AI and Climate-Smart Policies for Blue Economy:

Eco-Digital Futures

Altuğ Günar
https://orcid.org/0000-0001-8094-624X
Bandırma Onyedi Eylül University, Turkey

Didem Saygın
https://orcid.org/0000-0002-2832-351X
Çanakkale Onsekiz Mart University, Turkey

Çağatay Başarır
Bandirma Onyedi Eylul University, Turkey

IGI Global Scientific Publishing
Publishing Tomorrow's Research Today

Vice President of Editorial — Melissa Wagner
Director of Acquisitions — Mikaela Felty
Director of Book Development — Jocelynn Hessler
Production Manager — Mike Brehm
Cover Design — Jose Rosado

Published in the United States of America by
IGI Global Scientific Publishing
701 East Chocolate Avenue
Hershey, PA, 17033, USA
Tel: 717-533-8845 | Fax: 717-533-7115
Website: https://www.igi-global.com E-mail: cust@igi-global.com

Library of Congress Cataloging-in-Publication Data

Names: Günar, Altuğ, 1985- editor | Saygın, Didem, 1981- editor | Başarır, Çağatay, 1980- editor
Title: AI and climate-smart policies for blue economy : eco-digital futures / edited by Altuğ Günar, Didem Saygın, Çağatay Başarır.
Other titles: Artificial intelligence and climate-smart policies for blue economy
Description: Hershey, PA : IGI Global Scientific Publishing, [2026] | Includes bibliographical references and index. | Summary: "This book examines the sustainable use of marine resources and their ability for green financing, accounting for economic and security dimensions. It further provides a comprehensive overview of climate change and the environmental challenges it presents. It also examines how AI and digital technologies are driving transformation models for the blue economy and sustainability"-- Provided by publisher.
Identifiers: LCCN 2025030198 (print) | LCCN 2025030199 (ebook) | ISBN 9798337352381 hardcover | ISBN 9798337352398 paperback | ISBN 9798337352404 ebook
Subjects: LCSH: Environmental economics | Finance--Environmental aspects | Green technology | Artificial intelligence--Financial applications
Classification: LCC HC79.E5 A3675 2026 (print) | LCC HC79.E5 (ebook)
LC record available at https://lccn.loc.gov/2025030198
LC ebook record available at https://lccn.loc.gov/2025030199

British Cataloguing in Publication Data
A Cataloguing in Publication record for this book is available from the British Library.

Table of Contents

Detailed Table of Contents

This chapter explores the complex relationship between artificial intelligence (AI) and climate change by offering a comprehensive analysis of AI's role in understanding, mitigating, and adapting to climate-related challenges. In this regard, the first section examines the significant role of AI's in analyzing climate data. The second section explores the growing role of AI in addressing environmental challenges and promoting sustainability across various sectors. It discusses how AI can help agriculture and forestry transition to more sustainable practices by analyzing data to reduce emissions, enhance food security, and improve carbon sequestration. The final section examines how AI can help shape future climate solutions, particularly in increasing renewable energy production and building climate resilience. However, it mostly raises concerns about the growing energy demands of AI. As AI models get more complex, they require more energy, which could counteract the sustainability goals they're meant to support.

This paper integrates studies on AI deployment in the Circular Economy through the lens of waste management, resource beneficiation, and product life cycle analysis. Examination of more than sixty studies in machine learning, digital twining, and IoT reveals that AI reduces wastage, boosts recycling coverage and accuracy, and enhances efficiency—up to 25% and 40% respectively. AI-based systems incorporating life cycle studies and real-time monitoring deliver closed-loop production and robust supply chains. However, the barriers exist in data quality and energy consumption by the AI systems and in the realms of ethics and scalability through gaps in policies

The Blue Economy represents a critical pathway toward sustainable development, balancing economic growth with the conservation of marine ecosystems. This chapter explores the transformative role of Artificial Intelligence (AI) in advancing sustainable practices within the Blue Economy through a series of global case studies. It examines how leading companies and initiatives—such as Sinay, Saildrone, Sea.AI, Buffalo Automation, IBM, Microsoft, Ocean Cleanup, and Veolia Group—leverage AI-driven solutions for real-time environmental monitoring, predictive climate modeling, autonomous navigation, and circular waste management. Key insights reveal that AI enables data-driven decision-making, reduces carbon emissions, enhances operational efficiency, and supports closed-loop recycling systems, while confronting challenges related to data integration, regulatory frameworks, and economic viability. Policy implications highlight the urgent need for adaptive regulations, open data ecosystems, and public-private partnerships to scale smart marine ecosystems globally.

Artificial intelligence (AI) is increasingly utilized to predict climate change–related risks in marine systems. This analysis establishes a typology of four model classes: deep learning, probabilistic and ensemble models, reinforcement learning, and hybrid physics-informed approaches. These classes are systematically mapped to key marine-climate applications, including extreme events, ecosystem change, and adaptive resource management. The results indicate that AI can significantly improve short-term risk detection and scenario exploration, whereas long-term projections benefit from hybrid models that incorporate physical constraints and uncertainty quantification. Nevertheless, structural data gaps, regional biases, and limited interpretability restrict the generalizability and policy relevance of many AI models. The chapter contends that the value of AI for marine climate risk assessment depends not only on predictive accuracy but also on transparent model design, robust data foundations, and alignment with governance requirements.

This study investigates the global challenge of accessing safe freshwater through the perspectives of climate change and artificial intelligence (AI). Freshwater, only 2.5% of global water, is vital for life, yet over 2.1 billion people lack safe access. Climate change disrupts water cycles via rising temperatures, altered precipitation, glacial melt, and extreme weather, worsening scarcity and quality issues across regions like South Asia, Africa, and Europe. AI offers transformative solutions in real-time monitoring, efficient treatment, optimized irrigation, and disaster management, with successful cases in Singapore, India, and Africa. However, high costs, data dependency, ethical concerns, and limited local capacity hinder AI adoption. The study debates water as a human right versus a commodity, advocating a balanced approach to ensure equitable access and environmental sustainability. Inclusive, ethical AI policies and global cooperation are essential to address this crisis, aligning with Sustainable Development Goals for social justice and sustainable water management.

This chapter explores the intersection of maritime security and the sustainable blue economy, emphasizing the critical role of secure seas in advancing ecological health, economic prosperity, and global stability. It highlights how maritime threats—ranging from piracy and smuggling to climate-driven risks and cyber vulnerabilities—undermine efforts to build resilient ocean economies. Drawing on regional case studies such as the Red Sea, Indo-Pacific, Arctic, and African maritime zones, the discussion underscores the interplay between security frameworks, governance mechanisms, and sustainable development. The chapter also examines the role of international law, regional cooperation, and technological innovation in mitigating risks. By linking security with sustainability, it argues that safeguarding maritime spaces is not only a matter of protecting trade routes and national interests but also a prerequisite for ensuring long-term ecological and economic resilience.

Oceans cover 72% of the Earth's surface, constituting 99% of the living space by volume. By absorbing almost 25% of the carbon dioxide, it tries to prevent global warming. Oceans and maritime resources are becoming key players in international economic interactions, which has developed the idea of the blue economy. However, the increased economic interactions cause various insecurities, like interstate disputes, maritime zone confusions, piracy, terrorism, pollution, and illegal and unregulated exploitation of marine resources. This necessitated the advancement of maritime law with more possibilities and enforcement mechanisms. The interplay between the blue economy and maritime law is essential to realize Sustainable Development Goal No. 14. Their mutual dependence and empowerment are vital for conserving and promoting marine resources. From this perspective, this chapter, using qualitative methodology's critical analytical method, explores the possibilities of converging AI and climate-smart policies with the blue economy and maritime law framework to realize a sustainable future.

Artificial intelligence (AI) systems increasingly shape decisions in health care, finance, employment, and public services—often by learning from vast, sensitive datasets. This chapter examines the ethical and compliance landscape for data privacy in AI. We ground the discussion in major ethical theories (utilitarianism, deontology, virtue ethics, and care ethics) and connect them to practical challenges such as algorithmic bias, surveillance, and opacity. We then synthesize global legal regimes (GDPR, CCPA/CPRA, LGPD, PIPL, PDPA) and extract the core protection principles that matter most for AI builders. Governance tools including Data Protection Impact Assessments (DPIAs), privacy-by-design/default, and post-deployment monitoring are translated into concrete practices. Four case studies illustrate recurring pitfalls and lessons learned. Finally, we survey privacy-enhancing technologies (federated learning, differential privacy, homomorphic encryption, and secure multi-party computation) and outline emerging trends.

Renewables have moved from the sidelines to the center of policy and investment. Evidence from IEA, IRENA, REN21, and recent studies shows steep cost drops in solar and wind, record capacity growth, and more jobs. Environmental gains (lower GHGs, cleaner air and water, smarter land use) come with economic and social upsides (jobs, energy security, rural access via decentralized systems). But real constraints persist: grid congestion, curtailment, flexibility gaps; high financing costs and regulatory uncertainty in emerging markets; supply chain risks; and community concerns over siting and land rights. The challenge has shifted from tech costs to system integration and governance—durable policies, markets that value flexibility, quicker permitting, and just transition tools. Priorities ahead: operating high VRE under extremes, long-duration storage and sector coupling (hydrogen, heat, industry), critical minerals and circularity, resilience to climate and cyber risks, and effective de-risking in the Global South.

This study evaluated corporate alignment with the 2050 Net Zero Emissions targets, a cornerstone of the global transition to a low-carbon economy as mandated by the Paris Agreement. Utilizing the Transition Pathway Initiative (TPI) framework, as endorsed by OECD reports, we analyzed the carbon performance of 2,028 firms across 24 sectors and 28 countries. Companies are scored on a 0–5 scale, with 5 indicating the highest alignment with net-zero goals. Through descriptive statistics, statistical analyses, and hierarchical clustering, we assessed performance variations by firm size, sector, and countries, comparing their alignment with Paris-Compliant Pathways. Results reveal that larger firms, sectors such as telecommunications and oil & gas, and European countries like Denmark and France exhibit superior carbon management, while smaller firms, carbon-intensive sectors like coal mining, and regions like China and the Middle East lag. These findings reveal sectoral and

Sustainability awareness has led to green marketing practices, which businesses have either voluntarily adopted or been compelled to embrace by competitors, consumers, and governments.To ensure the effective functioning of this green marketing approach holistically, driven by a firm's own green mission, eco-conscious consumers, green-marketing competitors, and imposed green standards, artificial intelligence (AI) has become the most crucial tool. This study investigates and discusses the driving and assisting role of AI within the complex nexus of green competition, eco-consumers, and green standards as applied in a holistic green marketing approach in businesses. Through its integration with techniques such as algorithms and machine learning, as well as systems like blockchain, AI assumes a supportive and leveraging role in businesses' green marketing strategies.

The transition to a sustainable ocean economy requires innovative approaches that integrate environmental protection with technological advancement. This chapter investigates the transformative role of Artificial Intelligence (AI) and climate-smart policies in advancing the Blue Economy and supporting the achievement of the United Nations Sustainable Development Goals (SDGs), especially SDGs 13, 14, and 9. It highlights the use of AI in marine biodiversity monitoring, climate risk modeling, and sustainable resource management. The discussion includes real-world examples from Indonesia and small island developing states (SIDS), emphasizing multisectoral collaboration and ethical challenges such as data inequality, algorithmic bias, and the digital divide. By outlining strategic pathways for AI adoption in ocean governance, the chapter offers policy and innovation recommendations to promote inclusive, eco-digital futures in the maritime sector.

Preface

Awareness of climate and environmental issues on a global scale dates back to the 1960s. Following the Second World War, globalisation gained momentum, leading to significant technological and production advances and a major economic revolution. The communications revolution has not only made the world a smaller place, but also led to a substantial increase in production. In this new era, multinational companies have established global production networks to meet the demands of the global market. The United Nations took the first step towards raising global awareness of climate and environmental issues in 1972. From an academic perspective, Garet Hardin's 'tragedy of the commons' is a key concept in this context. According to Hardin, the misuse of natural resources leads to their excessive use and eventual depletion, leaving farmers unable to find resources to feed their animals (Dyer, 2018, p. 84).

Following the global environmental conference organised by the UN in 1972, the first action was taken by the nations of the world in response to climate and environmental issues, and World Day was first celebrated in 1970. The importance of this conference is evident in the context of the 'Stockholm Report'. This report focused on global environmental and climate issues (Global Sustainable Development Report, 2013, p. 9).

In terms of global organisations and academic studies, no discussion on climate, the environment, or humankind's world has had as much impact as the 'Great Extinction Debate'. The diversification of economic growth and development processes, and their impact on the natural world, has prompted a group of academics to publish a report highlighting the urgent need to address environmental degradation. The 'Club of Rome' report 'Limits to Growth' led to global awareness transcending the nation state, resulting in the beginning of the 'Great Debate' (Spretnak and Capra, 1985, p. 163; Richardson, 2006, p. 4). In 1968, scientists gathered in Rome to launch the 'Club of Rome' initiative, which aimed to raise awareness of environmental and climate issues and encourage decision-makers to take action by presenting them with a scientific report on the consequences of environmental and climate damage.

This report, often referred to as 'Forecasting the State of Humanity', aimed to explain why humanity is so destructive and address the root causes of the problems they identified as 'World's Problems', thereby initiating a global project (Meadows et al., 1972, p. 9).

The first Climate Conference, held in 1979, saw the first concrete steps taken on climate change. At this conference, climate change was recognised as a serious issue for the first time. The impact of climate change on humanity was discussed and world governments were encouraged to take action (United Nations Environment Programme, 2000). Against this backdrop, the 1985 conference — the 'Vienna Convention for the Protection of the Ozone Layer' — focused on actions and principles to protect the ozone layer. It came into force in 1988 and had been accepted globally by 2009. Its primary objective was to demonstrate and discuss the impact of human activities on the ozone layer and encourage international collaboration (UN Environment Programme Ozone Secretariat, 2020). Following the establishment of the Intergovernmental Panel on Climate Change (IPCC) in 1988, scientific action on climate change was initiated and research was conducted into its environmental and socio-economic effects, as well as new strategies to address climate change issues (IPCC, 2010, p. 4). In 1992, the 'United Nations Framework Convention on Climate Change' brought states together to prevent human intervention in the climate system, introducing the concept of 'differentiated responsibility' to place a greater burden of responding to climate change on states (United Nations Framework Convention on Climate Change, n.d.).

One of the most significant steps taken in the fight against climate change was the Kyoto Protocol, adopted in 1997. It came into force on 16 February 2005, and its significance lies in the fact that it was ratified by 192 countries. It has set limits on and reduced greenhouse gas emissions for industrialised countries and transition economies. It has requested regular reporting from the participating countries and, in line with the principle of differentiated responsibility, placed a greater burden on developed countries. It has also set emission reduction targets for the European Union. The most important developments have occurred since 2012. The Doha Amendment to the Kyoto Protocol was adopted in 2012, and in December 2020, the parties agreed to changes for the second commitment period, which began in 2013 and will continue until 2020. The parties agreed to reduce emissions by at least 18% below 1990 levels (United Nations Framework Convention on Climate Change, n.d.).

One of the most significant climate-related developments was the 2015 Paris Climate Agreement. Adopted in December 2015, it came into force in November 2016. Its primary objective is to keep global temperatures below 2°C and limit them to 1.5°C above pre-industrial levels. To combat climate change, it has been determined that reducing greenhouse gas emissions by 43% by 2030, after peaking before 2025, is crucial to achieving the 1.5°C target (United Nations Framework Convention on Climate Change, n.d.).

Examining the series of actions from the 1960s to the present day reveals that global emissions have decreased, particularly after the Paris Agreement. Although the latest data indicates an increase in emission rates, it is estimated that the rate of increase has fallen from 2% to 0.6%. This positive trend is being driven by the gradual phase-out of coal and fossil fuels, with almost all European Union countries having agreed to eliminate coal entirely. The impact of low-emission energy sources on the global energy transition is said to have increased to 40% by 2015, with significant steps being taken towards a global transition to clean energy. 2025 is said to be a turning point in the energy transition, when solar and wind energy production will surpass that of coal. This could double the use of renewable energy sources by 2030, increase the proportion of energy derived from solar power, reduce costs and boost investment in clean energy tenfold to 2.2 trillion dollars. It is estimated that climate change and environmental initiatives, ongoing for 70 years, have gained momentum in the last two years. Between 2015 and the Paris Agreement, solar energy usage doubled (Climate Council, 2025).

The concept of the blue economy has evolved from the long-standing view of the seas and oceans as inexhaustible sources of raw materials, to a more holistic approach that considers the integrity of ecosystems. Historically, coastal communities have sustained themselves through fishing, maritime trade, and port activities. However, these practices have largely been short-term, profit-driven, and have not considered the capacity of nature to regenerate itself. The seas have been used, but not managed. The Industrial Revolution completely disrupted this balance. The advent of steam power, industrial fishing techniques and the expansion of global trade networks has dramatically increased demand for seafood, fossil fuels and seabed minerals. This has not only resulted in economic expansion, but also in overfishing, the fragmentation of ecosystems, the decline of biodiversity and chronic marine pollution. The seas have effectively become the production yard. The modern form of the blue economy has emerged through certain turning points. Rising environmental awareness in the second half of the 20th century brought the protection of the seas from the periphery to the centre of political agendas. The designation of marine protected areas and the establishment of pollution control regimes and coastal management approaches represented the first systematic efforts to protect marine ecosystems. This transformation became more institutionalised with the emergence of the sustainable development paradigm at the end of the 20th century. The idea that economic growth must be considered alongside environmental limits and social justice has also transformed ocean governance. Consequently, the blue economy is now defined not only as a means of generating income from the seas, but also as a means of fostering prosperity by safeguarding the continuity of marine ecosystems.

This perspective is based on the idea that the seas are strategic partners that can be managed rationally and are not resources to be exploited (Chiorino, 2024).

Today, climate change is a major global issue affecting production and sovereignty relations and posing a significant challenge to the world. During this period of change, the seas and oceans, or the 'blue world' as we might call them, are the most vulnerable areas, i.e. the sources of our resources. In this context, the blue economy is an important concept to discuss in relation to resource use and reconfiguring production using these resources to offer to societies in this era of climate crisis. Due to their central role in energy, food and transport systems, the seas are areas where the ecological losses and inequalities caused by climate change are particularly evident. However, as with everything else, these processes are subject to the transformative effects of artificial intelligence. AI's advanced processing capacity and data processing power have the potential to be a turning point in the fight against climate change, as well as in the development of new applications and policies to protect the planet and the blue economy.

In 2024, Geoffrey Hinton and John Hopfield made revolutionary breakthroughs in artificial intelligence by demonstrating the potential of artificial neural networks and machine learning. Their work has also paved the way for the development of climate change models. The emergence of AI is set to transform the economy, politics, and society as we know them. As Hinton put it, 'In the Industrial Revolution, we made human labour insignificant. Now, we are rendering human intelligence irrelevant, and this is very scary" (Lee, 2024). This highlights the potential dangers and power of AI. This study examines the relationship between climate change, the blue economy and artificial intelligence, three of the most discussed topics of recent years, from different perspectives.

Climate change has evolved from being an environmental issue to a defining structural force reshaping economic systems, governance, social relations and geopolitical dynamics. The scale and intensity of climate-related disruptions — ranging from extreme weather events and ecosystem degradation to resource scarcity and economic volatility — have exposed the limitations of traditional policy approaches based on sectoral silos, static regulatory tools and reactive governance. Meanwhile, the rapid development of artificial intelligence (AI) and data-driven technologies has transformed the way societies generate knowledge, assess risks and make collective decisions. These parallel transformations have created both unprecedented opportunities and significant challenges, raising a fundamental question at the heart of this book: How can artificial intelligence be meaningfully integrated into climate-friendly policies to support sustainable, resilient and equitable development, particularly within ocean-related economic systems? This book, AI and Climate-Smart Policies for the Blue Economy: Eco-Digital Futures, recognises that climate change, digital transformation and the blue economy are now inextricably linked.

They form an interconnected system, meaning decisions made in one area have repercussions in others. Oceans play a pivotal role in regulating the global climate by absorbing heat and carbon, while also sustaining the biodiversity and livelihoods of millions of people. At the same time, however, marine and coastal areas are also becoming focal points for economic expansion, technological innovation, and geopolitical competition. Without carefully designed governance frameworks, however, this expansion could intensify ecological degradation and social inequality. This book addresses this challenge by presenting artificial intelligence not just as a technological solution, but as a vital element of climate-friendly governance that must be aligned with legal, ethical, economic and social factors.

A defining feature of contemporary climate governance is the growing dependence on data, modelling and predictive analytics to inform decision-making processes. Artificial intelligence systems are being used more and more to predict climate risks, optimise resource use, monitor emissions and support early warning mechanisms. These tools have transformed the epistemic foundations of climate policy by expanding our capacity to analyse complex systems and anticipate future scenarios. However, this transformation also raises important questions about transparency, accountability, and power. Algorithms determine what is visible and actionable, yet their underlying assumptions and limitations often remain opaque to policymakers and the public. Furthermore, the computational infrastructure that supports AI has its own environmental impact, particularly in terms of energy consumption. While acknowledging AI's transformative potential, this book takes a critical yet constructive approach, emphasising the need for governance structures that enable its responsible, inclusive and sustainable use.

CHAPTER OVERVIEW

The first chapter establishes a conceptual foundation for understanding AI's role in climate solutions. It examines the complex relationship between AI and climate change, providing a comprehensive analysis of AI's role in understanding, mitigating and adapting to climate-related challenges. In this context, the chapter explores the important role of AI in analysing climate data.

The second chapter focuses on the transformative power of AI and its unique ability to optimise and scale circular economy applications, and integrates studies on AI deployment in the circular economy through the lens of waste management, resource beneficiation, and product life cycle analysis. Examination of more than sixty studies in machine learning, digital twining, and IoT reveals that AI reduces wastage, boosts recycling coverage and accuracy, and enhances efficiency—up to 25% and 40% respectively.

The third chapter critically examines its foundations, key sectors, environmental challenges, governance, and future pathways through a structured literature review and case study synthesis drawing on over 130 sources, including UNCLOS, SDG 14, Seychelles' Blue Bond, and EU Marine Spatial Planning. Findings show substantial economic potential, estimated at over $3 trillion annually by 2030, yet persistent issues of overfishing, pollution, and fragmented governance. While technological and financial innovations are advancing, they require safeguards to ensure equity and accountability.

The fourth chapter explores the deep and increasingly vital connections between AI, climate resilience, and the future trajectory of the blue economy and argues that AI represents a transformative force, offering unprecedented capabilities for enhancing our understanding, monitoring, and adaptive management of marine and coastal ecosystems.

The fifth chapter chapter examines the development of the blue economy in Türkiye, highlighting its strategic potential and structural constraints. Türkiye's maritime sectors, including shipping, fishing, aquaculture, tourism and renewable energy, are presented as critical drivers of sustainable growth given the country's position at the crossroads of global maritime trade.

The sixth chapter investigates the transformative impact of AI on the blue economy, focusing on how AI adoption influences sustainability outcomes. Utilizing a quantitative research design, a sample of 289 respondents from various marine sectors was surveyed, with data analyzed using regression analysis, ANOVA, and correlation techniques. The findings reveal significant positive relationships between AI adoption and sustainability outcomes, highlighting the importance of supportive policy frameworks and robust technological infrastructure in driving eco-efficient practices.

The seventh explores the transformative role of AI in advancing sustainable practices within the blue economy through a series of global case studies. It examines how leading companies and start-ups such as Sinay, Saildrone, Sea.AI, Buffalo Automation, IBM, Microsoft, Ocean Cleanup and Veolia Group use AI-powered solutions for real-time environmental monitoring, predictive climate modelling, autonomous navigation and circular waste management.

The eighth chapter establishes a typology comprising four model classes: deep learning; probabilistic and community models; reinforcement learning; and hybrid physics-based approaches. These classes are systematically matched to fundamental ocean-climate applications, including extreme events, ecosystem change, and adaptive resource management. The results demonstrate that AI can significantly enhance short-term risk detection and scenario exploration, whereas long-term projections benefit from hybrid models incorporating physical constraints and uncertainty measurement.

The ninth chapter examines the global challenge of accessing safe freshwater from the perspectives of climate change and AI. Freshwater constitutes only 2.5% of the world's water and is vital for life, yet over 2.1 billion people lack safe access to it. Rising temperatures, changing precipitation patterns, melting glaciers and extreme weather events disrupt water cycles, exacerbating scarcity and quality issues in regions such as South Asia, Africa and Europe. AI offers transformative solutions in the form of real-time monitoring, efficient purification, optimised irrigation and disaster management, with successful examples in Singapore, India and Africa.

The tenth chapter explores the intersection of maritime security and the sustainable blue economy, emphasising the pivotal role secure seas play in promoting ecological health, economic prosperity, and global stability. It illustrates how maritime threats ranging from piracy and smuggling to climate-related risks and cyber security vulnerabilities undermine efforts to develop resilient ocean economies.

The eleventh chapter takes a critical analytical approach to qualitative methodology to explore the potential of combining AI and climate-friendly policies with the blue economy and maritime law framework to achieve sustainability. Oceans cover 72% of the Earth's surface, constituting 99% of the living space by volume. By absorbing almost 25% of the carbon dioxide, it tries to prevent global warming. Oceans and maritime resources are becoming key players in international economic interactions, which has developed the idea of the blue economy. However, the increased economic interactions cause various insecurities, like interstate disputes, maritime zone confusions, piracy, terrorism, pollution, and illegal and unregulated exploitation of marine resources. This necessitated the advancement of maritime law with more possibilities and enforcement mechanisms. The interplay between the blue economy and maritime law is essential to realize Sustainable Development Goal No. 14. Their mutual dependence and empowerment are vital for conserving and promoting marine resources.

The twelfth chapter examines the ethical and compliance environment surrounding data privacy in artificial intelligence. AI systems increasingly shape decisions in health care, finance, employment, and public services—often by learning from vast, sensitive datasets. Therefore the chapter open a discussion in major ethical theories (utilitarianism, deontology, virtue ethics, and care ethics) and connect them to practical challenges such as algorithmic bias, surveillance, and opacity. Then it synthesize global legal regimes (GDPR, CCPA/CPRA, LGPD, PIPL, PDPA) and extract the core protection principles that matter most for AI builders. Governance tools including Data Protection Impact Assessments (DPIAs), privacy-by-design/default, and post-deployment monitoring are translated into concrete practices.

The thirteenth chapter explores the transition of renewable energy from a marginal issue to the centre of global policy and investment agendas. Drawing on evidence from the IEA, IRENA, REN21 and recent academic studies, it highlights the sharp decline in the costs of solar and wind technologies, record-breaking growth in installed capacity and rapid employment growth in the sector.

The fourteenth chapter assesses corporate alignment with the 2050 Net Zero Emissions targets, which form the cornerstone of the global transition to a low-carbon economy as set out in the Paris Agreement. Utilizing the Transition Pathway Initiative (TPI) framework, as endorsed by OECD reports, it analyzed the carbon performance of 2,028 firms across 24 sectors and 28 countries. Through descriptive statistics, statistical analyses, and hierarchical clustering, it assessed performance variations by firm size, sector, and countries, comparing their alignment with Paris-Compliant Pathways.

Chapter 15 analyses the impact of green finance and blue economy concepts shaped in line with sustainable development goals on economic structures. Green finance encompasses financial instruments that reduce environmental risks and support sustainable projects, while the blue economy aims to promote the sustainable use of marine and ocean resources for economic development. Chapter analyses the impacts of green financing mechanisms and blue economy models on economic development. It also discusses how artificial intelligence technologies support the green and blue economy in areas such as data analytics, resource management and climate risk prediction.

Chapter 16 aims to contribute to existing literature by taking a holistic view of smart city applications and matching them with climate change mitigation policies. As a result of the urbanization movements that started to increase after the Industrial Revolution, cities cause climate change and are also affected by the climate crisis caused by climate change. Especially overpopulated cities have implemented many smart city applications to combat climate change. The idea of smart cities aims to make cities more livable, resilient and sustainable with the help of technology.

Chapter 17 examines the role of AI in driving and supporting green competition, eco-consumers and green standards within a holistic green marketing approach in businesses. Through integration with techniques such as algorithms and machine learning, as well as systems such as blockchain, AI plays a supportive role in businesses' green marketing strategies.

Chapter 18 examines the transformative impact of AI and climate-smart policies on the advancement of the Blue Economy and the achievement of the United Nations Sustainable Development Goals (SDGs), particularly SDGs 13, 14 and 9. The chapter highlights the use of AI in marine biodiversity monitoring, climate risk modelling and sustainable resource management.

As editors, we extend our gratitude to the contributors whose expertise has enriched this volume and to the readers who will engage with its content. Together, let us explore the transformative journey of smart applications of AI in climate change and blue economy and its far-reaching implications, while striving to balance its creative potentials.

Yours Respectfully

Editors

Altuğ Günar

Didem Saygın

Çağatay Başarır

REFERENCES

Chiorino, L. (2024). Blue economy and offshore wind (Bachelor's thesis, Luiss Guido Carli University, Italy). LuissThesis. https://tesi.luiss.it/40359/1/273681_CHIORINO_LODOVICA.pdf

Climate Council. (2025, November 12). *Ten years of the Paris Agreement: what have we achieved?* Climate Council. https://www.climatecouncil.org.au/ten-years-of-the-paris-agreement-what-have-we-achieved/

Information Unit for Conventions, United Nations Environment Programme. (2000). Climate change information sheet 17: The international response to climate change: A history. UNFCCC. https://unfccc.int/cop3/fccc/climate/fact17.htm

Intergovernmental Panel on Climate Change. (2010). *Understanding climate change: 22 years of IPCC assessment* (Broşür). IPCC. https://www.ipcc.ch/site/assets/uploads/2018/04/ipcc-brochure_understanding.pdf

Lee, T. B. (2024, November 21). *Seven big advantages human workers have over AI.* Understanding AI. https://www.understandingai.org/p/seven-big-advantages-human-workers

United Nations Environment Programme. (n.d.). Vienna Convention for the Protection of the Ozone Layer. Ozone Secretariat. https://ozone.unep.org/treaties/vienna-convention

United Nations Framework Convention on Climate Change. (n.d.). The Kyoto Protocol. UNFCCC. https://unfccc.int/process-and-meetings/the-kyoto-protocol

United Nations Framework Convention on Climate Change. (n.d.). The Paris Agreement. UNFCCC. https://unfccc.int/process-and-meetings/the-paris-agreement

United Nations Framework Convention on Climate Change. (n.d.). *United Nations Framework Convention on Climate Change.* UNFCCC. https://unfccc.int/process-and-meetings/united-nations-framework-convention-on-climate-change

Acknowledgement

This book would not have come to fruition without the contributions of many individuals. First and foremost, I would like to thank the authors of the chapters. These remarkable professionals are dedicated to leading and genuinely collaborating to make the world a better place through meaningful work. My deepest gratitude extends IGI Global team whose has provided unwavering support and shown patience throughout this book project's many changes.

Altuğ Günar

Didem Saygın

Çağatay Başarır

Chapter 1

Understanding, Impact, and Upcoming Innovations in the Application of AI to Climate Solutions

Begüm Özel
https://orcid.org/0009-0004-8017-6811
Veritas AG, Turkey

Murat Yorulmaz
https://orcid.org/0000-0002-3664-0451
Trakya University, Turkey

ABSTRACT

This chapter explores the complex relationship between artificial intelligence (AI) and climate change by offering a comprehensive analysis of AI's role in understanding, mitigating, and adapting to climate-related challenges. In this regard, the first section examines the significant role of AI's in analyzing climate data. The second section explores the growing role of AI in addressing environmental challenges and promoting sustainability across various sectors. It discusses how AI can help agriculture and forestry transition to more sustainable practices by analyzing data to reduce emissions, enhance food security, and improve carbon sequestration. The final section examines how AI can help shape future climate solutions, particularly in increasing renewable energy production and building climate resilience. However,

DOI: 10.4018/979-8-3373-5238-1.ch001

it mostly raises concerns about the growing energy demands of AI. As AI models get more complex, they require more energy, which could counteract the sustainability goals they're meant to support.

INTRODUCTION

The role of AI in addressing climate change has become increasingly crucial as the world faces the urgent challenges of global warming and environmental degradation. As climate patterns grow more unpredictable and extreme, AI provides powerful tools to help understand, manage, and reduce the impacts of climate change. By utilizing advanced algorithms and machine learning (ML), AI helps analyze large amounts of climate data, monitor environmental changes, and make more reliable predictions about the future of our planet. AI has made many significant contributions, particularly in the fight against climate change. One of its most valuable contributions is its ability to process complex climate data, helping better understand the detailed systems that influence climate change. By analyzing these data sets, AI improves climate models, making it possible for researchers to spot patterns that were once hard to detect. On top of that, AI enables real-time monitoring of environmental changes, offering crucial insights into issues like deforestation, ocean temperatures, and the rate at which ice caps are melting. AI is also transforming weather forecasting and climate predictions, helping to minimize uncertainty and improve the accuracy of future climate scenarios. By using predictive models and analyzing data, AI can provide early warnings for extreme weather events like hurricanes, droughts, or floods, potentially saving lives and reducing the damage they cause.

Looking ahead, AI's role in tackling climate challenges will continue to grow. New AI technologies hold great promise for improving climate adaptation, such as making renewable energy systems more efficient and optimizing resource management. However, as AI advances, it also brings concerns about its rising energy demands. The powerful computational resources required by complex AI models can lead to significant energy consumption, which, if not carefully managed, could undermine the very sustainability goals AI aims to support. That said, AI's potential to provide innovative solutions for both climate mitigation and adaptation remains crucial in shaping how we respond to the climate crisis providing that we can balance its energy use with the need for greater efficiency and sustainability. In summary, this chapter explores how AI helps understand climate patterns, monitor environmental changes, and predict future climate events. Through this work, a deeper understanding of AI's potential will be gained not only to advance our climate knowledge but also to address its environmental impact and contribute to sustainable solutions for the future.

The aim of the study is to analyze the complex relationship between artificial intelligence (AI) and climate change by offering a comprehensive analysis of AI's role in understanding, mitigating, and adapting to climate-related challenges. In this regard, the first section examines the significant role of AI's in analyzing climate data. By processing large amounts of data, AI helps spot complex environmental patterns, monitor changes in ecosystems, and improve climate models. These capabilities are essential for making weather and climate predictions more accurate, allowing future climate scenarios to be better understood and assessed. The second section explores the growing role of AI in addressing environmental challenges and promoting sustainability across various sectors. It discusses how AI can help agriculture and forestry transition to more sustainable practices by analyzing data to reduce emissions, enhance food security, and improve carbon sequestration. AI's impact is not limited to these industries. It extends to energy, transportation, and urban sustainability, where AI optimizes systems, reduces waste, and supports renewable energy integration. The section also highlights AI's potential in minimizing energy consumption, reducing greenhouse gas emissions, and supporting smarter energy management. Moreover, the section calls for faster integration of AI into key sectors to fully unlock its environmental benefits and address the urgent need for climate action. The final section examines how AI can help shape future climate solutions, particularly in increasing renewable energy production and building climate resilience. However, it mostly raises concerns about the growing energy demands of AI. As AI models get more complex, they require more energy, which could counteract the sustainability goals they're meant to support. The section stresses the importance of developing more energy-efficient AI technologies that strike a balance between climate action and their environmental impact, ensuring AI continues to be a powerful tool in the fight against climate change without adding to its own carbon footprint. In the end, it calls for the responsible development of AI technologies that make a positive impact on global climate action and sustainability.

AI'S ROLE IN UNDERSTANDING CLIMATE PATTERNS

Recent studies highlight the urgent reality of the climate crisis, emphasizing the need for immediate action. Experts warn that if the course is not changed, extreme weather events such as prolonged heatwaves, severe water shortages, and destructive weather patterns will become even more intense in the coming decades. Addressing this crisis requires a collective effort across all sectors, including politics, business, science, and civil society. In this regard, the growing role of AI holds significant potential in supporting climate action. This section explores how AI contributes to analyzing climate data, monitoring environmental changes, and improving weather and climate predictions.

How AI Helps Analyze Climate Data

AI helps analyze climate data. It does it by processing large and complex datasets that traditional methods often struggle with. Climate change is a global challenge that requires action on two fronts: reducing carbon emissions and adapting to its impacts. AI plays a key role in both by helping scientists analyze vast amounts of data quickly and accurately. With AI's machine learning algorithms, researchers can now identify patterns and connections in environmental data like temperature changes and atmospheric conditions that were once hard to detect. This leads to more accurate predictions and better decision-making when it comes to climate policies and solutions. As global temperatures continue to rise due to human-caused greenhouse gas emissions since the industrial revolution, AI helps deepen understanding of these shifts, offering valuable insights that guide meaningful action to combat climate change.

The use of AI systems has become increasingly widespread today and impacts many areas of daily life. As a result, AI technologies have become an indispensable necessity in modern times, while providing innovative solutions across various sectors. AI's ability to analyze vast amounts of data and identify patterns in several areas holds enormous potential to revolutionize our approach to climate change. By using ML algorithms, advanced climate models can be created to simulate different scenarios and predict how decisions or technological innovations might impact the climate (Butler et al., 2024). This deep analysis provides clearer insights into the effectiveness of various mitigation and adaptation strategies by making AI a crucial tool for addressing the challenges of climate change.

AI contribution to climate data collection is significant. It does it by analyzing satellite images and tracking climate phenomena such as hurricanes, wildfires, and ice sheet melt. However, challenges related to the availability and quality of climate data still exist, especially concerning non-financial entities like companies, governments, and households. These gaps hinder the proper financial assessment of climate risks (Changlani & Thakore, 2023). AI has the potential to address these gaps as well by offering more comprehensive climate projections, better forecasts, and more accurate disclosures, which provide a clearer view of financial risks linked to climate change. Additionally, AI-powered remote sensing tools, such as satellites and drones, are essential in filling these data gaps, helping to improve data collection with the benefit of local community collaboration. AI is already being used to monitor the effects of climate change on specific environmental features, such as analyzing shallow-water reefs and gathering temperature, humidity, and carbon dioxide data to assess forest health (Stein, 2020). AI's ability to fill data gaps and provide precise climate forecasts is crucial for making informed decisions on climate action. This technology is key to improving our understanding and response to climate challenges in the future.

Similarly, AI is making a big impact in areas like creating more sustainable energy systems, improving disaster preparedness, and optimizing how resources are allocated to tackle climate change. AI-powered tools help in planning energy networks and managing resources more efficiently, while natural language processing allows climate researchers to sift through vast amounts of unstructured data such as academic papers and policy reports to uncover valuable insights. As AI continues to harness the potential of big data, it plays a crucial role in shaping effective climate policies, offering innovative solutions to climate challenges, and heilping design of more resilient and sustainable environmental strategies (UNIDO, 2024).

The era we are in is called the age of "big data" (Kumar, Marchena, Awlla, Li & Abdalla, 2024). The age of "big data" has significantly contributed to the increasing use of AI across various fields of Earth science. In this context, AI techniques, such as ML and deep learning (DL), allow machines to process large amounts of datasets and identify complex, nonlinear patterns that would be challenging for humans to detect. These AI algorithms are capable of analyzing vast environmental data, revealing subtle relationships between variables that traditional physics-based models might miss. For example, AI's ability to utilize remote sensing data and analyze satellite images enables researchers to monitor deforestation, track land use changes, and assess natural disaster damage more effectively. Moreover, AI can uncover hidden insights within environmental datasets, such as temperature shifts, carbon dioxide levels, and rainfall patterns, helping to improve climate models and forecast future environmental changes (Materia et al., 2024). The evolution of AI in environmental sciences not only paves the way for more accurate predictions but also has the potential to fill gaps in our understanding, enabling more adaptive and proactive solutions to the growing challenges of climate change.

AI is being applied in various areas of climate change. These applications include predicting global temperature changes and simulating atmospheric conditions, providing essential insights into climate and oceanic events (UNFCCC, 2024). Additionally, AI is used to understand the structure and functioning of weather systems, especially in specific localities, and to assess water demand in particular regions. It plays a key role in identifying losses and damages caused by extreme weather events and climate-related disasters, as well as monitoring climate-induced migrations and developing appropriate social policies (Mehryar, Yazdanpanah & Tong, 2024). AI is also crucial for managing electricity grids, enhancing energy efficiency in buildings, and calculating the carbon footprint of construction materials. Furthermore, it helps develop and manage smart transportation systems in cities, forecast the social and economic impacts of climate policies, and even assess the psychological effects of climate change on communities (Tuğaç, 2023). Through these applications, AI provides critical tools to better understand and mitigate the impact of climate change across multiple sectors.

For instance, in recent years, AI has made significant strides in both technology and climate action. A landmark achievement in AI occurred in late 2017, when Google DeepMind's AlphaZero, trained purely through AI rather than human-derived strategies, defeated the renowned chess engine, Stockfish (BBC News, 2017). This breakthrough demonstrated AI's potential to outperform established systems by learning through its own experience. Similarly, OpenAI's GPT-3, introduced in 2020, revolutionized natural language processing by generating human-like text across a vast range of topics, providing valuable applications in a variety of sectors, including climate science (Kissinger et al.,, 2021). These advancements demonstrate that AI holds great potential, not only in technology but also in addressing global challenges like climate change. AI will be a crucial tool in providing more effective solutions and decision-making systems in the future.

In this context, AI's capabilities in climate science are becoming increasingly instrumental, as seen in initiatives like Climate Trace, supported by Google, which uses AI to improve the transparency and accuracy of global greenhouse gas emission records (Climate Trace). Traditional emissions data, often self-reported and unreliable, is now complemented by AI-powered algorithms that analyze data from over 300 satellites and 11,000 sensors to track emissions across diverse sectors. This technology has uncovered that emissions are higher than previously reported, offering critical data for more effective climate policy (Dannouni et al., 2023). This advancement in AI technology is a game-changer, as it not only improves the accuracy of emissions tracking but also provides more reliable data to inform climate policy, ultimately leading to more effective actions against global warming.

In practice, AI is also being used to address climate change impacts in vulnerable regions. In Saint Kitts and Nevis, for example, AI-driven drought risk modeling is helping the island navigate the severe effects of climate change, including decreasing rainfall and the threat of more intense tropical storms. By collaborating with the Climate Technology Centre and Network (CTCN), the nation has integrated AI-based systems for water resource management and drought forecasting (UNIDO, 2024). These tools provide valuable insights for both climate adaptation and decisionmaking, highlighting AI's critical role in mitigating climate impacts and managing natural resources.

Defining climate change is important, as it helps understand what needs to be addressed. Climate change is described in the United Nations Framework Convention on Climate Change (UNFCCC) as the alteration in the climate over an extended period, resulting from both natural variations and human activities that directly or indirectly disturb the composition of the global atmosphere (Tuğaç, 2023). As can be understood, this issue is not solely environmental; it also has significant economic and social repercussions that must be addressed. Addressing climate change requires not only environmental solutions but also socio-economic adaptations to mitigate its widespread effects. As we move forward, it is essential that technological innovations, such as AI, work in tandem with global policy efforts, local community initiatives, and international cooperation to build a sustainable future. Without a multi-faceted approach, the challenges posed by climate change will only intensify, demanding urgent and coordinated action from all sectors of society.

TRACKING ENVIRONMENTAL CHANGES WITH AI

As ecological challenges grow, keeping track of and managing the environment is more important than ever. As technology advances, AI emerges as a key tool to tackle these issues. AI has the ability to make a huge impact on forest management, helping to prevent illegal logging, track deforestation, and aid in reforestation efforts (Masterson, 2024). These efforts are essential for protecting biodiversity, boosting carbon capture, and guiding smart policy decisions. AI also plays a key role in ensuring compliance with regulations that promote sustainable supply chains, like the European Union's (EU) deforestation law. Often, this work involves collaborating with leading organizations, such as the World Resources Institute (WRI), to make the biggest impact possible (Butler et al., 2024). AI's ability to monitor environmental changes in real-time provides a unique opportunity to tackle ecological threats quickly and accurately. As AI and similar technologies continue to evolve, they can significantly improve forest management and help create more effective, data-driven strategies for global climate adaptation and conservation efforts.

Regarding deforestation, AI is being used to track patterns and support conservation efforts. For instance, neural networks that process spatial and temporal data are used to predict gradual deforestation and estimate future loss rates. This enables continuous updates to models and more accurate risk assessments, helping to identify areas at risk (UNIDO, 2024). In projects like the Guacamaya Project in the Amazon, AI technologies are combined with satellite images, camera trap data, and bioacoustic information to monitor deforestation in real-time (Sadhukhan, 2019). These AI systems help quickly identify critical deforestation zones, guiding immediate intervention and supporting reforestation and carbon sequestration programs.

The effects of climate change are impossible to ignore. Agriculture and forestry are feeling the impact directly, as rising temperatures, droughts, and floods take a toll on crop yields, forest health, and livestock productivity (Sleeman, 2023). These changes not only threaten ecosystems but also put global food security and carbon cycling at risk. AI plays a key role in climate mitigation by improving forest mapping and monitoring through data-driven methods. For example, projects like the FAIR Forward initiative in Indonesia use AI to better classify forests, identifying those with high carbon storage potential, which helps support efforts to prevent deforestation and reduce its impact (Butler et al., 2024). AI plays a crucial role in improving forest management, providing more efficient and targeted solutions to protect ecosystems and promote sustainability in the fight against climate change.

AI has great potential to optimize other various sectors for climate mitigation as well, such as transportation, waste management, and energy production. For example, AI boosts urban sustainability by improving traffic management and reducing transportation emissions. It also supports off-grid, zero-carbon electrification through microgrid systems, making electricity more affordable and accessible (Stein, 2020). In terms of emissions monitoring, AI enhances climate modeling and predictions by providing real-time data on greenhouse gas reductions and carbon sequestration (Tuğaç, 2023). This ability to track emissions at various administrative levels allows for more effective policy adjustments and the identification of further mitigation strategies regarding environment.

Human-driven climate change has caused extreme weather events that directly impact both nature and human activities. To tackle this, it's essential to monitor and understand the sources of greenhouse gas emissions (Demirci, 2024). Understanding where, how, and to what extent greenhouse gases are being emitted is essential. Without continuous tracking of these emissions and their sources, it becomes impossible to address the issue without truly understanding it. This is where technology becomes crucial. Modern advancements like the "Artificial Internet of Things" (AIoT), Earth observation satellites, cloud computing, and AI are making it possible to collect detailed data on greenhouse gas emissions, offering near-instant updates on their progress (Singh, 2024). That new concept, AIoT, has even emerged to describe the technologies utilized for climate change. Through the work of organizations like Climate Trace, AI is revolutionizing emission tracking by gathering data from satellites and sensors worldwide, allowing for real-time monitoring and enabling governments and organizations to immediately detect and address illegal polluters (TkinkTech). This combination of AI and advanced technology offers a promising path toward more efficient climate action. By providing real-time, accurate data on emissions, it empowers stakeholders to act swiftly and make informed decisions, ultimately enhancing global efforts to combat climate change.

USING AI FOR WEATHER AND CLIMATE PREDICTIONS

AI is transforming weather and climate predictions by making forecasting models more accurate and efficient. Traditional methods often struggle with the complexity and unpredictability of atmospheric data. But AI's advanced ML algorithms analyzes vast amounts of data from satellites, weather stations, and other sources, identifying patterns and making more reliable predictions. By improving climate models and weather forecasts, AI helps anticipate extreme weather events, giving valuable lead time for disaster preparedness and climate risk management.

Weather predictions are able to focus on specific extreme events, but climate predictions can only aggregate these events over time, which leads to a reduced number of observational samples available for climate forecasting. AI presents a powerful opportunity to improve the accuracy of predicting climate extremes across various timescales, which could significantly enhance both our scientific knowledge and the preparedness of societies (Lucas, 2024). Using advanced, data-driven techniques, AI helps uncover the connections between climate drivers and processes, such as local and distant interactions, the link between the stratosphere and troposphere, and the relationship between oceans and the atmosphere. AI algorithms are capable of analyzing vast climate datasets and identifying complex patterns that influence extreme events in a nonlinear way, without depending on potentially biased numerical models of Earth systems (Materia, 2024). As AI continues to improve the accuracy of climate predictions, it's crucial to implement effective policies that support its integration into climate action efforts. A holistic approach, where AI research and climate change policies work together, will be crucial for fostering meaningful and sustained progress.

Effective policymaking in both AI and climate change will be crucial for guiding incentives and driving meaningful progress. Supporting AI research, development, and application in ways that benefit the climate encourages interdisciplinary collaboration between computer science and fields related to climate action such as engineering, economics, and urban planning (Lucas, 2024). It helps pave the way for improving the technical readiness of AI applications for climate change mitigation and adaptation through research, development, and demonstration efforts. Additionally, it lowers regulatory barriers that could slow the adoption of AI technologies in key sectors, like energy, that are crucial for tackling climate change (Kaack et al., 2020). Therefore, policies that encourage innovation and eliminate obstacles will be crucial in unlocking AI's full potential to combat climate change. This will enable AI to play a significant role in improving climate predictions and guiding decisions for a more sustainable future.

AI can provide more precise forecasts of future climate trends by analyzing large datasets and identifying key patterns. These advanced predictions help in understanding how the climate might evolve under different scenarios (UNFCCC, 2024). Additionally, AI plays a pivotal role in combating climate change by enhancing resource efficiency and energy use, enabling businesses, governments, and individuals to make more climate-conscious decisions (Butler et al., 2024). By optimizing resource usage, AI supports efforts to mitigate and adapt to climate impacts. With more accurate predictions of climate trends and weather patterns, AI enables better decision-making for climate adaptation and resource management (Lucas, 2024). This ability to enhance predictions also extends to weather patterns, where AI-driven analyses of both past and present data can significantly improve the accuracy of meteorological models.

Analyzing both past and present weather data through machine learning algorithms leads to the creation of more accurate meteorological models. These models, in turn, provide better forecasts for various climate factors like temperature, precipitation, and wind speed (Chen et al., 2023). By making these predictions more precise, AI not only enhances weather forecasts but also enables a more informed approach to managing environmental and agricultural challenges.

Environmental and agricultural challenges can be listed as transportation and wrong farming practices. AI-powered autonomous electric vehicles set an example to improve transportation efficiency while cutting emissions. Additionally, AI is used in traffic management and route optimization to boost fuel efficiency (Miller et al., 2024). Beyond transportation, AI is also transforming other sectors like agriculture, where it is used to optimize farming practices, improving crop yields and cutting emissions associated with agriculture. Through precision farming, AI enables more efficient use of resources like water, fertilizers, and pesticides, lowering the carbon footprint (Changlani & Thakore, 2023). Additionally, AI-powered early warning systems help predict crop yields and provide recommendations to address issues like crop failure and malnutrition, particularly for smallholder farmers who form the backbone of agricultural production in Kenya (Butler et al., 2024). By offering timely predictions and advice, AI helps farmers better prepare for adverse weather conditions and mitigate the effects of climate change.

Apparently, the vast amount of data collected from observation satellites, along with the complexity of climate models, has made AI essential for weather forecasting. AI is increasingly used to analyze large datasets and uncover new climate patterns, helping to reduce prediction biases and improve forecasting accuracy (Liu et al., 2023). AI integrates data, algorithms, and computational technologies, to provide valuable insights for tackling climate change. It contributes significantly to areas like developing environmental monitoring systems, detecting carbon sources, weather predictions, assessing the effects of extreme weather events across various regions, and understanding the socio-economic impacts of such events. In this regard, AI supports the creation of more reliable climate policies by providing essential data for national documents like Nationally Determined Contributions (NDCs) and National Climate Change Reports under the UNFCCC and the Paris Agreement. These contributions help countries take more informed actions in reducing greenhouse gas emissions and adapting to the changing climate (Tuğaç, 2023). By enhancing the accuracy and reliability of climate predictions, AI not only mitigates the impacts of extreme weather but also plays a crucial role in shaping effective climate policies. This fosters more informed global action to combat climate change and build a more resilient future.

AI'S ENVIRONMENTAL IMPACT

AI offers transformative solutions for environmental issues. However, the development and use of AI technologies have various environmental impacts, including energy consumption, carbon emissions, and other environmental implications. AI applications, especially in areas like big data processing, ML, and DL, require significant amounts of energy, which can have negative effects on the environment. To address the environmental impact of AI, agriculture and forestry must adopt more sustainable and resilient practices. AI supports this transformation by analyzing data and providing insights into how these sectors can reduce emissions, enhance food security, and improve carbon sequestration. By offering targeted solutions, AI helps these industries adapt to changing climate conditions while promoting sustainability (Butler et al., 2024).

AI is increasingly positioned as a cornerstone of economic competitiveness, public service delivery, and scientific advancement. From large language models to computer vision systems and predictive analytics, AI applications are expanding rapidly across sectors. However, this expansion carries a growing environmental footprint that is often underexamined in policy debates and corporate strategies. Understanding AI's environmental impact requires a balanced assessment of its energy consumption, resource use, and emissions—alongside its potential to enable environmental solutions (Bolon-Canedo et al., 2024). While AI holds great promise in supporting sustainable practices, the scale of transformation required in agriculture and forestry to truly mitigate environmental impact remains a major challenge. Without significant, systemic changes, the full potential of AI can not be realized in these critical sectors. Therefore, this section aims to address both the environmental impact of AI, energy efficiency, and sustainability challenges, while also highlighting the positive contributions AI can make in protecting the environment, enhancing energy efficiency, and promoting sustainable practices.

ENERGY DEMANDS OF AI TECHNOLOGIES

The demand for energy is set to grow steadily as long as humanity continues to progress and technological advancements take place. As the need for energy increases, the challenge of resource depletion becomes more pressing, pushing the energy sector to focus on maximizing the efficiency of available resources. At the same time, the growing energy consumption is worsening global challenges like climate change and energy security concerns. Recently, AI has become a key player in the energy sector, helping to improve system efficiency, lower costs, reduce carbon footprints, and ensure sustainable, reliable energy production (Çetinkaya & Genç, 2024). As AI continues to evolve, it holds the potential to drive more efficient, cleaner, and smarter energy systems across the globe. The thing is, in order to achieve all of this, AI technologies also come with their own energy demands. As AI technologies continue to advance, their energy demands are becoming an increasing concern. ML models and complex algorithms require significant computational power. This section explores the substantial energy consumption involved in training and running AI systems, as well as the environmental impact of these high energy needs.

AI plays a crucial role in transforming various sectors by enabling systems to perform complex tasks traditionally done by humans. AI technologies, such as ML and DL, are increasingly applied in areas like data analysis, predictive forecasting, decision making tools, and optimizing energy markets (UNFCCC, 2024). By simulating human-like decision making, AI systems have the potential to manage intelligent energy infrastructures autonomously, without the need for human intervention. ML enables computers to learn from data, improving decision-making capabilities over time. These systems can surpass older models, like linear regression, by recognizing non-linear patterns and managing complex datasets. Even in the presence of noise or anomalies, AI-driven models are still capable of delivering accurate predictions (Çetinkaya, 2024). As AI continues to evolve, its ability to optimize energy usage across various sectors has the potential to drastically change how resources are managed, bringing humanity closer to more efficient and sustainable energy systems. However, the increasing energy demands of AI present a significant challenge. As AI technologies advance and become more powerful, their energy consumption is also rising, which could potentially counteract some of the benefits they offer in terms of energy efficiency and sustainability.

While AI has tremendous potential to help fight climate change issues, there are some key challenges to its sustainable development and use. One major concern is its environmental impact. AI systems can produce high levels of carbon dioxide emissions and require significant material resources. These costs always need to be weighed against the benefits AI can bring to climate efforts. On top of that, effectively deploying AI comes with complex, mostly technical demands. This is especially true in climate-related work, which often depends on very specific and hard-to-access data (Butler et al., 2024). As AI continues to grow and become more widely adopted, it's becoming increasingly important to pay attention to how much energy it actually uses. Recent studies are already shedding light on just how big of an impact this could have.

Recent studies on the widespread use of AI systems have revealed some concerning findings: the energy consumed by AI has increased by 300,000 times since 2012. Today, data centers running massive AI software are responsible for 2% of global carbondioxyde emissions and up to 5-9% of the world's electricity consumption. Experts predict that information and communications technology will account for 14% of greenhouse gas emissions by 2040 (Hidalgo et al., 2023). The International Energy Agency estimates that in 2023, AI accounted for 1.5% of the world's energy consumption. However, AI technology is still in its early stages and is currently used by only a few companies. As adoption increases, it's expected that its energy consumption will quadruple in the coming years (Cojocaru & Cojocaru, 2024). This rise in energy consumption is mainly driven by the rapid development of new AI-specific architectures aimed at enhancing the performance of AI models.

The development, deployment, maintenance, and disposal of AI technologies have a significant impact on energy consumption and contribute to greenhouse gas emissions. One of the direct effects is the high energy usage, which leads to substantial carbon emissions, especially when data centers rely on non-renewable energy sources. The growing use of AI could also have unintended side effects, such as a higher demand for water, electricity, and other resources. Additionally, it requires the construction of infrastructure and the production of hardware components (UNFCCC, 2024). Over time, these factors can lead to broader, systemic consequences for both the environment and society, particularly for the developing countries. Unlike wealthier nations, developing countries face unique challenges and risks that could hinder the effectiveness and fairness of AI technologies when used for climate action. There are several factors contributing to these difficulties, and understanding them is key to developing strategies that maximize the positive impact of AI while minimizing any negative effects or harmful practices (UNIDO, 2024). This ensures that AI technology can be used responsibly, ethically, and equitably by all.

When AI is applied sensibly, fairly, and morally, AI plays a valuable role in fighting climate change. For example, by helping reduce emissions or predict extreme weather events, as it has seen in projects like Fair Forward's forest mapping in Indonesia and Sunny Lives AI. However, at the same time, training AI models comes with a significant carbon footprint. Each model can produce up to 300,000 kg of carbondioxyde, largely due to the massive computing power and physical infrastructure required. That's why it's so important to estimate and monitor the carbon impact of AI using tools like ML carbon dioxide impact or code carbon, and to weigh those emissions against the potential climate benefits, both short and long term (Butler et al., 2024). In considering AI's impact on climate change, it's also crucial to examine how its infrastructure and lifecycle contribute to its overall environmental footprint. By understanding the different stages, from software development to hardware setup, we can find ways to cut down on energy use and reduce carbon emissions.

The AI lifecycle is considered in two main stages: software and hardware. In the software realm, this lifecycle includes phases such as collecting and preparing data, developing models, training them, validating their accuracy, deploying them for use, performing inferences, and maintaining and eventually retiring the system. For hardware, it involves the creation of essential computer chips, particularly graphical processing units that are critical for training and inference processes, alongside the construction and operation of data centers (UNEP, 2024). Data centers are the most energy consumers. Three main factors drive the energy consumption of data centers: cooling, storage, and processing. As AI models grow more complex, they require larger datasets and faster access times, which in turn increases storage needs and boosts power consumption. In fact, nearly 40% of a data center's electricity is spent on cooling, making it a major contributor to overall energy use. Traditional air-based cooling methods can't handle the high-density demands of AI and cloud workloads, so more efficient solutions like liquid and immersion cooling have become crucial. While these advanced cooling techniques improve thermal efficiency, they do come with high upfront costs for the specialized equipment and infrastructure required (Chauhan, 2024). Despite these energy demands, AI's impact on industries goes beyond just its power consumption. As AI continues to evolve, it is increasingly filling the gaps left by traditional hardware sensors, offering the potential to optimize processes and improve energy efficiency across sectors. This makes it an important tool for achieving long term environmental goals.

AI's ability to fill the gaps left by traditional hardware sensors is revolutionizing the industrial sector by providing deeper insights and streamlining processes. This not only boosts energy efficiency but also plays a vital role in cutting emissions. As industries increasingly adopt AI, the potential for significant environmental benefits, like reduced carbon footprints and more sustainable production practices, becomes clearer. Despite its own energy demands, AI is becoming a key player in shaping a more energy efficient and ecofriendly industrial landscape.

THE FUTURE OF AI AND CLIMATE SOLUTIONS

In late 2017, a major breakthrough occurred when Google DeepMind's AI program, AlphaZero, defeated Stockfish, the world's most powerful chess engine at the time. What made AlphaZero unique was that it didn't rely on pre-programmed moves or human strategies. Instead, it developed its own approach entirely through AI training. Then, in 2020, OpenAI introduced GPT-3, a model capable of generating human-like text based on a prompt. GPT-3 can complete sentences, answer questions, create drafts, and even simulate potential conversations. It generates responses on any topic by analyzing vast amounts of online text. The success of AlphaZero and the abilities of GPT-3 highlight AI's increasing potential, not just to mimic human intelligence, but to go beyond traditional problemsolving methods (Kissinger et al.,, 2021). This highlights AI's immense potential to revolutionize various sectors, including solutions for climate change.

AI has the potential to reshape how we understand and address climate change. By utulizing AI techniques, humanity can speed up problemsolving, gain deeper insights into the causes of climate change, improve responses to its impacts, and create more effective solutions (Lewis, Toney & Shi, 2024). This progress not only offers a deeper understanding of climate-related challenges but also paves the way for more efficient, data-driven strategies to address them.

AI capabilities determine how effectively an organization can develop and implement AI technologies. To fully unlock AI's potential, substantial investment in research and development is crucial to strengthen these capabilities. This investment is key to staying competitive and driving growth. Companies that prioritize AI development are better positioned to use AI to boost efficiency, make smarter business decisions, and ultimately benefit various industries while contributing to climate solutions (Mondal et al., 2024). However, despite these opportunities, the urgency of addressing climate change often gets sidelined in political discussions, where more immediate concerns tend to take priority.

In light of this, The 2020 White Paper on Artificial Intelligence, published by the European Commission, emphasizes how AI systems can make a significant impact not only through their various benefits but also by helping reduce greenhouse gas emissions and supporting efforts to adapt to climate change (European Commission, 2020). AI is recognized as a key factor in achieving the United Nations Sustainable Development Goals (SDGs), particularly SDG 13: Climate Action. The European Green Deal, announced in December 2019, sets the goal for Europe to become the first climate-neutral continent by 2050, highlighting AI's crucial role in reaching this target. Additionally, the White Paper stresses the need to ensure that AI systems have a minimal environmental impact throughout their entire lifecycle, from production to algorithm training (Tuğaç, 2023). As AI continues to evolve, it offers innovative solutions for both mitigating and adapting to climate change, further underscoring its potential in building a sustainable future.

In order to fully utulize those AI capabilities, a strong digital framework is essential. This includes cloud computing, storage, and networking capabilities that support the collection, processing, and analysis of the vast amounts of data needed to train AI models. By making substantial investments in digital infrastructure, organizations can ensure their AI systems run efficiently, helping them tackle climate-related challenges and implement effective AI-driven solutions to combat climate change (Mondal et al., 2024). To make the most of these capabilities, it's vital to ensure that the infrastructure is not only robust but also adaptable to support the growth and scalability of AI solutions.

The global growth of AI is driven by advancements in algorithms, the availability of inexpensive computing power, and increasing data. This rapid development, particularly in ML, is mostly concentrated in countries like the United States and China. Predicting the pace of AI's progress is difficult, but it's clear that its capabilities will continue to expand significantly. Whether in five, ten, or twenty-five years, AI applications will become more efficient, cost-effective, and integrated into daily life. Looking ahead, it's reasonable to expect processing power to grow up to one million times faster, potentially creating neural networks comparable in size to the human brain (Kissinger et al., 2021). As AI Technologies continue to evolve, its transformative potential will extend to addressing some of the most pressing challenges of our time, such as climate change.

Emerging AI technologies offer significant potential for improving climate adaptation strategies. From enhancing resource management to boosting the resilience of infrastructure, AI is bringing forward innovative solutions to help communities adjust to shifting environmental conditions. In addition, AI plays a key role in climate mitigation efforts by predicting extreme weather events, promoting energy-efficient buildings, optimizing agricultural practices, supporting sustainable forest management, and improving waste management. These advances are helping create smarter, more sustainable solutions for addressing the challenges posed by climate change (Liu et al., 2023). Through these applications, AI can help create more resilient and sustainable environments.

Looking ahead to the next fifteen to twenty years, it's reasonable to expect that processing power could grow at a rate up to one million times faster, which may lead to the creation of neural networks comparable to the size of the human brain. As AI continues to advance, this rapid growth in computational power will only enhance AI's ability to tackle climate change. With such powerful systems, it is likely to see even more efficient solutions in areas like energy management, carbon capture, and environmental monitoring, ultimately driving a greater impact on global efforts for climate mitigation and adaptation.

EMERGING AI TECHNOLOGIES FOR CLIMATE ADAPTIONS

The planet is facing a serious threat from climate change, which is mainly caused by human activities such as burning fossil fuels, deforestation, and industrial emissions. The impacts of this crisis are already being felt, with rising global temperatures, more frequent extreme weather events, rising sea levels, and disruptions to ecosystems (Changlani & Thakore, 2023). Human activities are the primary cause of greenhouse gas emissions, with carbon dioxide accounting for roughly two-thirds of these emissions, primarily from burning fossil fuels and destroying forests. Methane, which is even more harmful, is mostly released from livestock farming, energy production, and heating. Nitrous oxide, fluorine, and other greenhouse gases are released from agricultural and industrial processes (ThinkTech). Addressing these challenges requires both reducing the damage already done and adjusting to the new realities of our environment. To effectively tackle these issues, it's crucial to harness innovative technologies, including artificial intelligence.

In this regard, AI is increasingly being used to help adapt to and build resilience against the impacts of climate change, especially in tracking long-term changes like rising sea levels and predicting extreme weather events such as hurricanes and droughts. As an integral part of modern society, AI is uniquely equipped to tackle the complex challenges caused by climate change. With tools like machine learning and predictive analytics, AI can analyze vast amounts of data, spot patterns, and offer valuable insights that help reduce the impact of climate change (Stein, 2020). These technologies not only support immediate response efforts but also play a crucial role in shaping long-term strategies for a more sustainable future.

Building on this potential, the use of AI in combating climate change presents significant opportunities for both mitigation and adaptation efforts (Climate Change AI, 2021). One of the main reasons AI is becoming so widely used in tackling climate change is because of how unpredictable and complex the issue is. Traditional methods often struggle to keep up with the constantly shifting nature of climate systems. AI, on the other hand, is especially good at handling uncertainty. It can process huge amounts of data and uncover insights that lead to smarter, more informed decisions (Tuğaç, 2023). Continuing to face increasingly urgent climate challenges, AI offers a powerful tool to guide more effective and timely responses.

In line with this, AI-powered climate solutions refer to the application of AI technologies specifically designed to address environmental challenges (Mondal et al., 2024). To avoid the worst effects of a 2°C rise in global temperatures, the Intergovernmental Panel on Climate Change (IPCC) has called for cutting greenhouse gas emissions by 45% by 2030 and reaching net-zero emissions by 2050. AI can play a key role in reaching these goals by helping optimize energy use, boost resource efficiency, and drive the development of clean Technologies (Stein, 2020).

Realizing the full potential of AI will take more than just technological innovation. It also requires smart policies and strong collaboration across different sectors.

To make this integration successful, it is widely believed that policymakers must take action in three key areas. First, they should encourage the development and deployment of AI technologies specifically designed to combat climate change. Second, it's important to manage the impact of AI-driven technologies that are already in use or emerging across different sectors. Finally, strengthening the public sector's ability to both regulate and innovate at the intersection of AI and climate change will be key to ensuring successful implementation (Kaack et al., 2020). By taking these steps, policymakers can create a framework that allows AI to play a pivotal role in addressing climate change.

A great example of this is SEEDS in India, a non-profit organization that has developed an AI-powered system to predict natural disaster risks by analyzing local factors. This system improves disaster response and speeds up the delivery of humanitarian aid. Their Sunny Lives AI model considers elements like water bodies, rivers, proximity to road networks, and landslide risks, helping to create more effective strategies for reducing the impact of climate-related disasters. This innovative approach underscores the crucial role AI plays in enhancing disaster preparedness (Butler et al., 2024). This innovative approach highlights the essential role AI plays in improving disaster preparedness, demonstrating how it can greatly enhance our ability to respond to and reduce climate-related risks.

Building on this, global initiatives like the United Nations's (UN) Early Warnings for All, in partnership with Microsoft and Ethiopia's Ministry of Irrigation, are using AI, satellite imagery, and predictive modeling to identify at-risk communities and improve disaster preparedness (UNIDO, 2024). AI also plays a crucial role in minimizing the environmental impacts of agrochemical use. By analyzing agricultural data, farmers can optimize resource use, reduce chemical applications, and improve crop yields in more sustainable ways. These efforts are part of a broader strategy to harness AI for enhancing environmental sustainability, from disaster risk reduction to more sustainable farming practices (Liu et al., 2023). This highlights the growing potential of AI in driving sustainable solutions across various sectors.

AI is also instrumental in developing adaptive strategies for climate change. It helps predict extreme weather events, track sea-level rise, and monitor environmental data in real-time, allowing communities to better prepare for climate impacts and respond more effectively (Changlani & Thakore, 2023). Several key initiatives are using AI to combat climate change. Programs like the UN's AI for Earth and Microsoft's AI for Good are focused on reducing disaster risks, improving environmental monitoring, and building climate resilience through the use of AI (Changlani & Thakore, 2023). This proactive approach becomes even more effective when AI is integrated into climate strategies, boosting resource efficiency and improving environmental monitoring across various sectors.

AI technologies are key to enhancing resource efficiency and improving environmental monitoring. By integrating AI into climate strategies, significant progress can be made in sectors like agriculture, energy, and disaster management, helping to reduce emissions, improve sustainability, and foster resilience to climate change (Mondal et al., 2024). AI-driven climate solutions are not just technological advancements, yet they are still among the most efficient methods for addressing climate change. By optimizing energy use, reducing emissions, and enabling smarter resource management, AI plays a pivotal role in climate mitigation and adaptation efforts (Pimenow et al., 2024). Ultimately, the integration of AI into climate strategies represents a transformative step toward a more sustainable and resilient future, where technology acts as a critical enabler in the fight against climate change.

In this context, the global population is projected to reach nearly 10 billion by 2050, significantly increasing the demand for food while climate change exacerbates vulnerabilities in agricultural systems. This makes food security a critical issue for both national and global stability (Kılavuz, 2024). Therefore, emerging AI technologies are crucial in the fight against climate change. They help with disaster preparedness, managing water resources more efficiently, and improving farming practices. ML in particular, is great at analyzing large amounts of data and making accurate predictions. It's becoming a key tool in climate science, helping spot patterns, forecast impacts, and find smarter solutions to climate-related challenges (Stein, 2020). There are several key initiatives looking at how AI can help tackle climate challenges. For instance, the AI for Climate Action Initiative and the International Telecommunication Union's AI for Good and Green Digital Action programs bring together experts and organizations to develop real-world solutions that support the UN's SDGs especially climate action. The International Energy Agency also runs programs like Energy for AI and AI for Energy, focusing on how AI can improve energy systems and meet growing energy demands. Meanwhile, the Global Partnership on AI, in collaboration with the Organisation for Economic Cooperation and Development (OECD), promotes the responsible use of AI while addressing its environmental impact. And organizations like Climate Change AI, a non-profit, are using ML to support climate mitigation, adaptation, and resilience all while working to minimize AI's own carbon footprint (UNIDO, 2024). AI has the potential to play a big role in addressing major environmental sustainability challenges. It can help close knowledge gaps and reduce human bias, leading to more informed, evidence-based decisions and policies. As the demand for renewable energy continues to grow, AI is also proving valuable in optimizing how we use these resources by analyzing data and predicting energy production, it helps tackle key issues in energy generation and management (Anavatan, 2023). The UN

Environment Programme (UNEP) plays a leading role in shaping global environmental policies and promoting sustainable development. It also looks closely at the environmental impact of digital technologies exploring how they can support sustainability goals while also examining their broader effects on the planet unep, 2024). As AI continues to evolve, its thoughtful and sustainable application will be key to shaping a future where technology serves both people and the planet. With the right guidance, AI can become one of the most powerful tools in building a more resilient, sustainable world.

Ultimately, AI holds incredible promise in the fight against climate change. It's becoming an essential part of how we build climate resilience and move toward a more sustainable future. But as we embrace AI's potential, we also need to be mindful of its own environmental impact, especially its rising energy consumption. The tools we use to protect the planet should be sustainable themselves. With global cooperation and a thoughtful, responsible approach, it is obvious that AI becomes a force for good where technology and nature grow together, creating a better, fairer world for everyone.

CONCLUSION

AI is becoming a key player in the fight against climate change, providing innovative solutions across research, engineering, and policy-making for both mitigating and adapting to climate impacts. As the world increasingly focuses on tackling the climate crisis, AI's ability to optimize processes, predict climate outcomes, and support better decision making will be crucial. How AI is used in the coming years will have a significant impact on our progress toward climate goals, with its flexibility either speeding up or slowing down efforts, depending on how it's applied. Creating strong policy frameworks to guide AI's role in supporting climate strategies will be essential to ensure it helps build a more sustainable and equitable future for all.

AI has huge potential to drive climate action, with applications covering areas like monitoring, data collection, climate forecasting, resource management, energy efficiency, and disaster risk reduction. Research shows that AI is playing a crucial role in tackling climate challenges. Its ability to predict and model climate impacts can significantly improve how we prepare for and respond to these challenges. A great example of this is AI's impact on weather forecasting, where analyzing large volumes of data directly enhances accuracy. This growing capability helps reduce disaster risks and optimize land use, showing just how powerful AI can be in supporting climate resilience.

Moreover, AI is a crucial part of climate adaptation efforts. By using techniques like supervised learning and reinforcement learning, it becomes easier to navigate the complexities of adaptation decision-making. For example, ML is playing a key role in advancing technologies such as smart grids, disaster forecasting, and risk assessment all of which are essential for cutting emissions and improving disaster preparedness. In addition, AI's ability to optimize industrial processes and boost energy efficiency is central to lowering carbon emissions and reaching carbon neutrality. It's already being applied in sectors like agriculture, transportation, and energy use, helping make these industries more sustainable.

The success of AI in climate strategies also relies on having strong digital infrastructure in place. For AI to effectively analyze climate data and inform policy decisions, it needs the ability to collect, process, and share vast amounts of information. A robust digital infrastructure will allow AI to drive more informed, data-driven decisions, ultimately advancing climate action more effectively.

Finally, AI-powered solutions have tremendous potential to help emerging markets shift toward net-zero carbon economies. By improving energy efficiency, automating carbon emission tracking, and supporting scalable, flexible solutions, AI can enable these economies to skip traditional development stages while promoting sustainable growth. Its ability to predict climate impacts and automate policy decisions boosts resilience, ensuring more effective responses to the ongoing challenges of climate change.

In conclusion, AI is proving to be a powerful ally in the fight against climate change helping build cleaner technologies, use energy more efficiently, improve climate predictions, and adapt to changing conditions. However, as reliance on AI grows, it's important to remember that it also comes with its own environmental cost. Training large models and processing huge amounts of data require significant energy, and if not carefully managed, that could work against the very sustainability goals that are being tried to achieve. That's why it's crucial to shape AI development with smart policies that not only steer it toward climate solutions but also make sure it's used in an energy-efficient, responsible way. With the right planning and strong digital infrastructure, AI can continue to support a more sustainable, climate-resilient future for everyone.

REFERENCES

Anavatan, A. (2023). Yapay Zekâ ve Sürdürülebilirlik. In Türk, B. (Ed.), *İyisiyle Kötüsüyle Sürdürülebilirlik: Yaşasın İyilik* (pp. 117–138). Nobel Akademik Yayıncılık.

BBC News. (2017). Google's 'superhuman' DeepMind AI Claims Chess Crown. 6.12.2027, https://www.bbc.com/news/technology-42251535

Bolon-Canedo, V., Moran-Fernandez, L., Cancela, B., & Alenzo-Betanzos, A. (2024). A review of green artificial intelligence: Towards a more sustainable future. *Neurocomputing*, *599*, 128096. DOI: 10.1016/j.neucom.2024.128096

Butler, K., Harzbecher, M., Egbers, S. M., Voigt, L. & Lehmann, E. (2024). AI for Climate Action Applications and Principles in Development Cooperation. *Deutsche Gesellschaft für Internationale Zusammenarbeit (GIZ)*, 2-15.

Çetinkaya, E. S., & Genç, G. (2024). Enerji Sektöründe Yapay Zekâ Uygulamaları. In Özsoylu, A. F., & Kılavuz, E. (Eds.), *Sürdürülebilir Kalkınmada Dijital Dönüşüm ve Yapay Zekâ Uygulamaları* (pp. 179–212). Akademisyen Yayınevi Kitabevi.

Changlani, K., & Thakore, R. (2023). Artificial Intelligence for Climate Action. https://www.researchgate.net/publication/374899918_Artificial_Intelligence_for_Climate_Action

Chauhan, S. (2024). The Growing Energy Demand of Data Centers: Impacts of AI and Cloud Computing. *International Journal for Multidisciplinary Research*, *6*(4), 1–10.

Chen, L., Chen, Z., Zhang, Y., Liu, Y., Osman, A. I., Farghali, M., Hua, J., Al-Fatesh, A., Ihara, I., Rooney, D. W., & Yap, P. S. (2023). Artificial Intelligence-based Solutions for Climate Change: A Review. *Environmental Chemistry Letters*, *21*(5), 2525–2557. DOI: 10.1007/s10311-023-01617-y

Climate Change, A. I. the Centre for AI and Climate, the Global Partnership on Artificial Intelligence's Committee on Climate Action and Biodiversity Preservation. (2021). *Climate Change & AI: Recommendations for Government*. Retrieved from: https://gpai.ai/projects/ responsible-ai/environment/climate-change-and-ai.pdf

Climate Trace. Retrieved from: https://www.climatetrace.org/team

Cojocaru, S. S. T. & Cojocaru, C. M. (2024). The Influence of Artificial Intelligence on Energy Consumption in Data Centers: Bibliometric Research. *EasyChair*, 13792.

Dannouni, A., Deutscher, S. A., Dezzaz, G., Elman, A., Gawel, A., Hanna, M., Hyland, A., Kharij, A., Maher, H., Patterson, D., Jones, E. R., Rothenberg, J., Tber, H., Texier, M., & Ziat, A. (2023). Accelerating Climate Action with AI. Boston Consulting Group, https://web-assets.bcg.com/72/cf/b609ac3d4ac6829bae6fa88b8329/bcg-accelerating-climate-action-with-ai-nov-2023-rev.pdf

Demirci, M. (2024). İklim Değişikliği ve Yapay Zekâ Teknolojisinde Yaşanan Gelişmeler Işığında İlerleme Düşüncesinin Evrimi. *Dumlupınar Üniversitesi Sosyal Bilimler Dergisi, 80*(80), 249–269. DOI: 10.51290/dpusbe.1448966

European Commission. White Paper: On Artificial Intelligence - A European Approach to Excellence and Trust. https://commission.europa.eu/system/ files/2020-02/commission-white-paper-artificial-intelligence-feb2020_en.pdf da Hidalgo, J. I., Vega, V. V., Ceberio, J., Garnica, O., Velasca, J. M., Cortes, J. J., Villanueva, R. J. & Alvarez, J. D. (2023). Sustainable Artificial Intelligence Systems: An Energy Efficiency Approach. Authorea Preprints, https://www.semanticscholar.org/paper/Sustainable-Arti%EF%AC%81cial-Intelligence-Systems%3A-An-Hidalgo-ega/8d53f553e232f5423e74ac8d60a2174f93f7418a

Kaack, L. H., Donti, P. L., Emma, S. & David, R. (2020). Artificial Intelligence and Climate Change: Opportunities, Considerations, and Policy Levers to Align AI with Climate Change Goals. *Henrich Böll Stiftung Ecology*, 1-15.

Kılavuz, E. (2024). Dijital Gelişmeler Çerçevesinde Tarım Sektörünün Analizi. In Özsoylu, A. F., & Kılavuz, E. (Eds.), *Sürdürülebilir Kalkınmada Dijital Dönüşüm ve Yapay Zekâ Uygulamaları* (pp. 109–142). Akademisyen Yayınevi Kitabevi.

Kissinger, H., Schmidt, E., & Huttenlocher, D. (2021). *The Age of AI: And Our Human Future.* Little, Brown and Company.

Kumar, Y., Marchena, J., Awlla, A. H., Li, J. J., & Abdalla, H. B. (2024). The AI-Powered Evolution of Big Data. *Applied Sciences (Basel, Switzerland), 14*(22), 1–41. DOI: 10.3390/app142210176

Lewis, J. I., Toney, A., & Shi, X. (2024). Climate Change and Artificial Intelligence: Assessing the Global Research Landscape. *Discover Artificial Intelligence, 4*(64), 64. DOI: 10.1007/s44163-024-00170-z

Lucas, M. (2024). The Role of AI in Climate Change Mitigation and Environmental Monitoring. Retrieved from: https://www.researchgate.net/publication/377762233_The_ Role_of_AI_in_ Climate_Change_Mitigation_and_Environmental_Monitoring

Masterson, V. (2024). 9 Ways AI is Helping Tackle Climate Change. World Economic Forum, Retrieved from: https://www.weforum.org/stories/2024/02/ai-combat-climate-change/

Materia, S., García, L. P., & Straaten, C. V. (2024). Artificial Intelligence for Climate Prediction of Extremes: State of the Art, Challenges, and Future Perspectives. *Wiley Interdisciplinary Reviews: Climate Change*, *15*(6), 1–31. DOI: 10.1002/wcc.914

Mehryar, S., Yazdanpanah, V., & Tong, J. (2024). AI and Climate Resilience Governance. *iScience*, *27*(6), 1–12. DOI: 10.1016/j.isci.2024.109812 PMID: 38784017

Miller, T., Durlik, I., Kostecka, E., Łobodzińska, A., & Matuszak, M. (2024). The Emerging Role of Artificial Intelligence in Enhancing Energy Efficiency and Reducing GHG Emissions in Transport Systems. *Energies*, *17*(24), 1–31. DOI: 10.3390/en17246271

Mondal, S., Das, S., & Vrana, V. G. (2024). Exploring the Role of Artificial Intelligence in Achieving a Net Zero Carbon Economy in Emerging Economies: A Combination of PLS-SEM and fsQCA Approaches to Digital Inclusion and Climate Resilience. *Sustainability (Basel)*, *16*(23), 1–35. DOI: 10.3390/su162310299

Pimenow, S., Pimenowa, O., & Prus, P. (2024). Challenges of Artificial Intelligence Development in the Context of Energy Consumption and Impact on Climate Change. *Energies*, *17*(23), 1–34. DOI: 10.3390/en17235965

Sadhukhan, A., Kunhu, A., Karmakar, A., & Acharya, S. (2019). Illegal Hunting and Poaching Detection Using AI in Amazon Rainforest Birds. In Benavente-Peces, C., Slama, S. B., & Zafar, B., (Eds.), *Proceedings of 1st International Conference on Smart Technology for Emerging Problems*, (pp. 1-6). Springer.

Singh, M. (2024). How AI and Earth Observation can Help Life on Our Planet. World Economic Forum. Retrieved from: https://www.weforum.org/stories/2024/01/ai-earth-observation-technology/

Sleeman, J. (2023). Using Artificial Intelligence to aid Scientific Discovery of Climate Tipping Points. Retrieved from: https://www.researchgate.net/publication/368507547_ Using_Artificial_ Intelligence_to_aid_Scientific_Discovery_of_Climate_Tipping_Points

Stein, A. L. (2020). Artificial Intelligence and Climate Change. *Yale Journal on Regulation*, *37*(890), 891–934.

ThinkTech Stem Teknoloji Düşünce Merkezi. Küresel İklim Değişikliği ile Mücadelede Yeni Nesil Teknolojiler. Retrieved from: https://thinktech.stm.com .tr/uploads //docs/1638270030_ stmblogkureseliklimdegisikligiilemucadele.pdf?v=1643068800028

Tuğaç, Ç. (2023). Birleşmiş Milletler Sürdürülebilir Kalkınma Amaçlarının Gerçekleştirilmesinde Yapay Zekâ Uygulamalarının Rolü. *Sayıştay Dergisi*, *34*(128), 73–99. DOI: 10.52836/sayistay.1245051

United Nations Environment Programme (UNEP). (2024). Artificial Intelligence (AI) end-to-end: The Environmental Impact of the Full AI Lifecycle Needs to be Comprehensively Assessed. Retrieved from https://wesr.unep.org/sites/default/files/ 2024-09/AI%20Environmental%20Impact%20Issues%20Note%20%281%29.pdf

United Nations Framework Convention on Climate Change (UNFCCC). (2024). Draft Technical Paper on AI for Climate Action, Technology Executive Committee, Twenty-nineth Meeting. Retrieved from: https://unfccc.int/ttclear/misc_/StaticFiles/ gnwoerk_static/tn_ meetings/0ec396b0ba7b4d0d853b77c7b83dc172/3ebbf2e8e-7834a7f873b0ae9a86262f7.pdf

United Nations Industrial Development Organization (UNIDO). (2024). Artificial Intelligence for Climate Action in Developing Countries: Opportunities, Challenges and Risks. Retrieved from: https://unfccc.int/ttclear/misc_/StaticFiles/gnwoerk _static/ AI4climateaction/28da5d97d7824d16b7f68a225c0e3493/a4553e8f70f-74be3bc37c929b73d9974.pdf

Chapter 2
Artificial Intelligence in the Circular Economy and Its Impact on Climate Change:
A Review of Emerging Applications and Challenges

Lokeshwari D. V.
https://orcid.org/0000-0002-6112-3173
Dayanand Sagar College of Arts, Science, and Commerce, India

ABSTRACT

This paper integrates studies on AI deployment in the Circular Economy through the lens of waste management, resource beneficiation, and product life cycle analysis. Examination of more than sixty studies in machine learning, digital twining, and IoT reveals that AI reduces wastage, boosts recycling coverage and accuracy, and enhances efficiency—up to 25% and 40% respectively. AI-based systems incorporating life cycle studies and real-time monitoring deliver closed-loop production and robust supply chains. However, the barriers exist in data quality and energy consumption by the AI systems and in the realms of ethics and scalability through gaps in policies and infrastructural support. The studies underscore the transformative power of AI while noting the imperative for overcoming socio-technical barriers to successful integration in circular economy transformations.

DOI: 10.4018/979-8-3373-5238-1.ch002

INTRODUCTION

Research on the integration of Artificial Intelligence (AI) in the Circular Economy (CE) has received notable interest due to its potential to resolve the world's most burning environmental challenges through greater efficiency and reduced wastage. The transition from conventional linear "take-make-dispose" systems to restorative CE systems continues to be supported by digital technologies such as AI, Internet of Things (IoT), and big data analytics (Kio et al., 2022; Pathan et al., 2023). The last few years have shown radical potential held by AI in the enhancement of supply chains, lifecycles of products, and trash processing and consequently contributing to the achievement of sustainability objectives and economic prosperity (Singh, 2024; Mamudu et al., 2024). As global trash production continues to be predicted at 3.4 billion tons by the year 2050, the importance of AI centric solutions to mitigate environmental impacts and enhance circularity continues to be pronounced (Fang et al., 2023; Machine Learning and Artificial Intelligence, 2022).

Despite the growing scrutiny, major barriers continue to exist in the implementation of artificial intelligence in circular economy aspects more particularly connected to waste management, optimisation of resources and life cycle assessment. Recent studies offer fragmented findings on the practical implementation and viability of AI in overcoming systemic inefficiencies (Pathan et al., 2023; Bashynska & Prokopenko, n.d.; Arun et al., 2024). Even environmental repercussions of AI deployment in the form of the excessive energy required in model training that negates gains in the context of sustainability have also been voiced (Jaiswal & Pandey, 2024; Bashynska & Prokopenko, n.d.). Lastly, the socio-economic consequences—like job losses and data security concerns—make widespread adoption challenging (Mankar et al., 2024; Olawade et al., 2024). These factors hinder the overarching achievement of AI's potential for enabling sustainable industrial change (Alamelu et al., 2024; Ibrahim, 2024).

The conceptual model presented in this review integrates the values of Circular Economy (CE)—reduction, reuse, and recycling—concepts and the AI technologies fostering data-informed decision-making, predictive forecasting, and automations (Noman et al., 2022; Popolo et al., 2024; Daramola et al., 2023). The model depicts the interdependent relationship whereby AI facilitates CE by optimizing the flows of resources and prolonging the lifetimes of the products and thereby facilitating closed loops (Kumar et al., 2024; Moorthy et al., 2024). By utilizing the values of sustainability science and the Industry 4.0 worldviews, the review recognizes AI as the key change agent in the transformation to regenerative economies (Keivanpour, 2023; Agrawal et al., 2021).

Thus, the present systematic review aims to review the application of AI in managing wastage, optimizing the use of resources, and lifecycle evaluation. Empirical studies and technological advancements, the work clarifies the role of artificial intelligence in managing challenges pertinent to circular economy while highlighting research gaps that hamper widespread application. By integrating knowledge on AI-based circular practices and providing insights for future research and practical implementation and policy formulation (Besigomwe, 2024; Bashynska & Prokopenko, n.d.), the research contributes to knowledge development.

The research adopts a systematic literature synthesis method that incorporates peer-reviewed articles and case studies carried out on fields applicable to AI CE. The analysis applies thematic categorization and critical analysis of the influences of technologies, potential opportunities and constraints. The findings are structured in the context of waste management, resource enhancement, and lifecycle assessment and then the integration challenges and future trends (Farshadfar et al., 2024; Rusch et al., 2022).

RESEARCH METHODOLOGY

The deployment of Artificial Intelligence in the Circular Economy context, highlighting areas including waste management, resource optimisation, and product lifetime analysis. Paper aims to synthesise available knowledge, discover technological trends and gauge the effectiveness of AI enabled solutions in realising CE goals. The relationship between AI and sustainability and provides evidence for the potential intelligent systems to reduce waste and prolong the lifetimes of products through the optimisation of resources. By synthesising knowledge from various industries and research studies, this review offers an in depth understanding of the potential role of AI in facilitating CE transformations and provides useful insights for future studies, policy frameworks and implementation practices. The precise objectives focus on the assessment of the deployment of AI in garbage handling, the comparison of multiple AI-enabled optimisation techniques for resources, the identification of AI techniques for lifetime analysis and sustainability evaluation, the evaluation of the effectiveness of AI formulations in recycling and material retrieval processes and the surveying the challenges and opportunities inherent in the integration of CE and AI. The purpose of attaining the objectives and research question exploring key uses of Artificial Intelligence in the Circular Economy, highlighting waste management, optimisation and analysis of lifetimes and resources of products was methodically extended in concentrated sub questions to capture specialty studies and ensure that pertinent knowledge domain insights are not disregarded. The literature search used inclusion and exclusion criteria and concentrated on the last five years' peer-reviewed studies while excluding studies outside the CE and AI fields. The preliminary screening yielded 539 papers and by the application of citation chaining that included the backward tracing of lists of references to capture the roots studies and the forward tracing of works that cite them to capture the new debates emerging, a further 49 papers were obtained.

This procedure produced 588 studies in total, which were then subject to relevance scoring to verify consistency in the research scope and ultimately narrowed the papers down to 70 very relevant papers on which this review is based.

DISCUSSION

Discourse Most recent studies accentuate the diversified implementation of Circular Economy (CE) promotion by Artificial Intelligence (AI) in various industries. For manufacturing, energy, and logistics sectors, AI approaches such as generative AI and predictive maintenance have been shown to reduce wastage and energy intensity when complemented by Internet of Things (IoT), blockchain, and policy initiatives (Jaiswal & Pandey, 2024; Bashynska & Prokopenko, n.d.). Data-driven design, life cycle assessment (LCA), and material flow analysis have also advanced CE processes; however, customized approaches remain imperative (Kio et al., 2022). Examples in closed-loop manufacturing demonstrate that machine learning (ML) and reinforcement learning (RL) can achieve measurable effectiveness gains including a 12% decrease in raw materials and a 25% enhancement in the recovery of materials despite the complexity inherent in dynamic decision-making (Besigomwe, 2024). Solid and e-waste management can benefit from ML techniques such as support vector machine (SVM), random forest (RF), convolutional neural network (CNN), and artificial neural network (ANN) in improving recycling rates and classification accuracy while struggling with data veracity (Singh et al., 2024). For product life cycle management, ML-fortified life cycle analytics and predictive control reduce production inefficiency while the viability of implementation represents a major obstacle (Shipilova & Bobkov, 2024). CE supply chains, predictive analytics through AI and automations using robots represent real time savings and environmental impacts reductions while security and fairness through data and physical infrastructural challenges continue to pervade (Mankar et al., 2024; Singh, 2024).

Research indicates that the contexts of Industry 5.0 and smart cities illustrate that artificial intelligence and the Internet of Things facilitate the accurate allocation of resources, adaptive energy consumption, and real-time decision-making support, thus improving sustainability metrics despite encountering challenges related to infrastructure, skill management, and integration (Shamsuddoha et al., 2024; Mamudu et al., 2024; Kumar et al., 2024). Specific applications in sectors such as construction, electronics, and fashion reveal that the amalgamation of AI with IoT, blockchain, or robotics enhances eco-design, traceability, recycling automation, and consumer-focused sustainability, even as significant costs, regulatory voids, and obstacles to adoption persist (Levchenko et al., 2024; Goel et al., 2024; Ramos et al., 2023; Cheng et al., 2024; Zhou & Zhao, n.d.). In the context of electronic waste recovery and disassembly operations, AI methodologies like AdaBoost, in conjunction with robotics and vision systems, significantly improve recycling efficiency and accelerate disassembly processes by as much as 35% quicker, while ongoing challenges include safety, scalability, and the complexities inherent in recycling practices (Ranganathan et al., 2024; Maheshwari & Gupta, 2024; Yeruva et al., 2024). Comprehensive reviews and bibliometric analyses further validate these observations and recognize AI as a pivotal factor in promoting a circular economy by facilitating supply chain tracking, waste monitoring, and sustainable product design, while highlighting persistent deficiencies concerning data quality, ethical considerations, integration, and practical implementation (Noman et al., 2022; Pathan et al., 2023; Rusch et al., 2022; Baumgartner et al., n.d.; Popolo et al., 2024).

A critical and comprehensive review of the available scholarly sources presents significant advantages and disadvantages connected to the application of Artificial Intelligence in the Circular Economy context. Technically, the variegated set of methodologies such as machine learning (ML), reinforcement learning (RL), predictive analysis, and digital twin models has shown promise in optimizing the use of resources and reducing wastage in general (Besigomwe, 2024; Singh, 2024; Zocco et al., n.d.). Systematic reviews and bibliometric studies further enhance this research by tracing trends and research avenues (Noman et al., 2022; Machine Learning and Artificial Intellig…, 2022). Most studies, however, depend on simulation or case studies lacking extensive real-life verifications and thereby limiting generalizability, while a scarcity of longitudinal data and weak integration with industry processes narrow the scope for practical applicability (Besigomwe, 2024; Neto et al., 2023; Valeriya et al., n.d.; Popolo et al., 2024). The quality of data presents both enabling and inhibiting functions: while AI takes advantage of the larger dataset for better categorization of wastes and improved lifecycle assessment and tracking of resources (Singh et al., 2024; Shennib et al., 2024; Dhanya et al., 2023), incompleteness or inconsistency in the dataset and gaps in the available infrastructural support and risk of infringement on the right to privacy question the reliability (Olawade et al., 2024; Arun et al., 2024; Cugurra et al., 2024). Environmentally speaking, the AI-aided optimizations yield lower material intensity and reduced wastage and energy intensity while sometimes enabling the achievement of net-zero emissions (Besigomwe, 2024; Singh, 2024; Rane et al., 2024), while the energy-intensive nature of training complex AI models—most noticeably the generative AI and deep learning—entails environmental trade-offs remaining inadequately examined (Jaiswal & Pandey, 2024; Pratap & Venkatesh, 2024; Durai et al., 2024).

In waste management, AI improves sorting, recycling mechanization, and material recovery and gains measurable efficiencies (Singh et al., 2024; Cheng et al., 2024; Maheshwari & Gupta, 2024), while scalability suffers due to high deployment costs, excessive technical complexity, and complexity in the handling of heterogenic waste streams (RS & Kumar, 2024; Rani et al., 2025). Similarly, AI-fomented lifecycle analysis and resource optimisation support predictive maintenance, quality monitoring and closed-loop production (Shipilova & Bobkov, 2024; Keivanpour, 2023; Aroul, 2024), while a lack of standardised measures of sustainability and integration barriers to legacy systems hamper wider applicability (Rusch et al., 2022; Baumgartner et al., n.d.). Ethical, social, and policy considerations lag unevenly; while other studies focus on the imperative for responsible AI, on regulative frameworks and on the adaptation of the workforce (Jaiswal & Pandey, 2024; Sánchez-García et al., 2023), doubts regarding the presence of algorithmic bias, job displacement and the lack of effective policy coverage prevail (Olawade et al., 2024; Ibrahim, 2024; Cugurra et al., 2024). Lastly, technology integration holds hope through interoperations among IoT and blockchain and_big data analytics and the support for scalable CE solutions and Industry 5.0 systems (Bashynska & Prokopenko, n.d.; Shamsuddoha et al., 2024; Valeriya et al., n.d.), while isolated uses, interoperational complexities and infrastructural costs continue to limit scalability (Cheah et al., 2022; Gulyamov, n.d.; Shennib & Schmitt, 2021).

THEMATIC REVIEW OF LITERATURE

Thematic analysis including 70 studied articles unveils some key trends. Most commonly observed among 38 articles is the theme of the implementation of artificial intelligence in the realm of garbage management when machine learning, computer vision, and robots are employed in order to enhance the process of sorting garbage, enhance the rate of recycling, and reduce the use of landfills by municipalities, electronics, and other industries (Singh et al., 2024; RS & Kumar, 2024; Olawade et al., 2024; Maheshwari & Gupta, 2024; Arun et al., 2024; Fang et al., 2023; Gulyamov, n.d.; Sleem & Elhenawy, n.d.). The deployment of resources through the help of artificial intelligence, observed in 33 articles, uses predictive analysis, real-time decision-making protocols, and digital twins to decrease the consumption of raw materials while optimizing the chain of supply and energy management (Bashynska & Prokopenko, n.d.; Besigomwe, 2024; Mankar et al., 2024; Mamudu et al., 2024; Ibrahim, 2024; Thakkar et al., 2024; Dhanya et al., 2023). The other notable theme deals with product life cycle analysis and sustainability assessment, as seen in 27 articles, in which predictive maintenance, life cycle assessment (LCA), and sustainable design help enhance the adoption of closed-loop production (Shipilova &

Bobkov, 2024; Rusch et al., 2022; Popolo et al., 2024; Zhou & Zhao, n.d.; Machine Learning and Artificial Intellig…, 2022).

In addition, 21 studies highlight AI application in circular supply chains and reverse logistics and IoT and blockchain-enhanced material recovery, reverse logistics, and visibility (Mankar et al., 2024; Farshadfar et al., 2024; Goel et al., 2024; Moorthy et al., 2024; Daramola et al., 2023). Accordingly, 19 articles position AI in Industry 4.0 and 5.0 revolutions using intelligent manufacturing, IoT-infused tracking, and human-AI symbiosis to drive sustainability and robustness (Shamsuddoha et al., 2024; Prajapati, 2024; Valeriya et al., n.d.; Jain, 2024; Keivanpour, 2023; Cheah et al., 2022). Sectoral innovation also comes through 15 studies in fashion, construction, packing, and e-waste, in which eco-design optimisation by AI, recycling technologies by AI, and conservation of resources takes place (Левченко et al., 2024; Ramos et al., 2023; Yeruva et al., 2024; Ashatkar, 2024; Johnson et al., 2022). Moral-ethical considerations and policy and socio-economic analysis appear in 12 studies centred on algorithmic bias, job disruption by AI, data protection and the necessity for the development of responsible AI using regulatory frameworks (Jaiswal & Pandey, 2024; Pratap & Venkatesh, 2024; Kumar et al., 2024; Pathan et al., 2023; Sánchez-García et al., 2023; Cugurra et al., 2024.

Moreover, 11 articles promulgate AI-enabled decision support systems using big data, federated learning, and machine learning for strategic planning and sustainability monitoring (Ranganathan et al., 2024; Dhanya et al., 2023; Onyeaka et al., 2023; Agrawal et al., 2021). Niche and growing areas are the application of artificial intelligence to reduce food wastage (7 articles), in which the prediction of crop yields, supply chain analysis, and redistribution mechanisms work towards reducing wastage and ensuring sustainability (Onyeaka et al., 2023). Robotic and autonomous systems (6 articles) utilizing AI vision and robotic handling allow for the effective disintegration, sorting, and recycling of complex wastestreams (Cheng et al., 2024; Zocco et al., n.d.; Maheshwari & Gupta, 2024). Overall, the three threads collectively establish AI's transformational capacity in CE transitions while at the very same time underscoring sectoral, methodological, and policy-oriented research lacunas.

CHRONOLOGICAL REVIEW OF LITERATURE

The existing body of literature pertaining to the integration of Artificial Intelligence (AI) within the Circular Economy (CE) has undergone substantial advancements, emphasizing key areas such as waste management, resource optimization, and product lifecycle evaluation. Early research, and particularly during the years 2021 and 2022, focused principally on foundation questions that employed bibliometric studies and systematic reviews to describe the profession and gauge the potential of AI to redesign supply chain functions, reduce waste, and permit lifecycle analysis. Early research studies also initiated the discussion on the integration of Circular Economy (CE) and Industry 4.0 technologies and yielded the foundation for the applied research that extended beyond. Up to the year 2023, research focus had shifted to the exploration of sectoral uses and policy implications evaluation through the example of prominent research aimed at the fashion sectors involvement with artificial intelligence, electronic waste recycling, and environmentally friendly product control. The period again emphasized the key strategic importance of digital transformation, business models built on the circular economy model, and the potential linkage of artificial intelligence with the achievement of the Sustainable Development Goals while at the same time accentuating the imperative for wider examination in respect to the diverse dimensions of data governance and ethical uses. The years ahead, namely 2024 to 2025, are expected to bring forth notable breakthroughs in technological innovation such that advanced artificial intelligence methods, such as machine learning, reinforcement learning, and federated learning, are predicted to be combined using Internet of Things (IoT), blockchain technologies, digital twins, and robots to resolve multiple challenges inherent in the circular economy through transdisciplinary approaches. Empirical case studies elucidate real-world improvements in the effectiveness of waste management, energy uses, and sustainability measures despite the continued challenges inherent in scalability, the ethic execution, and compatibility with standardization in regulatory measures.

CONVERGENCE AND DISAGREEMENT BETWEEN RESEARCHES

The research on the application of AI in the circular economy shows widespread consensus on various comparison parameters while the differences point to nuanced considerations. As far as coverage by application domain is concerned, the majority of the studies stress the implementation by the manufacturing, waste management, supply chain, and urban infrastructure domains and the enabling by AI of sustainable processes in e-waste, construction, fashion, and recycling of batteries (Jaiswal & Pandey, 2024; Bashynska & Prokopenko, n.d.; Singh et al., 2024; Munonye & Ajonye, 2024; Moorthy et al., 2024; Yeruva et al., 2024; Ashatkar, 2024; Zhou & Zhao, n.d.; Johnson et al., 2022). However, very few studies are particularly focused on individual fields such as the handling of food wastage (Onyeaka et al., 2023), the recycling of materials used in battery applications (Ranganathan et al., 2024), or the fashion sector (Ramos et al., 2023; Zhou & Zhao, n.d.). The difference mainly stems from the different scopes undertaken in the studies, ranging through wide reviews to focusing on specialized areas in order to obtain a deeper analysis.

There is consistency in the effectiveness delivered by artificial intelligence (AI) methods such that AI achieves waste reduction by minimizing what goes to landfills and optimizing the process to enhance the lifecycles by 12–25% in the recovery of materials and 7–15% in energy savings (Singh et al., 2024; Besigomwe, 2024; RS & Kumar, 2024; Valeriya et al., n.d.; Fang et al., 2023). Machine learning and deep learning are the most frequently identified effective methods (Sleem & Elhenawy, n.d.; Farshadfar et al., 2024). Other research lists the environmental quality of accessible data, the large amounts of energy to implement AI models, and scalability issues (Bashynska & Prokopenko, n.d. ; Jaiswal & Pandey, 2024; Olawade et al., 2024; Cugurra et al., 2024). Differences in effectiveness reported exist mainly as a function of the difference in research design, including simulation versus implementation in the real world, the maturity of the technology to be utilised, and the existence and locality of.

In the realm of integration frameworks, numerous studies suggest or critically review frameworks that integrate artificial intelligence and lifecycle assessments, digital twins, and sustainability metrics (Shipilova & Bobkov, 2024; Aroul, 2024; Baumgartner et al., n.d.; Popolo et al., 2024). New methodologies proposed are those of federated AI and product lifecycles management frameworks (Kumar et al., 2024; Popolo et al., 2024). Comparatively, other studies attempt to focus mainly on the application of AI lacking formal frameworks or thorough integration with lifecycles or sustainability metrics (Singh, 2024; Bashynska & Prokopenko, n.d.). These contrasts mainly represent the development state of the research area, whereby some studies represent the development of conceptual models while the rest focus on the importance of practical uses or literature reviews.

In the context of operational effectiveness improvements, many studies report gains ranging as far as 40% cost savings, 12% material savings, uptil 40% recycling rate increase and energy savings between 7% and 15% (Besigomwe, 2024; RS & Kumar, 2024; Vashishth et al., 2024; Fang et al., 2023; Gulyamov, n.d.). The application of AI-based predictive analytics and automation is highlighted as a way to enhance process effectiveness (Jaiswal & Pandey, 2024; Mankar et al., 2024; Farshadfar et al., 2024). However, the few studies available express doubts about the net savings by attributing them to the energy needs of AI technology, the cost of infrastructures and the barriers afforded by upfront investments (Jaiswal & Pandey, 2024; Gulyamov, n.d.; Durai et al., 2024). The controversies are informed by whether or not studies factor the whole lifecycle energy costs inherent to the AI systems and the specific economic environments the implementation occurs.

Finally, considering implementation challenges, there is general agreement on the challenge of data quality and accessibility, ethics, regulative constraints, and skills gaps (Bashynska & Prokopenko, n.d.;Olawade et al., 2024; Alamelu et al., 2024; Sánchez-García et al., 2023; Cugurra et al., 2024). Constant areas of concern are AI bias and data confidentiality (Ibrahim, 2024; Cugurra et al., 2024). However, there is divergence in the coverage of such challenges such that some researches are more concerned with technological constraints while other researches are concerned with socio-economic and policy issues depending on the focus (Sharma, 2023; Rane et al., 2024; Durai et al., 2024). Technically oriented researches are more concerned with algorithmic constraints while policy oriented researches are concerned with governance adaptation by the workforce and broad social considerations.

THEORETICAL AND PRACTICAL CONSIDERATIONS

Theoretical Implications

Consolidated evidence from the literature reviews confirms the theory that Artificial Intelligence (AI) acts as a major enabling factor for the evolution of circular economy (CE) values through the improvement of resource effectiveness, reducing wastage, and the enhancement of the management of the lifespan of products. This verifies the prevailing theory on the convergence of digital innovation and sustainable systems and shows how the potential of artificial intelligence can adequately seal material loops and enhance systemic resilience (Noman et al., 2022; Bashynska & Prokopenko, n.d; Rane et al., 2024). Additionally, AI techniques including machine learning, deep learning, and reinforcement learning are increasingly applied to respond to the challenging CE tasks such as the sorting of wastes, predictive repair, and lifespan examination, completing and extending theory on data-guided decision-making and adaptive systems for sustainable production and consumption (Besigomwe, 2024; Farshadfar et al., 2024; Zocco et al., n.d.).

The research findings also question the viability of conventional sustainability frameworks by highlighting the necessity for artificial intelligence-based real-time and dynamic optimisation models responsive to variable market and environmental dynamics. This adds to the theorization debate relevant to Industry 4.0 and 5.0 as transformational frameworks for the implementation of circular economy (CE) practices (Shamsuddoha et al., 2024; Prajapati, 2024; Valeriya et al., n.d.). Further, identified gaps in the theorised connection of AI and emerging technologies such as blockchain, Internet of Things (IoT), and digital twins catalyze the development of a multi-technology convergent framework to achieve CE objectives in entirety. Accordingly, the creation of expanded theory models incorporating cross-technology interactions and moral dimensions (Kumar et al., 2024; Aroul, 2024; Baumgartner et al., n.d.) takes centre-stage. The integration of AI in the optimisation of circular supply chains and reverse logistics presents high theory value since it catalyses the development of closed-loop systems optimising the movement of materials on minimum environmental feedback and augmenting supply chain sustainability-related concepts (Farshadfar et al., 2024; Ibrahim, 2024; Daramola et al., 2023). Finally, the absence of centrality for industry sectors such as electronic waste management and food supply chains attests to the theorisation gap in areas aforementioned and suggests the imperative for domain-specific AI-CE frameworks addressing industry Realted challenges and opportunities which highlights in (Noman et al., 2022; Yeruva et al., 2024; Onyeaka et al., 2023).

REAL-WORLD APPLICATIONS

The results underscore the concrete advantages that artificial intelligence offers to industry practitioners, especially regarding the augmentation of operational efficiency, the reduction of raw material consumption, and the enhancement of recycling rates in manufacturing, waste management, and supply chain operations. These revelations advocate for the integration of AI-enabled process automation and predictive analytics as means to realize quantifiable sustainability advancements (Besigomwe, 2024; Singh et al., 2024; RS & Kumar, 2024). Furthermore, policymakers have the opportunity to harness AI's dual capacity to advance sustainability alongside economic development, thereby crafting regulatory frameworks that strike a balance between promoting adoption and tackling issues related to energy consumption, data privacy, and ethical deployment of AI technologies. Such policy measures are essential for encouraging responsible innovation during transitions toward circular economies (Jaiswal & Pandey, 2024; Pratap & Venkatesh, 2024; Cugurra et al., 2024).

The integration of artificial intelligence and the Internet of Things and blockchain technologies offers promising opportunities for enhanced clarity, traceability, and real-time tracking in circular systems. This combination enables better resource allocation and management, waste handling and control, and lifecycle management in different industries, including smart cities and e-business (Bashynska & Prokopenko, n.d.; Shennib et al., 2024; Yu et al., 2024). It is recommended that product designers and manufacturers utilize artificial intelligence during the processes of eco-design and optimizing lifecycles and support the development of products whose recycling, reuse, and remanufacturing can be done more easily. This reduces environmental influences and enhances circular business models as per (Shipilova & Bobkov, 2024; Ramos et al., 2023; Popolo et al., 2024).

In addition, the deployment of robot systems with artificial intelligence and state-of-the-art waste sorting technologies provides feasible remedies to resolve labor scarcity and ineffectiveness in recycling plants and improve rates of material recovery and reduce landfill wastes (Cheng et al., 2024; Maheshwari & Gupta, 2024; Sleem & Elhenawy, n.d.). Additionally, business ventures concerned about sustainability, including startups, can exploit AI-enabled business model innovation and scale their businesses while ensuring maximum resource efficiency and environmental degradation reduction and thereby highlighting the prospective contribution of AI to sustainable entrepreneurship and the circular economy industry (Thakkar et al., 2024; Subramanyam et al., 2024).

SHORTCOMINGS OF CURRENT LITERATURE

The research literature available on the application of artificial intelligence (AI) in the case of the circular economy (CE) emphasizes several remarkable limitations for increased analysis. The case is the quality and the accessibility of the data since most studies reflect challenges arising from the scarcity or inadequacy of homogeneous or reliable datasets (Singh et al., 2024; Olawade et al., 2024; Fang et al., 2023; "Machine Learning and Artificial Intelligence", 2022). Another limitation entails energy consumption and the corresponding trade-offs in the case of sustainability; that is, the intensive energy requirements for training and applying AI models potentially eliminate the environmental advantage and decrease the overall positive impact on CE efforts (Jaiswal & Pandey, 2024; Bashynska & Prokopenko, n.d.; Pratap & Venkatesh, 2024).

Another limitation applies to the limited range of application, whereby research often focuses on specific industries or isolated processes, such as production or sorting garbage, and thus restricts broad knowledge on the application of AI across the whole CE value chain and reducing its overall applicability (Kio et al., 2022; Noman et al., 2022; Cheah et al., 2022). The absence of unified evaluation measures also hinders the conduct of comparative studies and benchmarking activities and thus inhibits the ability for uniform assessment of AI performance for CE systems (Singh et al., 2024; Arun et al., 2024; Chen & Chang, 2024). Moral and social aspects such as data protection issues, algorithmic biasing, job displacement, and violations of human rights are not sufficiently attended to and which hinders the proper application of AI and social acceptability (Bashynska & Prokopenko, n.d.; Pratap & Venkatesh, 2024; Cugurra et al., 2024).

Overreliance on single AI models also restricts both the strength of research and its real-world effectiveness since the widespread deployment of single models escalates the chances of overfitting in complex waste management contexts (Fang et al., 2023; "Machine Learning and Artificial Intellig…", 2022). Additionally, geographic and industry biases are evident since available research often concentrates on developed countries or single industries that reduce the transference of findings to limited-resource or new economies (Noman et al., 2022; Pathan et al., 2023; Yu et al., 2024). Finally, the scarcity of longitudinal studies hinders understanding the long-term impacts of AI on circular economy (CE) outcomes and thus hinders the assessment of durability, scalability, and the dynamic challenges in the long term (Valeriya et al., n.d.; Subramanyam et al., 2024).

FOUND GAPS AND FUTURE RESEARCH PATHS

The studies on AI deployment in the circular economy (CE) report some major areas for research requiring specialized focus. Among them is the high energy intensity of the AI model, and the generative model in particular, which cancels the environmental benefits of the circular intervention. The future research needs to be directed at the development of energy-efficient technologies and algorithms, the estimation of the carbon intensity of AI, and the exploration of the integration of renewable energy sources in order to obtain net positive environmental outcomes (Jaiswal & Pandey, 2024; Pratap & Venkatesh, 2024; Durai et al., 2024). Data quality and accessibility while dealing with waste management are other major areas of concern since sets of missing or inconsistent data put the accuracy and scalability of AI models at risk; what are needed are standard operating protocols for data collection, noisy data robust models, and open platforms for the sharing and release of data (Singh et al., 2024; Olawade et al., 2024; Shennib et al., 2024; Shennib & Schmitt, n.d.).

Few studies integrate AI and chemical and microbiological processes to enhance material recovery and degradation of waste. More studies on hybrid AI-chemical-microbiological systems and industrial corroboration should be conducted to enhance recovery yields and environmental sustainability (Chittalury, 2025; Anusha et al., 2025; Arun et al., 2024). The absence of standard frameworks for cross-sector comparability in AI-enabled product lifecycle analysis needs to be addressed through the development of common lifecycle assessment tools and interoperable tools validated by longitudinal studies (Shipilova & Bobkov, 2024; Rusch et al., 2022; Popolo et al., 2024). Social and moral dimensions such as algorithmic bias, data security, job displacement, and human rights also weren't adequately examined and call for thorough ethical effect studies besides regulative frameworks and workforce adjustment measures (Olawade et al., 2024; Ibrahim, 2024; Cugurra et al., 2024).

Another gap involves scalability and end-to-end interoperability for AI systems since most of the applications are siloed. Interoperable AI-IoT-blockchain architectures should be studied and pilot cross-sectors AI ecosystems be experimented on to realize the whole-of-circular economy advantage (Gulyamov, n.d.; Cheah et al.,

2022; Bashynska & Prokopenko, n.d.). Use of artificial intelligence in underserved agri-food supply chains dictates the necessity to develop additional innovation to minimize wastage, enhance logistics, and provide real-time tracking (Noman et al., 2022; Onyeaka et al., 2023). Absence of applicative testing and longitudinal research inhibits the attainment of long-term in-depth understanding on the impact of AI and the resulting value of research studies on the subject, long-term protection measures, and available data (Besigomwe, 2024; Valeriya et al., n.d.; Neto et al., 2023). Additionally, the challenge-seeking environments deprived of resources constitute barriers to the implementation of AI that require lightweight and low-cost solutions and socio-technical studies to offer inclusive CE outcomes (Shennib et al., 2024; Olawade et al., 2024). Finally, the exploitation of robots and automatic systems continues to be underserved by the complexity and the cost thereof and presents the opportunity to design and develop flexible AI-robot systems and approximate the economic viability to scale wastage sorting and recycling (Cheng et al., 2024; Zocco et al., n.d.; Maheshwari & Gupta, 2024).

FINAL INSIGHTS AND INTEGRATION

There exists a large body of research clearly establishing that Artificial Intelligence (AI) serves as a key enabling agent for the development of circular economy (CE) values, especially in the case of waste management, resource intensity, and product life cycle assessment. Technology fueled by AI, such as machine learning, deep learning, and robotic process automation, helps enhance the effectiveness of operations across various sectors, including manufacturing, smart cities or intelligent urban environments, e-waste recycling, fashion industry, and supply chain logistics. Such technologies enhance the effectiveness and accuracy of sorting and grading wastage, optimize the distribution and allocation of resources, and offer predictive maintenance assistance, all contributing to reduced exploitation of raw materials, increased recycling rates, and extended product longevity. The combination of AI and other digital technologies such as the Internet of Things (IoT), blockchain technology, and digital twins helps multiply such benefits through the enabling of real-time monitoring and tracking, ensuring end-to-end visibility and exchange of data in the process and fostering adaptive decision-support systems closely aligned to CE objectives.

Life cycle management and resource management areas have witnessed significant assistance through AI-empowered analytics in order to gauge sustainability, manage control quality, and sustain closed-loop production systems. The outcomes are quantitative and range from lower generation of wastes and energy consumption and emissions to greater cost savings and supply chain resiliency. Examples of successful execution of material recovery and process effectiveness through the automated execution of the complex recycling processes such as disassembly of electronic wastes and sorting of beverage cans represent the execution through AI to meet the outcomes.In addition, new trends such as the federated model of AI and interdisciplinary approaches merging AI and microbiology or green chemistry represent promising future areas for development in managing sustainable resources.

Despite such strengths, the study identifies significant challenges that continue to hamper the achievement of artificial intelligence in circular economy transformations. Data quality concerns such as incompleteness, inconsistency, or corruption of datasets negatively affect model validity when proper resources are not available. The environmental effect of the AI systems, particularly the energy-intensive nature of the deployment of deep learning models for the purpose of training, offers a sustainability trade-off requiring balanced policy and technological mitigation. Hoplessness regarding the ethics such as data privacy, algorithmic bias, disruption to the job market, and the necessity for robust regulative measures are not adequately alleviated and may impair socially beneficial AI deployment. Finally, the expenses for the infrastructural support, interoperability challenges, and piecemeal implementation among industry limit the scalability and end-to-end integration of the AI-enabled circular solutions. Overall, the evaluation holds that while artificial intelligence provides the key driver in the achievement of circular economy goals, implementation thereof hinges on overcoming numerous complex barriers through the development of collective policy, multi-disciplinary research, and technological innovation. Future initiatives should be directed at reinforcing the data governance, minimizing environmental impacts of AI, fostering the right uses of AI by means of ethics, and the creation of model frameworks to help aggregate AI uses in conjunction across several industries. Attentive focus on such areas allows AI to meet the potential as a key instrument for sustainable development, preservation of the resources, and global transformation to the circular economy.

REFERENCES

Agrawal, R., Wankhede, V. A., Kumar, A., Luthra, S., Majumdar, A., & Kazancoglu, Y. (2021). An exploratory state-of-the-art review of artificial intelligence applications in circular economy using structural topic modeling. Operations Management Research, 1-18. https://doi.org/.DOI: 10.1007/s12063-021-00212-0

Alamelu, R., Sudha, M., & Raja, P. (2024). Adoption of circular economy strategies with artificial intelligence. Advances in logistics, operations, and management science book series, 179-204. https://doi.org/.DOI: 10.4018/979-8-3693-6205-1.ch007

Anusha, S., Yamini, B., Udhayashankar, S., Priya, V., Ezhilvendan, M., Vinoth, R., & Subramanıan, R. S. (2025). AI and applied microbiology in sustainable waste recycling and the circular economy. Advances in environmental engineering and green technologies book series, 83-100. https://doi.org/DOI: 10.4018/979-8-3693-7282-1.ch005

Aroul, G. G. (2024). *Enabling circular economy through integrated digital twins: A framework for ai-driven business transformation*. Indian Scientific Journal Of Research In Engineering And Management., DOI: 10.55041/IJSREM28351

Arun, M., Barik, D., & Chandran, S. S. R. (2024). Exploration of material recovery framework from waste -a revolutionary move towards clean environment. *Chemical Engineering Journal Advances*, *18*, 100589–100589. DOI: 10.1016/j.ceja.2024.100589

Arun, V., Patro, E. K. R., Devi, V. A., Nagpal, A., Chandra, P. K., & Albawi, A. (2024). AI based prediction algorithms for enhancing the waste management system: A comparative analysis. E3S web of conferences, 552, 01052-01052. https://doi.org/DOI: 10.1051/e3sconf/202455201052

Ashatkar, A. (2024). Revolutionizing plastic waste management in the packaging industry An indepth exploration of ai-driven circular economy strategies. *International Journal for Research in Applied Science and Engineering Technology*, *12*(3), 370–375. Advance online publication. DOI: 10.22214/ijraset.2024.58806

Bashynska, I., & Prokopenko, O. (n.d.). Innovative technologies and digital platforms: Ai's role in a sustainable circular economy. https://doi.org/DOI: 10.19192/wsfip.sj3.2024.10

Bashynska, I., & Prokopenko, O. (n.d.). Leveraging artificial intelligence for circular economy Transforming resource management, supply chains, and manufacturing practices. https://doi.org/DOI: 10.19192/wsfip.sj2.2024.13

Baumgartner, R. J., Berger, K., & Schöggl, J. (n.d.). Digital technologies for sustainable product management in the circular economy. DOI: 10.1007/978-3-031-61749-2_7

Besigomwe, K. (2024). Ai-driven process design for closed-loop manufacturing. Cognizance journal, 4 (12), 372-380. DOI: 10.47760/cognizance.2024.v04i12.035

Cheah, C. G., Chia, W. Y., Lai, S. F., Chew, K. W., Chia, S. R., & Show, P. L. (2022). Innovation designs of industry 4.0 based solid waste management: Machinery and digital circular economy. *Environmental Research*, *213*, 113619–113619. DOI: 10.1016/j.envres.2022.113619 PMID: 35700763

Chen, C., & Chang, Y. (2024). Leveraging circular economy metrics for data-driven forecasting of solid waste production in europe. *Sustainability (Basel)*, *16*(3), 1017. Advance online publication. DOI: 10.3390/su16031017

Cheng, T., Kojima, D., Hu, H., Onoda, H., & Pandyaswargo, A. H. (2024). Optimizing waste sorting for sustainability: An ai-powered robotic solution for beverage container recycling. *Sustainability (Basel)*, *16*(23), 10155–10155. DOI: 10.3390/su162310155

Chittalury, D. (2025). Chemical breakthroughs facilitating superior material reclamation techniques. Advances in environmental engineering and green technologies book series, 101-112. https://doi.org/DOI: 10.4018/979-8-3693-7282-1.ch006

Cugurra, M. D. B. I., Koussouris, S., Cutrona, V., & Canetta, L. (2024). A human rights-driven AI assessment of ethical industry 5.0 transition and sustainability in circular value chains. https://doi.org/DOI: 10.1109/ICE/ITMC61926.2024.10794210

Daramola, O. M., Apeh, C. E., Basiru, J. O., Onukwulu, E. C., & Paul, P. O. (2023). Optimizing reverse logistics for circular economy: Strategies for efficient material recovery and resource circularity. *International Journal of Social Science Exceptional Research*, *2*(1), 16–31. DOI: 10.54660/IJSSER.2023.2.1.16-31

Dhanya, D., Kumar, S., Thilagavathy, A., Prasad, D. V. S. S. S. V., & Boopathi, S. (2023). Data analytics and artificial intelligence in the circular economy. Advances in civil and industrial engineering book series, 40-58. https://doi.org/DOI: 10.4018/979-8-3693-0044-2.ch003

Durai, S., Manoharan, G., & Ashtikar, S. P. (2024). Harnessing artificial intelligence. Advances in finance, accounting, and economics book series, 89-117. https://doi.org/DOI: 10.4018/979-8-3693-2881-1.ch003

Fang, B., Chen, Z., Osman, A. I., Farghali, M., Ihara, I., Hamza, E. H., Rooney, D., & Yap, P. (2023). Artificial intelligence for waste management in smart cities: A review. *Environmental Chemistry Letters*, *21*(4), 1–31. DOI: 10.1007/s10311-023-01604-3 PMID: 37362015

Fang, I., Yen, F., & Hsu, Y. (2023). Using artificial intelligence for recycling-a case study in taiwan's resource recycling industry. https://doi.org/DOI: 10.1145/3624875.3624897

Farshadfar, Z., Mucha, T., & Tanskanen, K. (2024). Leveraging machine learning for advancing circular supply chains: A systematic literature review. *Logistics (Basel)*, *8*(4), 108–108. DOI: 10.3390/logistics8040108

Goel, R., Singla, J., & Kumar, S. (2024). The digital circular economy. Advances in e-business research series, 441-480. https://doi.org/DOI: 10.4018/979-8-3693-9699-5.ch019

24. Gulyamov, S. (n.d.). Intelligent waste management using iot, blockchain technology and data analytics. E3S web of conferences . https://doi.org/DOI: 10.1051/e3sconf/202450101010

Ibrahim, M. (2024). Unleashing the potential of artificial intelligence for dynamic supply chain optimization and sustainability. *Practice, Progress, and Proficiency in Sustainability*, •••, 381–410. DOI: 10.4018/979-8-3693-2827-9.ch013

Jain, A. (2024). Fostering sustainable innovation through digital transformation. Advances in finance, accounting, and economics book series, 353-372. https://doi.org/DOI: 10.4018/979-8-3373-0578-3.ch016

Jaiswal, S., & Pandey, S. (2024). Exploring generative AI as a catalyst for sustainability Strategies for waste and energy reduction. *Management Insight*, *20*(01), 14–26. DOI: 10.21844/mijia.20.1.2

Johnson, M., Khatoon, A., & Fitzpatrick, C. (2022). Application of AI and machine vision to improve battery detection and recovery in e-waste management. https://doi.org/DOI: 10.1109/ICECCME55909.2022.9988629

Keivanpour, S. (2023). Industry 4.0 driven circular economy—an introduction. Synthesis lectures on sustainable development, 133-148. https://doi.org/DOI: 10.1007/978-3-031-44652-8_7

Kio, P., Anumba, C. J., & Ali, A. K. (2022). Circular economy trends – potential role of emerging technologies. *IOP Conference Series. Earth and Environmental Science*, *1101*(6), 062005–062005. DOI: 10.1088/1755-1315/1101/6/062005

Kumar, P., Kumar, M., & Gautam, M. P. (2024). Synergizing federated AI systems with circular economy principles: A framework for sustainable and resilient data science. *Indian Scientific Journal Of Research In Engineering And Management*, *08*(12), 1–7. DOI: 10.55041/IJSREM39516

Levchenko, N., Nazarenko, O., Berezovska A., Bielousova, P., Shchemeliev, M., & Hadzhiverdiyev, A. (2024). Innovative technologies in construction waste management in accordance with circular economy. https://doi.org/DOI: 10.33042/2311-7257.2024.110.1.9

Maheshwari, A., & Gupta, S. (2024). Ai-powered robotic systems for disassembly and recycling of complex electronic devices. https://doi.org/DOI: 10.52458/9788197112492.nsp.2024.eb.ch-21

Mamudu, U. U., Obasi, C. D., Awuye, S. K., Danso, H., Ayodele, P., & Akinyemi, P. (2024). Circular economy in the manufacturing sector: Digital transformation and sustainable practices. *International Journal of Science and Research Archive*, *12*(2), 129–141. DOI: 10.30574/ijsra.2024.12.2.1217

Mankar, V. A., Ali, A. J., Kamlani, R., Sure, Y., Kediya, S., & Gudadhe, A. (2024). The role of AI in circular economy supply chains: A comparative analysis of industry practices. https://doi.org/DOI: 10.1109/IDICAIEI61867.2024.10842918

Moorthy, K., Balakrishnan, G., Kumar, S., Raja, L., & Vijayalakshmi, A. (2024). Embracing circular economy principles for sustainable green supply chain management in manufacturing industries. Advances in human resources management and organizational development book series, 85-110. https://doi.org/DOI: 10.4018/979-8-3693-1343-5.ch005

Munonye, W. C., & Ajonye, G. O. (2024). Redesigning urban infrastructure for circularity: The role of smart cities in reducing waste. Journal of environmental science studies, 7 (2), 29-29. https://doi.org/DOI: 10.20849/jess.v7i2.1468

Neto, G. C. D. O., Araújo, S. A. D., Araújo, S. A. D., Flausino, F. R., & Silva, C. (2023). Simulation of electronic waste reverse chain for são paulo circular economy: An artificial intelligence-based approach for economic and environmental optimization. https://doi.org/DOI: 10.20944/preprints202307.1685.v1

Noman, A. A., Akter, U. H., Pranto, T. H., & Haque, A. B. (2022). Machine learning and artificial intelligence in circular economy: A bibliometric analysis and systematic literature review. Annals of emerging technologies in computing., abs/2205.01042, . https://doi.org/DOI: 10.33166/AETiC.2022.02.002

Olawade, D. B., Fapohunda, O., Wada, O. Z., Usman, S. O., Ige, A. O., Ajisafe, O., & Oladapo, B. I. (2024). Smart waste management: A paradigm shift enabled by artificial intelligence. https://doi.org/DOI: 10.1016/j.wmb.2024.05.001

Onyeaka, H., Nwauzoma, U., Miri, T., Nwaiwu, O., & Akinsemolu, A. A. (2023). Using artificial intelligence to tackle food waste and enhance the circular economy: Maximising resource efficiency and minimising environmental impact: A review. *Sustainability (Basel), 15*(13), 10482–10482. DOI: 10.3390/su151310482

Pathan, M. S., Richardson, E., Galvan, E., & Mooney, P. (2023). The role of artificial intelligence within circular economy activities—A view from ireland. *Sustainability (Basel), 15*(12), 9451–9451. DOI: 10.3390/su15129451

Popolo, V., Vespoli, S., Gallo, M., & Grassi, A. (2024). Revolutionizing end-of-life product recovery with product 4.0: An examination of intelligent products in industry 4.0. *Sustainability (Basel), 16*(24), 11017–11017. DOI: 10.3390/su162411017

Prajapati, N. (2024). Industry 4.0 and sustainability: Towards vision and ideas. International journal of innovative science and research technology, 2817-2831. https://doi.org/DOI: 10.38124/ijisrt/IJISRT24OCT1894

Pratap, S., & Venkatesh, K. (2024). The role of AI in enhancing green management and advancing digital lean practices for sustainable efficiency. *ShodhKosh Journal of Visual and Performing Arts, 5*(6). Advance online publication. DOI: 10.29121/shodhkosh.v5.i6.2024.1725

Ramos, L., Rivas-Echeverria, F., Casas, E., & Pérez, A. G. (2023). The role of artificial intelligence in enhancing sustainability in the fashion industry: A 2012-2022 review. https://doi.org/DOI: 10.1109/MCSI60294.2023.00022

Rane, N., Kaya, Ö., & Rane, J. (2024). Advancing the sustainable development goals (sdgs) through artificial intelligence, machine learning, and deep learning. https://doi.org/DOI: 10.70593/978-81-981271-8-1_4

Ranganathan, C. S., Riazulhameed, A. A. M. A., Saravanan, S. K., Mohankumar, N., Rajmohan, M., & Murugan, S. (2024). Enhancing circular economy models for battery material recovery using iot data and adaboost algorithm. 2020 4th International Conference on Electronics, Communication and Aerospace Technology (ICECA), 25-30. https://doi.org/DOI: 10.1109/ICECA63461.2024.10800942

Rani, M., Rani, M., Srivastava, R., & Prakash, S. (2025). Intelligent approaches to promote recycling aligned with sustainability objectives. Advances in environmental engineering and green technologies book series, 295-318. https://doi.org/DOI: 10.4018/979-8-3693-7282-1.ch014

RS, G., & Kumar, S. (2024). Smart waste: Ai's impact on solid waste management. Open access journal of waste management & xenobiotics, 7 (4), 1-2. https://doi.org/DOI: 10.23880/oajwx-16000201

Rusch, M., Schöggl, J., & Baumgartner, R. J. (2022). Application of digital technologies for sustainable product management in a circular economy: A review. *Business Strategy and the Environment*, *32*(3), 1159–1174. DOI: 10.1002/bse.3099

Sánchez-García, E., Martínez-Falcó, J., Marco-Lajara, B., & Manresa-Marhuenda, E. (2023). Revolutionizing the circular economy through new technologies: A new era of sustainable progress. https://doi.org/DOI: 10.1016/j.eti.2023.103509

Shamsuddoha, M., Kashem, M. A., Nasir, T., & Hossain, N. U. I. (2024). From industry 4.0 to industry 5.0. https://doi.org/DOI: 10.1201/9781003438748-13

Sharma, D. (2023). *Novel technological approaches in waste management for environmental conservation and resource recovery*. Indian Scientific Journal Of Research In Engineering And Management., DOI: 10.55041/IJSREM27793

Shennib, F., Eicker, U., & Schmitt, K. A. (2024). Openwasteai—Open data, iot, and AI for circular economy and waste tracking in resource-constrained communities. *IEEE Technology and Society Magazine*, *43*(1), 39–53. DOI: 10.1109/MTS.2024.3372610

Shennib, F., & Schmitt, K. (2021, October 28). Data-driven technologies and artificial intelligence in circular economy and waste management systems. *RE:view*, 1–5. Advance online publication. DOI: 10.1109/ISTAS52410.2021.9629183

Shipilova, T., & Bobkov, A. (2024). Assessment of the possibility of using AI to manage the product lifecycle in a closed-loop economy. Moscow Economic Journal, 286-298. https://doi.org/DOI: 10.55186/2413046X_2024_9_6_292

Singh, A. (2024). Ai-driven innovations for enabling a circular economy. Advances in finance, accounting, and economics book series, 47-64. https://doi.org/DOI: 10.4018/979-8-3373-0578-3.ch003

Singh, P., Azad, M. L., & Mishra, V. (2024). Leveraging AI for sustainable waste management: Enhancing recycling and reducing landfill waste in circular economy models. https://doi.org/DOI: 10.1109/IC3I61595.2024.10829161

Sleem, A., & Elhenawy, I. (n.d.). Intelligent waste management system for recycling and resource optimization. https://doi.org/DOI: 10.54216/JISIoT.010205

Subramanyam, N., Bhavitha, Y., Pallavi, V., Jain, D. K., & Seshu, G. Y. (2024). Ai-driven scaling of sustainable startups: Business models and environmental impact. *Deleted Journal*, *4*(6), 35–38. DOI: 10.62225/2583049X.2024.4.6.3391

Thakkar, A., Zaveri, B., & Mathewos, T. (2024). Ecotech harmony. Advances in logistics, operations, and management science book series, 123-145. https://doi.org/DOI: 10.4018/979-8-3693-2346-5.ch009

Valeriya, G., Kirola, M., Gupta, M., Bharathi, P., & Acharya, P. (n.d.). Evaluating the impact of ai-based sustainability measures in industry 5.0: A longitudinal study. BIO web of conferences, . https://doi.org/DOI: 10.1051/bioconf/20248601058

Vashishth, T. K., Sharma, V., Sharma, K. K., Kumar, B., Chaudhary, S., & Panwar, R. (2024). Waste-to-energy solutions harnessing iot and ML for sustainable power generation in smart cities. Advances in computational intelligence and robotics book series, 125-146. https://doi.org/DOI: 10.4018/979-8-3693-1062-5.ch007

Yeruva, J. R., Guduru, A. K., & Saripalli, B. P. (2024). Technological advancements in e-waste management and resource recovery. *Practice, Progress, and Proficiency in Sustainability*, •••, 275–288. DOI: 10.4018/979-8-3693-1018-2.ch018

Yu, P., Zhang, Z., Wong, S. K. M., Lu, S., Chen, Y., & Wang, Y. (2024). The role of technologies in facilitating circular economy of china's e-commerce section. Advances in finance, accounting, and economics book series, 151-174. https://doi.org/DOI: 10.4018/979-8-3693-6392-8.ch008

Zhou, Y., & Zhao, J. (n.d.). Leveraging computational systems for lifecycle management and enhancement of circular economy in fashion: A study on tracking, recycling, and reuse technologies. https://doi.org/DOI: 10.54254/2755-2721/87/20241531

Zocco, F., Haddad, W. M., Corti, A., & Malvezzi, M. (n.d.). A unification between deep-learning vision, compartmental dynamical thermodynamics, and robotic manipulation for a circular economy. https://doi.org/DOI: 10.1109/ACCESS.2024.3493755

Chapter 3
The Blue Economy:
Growth, Sustainability, and Economic Potential

Mesut Savrul
https://orcid.org/0000-0003-1513-1874
Canakkale Onsekiz Mart University, Turkey

ABSTRACT

The Blue Economy has emerged as a strategy to align ocean-based growth with sustainability and equity. This study critically examines its foundations, key sectors, environmental challenges, governance, and future pathways through a structured literature review and case study synthesis drawing on 110 sources, including UNCLOS, SDG 14, Seychelles' Blue Bond, and EU Marine Spatial Planning. Findings show substantial economic potential, estimated at over $3 trillion annually by 2030, yet persistent issues of overfishing, pollution, and fragmented governance. While technological and financial innovations are advancing, they require safeguards to ensure equity and accountability. The paper offers six policy recommendations for inclusive and sustainable outcomes, concluding that the Blue Economy is both a major opportunity and a governance challenge, demanding integrated and ethically grounded strategies.

INTRODUCTION

Coastal and island nations, especially Small Island Developing States and Least Developed Countries, have increasingly embraced the Blue Economy as a pathway to economic resilience and sustainable livelihoods (World Bank, 2017). Innovative policy and financial approaches illustrate this shift. Seychelles' issuance of a sovereign blue bond to finance sustainable fisheries management demonstrates how economic

DOI: 10.4018/979-8-3373-5238-1.ch003

and environmental objectives can be aligned through novel financial instruments (World Bank, 2018). At the regional level, the European Union's Integrated Maritime Policy reflects efforts to coordinate sustainable ocean development across member states, with emphasis on renewable energy, marine biotechnology, and sustainable fisheries (European Parliament, 2025).

Despite its promise, the Blue Economy faces significant and intensifying challenges. Human activity in marine spaces is accelerating, a process described as the "Blue Acceleration," driven by the rapid expansion of shipping, offshore energy infrastructure, aquaculture, and extractive industries (Jouffray et al., 2020). These developments often generate trade-offs between economic growth and ecosystem health. Overfishing continues to threaten marine biodiversity and food security, while pollution from plastics, chemicals, and oil spills further degrades marine environments (Tedsen et al., 2014). Climate change compounds these pressures through sea-level rise, ocean warming, and acidification, increasing risks for coastal communities and undermining the resilience of critical infrastructure (Oppenheimer et al., 2019).

Equity has emerged as a central concern within Blue Economy debates. Critics argue that prevailing narratives often privilege industrialized countries and large-scale corporate actors, while marginalizing small-scale fishers and Indigenous communities who depend most directly on marine resources (Bennett et al., 2019). Without inclusive governance frameworks that ensure participation and fair benefit sharing, the Blue Economy risks reinforcing existing inequalities rather than promoting sustainable and equitable development.

Given its interdisciplinary nature, the Blue Economy offers fertile ground for academic and policy analysis. This chapter provides a comprehensive assessment of the Blue Economy's growth trajectory, sustainability challenges, and economic potential. By examining key sectors alongside governance mechanisms and financial innovations, it seeks to identify pathways for aligning ocean-based economic expansion with ecological resilience and social equity, contributing to broader debates on moving beyond extractive ocean economies toward more sustainable Blue Economy models.

KEY SECTORS DRIVING GROWTH

The Blue Economy encompasses a diverse set of ocean-based industries that collectively drive economic growth, support livelihoods, and shape pathways toward sustainability. Among these, fisheries and aquaculture, maritime transport, offshore renewable energy, marine biotechnology, and coastal tourism represent the principal engines of Blue Economy expansion. Together, these sectors illustrate both the opportunities and tensions inherent in reconciling economic development with long-term ecological stewardship.

Fisheries and aquaculture remain a cornerstone of the Blue Economy, underpinning global food security and coastal livelihoods. Global production reached a record 214 million tonnes in 2020, valued at USD 141 billion, making it the fastest-growing food production sector worldwide (FAO, 2022). Aquaculture now supplies more than half of global fish consumption, reflecting rising demand and increasing pressure on wild stocks (FDA, 2023). The sector contributes billions to global GDP and supports millions of people, particularly in coastal regions where marine ecosystem services are essential for income and food security (Sarker et al., 2019). However, sustainability challenges persist, as approximately 35% of global fish stocks are exploited beyond biologically sustainable levels due to overfishing, habitat degradation, and climate change (FAO, 2025a). In response, innovations such as integrated multi-trophic aquaculture, alongside AI-based monitoring, blockchain-enabled traceability, selective breeding, and advances in marine biotechnology, are reshaping production practices and reducing environmental impacts (Loayza et al., 2023; Bhanja, 2024; Kamanzi, 2025).

Maritime transport is another foundational pillar, accounting for over 80% of global trade by volume and serving as a critical driver of globalization and economic integration (UNCTAD, 2022). Encompassing shipping, port logistics, and shipbuilding, the sector contributes trillions of dollars to global commerce, with Asia—particularly China, Singapore, and South Korea—dominating infrastructure and manufacturing capacity. Efficient maritime transport lowers trade costs, strengthens supply chain resilience, and supports industrial development (UNCTAD, 2024). Yet the sector also poses major sustainability challenges, as international shipping generated nearly 3% of global greenhouse gas emissions in 2018, a share projected to increase without intervention. Decarbonization has therefore become central to Blue Economy strategies, with the International Maritime Organization's 2050 targets reshaping investment and innovation priorities (Bonnet-Cantalloube, 2024). Emerging solutions include alternative fuels, electrification for short-sea shipping, wind-assisted propulsion, and smart ports leveraging AI and IoT to improve efficiency and reduce emissions (Vidović et al., 2023).

Offshore renewable energy represents one of the most rapidly expanding Blue Economy sectors. According to the 2023 Global Offshore Wind Report, more than 380 GW of additional offshore wind capacity is projected by 2032, bringing the global total to 447 GW, with Europe, China, and the United States leading deployment and growing investment in Africa and Latin America (GWEC, 2023). Beyond offshore wind, wave and tidal energy hold long-term potential despite current technological and financial barriers. Integrating offshore renewable energy into coastal grids can diversify energy systems, reduce fossil fuel dependence, and stimulate regional development, as demonstrated by job creation and investment in countries such as Portugal and the United Kingdom (Majidi et al., 2024). However, offshore energy infrastructure can affect marine ecosystems, sediment dynamics, and migratory species, underscoring the need for marine spatial planning and advances in turbine and anchoring design (Galparsoro et al., 2022). Its strong linkages with shipping and marine biotechnology further position offshore renewable energy as both a pillar of the energy transition and a catalyst for cross-sectoral growth (Durr & Shumway, 2024).

Marine biotechnology is an emerging frontier with applications spanning pharmaceuticals, nutraceuticals, cosmetics, biofuels, and biomaterials. The ocean's biodiversity provides a rich source of genetic and biochemical resources, with marine microorganisms already yielding novel antibiotics, anti-cancer, and anti-inflammatory compounds (Kumar et al., 2025). The global marine biotechnology market is valued at USD 6.2 billion, driven largely by pharmaceuticals and industrial enzymes (Ken Research, 2025). Beyond health applications, the sector supports aquaculture through improved disease management and feed efficiency and contributes to renewable energy through algae-based biofuels. Nevertheless, concerns over the overexploitation of marine genetic resources and ethical issues related to access and equity highlight the importance of coordinated governance and partnerships among research institutions, industry, and governments.

Coastal and marine tourism constitutes another major driver of Blue Economy growth. Tourism generates trillions of dollars globally and supports around one in ten jobs worldwide, with coastal and marine tourism providing over 6.5 million jobs and ranking second only to industrial fishing within the ocean economy. With annual growth projected to exceed 3.5%, it is expected to become the largest value-adding ocean sector by 2030, accounting for 26% of total value (Brumbaugh & Patil, 2017). Tourism is also a critical source of foreign exchange, particularly for small island and coastal states, where it can account for more than half of total exports (Badrinarayanan & Akhter, 2025). However, unmanaged tourism can intensify coastal erosion, pollution, and biodiversity loss (Ebarvia, 2016). Ecotourism has therefore emerged as a more sustainable alternative, promoting conservation, supporting local communities, and reducing environmental pressures.

Although each sector exhibits distinct dynamics, their interconnections amplify overall Blue Economy growth. Fisheries support tourism through cultural and culinary linkages, maritime transport underpins global trade, offshore energy advances decarbonization, and biotechnology drives innovation across industries. These cross-sectoral linkages underscore the integrated nature of ocean-based economies and the need for governance frameworks capable of leveraging synergies while managing trade-offs.

SUSTAINABILITY CHALLENGES

The Blue Economy is widely promoted as a pathway for inclusive economic growth, employment creation, and innovation across ocean-based sectors. However, its rapid expansion has intensified concerns over sustainability, ecological limits, and governance capacity. Without precautionary approaches and robust institutional frameworks, scholars warn that the ocean economy risks replicating patterns of unsustainable exploitation rather than evolving into a regenerative development model (Jouffray et al., 2020). Central challenges shaping the long-term trajectory of the Blue Economy include overfishing and biodiversity loss, climate change impacts, marine pollution, and persistent governance fragmentation.

Overexploitation of marine resources remains one of the most critical threats. Although 64.5% of global fish stocks are still harvested within biologically sustainable limits, 35.5% are overfished, a proportion that has steadily increased over the past five decades (FAO, 2025a). Overfishing not only threatens food security but also disrupts trophic relationships and weakens ecosystem resilience. These pressures are intensified by industrial fishing, weak quota enforcement, and insufficient monitoring (Simataa et al., 2025). In West Africa, illegal, unreported, and unregulated fishing accounts for up to 37% of total catches, resulting in severe economic losses for coastal communities and accelerating ecosystem degradation (Environmental Justice Foundation, 2012).

Habitat destruction further compounds biodiversity loss. Coral reefs, mangroves, and seagrass beds have suffered extensive degradation due to human activities and climate-related stressors. More than 11% of coral reefs have been permanently destroyed, with an additional 16% lost during the 1998 bleaching event alone, and up to 30% may be severely degraded within decades without stronger protection. Similarly, over half of the world's mangrove forests and large areas of seagrass beds have been cleared, despite their vital role as nurseries for marine species and buffers against coastal erosion (Talbot & Wilkinson, 2001). These losses directly threaten key Blue Economy sectors, including fisheries, tourism, and marine biotechnology.

Climate change represents a systemic challenge that amplifies existing vulnerabilities. Rising sea surface temperatures are altering fish stock distributions, forcing fishers to adapt practices and increasing competition over resources. Poleward species migration is disrupting traditional fisheries and intensifying geopolitical tensions (Gill et al., 2023). Ocean acidification further undermines marine ecosystems by reducing calcification in corals and shell-forming organisms, weakening marine food webs and threatening aquaculture and tourism (Smithsonian Ocean, 2019). Sea level rise and extreme weather events compound these risks by damaging coastal infrastructure, ports, and settlements, particularly in low-lying coastal cities and small island developing states (Oppenheimer et al., 2019).

Marine pollution is another major constraint on sustainable development. Wastewater discharges and microplastic pollution degrade habitats, reduce fish yields, compromise seafood safety, and diminish the attractiveness of coastal tourism, resulting in significant economic losses (Anuar et al., 2025). The cumulative effects of pollutants also pose complex risks to marine life and human health, underscoring the need for circular economy approaches and stronger international pollution controls.

Underlying these challenges is persistent governance fragmentation. While UNCLOS provides a foundational legal framework, uneven implementation and overlapping jurisdictions hinder coordinated ocean management (United Nations, 1982; Blasiak & Claudet, 2024). Regional Fisheries Management Organizations show mixed performance, and the rapid spread of blue growth strategies has raised concerns over blue washing (OECD, 2009; Janisch et al., 2020). Although innovative tools such as Blue Bonds offer promise, their scale remains limited. Realizing the Blue Economy's potential therefore requires integrated, transparent, and inclusive governance that aligns economic ambition with ecological limits and the Sustainable Development Goals (OECD, 2024; UNCTAD, 2025).

ECONOMIC POTENTIAL AND OPPORTUNITIES

The ocean is increasingly recognized as a critical economic frontier, offering substantial opportunities for growth, employment, and innovation while contributing to the resolution of global challenges such as food security, climate change, energy demand, and healthcare. In 2010, the ocean economy generated approximately USD 1.5 trillion, representing around 2.5% of global gross value added. Offshore oil and gas accounted for roughly one-third of this value, followed by maritime and coastal tourism, maritime equipment, and port activities. The sector also supported approximately 31 million full-time jobs, with industrial fisheries and tourism together accounting for most of the employment (OECD, 2016). Realizing the ocean economy's full potential, however, requires balancing economic expansion with ecological stewardship and inclusive governance.

The contemporary Blue Economy extends well beyond traditional sectors such as fishing and shipping. By 2023, ocean-related exports reached USD 2.2 trillion, with services accounting for nearly 60% of the total. The sector underpins a diverse range of industries, including tourism, pharmaceuticals, biotechnology, and renewable energy, and supports more than 600 million jobs globally. Since 1995, the ocean economy has expanded by a factor of 2.5, growing faster than the global economy overall, with much of this growth driven by developing countries (UNCTAD, 2025). The World Bank (2017) similarly emphasizes the ocean economy's importance for sustainable development, particularly in coastal and island states.

Shipping and maritime transport remain indispensable, carrying more than 80% of global trade by volume (UNCTAD, 2024). Coastal and marine tourism also plays a central economic role, especially in Small Island Developing States (SIDS), where it serves as a primary source of income, employment, and foreign exchange. Between 2000 and 2013, international tourist arrivals in SIDS increased significantly, while tourism exports more than doubled (UNWTO, 2019). At the same time, aquaculture has become the fastest-growing food production system globally, surpassing capture fisheries for human consumption in 2014 (FAO, 2022).

Beyond its economic contributions, the Blue Economy delivers critical ecological functions. The ocean acts as a major carbon sink, regulates the global climate system, and provides ecosystem services essential for long-term resilience. However, mounting climate pressures and intensifying human activities underscore the urgency of adopting sustainable investment and management practices to ensure that economic benefits do not come at the expense of ecosystem degradation (CINEA, 2021). Whether the Blue Economy advances sustainability or accelerates environmental decline ultimately depends on how ocean-based activities are governed.

Employment generation is one of the Blue Economy's most significant contributions. Fisheries and aquaculture directly employ 58.5 million people and support the livelihoods of nearly 600 million when related activities are included, particularly in developing countries where small-scale fisheries are vital for food security and social resilience (FAO, 2022). Coastal tourism is similarly labour-intensive, accounting for more than half of total employment in some SIDS (UNWTO, 2019). Sustainable tourism and ecotourism offer pathways to protect biodiversity while generating long-term employment, emphasizing balanced economic, environmental, and socio-cultural outcomes and empowering local communities (Mortelliti & Meng, 2025).

Looking ahead, ocean-based industries are projected to generate nearly 40 million full-time equivalent jobs by 2030, with rapid growth expected in offshore wind energy, marine aquaculture, fish processing, and port activities (OECD, 2016). However, these opportunities remain unevenly distributed. Large corporations dominate many sectors, while small-scale fishers, coastal communities, and Indigenous groups often face barriers related to access, rights, and market participation. Inclusive growth will therefore require targeted policies that promote equity, education, capacity building, and stronger labour protections.

A defining feature of recent Blue Economy development is the rise of blue finance, which seeks to align investment flows with sustainability objectives. Blue finance includes a range of instruments designed to support ocean conservation and sustainable marine industries, demonstrating flexibility across debt and investment formats (Shiiba et al., 2022). The Seychelles' pioneering sovereign blue bond illustrates how capital markets can be leveraged to fund sustainable fisheries and marine conservation (World Bank, 2018), while Belize's debt-for-nature swap highlights innovative approaches to debt restructuring and marine protection (Jiang & Cao, 2024). Growing private investment in marine biotechnology, aquaculture technology, and digital monitoring—supported by tools such as satellite tracking, AI, and genetic techniques—further enhances governance and efforts to curb illegal fishing (Beukema et al., 2023). Scaling such investment will require addressing risk perceptions, strengthening governance frameworks, and expanding blended finance mechanisms to mobilize public, private, and philanthropic capital (Swift & Baines, 2024).

Regional Perspectives

The economic potential of the Blue Economy varies considerably across regions, reflecting differences in resource endowments, governance capacity, and development priorities. In Africa, the African Union's 2050 Integrated Maritime Strategy frames the ocean as a strategic frontier for economic diversification, emphasizing fisheries, shipping, offshore energy, aquaculture, and maritime security (AU, 2014; Nagy & Nene, 2021). Several countries are advancing national initiatives in this direction, with Nigeria developing a National Policy on Marine and Blue Economy to address illegal, unreported, and unregulated fishing, and Kenya piloting governance frameworks for the sector (Karani et al., 2022). Island states such as Mauritius and Seychelles are comparatively more advanced, having established dedicated institutional arrangements and marine financing mechanisms (Rustomjee, 2018). Despite this progress, weak enforcement, limited maritime capacity, fragmented jurisdictions, and persistent IUU fishing continue to constrain the realization of Africa's Blue Economy potential.

The Asia-Pacific region represents a global hub of ocean-based production and innovation. It hosts the world's largest aquaculture industries, led by China, Indonesia, and Vietnam, with the top seven producers accounting for approximately 86% of global output (FAO, 2024). China has also emerged as a leader in offshore renewable energy, rapidly expanding offshore wind capacity from under 5 GW in 2018 to about 42.7 GW by early 2025 (Global Energy Monitor, 2025). Across the Pacific, Small Island Developing States are increasingly leveraging marine conservation, eco-tourism, and sustainable ocean financing instruments, including blue bonds, to support biodiversity protection, livelihoods, and climate resilience (UNDP, 2022).

Within the European Union, the Blue Growth Strategy seeks to integrate maritime industries into a broader agenda of sustainability and innovation, with a strong emphasis on marine spatial planning and smart governance (European Commission, 2012). This approach aligns closely with the European Green Deal and positions oceans and seas as engines of sustainable growth. Offshore wind energy is a central pillar of this strategy, with plans to expand installed capacity from roughly 12 GW to at least 60 GW by 2030 and 300 GW by 2050, supporting long-term decarbonization and renewable energy objectives (European Commission, 2020a; Global Energy Monitor, 2025; Wind Power Monthly, 2021).

In the Caribbean and Latin America, Blue Economy strategies reflect both vulnerability and opportunity. Caribbean economies remain heavily dependent on tourism and are increasingly adopting innovative financial instruments, such as blue bonds and debt-for-nature swaps, to enhance climate resilience and finance marine conservation (IMF, 2023; James & Clarke, 2025). In Latin America, countries including Chile and Peru are prioritizing fisheries and aquaculture development while seeking to balance productivity gains with sustainability and environmental protection (Anbleyth-Evans et al., 2024; FAO, 2025b). Collectively, these regional experiences underscore the diversity of Blue Economy pathways, ranging from resource-oriented development to conservation-led growth models.

Emerging Frontiers

Several frontier sectors within the Blue Economy present transformative economic opportunities, driven by technological innovation and growing sustainability imperatives. Marine biotechnology is emerging as a particularly dynamic field, as advances in transcriptomics, proteomics, and bioprospecting reveal marine organisms as valuable sources of pharmaceuticals and novel biomaterials (Moutinho Cabral et al., 2024). Beyond medical applications, marine-derived compounds are increasingly explored for biodegradable plastics and bio-adhesives, while the genomic potential of symbiotic bacteria enables more sustainable production pathways for scarce marine natural products (World Ocean Review, 2021).

Offshore renewable energy represents another rapidly expanding frontier. Global offshore wind capacity is projected to grow at least fifteenfold by 2040, effectively transforming the sector into a trillion-dollar industry. Annual market growth is estimated at around 13%, with new installations expected to exceed 20 GW per year by 2030 (IEA, 2019). Although wave and tidal energy technologies remain at early stages of development, they hold considerable long-term potential, particularly for coastal and island nations seeking to diversify their energy systems.

Circular economy approaches to waste management are also creating new value chains while addressing marine pollution. Innovations in advanced recycling technologies allow marine plastic waste to be converted back into usable raw materials, while emerging biodegradable polymers offer alternatives that reduce the long-term persistence of plastics in marine ecosystems. In parallel, increasingly sophisticated clean-up technologies are being deployed to intercept plastics both before and after they enter aquatic environments, contributing to pollution mitigation and resource recovery (Schmaltz et al., 2020; RIKEN, 2025).

Digital blue technologies further expand the scope of the Blue Economy by enhancing efficiency, transparency, and governance across ocean-based industries. The integration of artificial intelligence, the Internet of Things, and blockchain technologies significantly improves the reliability and traceability of fisheries and seafood supply chains. These systems enable real-time data collection, immutable records, and greater transparency throughout supply chain operations, strengthening monitoring and accountability (Alsharabi et al., 2024; Liu et al., 2025).

Together, these frontier sectors extend the Blue Economy well beyond traditional activities, positioning it as a catalyst for wider transitions toward a bioeconomy and a digital economy.

Risks and Conditions for Realizing Potential

The Blue Economy holds substantial potential for economic and social development but carries significant risks if shaped by extractive practices and weak governance. Without moving beyond business-as-usual approaches, it may intensify resource depletion, environmental degradation, and social inequality (Cisneros-Montemayor et al., 2022; Bennett et al., 2021). A key barrier to sustainable investment is the lack of reliable data, consistent standards, and robust scientific baselines, which increases costs and heightens the risk of greenwashing or bluewashing, undermining investor confidence (OECD, 2025a). Realizing the Blue Economy's potential requires strong governance and marine spatial planning, equity for small-scale fishers and coastal communities, science-driven innovation, and enhanced global cooperation to address transboundary challenges such as climate change, pollution, and illegal fishing.

POLICY AND GOVERNANCE DIMENSIONS

The governance of the Blue Economy is as vital as its economic and ecological dimensions. Without strong institutions, legal frameworks, and policies, the growth of ocean industries risks replicating patterns of unsustainable exploitation, inequality, and environmental degradation (Voyer et al., 2022:1). In this context, governance extends beyond regulation; it encompasses the institutions, norms, and practices that shape how marine resources are accessed, managed, and distributed.

Multi-Level Governance Architecture of the Blue Economy

Global and regional governance frameworks together form the institutional backbone of the Blue Economy, shaping how ocean resources are accessed, managed, and conserved across scales. At the global level, the United Nations Convention on the Law of the Sea (UNCLOS) remains the cornerstone of international ocean governance, defining maritime zones and clarifying states' rights and responsibilities regarding navigation, resource exploitation, marine scientific research, and dispute settlement (UNCLOS, 1982). While UNCLOS provides a durable legal foundation, it predates emerging challenges such as climate change impacts and deep-sea mining, necessitating complementary instruments to address these gaps. The governance of the Blue Economy is therefore embedded within a wider system of global agendas, most notably the United Nations Sustainable Development Goals. SDG 14 reinforces sustainable fisheries management, pollution reduction, and marine conservation, directly aligning with Blue Economy objectives (United Nations, 2025). Sectoral and cross-cutting frameworks, including the FAO Code of Conduct for Responsible Fisheries and international climate agreements, further shape governance by influencing fisheries practices, shipping decarbonization, and offshore energy development, with the International Maritime Organization linking maritime regulation to global climate commitments (European Commission, 2025).

While global frameworks establish overarching principles, much operational governance occurs at the regional level, where policies are adapted to ecological and socio-economic contexts. The European Union represents one of the most advanced regional models through its Integrated Maritime Policy and Maritime Spatial Planning Directive, which promote coordinated, ecosystem-based allocation of marine space and help manage trade-offs among competing uses (Schäfer, 2009; Zaucha, 2025; European Parliament, 2025). In Africa, the African Union's 2050 Integrated Maritime Strategy frames the ocean as a driver of continental development, though uneven implementation persists due to capacity and funding constraints (AU, 2014; AU-IBAR, 2019). In the Asia-Pacific, cooperative mechanisms address shared challenges such as illegal fishing, with ASEAN initiatives and Pacific regional institutions supporting fisheries management, enforcement, and emerging blue finance tools (ASEAN RPOA-IUU Secretariat, 2007; Forum Fisheries Agency, 2024).

Across regions, Regional Fisheries Management Organizations play a critical role in governing migratory fish stocks, enabling collective action despite persistent challenges related to compliance and transparency (Ewell et al., 2020; Haas, 2019). Together, these global and regional arrangements illustrate a multi-level governance architecture in which effectiveness depends on coherence, cooperation, and the capacity to address transboundary ocean challenges.

National Policy Innovations

Several countries have pioneered innovative Blue Economy policies, demonstrating different approaches to integrating governance with economic development.

Case Study 1: Seychelles' Blue Bond

Seychelles launched the world's first sovereign blue bond in 2018, raising $15 million to fund sustainable fisheries and marine conservation (World Bank, 2018). This instrument linked global capital markets with national ocean governance, creating incentives for sustainable practices. The bond proceeds supported marine protected areas, improved fisheries management, and community-based aquaculture. While relatively small in scale, it demonstrated how innovative finance can embed sustainability into national ocean policy.

Case Study 2: Kenya's Blue Economy Task Force

Kenya has positioned itself as a continental frontrunner by embedding the Blue Economy into its national policy framework. Executive Order No. 1 of 2016 expanded the mandate of the State Department for Fisheries and the Blue Economy, tasking it with coordinating initiatives across sectors such as aquaculture, maritime transport, tourism, and extractives (United Nations, n.d.-a). In the same year, Nairobi hosted the first global Sustainable Blue Economy Conference, which drew thousands of participants and elevated African ocean governance onto the global agenda (IISD, 2018). Yet, despite these achievements, significant challenges endure, particularly in combating IUU fishing and in safeguarding equitable access for artisanal fishers, whose livelihoods remain threatened by weak monitoring, enforcement gaps, and marginalization within policy processes (Kamau et al., 2021; United Nations Economic Commission for Africa, 2016).

Case Study 3: Indonesia's Archipelagic Strategy

With over 17,500 islands, 108,000 kilometres of coastline, and three-quarters of its territory covered by ocean, Indonesia's prosperity is closely tied to marine-based activities, including capture fisheries, aquaculture, coastal tourism, marine construction, and transportation (World Bank, 2021). Its blue economy framework prioritizes innovation in aquaculture and sustainable marine tourism, supported by digital monitoring systems and public–private partnerships to strengthen enforcement and environmental management. To finance these initiatives, Indonesia has turned to instruments such as blue bonds and sustainable finance, while fostering coordination across ministries and local communities to balance national security, food security, and ecological integrity. Yet institutional fragmentation, data limitations, and uneven local capacity remain significant challenges. Even so, Indonesia illustrates how a large developing archipelagic state seeks to harmonize strategic and sustainable objectives within its maritime domain (Wuwung et al., 2024). Indonesia's approach highlights how large developing states can strive to balance national security, food security, and environmental sustainability within ocean governance.

Case Study 4: EU Marine Spatial Planning in the North Sea

The EU's MSP Directive has been notably tested in the North Sea, where offshore wind development has expanded rapidly. By using MSP, regulators were able to designate areas for turbines while protecting vital shipping lanes and fishing grounds (European Commission, 2020b). Although not free from conflict, the process helped reduce disputes and established a governance model that is now emulated internationally.

These cases illustrate diverse governance pathways, from financial innovation to participatory planning, highlighting the adaptability of Blue Economy governance to national contexts.

Toward Integrated and Adaptive Blue Economy Governance

Despite notable progress, the governance of the Blue Economy continues to face significant structural and operational challenges. Fragmented institutional arrangements often result in overlapping or conflicting mandates across fisheries, shipping, energy, and environmental sectors, weakening coherent oversight and policy effectiveness (eMSP NBSR, 2024). Enforcement deficits remain particularly pronounced, with illegal, unreported, and unregulated fishing persisting in regions such as Southeast Asia and West Africa due to limited surveillance capacity and regulatory loopholes (Widjaja et al., 2020). Governance outcomes are further shaped by equity and power asymmetries, as industrial actors frequently dominate decision-making while small-scale fishers and coastal communities struggle for representation and access (Gerhardinger et al., 2022). These challenges are compounded by the inherently transboundary nature of marine issues, including climate impacts, pollution, and migratory species, where political tensions and sovereignty concerns often constrain effective cooperation (Mahon & Fanning, 2019).

Addressing these constraints requires a more integrated and adaptive governance approach. Strengthening science–policy integration is essential to ensure that decision-making reflects the best available knowledge and can adjust to environmental change (Zhang et al., 2023). Marine spatial planning offers a practical mechanism to reduce intersectoral conflicts and promote synergies among competing ocean uses (UNESCO-IOC, 2021). Equally important is participatory governance that meaningfully involves small-scale fishers, Indigenous peoples, and civil society to enhance legitimacy and equity (Winther & Dai, 2020). Transparent blue finance, reinforced by robust monitoring and verification, can prevent superficial sustainability claims

(OECD, 2025a), while enhanced regional cooperation and alignment with climate policy are critical for managing shared marine challenges and supporting resilient Blue Economy development (Garcia & Cochrane, 2005).

INNOVATION AND FUTURE PATHWAYS

The Blue Economy stands at a critical juncture, as mounting pressures on marine ecosystems coincide with growing expectations for ocean-based growth. Decisions made in the coming decades will determine whether the oceans become engines of sustainable prosperity or sites of escalating ecological and social crisis. Looking toward 2050, the Blue Economy presents a range of possible trajectories shaped by innovation, governance choices, and the capacity to manage risk responsibly. These dynamics highlight both the scale of the challenges involved and the profound responsibility inherent in pursuing ocean-based development.

Global interest in the Blue Economy is projected to expand substantially over the coming decades. Ocean-based industries are expected to contribute more than USD 3 trillion annually by 2030, with continued growth through mid-century, driven by sectors such as offshore renewable energy, aquaculture, coastal tourism, marine biotechnology, and shipping (OECD, 2016). While traditional activities like fisheries will remain essential for livelihoods, newer industries are expected to account for an increasing share of economic output (World Bank, 2017). However, these projections depend on the ability to address intensifying sustainability challenges. Climate change is already reshaping marine ecosystems, disrupting fisheries, and heightening risks for coastal communities through ocean warming, acidification, and sea-level rise (IPCC, 2022). As the OECD (2025b) notes, the future of the Blue Economy will ultimately be defined by how growth ambitions are balanced against ecological limits and supported by effective governance frameworks.

Technological innovation is expected to play a defining role in this transition. Advances in digital ocean governance are transforming how marine environments are monitored and managed. Satellite systems, remote sensing, and artificial intelligence enable more precise tracking of fishing fleets, improved identification of vessel activity, and real-time responses to changing marine conditions. At the same time, blockchain and distributed ledger technologies are being tested to enhance transparency and traceability in seafood supply chains, supporting efforts to curb illegal, unreported, and unregulated fishing while strengthening consumer confidence (Cromwell, 2025). Marine biotechnology further expands economic possibilities by unlocking marine genetic resources, with bioprospecting revealing novel compounds for pharmaceuticals and cosmetics derived from microbes and macroalgae adapted to extreme environments (Nurkolis et al., 2025). Algae-based systems are also gaining attention as platforms for biofuels and food ingredients due to their rapid growth, minimal reliance on land and freshwater, and carbon fixation potential (Chiriví-Salomón et al., 2024).

Technological progress is also reshaping maritime transport. Alternative fuels and propulsion systems, including hydrogen, ammonia, and electric power, offer pathways to significantly reduce emissions from shipping. Reflecting this shift, the International Maritime Organization has established targets to improve energy efficiency and reduce carbon intensity by at least 40% relative to 2008 levels by 2030 (International Maritime Organization, n.d.). In parallel, circular economy innovations are embedding waste reduction and resource recovery into marine value chains. Chemical recycling and advanced polymer regeneration technologies convert plastics retrieved from marine environments into reusable materials, transforming pollution into economic inputs (Kibria et al., 2023). Yet these innovations also raise governance concerns, as weak regulatory frameworks may exacerbate inequalities and ecological pressures, particularly in sensitive areas such as marine biotechnology and digital governance.

Policy and institutional innovation will therefore be as critical as technological progress. Marine spatial planning is expected to become a central framework for balancing competing ocean uses across national and transboundary waters, with the European Union's approach illustrating how spatial tools can enhance policy coherence and reduce conflict (Zaucha et al., 2025). Scaling such approaches globally requires capacity building, inclusive participation, and mechanisms to overcome jurisdictional fragmentation (Papageorgiou et al., 2024). Inclusive governance is especially vital for small-scale fisheries, which account for at least 40% of global capture fisheries production and support around 60.2 million livelihoods, yet remain marginalized in decision-making (FAO et al., 2023). Blue finance instruments, such as blue bonds and debt-for-nature swaps, can mobilize capital for sustainability, but their effectiveness depends on transparency and accountability, as illustrated by concerns surrounding blue washing (World Bank, 2018).

Ultimately, the Blue Economy faces divergent futures. A business-as-usual trajectory risks intensifying overfishing, habitat loss, biodiversity decline, and climate-related impacts that threaten billions of people dependent on marine systems (Le Gouvello et al., 2024; United Nations, n.d.-b). In contrast, a transformative pathway grounded in ecosystem-based management, circular economy principles, and inclusive governance offers the potential for regenerative development aligned with global climate goals and the Sustainable Development Goals (Le Gouvello et al., 2024; UNEP, 2024). Which path prevails will depend less on technology alone than on political will, institutional strength, and societal values, determining whether the Blue Economy becomes a driver of long-term prosperity or a continuation of extractive patterns.

CONCLUSION

This chapter has examined the Blue Economy through the interconnected dimensions of economic growth, sustainability, and governance, highlighting both its transformative potential and its inherent risks. Within global policy discourse, the Blue Economy is increasingly framed as a vehicle for economic transformation and a contributor to the United Nations Sustainable Development Goals, particularly SDG 14 on Life Below Water. However, as the analysis demonstrates, realizing this potential requires careful navigation of trade-offs between economic ambition, ecological limits, and social equity.

The Blue Economy encompasses significant economic opportunities alongside profound sustainability challenges. Projections indicate that ocean-based industries could generate more than USD 3 trillion annually by 2030, with fisheries, aquaculture, shipping, coastal tourism, offshore renewable energy, and marine biotechnology driving employment and innovation. Technological advances in digital monitoring, sustainable shipping fuels, and marine biotechnology further expand these opportunities. Yet these prospects are inseparable from mounting environmental pressures. Overfishing, habitat degradation, climate change, pollution, and fragmented governance structures continue to undermine marine ecosystem resilience and threaten the livelihoods of communities dependent on ocean resources. The long-term viability of the Blue Economy therefore depends on governance systems capable of aligning economic development with ecological protection.

Governance emerges as the central link between opportunity and sustainability. Global frameworks such as UNCLOS and SDG 14 provide essential normative foundations, while regional initiatives, including the European Union's Marine Spatial Planning Directive and the African Union's 2050 Integrated Maritime Strategy, demonstrate how these principles can be adapted to diverse contexts. At the national level, innovative approaches such as Seychelles' Blue Bond and Indonesia's archipelagic strategy illustrate how finance, conservation, and development can be integrated. Despite these advances, persistent gaps remain in addressing illegal, unreported, and unregulated fishing, ensuring equity for small-scale fisheries, and preventing blue washing in emerging financial markets.

Future pathways for the Blue Economy reveal starkly divergent outcomes. A business-as-usual trajectory risks accelerating ecological degradation and deepening social inequalities, while a transformative sustainability pathway envisions a regenerative Blue Economy grounded in ecosystem-based management, circular economy principles, and inclusive governance. The direction ultimately taken will depend less on technological capacity alone than on political will, institutional effectiveness, and societal values.

Based on the analysis, several priorities emerge. Strengthening global governance frameworks is essential, including updating legal instruments to address emerging challenges and reinforcing accountability under SDG 14. Expanding marine spatial planning and ecosystem-based management can help balance competing ocean uses while safeguarding ecological integrity. Ensuring equity for small-scale fisheries and coastal communities through participatory governance and co-management is critical for inclusive growth and food security. Transparent blue finance mechanisms, supported by rigorous monitoring and blended finance approaches, can mobilize investment while reducing the risk of blue washing. Integrating science, digital technologies, and adaptive governance can enhance transparency and enforcement, while stronger regional cooperation is needed to address transboundary challenges such as IUU fishing, marine pollution, and migratory species protection.

This chapter contributes to scholarship by synthesizing the Blue Economy's conceptual foundations, sectoral drivers, sustainability challenges, governance frameworks, and future pathways into a cohesive analytical perspective. It underscores that the Blue Economy is not merely a policy slogan, but an evolving paradigm reshaping how societies understand and govern the ocean. Ultimately, the Blue Economy represents both a transformative opportunity and a profound responsibility. Anchored in equity, transparency, and ecological stewardship, it can reconcile economic growth with sustainability; without such foundations, it risks becoming another arena of overexploitation, inseparable from the broader struggle for sustainable development.

REFERENCES

African Union – Inter-African Bureau for Animal Resources. (2019). *Africa blue economy strategy*. African Union. https://faolex.fao.org/docs/pdf/au228027.pdf

African Union. (2014). 2050 Africa's integrated maritime strategy (2050 AIM Strategy). https://au.int/sites/default/files/documents/30929-doc-2050_aim_strategy_eng_0.pdf

Alsharabi, N., Ktari, J., Frikha, T., Alayba, A., Alzahrani, A. J., Jadi, A., & Hamam, H. (2024). Using blockchain and AI technologies for sustainable, biodiverse, and transparent fisheries of the future. *Journal of Cloud Computing (Heidelberg, Germany)*, *13*, 135. Advance online publication. DOI: 10.1186/s13677-024-00696-8

Anbleyth-Evans, J., Araos Leiva, F., Gaymer, C. F., Alvarez Abel, R. R., Campos, L., & Hidalgo, C. (2024). From a brown to a blue economy in Chile. *Environmental Challenges*, *14*, 100846. https://doi.org/10.1016/j.envc.2024.100846

Anuar, S. T., Mohd Ali, A. A., Cordova, M. R., & Charoenpong, C. (2025). Editorial: Baselines, impacts and mitigation strategies for plastic debris and microplastic pollution in South East Asia. *Frontiers in Marine Science*, *12*. Advance online publication. https://doi.org/10.3389/fmars.2025.1586080

ASEAN RPOA-IUU Secretariat. (2007). Regional plan of action to promote responsible fishing practices including combating IUU fishing. Directorate General of Capture Fisheries, Ministry of Marine Affairs and Fisheries, Indonesia. https://imcsnet.org/sites/default/files/media_files/asean-rpoa-ıuu.pdf

Badrinarayanan, C., & Akhter, Z. (2023). *The cost of climate change: Assessing the impact on SIDS (Small Island Developing States)*. Blue Sky Analytics. https://blueskyhq.io/blog/the-cost-of-climate-changing-assessing-the-impact-on-sids-small-island-developing-states

Bennett, N. J., Blythe, J., White, C. S., & Campero, C. (2021). Blue growth and blue justice: Ten risks and solutions for the ocean economy. *Marine Policy*, *125*, 104387. https://doi.org/10.1016/j.marpol.2020.104387

Bennett, N. J., Cisneros-Montemayor, A. M., Blythe, J., Silver, J. J., Singh, G., Andrews, N., Calò, A., Christie, P., Di Franco, A., Finkbeiner, E. M., Gelcich, S., Guidetti, P., Harper, S., Hotte, N., Kittinger, J. N., Le Billon, P., Lister, J., López de la Lama, R., McKinley, E., & Rashid Sumaila, U. (2019). Towards a sustainable and equitable blue economy. *Nature Sustainability*, *2*(11), 991–993. https://doi.org/10.1038/s41893-019-0404-1

Beukema, P., Bastani, F., Wolters, P., Herzog, H., & Ferdinando, J. (2023). Satellite imagery and ai: a new era in ocean conservation, from research to deployment and impact (Preprint). *arXiv*. https://doi.org//arXiv.2312.03207DOI: 10.48550

Bhanja, A., Payra, P., Mandal, B., & Ghara, A. (2024). *Blockchain technology and artificial intelligence in fisheries management*. International Books & Periodical Supply Service. https://www.researchgate.net/publication/382640336_blockchain_technology_and_artificial_intelligence_in_fisheries_management

Blasiak, R., & Claudet, J. (2024). Energy and material demand: A review of models and scenarios. [*Annual Reviews*. https://doi.org/10.1146/annurev-environ-011023-022521]. *Annual Review of Environment and Resources*, *49*(1). Advance online publication.

Bonnet-Cantalloube, B. (2024). *Shipping forecasts: Scoring the decarbonisation proposals under the International Maritime Organisation's market-based measures (MBM) shipping strategies*. Carbon Market Watch. https://carbonmarketwatch.org/publications/shipping-forecasts-scoring-the-decarbonisation-proposals-under-the-international-maritime-organisations-market-based-measures-mbm-shipping-strategies/

Brumbaugh, R., & Patil, P. (2017, May 22). *Sustainable tourism can drive the blue economy: Investing in ocean health is synonymous with generating ocean wealth*. Voices – World Bank Blogs. https://blogs.worldbank.org/en/voices/Sustainable-Tourism-Can-Drive-the-Blue-Economy

Chiriví-Salomón, J. S., García-Huérfano, S., & Giraldo, I. A. (2024). Bioprospecting microalgae: A systematic review of current trends. *Phycology*, *4*(3), 508–532. DOI: 10.3390/phycology4030028

CINEA. (2021). *Sustainability criteria for the blue economy (ea-02-21-398-en-n)*. Publications Office of the European Union. https://cinea.ec.europa.eu/system/files/2021-05/Sustainability%20criteria%20for%20the%20blue%20economy%20.pdf

Cisneros-Montemayor, A. M., Croft, F., Issifu, I., Swartz, W., & Voyer, M. (2022). A primer on the "blue economy:" Promise, pitfalls, and pathways. *One Earth*, *5*(9), 982–986. DOI: 10.1016/j.oneear.2022.08.011

Cromwell, J., Turkson, C., Dora, M., & Yamoah, F. A. (2025). Digital technologies for traceability and transparency in the global fish supply chains: A systematic review and future directions. *Marine Policy*, *178*, 106700. DOI: 10.1016/j.marpol.2025.106700

Durr, S., & Shumway, S. (2024). *Biofouling prevention and management in offshore renewable energy infrastructure and construction: Best practices in biofouling management (Vol. 3)*. Liverpool John Moores University. http://researchonline.ljmu.ac.uk/id/eprint/25575/1/392045eng%20(1).pdf

Ebarvia, M. C. M. (2016). Economic assessment of oceans for sustainable blue economy development. *Journal of Ocean and Coastal Economics*, *2*(2), 7. Advance online publication. DOI: 10.15351/2373-8456.1051

eMSP NBSR. (2024). *Addressing the fragmentation of ocean governance across borders: Policy brief on ocean governance and maritime spatial planning in the North and Baltic Sea regions*. eMSP NBSR. https://www.emspproject.eu/wp-content/uploads/2024/01/Ocean-Governance-Policy-Brief-eMSP-NBSR-January-2024.pdf

Environmental Justice Foundation. (2012). *Pirate Fishing Exposed: The Fight Against Illegal Fishing in West Africa and the EU*. Environmental Justice Foundation. https://ejfoundation.org/resources/downloads/Pirate-Fishing-Exposed.pdf ejfoundation.

European Commission. (2012). *Blue growth: opportunities for marine and maritime sustainable growth (com (2012) 491 final)*. https://eur-lex.europa.eu/legal-content/en/txt/?uri=COM:2012:491:FIN

European Commission. (2020a). *Boosting offshore renewable energy for a climate neutral europe (press release)*. https://ec.europa.eu/commission/presscorner/detail/en/ip_20_2096

European Commission. (2020b). *An eu strategy to harness the potential of offshore renewable energy for a climate neutral future (com(2020) 741 final*). Publications Office of the European Union. https://eur-lex.europa.eu/legal-content/EN/TXT/?uri=CELEX%3A52020DC0741

European Commission. (2025). *Landmark agreement towards net-zero emissions from global shipping by 2050*. European Commission. https://blue-economy-observatory.ec.europa.eu/news/landmark-agreement-towards-net-zero-emissions-global-shipping-2050-2025-05-08_en/

European Parliament. (2025). *Integrated maritime policy of the European Union*. https://www.europarl.europa.eu/erpl-app-public/factsheets/pdf/en/FTU_3.3.8.pdf

Ewell, C., Cullis-Suzuki, S., Hocevar, J., Mitchell, E., Snowden, S., & Jacquet, J. (2020). An evaluation of Regional Fisheries Management Organization at-sea compliance monitoring and observer programs. *Marine Policy*, *115*, 103842. DOI: 10.1016/j.marpol.2020.103842

FDA (U.S. Food and Drug Administration). (2023). *Aquacultured seafood.* https://www.fda.gov/food/seafood-guidance-documents-regulatory-information/aquacultured-seafood

Food and Agriculture Organization (FAO). (2022). *The state of world fisheries and aquaculture 2022: towards blue transformation.* Food and Agriculture Organization of the United Nations., DOI: 10.4060/cc0461en

Food and Agriculture Organization (FAO). (2024). *Global fisheries and aquaculture production reaches a new record high.* Food and Agriculture Organization of the United Nations. https://www.fao.org/asiapacific/news/news-detail/fao-report--global-fisheries-and-aquaculture-production-reaches-a-new-record-high/en

Food and Agriculture Organization (FAO). (2025a). *FAO releases the most detailed global assessment of marine fish stocks to date.* Food and Agriculture Organization of the United Nations. https://www.fao.org/newsroom/detail/fao-releases-the-most-detailed-global-assessment-of-marine-fish-stocks-to-date/en

Food and Agriculture Organization (FAO). (2025b). *FAO and the World Bank highlight the importance of the blue transformation and blue economy in Latin America and the Caribbean projects.* https://www.fao.org/americas/news/news-detail/transformacion-azul/en

Food and Agriculture Organization of the United Nations (FAO), Duke University, & WorldFish. (2023). *Illuminating hidden harvests – the contributions of small-scale fisheries to sustainable development.* FAO. https://doi.org/DOI: 10.4060/cc4576en

Forum Fisheries Agency. (2024, November). *Operation Kurukuru 2024 successfully concludes.* Pacific Islands Forum Fisheries Agency. https://www.ffa.int/2024/11/operation-kurukuru-2024-successfully-concludes/

Galparsoro, I., Menchaca, I., Seeger, I., Nurmi, M., McDonald, H., Garmendia, J., Pouso, S., & Borja, Á. (2022). Mapping potential environmental impacts of offshore renewable energy. https://tethys.pnnl.gov/publications/mapping-potential-environmental-impacts-offshore-renewable-energy

Garcia, S. M., & Cochrane, K. L. (2005). Ecosystem approach to fisheries: A review of implementation guidelines. *ICES Journal of Marine Science*, *62*(3), 311–318. DOI: 10.1016/j.icesjms.2004.12.003

Gerhardinger, L. C., Holzkämper, E., de Andrade, M. M., Corrêa, M. R., & Turra, A. (2022). Envisioning ocean governability transformations through network-based marine spatial planning. *Maritime Studies*, *21*, 131–152. DOI: 10.1007/s40152-021-00250-1

Gill, D. A., Blythe, J., Bennett, N., Evans, L., Brown, K., Turner, R. A., Baggio, J. A., Baker, D., Ban, N. C., Brun, V., Claudet, J., Darling, E., Di Franco, A., & Epstein, G., Estradivari, Gray, N. J., Gurney, G. G., Horan, R. P., Jupiter, S. D., … Muthiga, N. A. (2023). Triple exposure: Reducing negative impacts of climate change, blue growth, and conservation on coastal communities. *One Earth*, *6*(2), 118–130. https://doi.org/10.1016/j.oneear.2023.01.010

Global Energy Monitor. (2025). *China's solar and onshore wind capacity reaches new heights while offshore wind shows promise.* https://globalenergymonitor.org/report/chinas-solar-and-onshore-wind-capacity-reaches-new-heights-while-offshore-wind-shows-promise/

GWEC (Global Wind Energy Council). (2023). *Global offshore wind report 2023.* GWEC. https://www.apren.pt/contents/publicationsothers/gwec-global-offshore-wind-report-2023.pdf

Haas, B., Haward, M., McGee, J., & Fleming, A. J. (2019). The influence of performance reviews on regional fisheries management organizations. *ICES Journal of Marine Science*, *76*(7), 2082–2089. DOI: 10.1093/icesjms/fsz088

IISD. (2018). *Summary report, 26–28 november 2018: sustainable blue economy conference*, Nairobi, Kenya. International Institute for Sustainable Development. https://enb.iisd.org/events/sustainable-blue-economy-conference/summary-report-26-28-november-2018

IMF. (2023). *Online annex 5: debt-for-nature swaps in latin america and the caribbean.* In Regional Economic Outlook – Western Hemisphere. International Monetary Fund. https://www.imf.org/-/media/Files/Publications/REO/WHD/2023/October/English/onlineannex-5-en.ashx

Intergovernmental Panel on Climate Change. (2022). *The ocean and cryosphere in a changing climate: special report of the intergovernmental panel on climate change.* Cambridge University Press., DOI: 10.1017/9781009157964

International Energy Agency. (2019). *Offshore wind outlook 2019 (world energy outlook special report).* https://iea.blob.core.windows.net/assets/495ab264-4ddf-4b68-b9c0-514295ff40a7/Offshore_Wind_Outlook_2019.pdf

International Maritime Organization. (n.d.). *IMO's work to cut GHG emissions from ships.* Retrieved August 21, 2025, from https://www.imo.org/en/mediacentre/hottopics/pages/cutting-ghg-emissions.aspx

James, K., & Clarke, J. T. (2025). *Securing our blue future: Caribbean leadership for global marine protection*. SDG Knowledge Hub. https://sdg.iisd.org/commentary/guest-articles/securing-our-blue-future-caribbean-leadership-for-global-marine-protection/

Janisch, C., Mander, M., Adams, J., Royden-Turner, S., & Amis, M. (2020). *Sustainable blue economy growth models (WRC Report No. TT 805/19)*. Water Research Commission.

Jiang, X., & Cao, H. (2024). Implementing the debt-for-nature swaps for marine protected areas: Case studies from Seychelles and Belize. *Humanities & Social Sciences Communications*, *11*, 331. DOI: 10.1057/s41599-024-02855-3

Jouffray, J. B., Blasiak, R., Norström, A. V., Österblom, H., & Nyström, M. (2020). The blue acceleration: The trajectory of human expansion into the ocean. *One Earth*, *2*(1), 43–54. DOI: 10.1016/j.oneear.2019.12.016

Kamanzi, N. G. (2025). Sustainable fisheries: Innovations in aquaculture. Eurasian Experiment Journal of Engineering, 5(1), 68–74. https://www.eejournals.org/public/uploads/1756137842_5787664b25b5d4e396b2.pdf

Kamau, J. N., Jacobs, Z. L., Jebri, F., Kelly, S., Kimani, E., Makori, A., Mwaluma, J., Mueni, E., Ong'anda, H., Palmer, M. R., Popova, E., Roberts, M. J., Taylor, S. F. W., Wihsgott, J. U., & Painter, S. C. (2021). Managing emerging fisheries of the North Kenya Banks in the context of environmental change. *Ocean and Coastal Management*, *209*, 105671. DOI: 10.1016/j.ocecoaman.2021.105671

Karani, P., Failler, P., Gilau, A. M., Ndende, M., & Diop, S. T. (2022). Africa blue economy strategies integrated in planning to achieve sustainable development at national and regional economic communities (RECs). *Journal of Sustainability Research*, *4*(3), e220011. DOI: 10.20900/jsr20220011

Ken Research. (2025, August). *Global marine biotechnology market 2024-2030*. https://www.kenresearch.com/global-marine-biotechnology-market

Kibria, M. G., Masuk, N. I., Safayet, R., Nguyen, H. Q., & Mourshed, M. (2023). Plastic waste: Challenges and opportunities to mitigate pollution and effective management. *International Journal of Environmental Research*, *17*(1), 20. https://doi.org/10.1007/s41742-023-00507-z

Kumar, A., Soratur, A., Kumar, S., & Venmathi Maran, B. A. (2025). A Review of Marine Algae as a Sustainable Source of Antiviral and Anticancer Compounds. *Macromol*, *5*(1), 11. DOI: 10.3390/macromol5010011 le Gouvello, R., & Simard, F. (2024). Towards a regenerative blue economy: Mapping the Blue Economy (IUCN). https://portals.iucn.org/library/sites/library/files/documents/2024-005-En.pdf

Liu, Z., Yu, X., Liu, N., Liu, C., Jiang, A., & Chen, L. (2025). Integrating AI with detection methods, IoT, and blockchain to achieve food authenticity and traceability from farm-to-table. *Trends in Food Science & Technology*, *158*, 104925. https://doi.org/10.1016/j.tifs.2025.104925

Loayza-Aguilar, R. E., Huamancondor-Paz, Y. P., Saldaña-Rojas, G. B., & Olivos-Ramirez, G. E. (2023). Integrated multi-trophic aquaculture (IMTA): Strategic model for sustainable mariculture in Samanco Bay, Peru. *Frontiers in Marine Science*, *10*, 1151810. DOI: 10.3389/fmars.2023.1151810

Mahon, R., & Fanning, L. (2019). Regional ocean governance: Polycentric arrangements and their role in global ocean governance. *Marine Policy*, *107*, 103590. DOI: 10.1016/j.marpol.2019.103590

Majidi, A. G., Ramos, V., Calheiros-Cabral, T., & Rosa Santos, P., das Neves, L., & Taveira-Pinto, F. (2024). Integrated assessment of offshore wind and wave power resources in mainland Portugal. *Energy*, *308*, 132944. DOI: 10.1016/j.energy.2024.132944

Mortelliti, C., & Meng, C. (2025) *Blue jobs for marine conservation in ASEAN*. Economic and Social Commission for Asia and the Pacific (ESCAP). https://hdl.handle.net/20.500.12870/8366

Moutinho Cabral, I., Gonçalves, C., Grosso, A. R., & Costa, P. M. (2024). Bioprospecting and marine 'omics': Surfing the deep blue sea for novel bioactive proteins and peptides. *Frontiers in Marine Science*, *11*, 1362697. Advance online publication. DOI: 10.3389/fmars.2024.1362697

Nagy, H., & Nene, S. (2021). Blue Gold: Advancing Blue Economy Governance in Africa. *Sustainability*, *13*(13), 7153. DOI: 10.3390/su13137153

Nurkolis, F. (2025). Marine bioactives: Pioneering sustainable solutions for advanced cosmetics and therapeutics. *Pharmacological Research*, *218*, 107868. DOI: 10.1016/j.phrs.2025.107868

OECD. (2009). *Strengthening regional fisheries management organisations*. OECD Publishing., DOI: 10.1787/9789264073326-

OECD. (2016). *The ocean economy in 2030*. OECD Publishing., DOI: 10.1787/9789264251724-

OECD. (2024). *The blue economy in cities and regions: A territorial approach (OECD Urban Studies)*. OECD Publishing., DOI: 10.1787/bd929b7d-

OECD. (2025a). *Promoting sustainable ocean economies: guidance for development co-operation (best practices in development co-operation)*. OECD Publishing., DOI: 10.1787/72055d7f-

OECD. (2025b). *The ocean economy and climate change nexus: The case of Portugal.* OECD Net Zero+ Policy Papers No. 16. https://www.oecd.org/content/dam/oecd/en/publications/reports/2025/06/the-ocean-economy-and-climate-change-nexus_2864ed2e/3d4949c2-en.pdf

Oppenheimer, M., Glavovic, B. C., Hinkel, J., van de Wal, R., Magnan, A. K., Abd-Elgawad, A., Cai, R., Cifuentes-Jara, M., DeConto, R. M., Ghosh, T., Hay, J., Isla, F., Marzeion, B., Meyssignac, B., & Sebesvari, Z. (2019). Sea level rise and implications for low-lying islands, coasts and communities. In Pörtner, H.-O., Roberts, D. C., Masson-Delmotte, V., Zhai, P., Tignor, M., Poloczanska, E., Mintenbeck, K., Alegría, A., Nicolai, M., Okem, A., Petzold, J., Rama, B., & Weyer, N. M. (Eds.), *IPCC special report on the ocean and cryosphere in a changing climate* (pp. 321–445). Cambridge University Press., DOI: 10.1017/9781009157964.006

Papageorgiou, M., Pozoukidou, G., Istoriou, T., & Kostopoulou, T. (2024). Inclusive maritime spatial planning: Stakes at the regional level. *Sustainability*, *16*(22), 10148. DOI: 10.3390/su162210148

RIKEN. (2025). *Plastic-like materials that dissolve in the sea.* https://www.riken.jp/en/news_pubs/ research_news/rr/20250327_1/

Rustomjee, C. (2018). *Green shoots for the African blue economy? (Policy Brief No. 132)*. Africa Portal. https://africaportal.org/wp-content/uploads/2023/06/PB_No.132.pdf

Sarker, S., Hussain, F. A., Assaduzzaman, M., & Failler, P. (2019). Blue economy and climate change: Bangladesh perspective. *Journal of Ocean and Coastal Economics*, *6*(2), 6. Advance online publication. DOI: 10.15351/2373-8456.1105

Schäfer, N. (2009). *European Marine Sand and Gravel Group - A wave of opportunities for the marine aggregates industry: The Integrated EU Maritime Policy and Maritime Spatial Planning – The Way Ahead.* EMSAGG Conference, Frentani Conference Centre, Rome, Italy. https://www.vliz.be/imisdocs/publications/ocrd/155807.pdf

Schmaltz, E., Melvin, E. C., Diana, Z., Gunady, E. F., Rittschof, D., Somarelli, J. A., Virdin, J., & Dunphy-Daly, M. M. (2020). Plastic pollution solutions: Emerging technologies to prevent and collect marine plastic pollution. *Environment International, 144*, 106067. DOI: 10.1016/j.envint.2020.106067

Shiiba, N., Wu, H. H., Huang, M. C., & Tanaka, H. (2022). How blue financing can sustain ocean conservation and development: A proposed conceptual framework for blue financing mechanism. *Marine Policy, 139*, 104575. Advance online publication. https://doi.org/10.1016/j.marpol.2021.104575

Silver, J. J., Gray, N. J., Campbell, L. M., Fairbanks, L. W., & Gruby, R. L. (2015). Blue Economy and competing discourses in international oceans governance. *Journal of Environment & Development, 24*(2), 135–160. DOI: 10.1177/1070496515580797

Simataa, C., Persendt, F., & Gomez, C. (2025). Illegal, unreported and unregulated (IUU) fishing in Africa: A systematic review of challenges and management strategies. *Aquaculture, Fish and Fisheries, 5.* Advance online publication. DOI: 10.1002/aff2.70107

Smithsonian Ocean. (2019). *Ocean acidification.* Smithsonian Ocean. https://ocean.si.edu/ocean-life/invertebrates/ocean-acidification

Swift, A., & Baines, J. (2024). *Closing the ocean funding gap is key to a sustainable ocean economy. World Resources Institute.* https://www.wri.org/technical-perspectives/closing-ocean-funding-gap-key-sustainable-ocean-economy

Talbot, F., & Wilkinson, C. R. (2001). *Coral reefs, mangroves and seagrasses*: A sourcebook for managers. Australian Institute of Marine Science. https://portals.iucn.org/library/efiles/documents/2001-055.pdf

Tedsen, E., Boteler, B., McGlade, K., & Srebotnjak, T. (2014*). Marine resource management and coastal livelihoods: an Atlantic perspective*. Future Ocean. https://www.vliz.be/imisdocs/publications/ocrd/340434.pdf

UNESCO-IOC. (2021). *MSPglobal policy brief: Ocean governance and marine spatial planning (IOC Policy Brief No. 5).* UNESCO. https://unesdoc.unesco.org/ark:/48223/pf0000375723

United Nations Conference on Trade and Development (UNCTAD). (2022). Review of maritime transport 2022: Navigating stormy waters (UNCTAD/RMT/2022). https://unctad.org/publication/review-maritime-transport-2022

United Nations Conference on Trade and Development (UNCTAD). (2024). *Review of maritime transport 2024* (UNCTAD/RMT/2024). https://unctad.org/publication/review-maritime-transport-2024

United Nations Conference on Trade and Development (UNCTAD). (2025). *Fast-growing trillion-dollar ocean economy goes beyond fishing and shipping*. United Nations. https://unctad.org/news/fast-growing-trillion-dollar-ocean-economy-goes-beyond-fishing-and-shipping

United Nations Development Programme (UNDP). (2022). *Demystifying green and blue bonds for the Pacific region*. https://www.undp.org/sites/g/files/zskgke326/files/2022-08/demystifying-green-and-blue-bonds-for-the-pacific.pdf

United Nations Economic Commission for Africa. (2016). *Africa's blue economy: A policy handbook*. United Nations. https://archive.uneca.org/sites/default/files/PublicationFiles/blue-eco-policy-handbook_eng_1nov.pdf

United Nations Environment Programme (UNEP). (2024). *Sustainable blue economy*. https://www.unep.org/topics/ocean-seas-and-coasts/ecosystem-based-approaches/sustainable-blue-economy

United Nations. (1982). *United Nations Convention on the Law of the Sea*. United Nations. https://www.un.org/depts/los/convention_agreements/texts/unclos/unclos_e.pdf

United Nations. (2025). *Goal 14: Conserve and sustainably use the oceans, seas and marine resources*. United Nations. https://www.un.org/sustainabledevelopment/oceans/

United Nations. (n.d.-a). Diving into the blue economy. Retrieved August 21, 2025, from https://www.un.org/fr/desa/diving-blue-economy

United Nations. (n.d.-b). Ocean impacts. Retrieved August 25, 2025, from https://www.un.org/en/climatechange/science/climate-issues/ocean-impacts

UNWTO. (2019). *Tourism in small island developing states (SIDS)*. UNWTO. https://webunwto.s3-eu-west-1.amazonaws.com/2019-07/unwtotourisminsidsa4wtables.pdf

Vidović, T., Šimunović, J., Radica, G., & Penga, Ž. (2023). Systematic overview of newly available technologies in the green maritime sector. *Energies*, *16*(2), 641. https://doi.org/10.3390/en16020641

Voyer, M., Benzaken, D., & Rambourg, C. (2022). Institutionalizing the blue economy: An examination of variations and consistencies among Commonwealth countries. Philosophical Transactions of the Royal Society B: *Biological Sciences*, 377(1854), 20210125. https://doi.org/DOI: 10.1098/rstb.2021.0125

Widjaja, S., Long, T., & Wirajuda, H. (2020). *Illegal, unreported and unregulated fishing and associated drivers (Blue Paper)*. High Level Panel for a Sustainable Ocean Economy / World Resources Institute. https://oceanpanel.org/wp-content/uploads/2022/05/Illegal-Unreported-and-Unregulated-Fishing-and-Associated-Drivers.pdf

Wind Power Monthly. (2021). *Europe's path to rolling out 300 GW of offshore wind by 2050*. https://www.windpowermonthly.com/article/1704375/europes-path-rolling-300gw-offshore-wind-2050

Winther, J. G., & Dai, M. (2020). *Integrated ocean management (blue paper). High level panel for a sustainable ocean economy*. World Resources Institute. https://oceanpanel.org/wp-content/uploads/2022/05/Integrated-Ocean-Management.pdf

World Bank. (2017). *The potential of the blue economy: increasing long-term benefits of the sustainable use of marine resources for small island developing states and coastal least developed countries*. World Bank. https://hdl.handle.net/10986/26843

World Bank. (2018). *Seychelles launches world's first sovereign blue bond*. https://www.worldbank.org/en/news/press-release/2018/10/29/seychelles-launches-worlds-first-sovereign-blue-bond

World Bank. (2021). *Oceans for prosperity: Reforms for a blue economy in Indonesia*. World Bank. https://documents.worldbank.org/en/publication/documents-reports/documentdetail/141421632525626590/oceans-for-prosperity-reforms-for-a-blue-economy-in-indonesia

World Ocean Review. (2021). Marine-derived active compounds: The ocean's genetic library for pharmaceuticals, cosmetics and biomaterials. World Ocean Review – The race for the oceans: Genetic diversity. https://worldoceanreview.com/en/wor-7/the-race-for-the-oceans-genetic-diversity/marine-derived-active-compounds/

Wright, G., Schmidt, S., Rochette, J., Shackeroff, J., Unger, S., Waweru, Y., & Müller, A. (2017). *Partnering for a sustainable ocean: The role of regional ocean governance in implementing SDG14*. PROG: IDDRI, IASS, TMG, & UN Environment. https://www.iddri.org/sites/default/files/import/publications/online_iass_report_report_170524.pdf

Wuwung, L., McIlgorm, A., & Voyer, M. (2024). Sustainable ocean development policies in Indonesia: Paving the pathways towards a maritime destiny. *Frontiers in Marine Science*, *11*. Advance online publication. https://doi.org/10.3389/fmars.2024.1401332

Zaucha, J., Gee, K., Ramieri, E., Neimane, L., Alloncle, N., Blažauskas, N., Calado, H., Cervera-Núñez, C., Marohnić Kuzmanović, V., Stancheva, M., Witkowska, J., Schütz, S. E., Ronco Zapatero, J., & Ehler, C. N. (2025). Implementing the EU MSP Directive: Current status and lessons learned in 22 EU Member States. *Marine Policy, 171*, 106425. https://doi.org/10.1016/j.marpol.2024.106425

Zhang, S., Butt, M. J., & Zulfiqar, K. (2023). Interconnected principles of ocean governance through science-policy integration under the judgments of the international dispute settlement bodies. *Science Progress, 106*(4). Advance online publication. DOI: 10.1177/00368504231205395

ADDITIONAL READING

Gerou, A., & Pantouvakis, A. (2025). The transition to a sustainable blue economy explored through frameworks and SDG alignment. *Discover Sustainability, 6*, 1039. DOI: 10.1007/s43621-025-01953-9

Heidkamp, M., Morrissey, K., & Duret, C. (Eds.). (2023). *Blue economy: People and regions in transitions*. Routledge.

Hutajulu, H. (2025). Blue Economy and Ocean Sustainability: A Bibliometric Review of Global Research. *The Es Economics and Entrepreneurship*, 3(03), 302 –. https://doi.org/DOI: 10.58812/esee.v3i03.505

Pauli, G. (2010). *The blue economy: 10 years, 100 innovations, 100 million jobs*. Paradigm Publications.

Saviano, M., Caputo, F., Fraccascia, L., & Pellegrini, M. M. (Eds.). (2024). *Handbook of sustainable blue economy*. Springer.

World Bank. (2016). Toward a blue economy: *A promise for sustainable growth in the Caribbean*. World Bank. http://hdl.handle.net/10986/25061

KEY TERMS AND DEFINITIONS

Blue Bond: These are bonds issued by governments, development banks, or other institutions to invest in maritime and ocean-based projects that have environmental, economic, and climate benefits.

Illegal, Unreported, and Unregulated (IUU) Fishing: It is a widespread and multifaceted problem occurring in all fisheries, encompassing various illegal activities from capture to use, and is sometimes linked to organized crime.

Marine Plastic Pollution: It is a type of marine pollution caused by plastics of varying sizes, from large original materials such as bottles and bags to microplastics formed by the breakdown of plastic materials.

Ocean Governance: It is the process of protecting and using marine and ocean resources in a sustainable, equitable, and science-based manner. This area is critical to combating global challenges such as climate change, marine pollution, overfishing, and habitat loss.

Regional Fisheries Management Organizations (RFMOs): It is a type of international organization dedicated to the sustainable management of fishery resources or species with high migration potential in a specific area of international waters.

SDG 14 (Life Below Water): It is about life below water and is one of the 17 Sustainable Development Goals established by the United Nations in 2015. The official statement is "Conserve and sustainably use the oceans, seas and marine resources for sustainable development."

UNCLOS (United Nations Convention on the Law of the Sea): It is an international agreement that defines the rights and responsibilities of nations regarding the use of the world's oceans, setting rules for international business, and managing businesses, the environment and marine natural resources.

Chapter 4
The Oceanic Nexus:
AI, Climate Resilience, and the Future of the Blue Economy

Amrik Singh
https://orcid.org/0000-0003-3598-8787
Lovely Professional University, India

ABSTRACT

The Oceanic Nexusexamines the profound and increasingly vital interconnections between Artificial Intelligence (AI), Climate Resilience, and the future trajectory of the Blue Economy. As global reliance on ocean resources intensifies and the impacts of climate change accelerate, the traditional approaches to marine resource management are proving insufficient. This chapter argues that AI represents a transformative force, offering unprecedented capabilities for enhancing our understanding, monitoring, and adaptive management of marine and coastal ecosystems. It highlights how AI can drive climate resilience by providing advanced predictive modeling for extreme weather events, optimizing sustainable aquaculture and fisheries, facilitating real-time monitoring of ocean health, and enabling more effective policy formulation for carbon sequestration and marine conservation. Furthermore, the abstract discusses how this AI-powered approach is crucial for unlocking the full potential of the Blue Economy fostering sustainable economic growth, creating new green jobs, and ensuring equitable benefits, all while safeguarding the ecological integrity of our oceans. Ultimately, the integration of AI is presented not just as a technological advancement but as a strategic imperative for building a resilient, sustainable, and prosperous Blue Economy in the face of a changing climate.

DOI: 10.4018/979-8-3373-5238-1.ch004

INTRODUCTION

The ocean is a vital component of Earth's climate system and a cornerstone of economic development for billions globally. As the climate crisis deepens and marine ecosystems face increasing threats, the role of Artificial Intelligence (AI) in shaping climate resilience and the future of the blue economy has become pivotal (Ahmed et al., 2018). The intersection of AI technologies, sustainable ocean practices, and adaptive strategies fosters oceanic climate resilience. It delves into how AI-driven innovations can advance the conservation, management, and economic utilization of ocean resources, ensuring sustainability, resilience, and equitable growth. The ocean covers more than 70% of Earth's surface and provides food, livelihoods, and ecosystem services for billions of people. However, climate change, overfishing, pollution, and habitat destruction threaten the health of marine ecosystems and the socioeconomic stability of coastal communities (Singh, 2024). The concept of the "blue economy" promotes sustainable use of ocean resources for economic growth, improved livelihoods, and ocean ecosystem health. Artificial Intelligence (AI) has emerged as a transformative force across sectors, and its integration into the blue economy presents unprecedented opportunities (Blasiak et al., 2022). From real-time ocean monitoring and fisheries management to predicting climate events and designing resilient coastal infrastructure, AI can enhance our capacity to understand, protect, and sustainably utilize marine resources.

Understanding the Blue Economy

The Blue Economy is a multifaceted concept that goes beyond simply harnessing ocean resources; it embodies a holistic approach to sustainable development that integrates economic growth, social inclusion, and environmental stewardship. At its core, it seeks to maximize the benefits derived from marine and coastal ecosystems while minimizing negative impacts and ensuring their long-term health. This paradigm shift recognizes the immense value of our oceans not just for traditional industries like fisheries and shipping, but also for emerging sectors such as renewable ocean energy, marine biotechnology, and sustainable tourism (Kaur & Singh, 2025). A Blue Economy is essentially a strategic framework aimed at maximizing the benefits derived from marine environments while actively safeguarding their inherent health and ensuring their long-term sustainability (Elston et al., 2024). This goes far beyond the traditional, often extractive, "ocean economy" by embracing a holistic and integrated approach to the management and development of ocean resources. It encompasses a wide array of interconnected sectors, including established industries like sustainable fisheries, responsible aquaculture, and maritime transport, alongside emerging fields such as renewable ocean energy (wave, tidal, and offshore wind), marine biotechnology, seabed mining (with strict environmental safeguards), and ecotourism. The core principle of the Blue Economy lies in recognizing the ocean as a critical engine for economic growth, innovation, and job creation, but always within the ecological carrying capacity of marine ecosystems (Supina & Singh, 2025; Singh, 2024). This necessitates robust governance frameworks, significant investment in marine scientific research, and the widespread adoption of innovative technologies to minimize environmental footprints. Key pillars include circular economy principles to reduce waste and pollution, climate change mitigation and adaptation strategies for coastal communities, and social equity to ensure that benefits are shared equitably and that vulnerable coastal populations are empowered (Gesami & Nunoo, 2024; Sharma & Singh, 2025). Ultimately, the Blue Economy is about fostering a symbiotic relationship between humanity and the ocean, ensuring that current and future generations can continue to thrive from its bounties without compromising its vitality (Supina & Singh, 2025). This is a vision for prosperity that sails in harmony with planetary health. The blue economy encompasses diverse ocean-based industries, including fisheries, aquaculture, maritime transport, coastal tourism, offshore renewable energy, and marine biotechnology. It emphasizes:

- Economic development without compromising ocean health
- Social inclusion and livelihood enhancement
- Innovation in marine conservation and governance

However, the blue economy is vulnerable to climate-induced challenges such as rising sea levels, ocean acidification, and intensified storms. Addressing these challenges requires integrated solutions AI being a promising ally.

AI in Oceanic Research and Climate Monitoring

Artificial Intelligence (AI) is rapidly transforming oceanic research and climate monitoring, offering unprecedented capabilities to analyze vast and complex datasets. The sheer volume of data generated by satellites, autonomous underwater vehicles (AUVs), smart buoys, and other sensors makes traditional manual analysis impractical. AI, particularly machine learning and deep learning, excels at identifying intricate patterns, anomalies, and correlations within this data that human researchers might miss. One of the most significant applications of AI in oceanic research is in marine biodiversity monitoring. AI-powered image and acoustic recognition systems can automatically identify and classify marine species from underwater video footage, photographs, and hydrophone recordings (Sharma & Singh, 2025). This allows scientists to track population dynamics, migratory patterns, and even individual animals, significantly accelerating biodiversity surveys and conservation efforts. For instance, AI can analyze hours of whale vocalizations to identify specific species and their movements, or rapidly assess coral reef health by analyzing visual data to detect bleaching or disease. Understanding the Blue Economy goes beyond merely utilizing ocean resources; it embodies a holistic approach to sustainability, focusing on economic growth, social equity, and ocean conservation. It recognizes that the ocean is a finite resource and that its exploitation must be balanced with its protection for future generations. Unlike traditional ocean economies, which have historically focused on extraction (e.g., fishing, oil and gas), the Blue Economy emphasizes sustainable practices across various sectors. This comprehensive framework includes established industries like sustainable fishing, responsible aquaculture, and eco-tourism, ensuring these activities operate within environmental limits (Hoosain et al., 2023; Kaur & Singh, 2025). It also encompasses emerging sectors that leverage ocean resources in innovative, non-extractive ways. This includes the development of renewable ocean energy, such as wave, tidal, and offshore wind power, which offers a clean alternative to fossil fuels. Marine biotechnology is another promising area, exploring the ocean's vast biodiversity for new medicines, materials, and industrial enzymes.

LITERATURE REVIEW

The Blue Economy considers sustainable shipping and port infrastructure, promoting greener vessels and efficient logistics to minimize pollution and carbon emissions. A crucial aspect of the Blue Economy is its commitment to environmental stewardship. This means prioritizing the health of marine ecosystems, combating pollution (especially plastic), addressing ocean acidification, and protecting vulnerable habitats like coral reefs and mangroves. It also involves investing in marine scientific research to better understand ocean processes and inform policy decisions. AI and advanced monitoring technologies play a pivotal role here, enabling real-time data collection and analysis. Moreover, the Blue Economy is inherently linked to social equity and inclusion. It aims to create "blue jobs" that are decent, safe, and contribute to local economies, particularly for coastal communities who depend directly on the ocean (Mustafa et al., 2019; Kaur & Singh, 2025). It encourages equitable access to resources and benefits, ensuring that marginalized groups are not left behind in this transition to a sustainable ocean economy. Strong governance and international cooperation are essential to achieve these goals, establishing clear regulations, combating illegal activities, and fostering shared responsibility for the global ocean commons.

Figure 1. Conceptual Framework

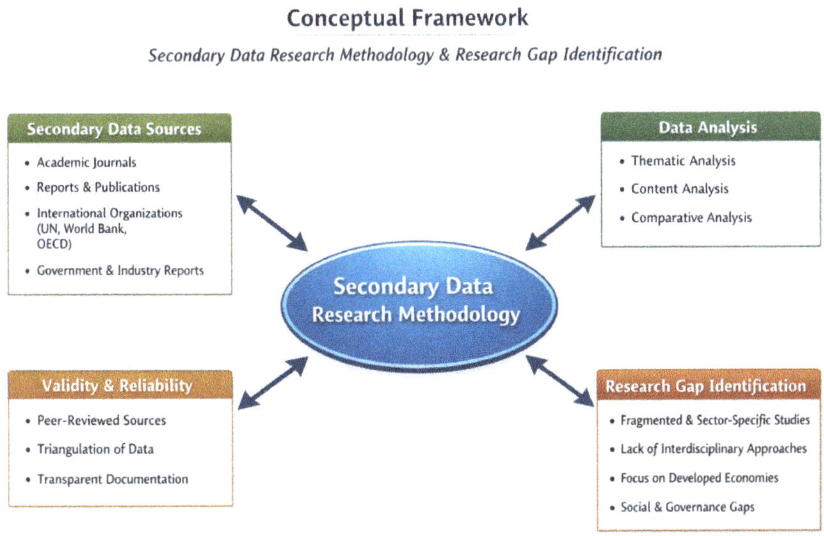

Sources: Author's own elaboration.

Artificial Intelligence in Ocean Observation and Monitoring

The Artificial Intelligence (AI) has become a revolution in ocean science, allowing scientists, policy makers and other environmental organizations to understand, predict and manage the systems in the oceans. Oceans produce vast amounts of complex information on satellites, sensors and autonomous platforms, which is way beyond the analysis capabilities of conventional techniques (Kaur & Singh, 2025). The use of AI methods, especially machine learning and deep learning is instrumental towards deriving meaningful patterns in this data, which may benefit ocean observation, modeling and biodiversity monitoring. The subsections below address some of the most important uses of AI in satellite and remote sensing, oceanographic modeling, and biodiversity monitoring.

Satellite and Remote Sensing Applications

The satellites and remote sensing technology offers constant, large-scale analysis of the oceanic condition, such as sea surface temperature, salinity, ocean color, chlorophyll content, sea level variations, and health of coral reefs. Nevertheless, these systems produce large amounts of data that can only be interpreted using some sophisticated ways of calculation. The accuracy and processability of satellite-derived ocean data are now much better due to AI-prompted analytics that can be utilized more easily now. Patterns, anomalies, and trends of multispectral and hyperspectral satellite imagery are identified using machine learning algorithms. As an example, AI models are able to detect minor changes in ocean color that give evidence of phytoplankton blooms, pollution or harmful algal bloom (Kaur & Singh, 2025). Likewise, deep learning methods can improve the process of detecting the coral bleaching and reef degradation using high-quality satellite and drone images, even in turbid or cloudy waters. One can point to the Copernicus Marine Environment Monitoring Service (CMEMS), which incorporates AI algorithms in order to enhance ocean forecasting and monitoring *Word Bank (2023)*. AI improves data integration by integrating both satellite measurements and in situ measurements to minimize uncertainties and close the data gaps created by the limitation of the sensors or unfavorable weather conditions. These enhancements can be AI-enhanced to make ocean currents, temperature variations, and biogeochemical cycles better predicted. Altogether AI-based satellite and remote sensing applications enhance the effectiveness of early warning systems and climate studies, as well as give practitioners action-oriented data on resource management of seas and coastal planning.s

Oceanographic Modeling

Oceanographic models are important to study the ocean circulation, the interaction of climate and extreme events like El Nino, marine heatwaves and cyclones. Although the classical numerical models are scientifically sound, they can be highly computationally expensive as well as suffer with uncertainties owing to incomplete information or noisy information. To meet the challenge of these issues, AI is able to refine the models using data-driven learning and real-time integration of different data streams (Kaur & Singh, 2025). AI systems combine endless data flows of the autonomous underwater vehicles (AUVs), moored buoys, gliders, and ARGO floats, which gather data on temperature, salinity, pressure, and ocean currents at various depths and locations. These data are analysed on machine learning algorithms in near real time making it possible to update a model adaptively and better estimate the parameters *OECD (2023)*. This is a hybrid model, which is a combination of physics-based models and AI, leading to more precise and computationally efficient ocean circulation models. Among the most serious contributions of AI enhanced oceanographic models, there is the enhanced prediction of large-scale climatic phenomena. As an example, simple cases can be used to illustrate how AI-based models can detect early warning signs of El Nino and La Nina events by learning complex interactions between subsurface heat content, atmospheric conditions, and sea surface temperatures. On the same note, AI enhances the prediction and observation of marine heat waves, which have been devastating to the coral reefs, fisheries, and the coastal ecosystems. The AI-assisted oceanographic models allow the enhancement of the preparedness, policy response, and climate adaptation measures due to the improved accuracy of the forecasts and the lead time.

Biodiversity Monitoring

The importance of biodiversity in the marine environment is in conservation, sustainability in fisheries management, and ecosystem stability. Nevertheless, the conventional biodiversity surveys are lengthy, costly and space and time constrained. Biodiversity monitoring based on AI provides non-invasive and scalable solutions to the problem through automated species detection and ecosystem measurements. Deep learning algorithms are commonly applied to classify marine life in underwater images of remotely operated vehicles (ROVs), AUVs, and stationary cameras. These models are able to identify fish, corals, marine mammals and other invertebrates quite effectively even during low visibility. Automated image analysis saves a lot of manual labour needed by taxonomic specialists and enables monitoring of marine environment on a long term basis. Besides visual information, AI recognizes the species using passive acoustic recordings, i.e., recognizing them by their vocalizations. Whale songs, dolphin clicks, and fish sounds are classified by machine learning models, and allow researchers to track the presence of the species, their migration patterns, and alterations in the behavior in the vast ocean regions. It is especially useful in observing difficult species, especially elusive or deep-sea species, which are hard to see visually (Gesami & Nunoo, 2024). AI-based biodiversity monitoring assists in conducting the overall ecosystem evaluation and decision-making on management by combining visual and audio information. These understandings are useful in determining the biodiversity hotspots, determining the effects of climate and human activities and in planning the effective marine protected areas. Finally, AI increases our ability to safeguard our marine life, by balancing the ecological sustainability and economic concerns (Pace et al., 2023). To conclude, ocean observation and management are changing with the available AI applications in satellite and remotely sensed observation, oceanographic modeling, and biodiversity monitoring. AI enhances scientific knowledge, enables climate resilience, and facilitates sustainable management of ocean ecosystems by converting huge and complicated data into valuable understanding.

AI and Fisheries Management

Artificial Intelligence (AI) is rapidly becoming an indispensable tool in modern fisheries management, addressing critical challenges like overfishing, illegal activities, and the need for more sustainable practices. By leveraging its ability to process and analyze vast datasets, AI provides unprecedented insights into fish populations, marine ecosystems, and fishing activities, ultimately supporting informed decision-making.

Enhancing Stock Assessment and Prediction

One of the most crucial areas where AI is making a significant impact is in fish stock assessment and prediction. Traditional methods for estimating fish populations are often labor-intensive, time-consuming, and prone to inaccuracies, leading to either overfishing or underutilization of resources (Singh, 2024). AI, particularly machine learning models, can analyze historical catch data, oceanographic conditions (like temperature, salinity, and chlorophyll concentrations), and even satellite imagery to provide more accurate and real-time predictions of fish stock levels. This allows fisheries managers to set more dynamic and sustainable quotas, aligning harvesting practices with the actual health and reproductive cycles of fish populations (Sharma & Singh, 2025). Predictive analytics can also forecast the impact of environmental changes, such as marine heatwaves, on fish distribution and abundance, enabling proactive management strategies.

Combating Illegal, Unreported, and Unregulated (IUU) Fishing

AI is proving to be a powerful weapon in the fight against Illegal, Unreported, and Unregulated (IUU) fishing, a major threat to global fish stocks and marine ecosystems. By integrating data from various sources, including satellite imagery, vessel tracking systems (like AIS - Automatic Identification System), and radar, AI algorithms can detect and analyze patterns of fishing vessel behavior (Novaglio et al., 2022). This allows for the identification of suspicious activities, such as vessels turning off their AIS transponders to evade detection or operating in marine protected areas. Organizations like Global Fishing Watch use AI and deep learning to map industrial fishing activity worldwide, even for vessels not publicly tracked, providing critical intelligence to authorities for enforcement. AI-powered drones and IoT sensors can also be deployed to monitor remote fishing zones, offering a cost-effective and efficient method for surveillance and compliance.

Optimizing Fishing Operations and Reducing Bycatch

Beyond monitoring and assessment, AI contributes to more efficient and sustainable fishing operations. Computer vision systems on board fishing vessels can automatically identify and classify fish species and their sizes as they're brought on deck. This real-time information allows fishers to make immediate decisions, for example, adjusting their gear or location to avoid catching unwanted or undersized fish, thereby significantly reducing bycatch (Sharma & Singh, 2025). AI can also optimize fishing routes by analyzing factors like weather conditions, water temperature, and species migration patterns, leading to reduced fuel consumption and operational costs. In aquaculture, AI algorithms analyze sensor data to maintain optimal water quality, temperature, and feeding schedules, leading to healthier fish, reduced feed waste, and early detection of diseases.

Identification of Research Gap

Despite the availability of extensive secondary literature, several research gaps are identified. First, much of the existing research is fragmented and sector-specific, with limited integrative or interdisciplinary frameworks. Second, there is an over-emphasis on technological or economic dimensions, while social, ethical, and governance perspectives remain underexplored. Third, many studies rely on data from developed economies, creating a geographical gap with limited contextual insights from developing countries, including India. Finally, empirical validation and longitudinal analysis are often missing, indicating a need for future studies that combine secondary insights with primary or mixed-method approaches.

RESEARCH METHODOLOGY

This study adopts a secondary data–based research methodology, which involves the systematic collection, evaluation, and analysis of existing data and published sources relevant to the research objectives. Secondary data research is particularly suitable for conceptual, exploratory, and review-oriented studies, especially in multidisciplinary areas such as sustainability, hospitality, tourism, blue economy, and technology-driven transformations, where extensive prior research and institutional data already exist.

Research Design

The research follows a descriptive and analytical design. Descriptive analysis helps in summarizing existing trends, patterns, and themes reported in prior studies, while analytical techniques enable critical interpretation, comparison, and synthesis of findings across multiple sources. This approach allows the study to build a comprehensive understanding of the topic without engaging in primary data collection.

Sources of Secondary Data

Secondary data are collected from reliable and authoritative sources, including peer-reviewed journals indexed in Scopus, Web of Science, and UGC-listed databases. In addition, reports and publications from international organizations such as the United Nations (UN), World Bank, OECD, FAO, and UNESCO are used to capture policy perspectives and global trends. Books, edited volumes, conference proceedings, government publications, and reputable industry reports further supplement the data. To ensure academic rigor, only sources published within a relevant time frame and aligned with the study objectives are considered.

Data Collection and Screening

A structured search strategy is employed using academic databases such as Google Scholar, ScienceDirect, SpringerLink, Taylor & Francis, and Emerald Insight. Keywords and Boolean operators relevant to the research theme are used to retrieve literature. The collected studies are screened based on predefined inclusion and exclusion criteria, such as relevance to the research objectives, methodological clarity, and credibility of the source. Duplicates and non-scholarly materials are excluded to maintain data quality.

Data Analysis Technique

The selected literature is analyzed using thematic analysis and content analysis. Key concepts, variables, challenges, opportunities, and outcomes discussed in previous studies are coded and grouped into themes. Comparative analysis is applied to identify convergences, divergences, and evolving perspectives across regions, sectors, and time periods. This method supports the development of a coherent narrative and theoretical insights grounded in existing knowledge. The reliability of the study is enhanced by relying on peer-reviewed and institutional sources, while validity is ensured through triangulation of data from multiple authors, disciplines, and organizations. Transparent documentation of sources and analytical procedures further strengthens the credibility of the methodology.

DISCUSSION AND INTERPREATION

Despite the immense potential, the integration of AI in fisheries management faces several challenges. Data availability and quality are paramount; AI models require large, diverse, and well-annotated datasets for effective training. There are also economic barriers and a lack of expertise in AI and data science within many fisheries organizations. Ethical considerations surrounding data privacy and the potential impact of automation on traditional fishing livelihoods also need careful navigation (Sharma & Singh, 2025). However, as AI technology continues to evolve and become more accessible, its role in ensuring the long-term sustainability of our oceans will only grow. Collaborative efforts between scientists, governments, the fishing industry, and AI developers are crucial to overcoming these challenges and fully harnessing AI's potential to create a more responsible, efficient, and equitable future for fisheries worldwide. Sustainable fisheries are vital to food security (Novaglio et al., 2022; Supina & Singh, 2025; Singh, 2024). Artificial Intelligence is revolutionizing fisheries management by significantly enhancing monitoring and surveillance capabilities. For instance, computer vision systems and AI-powered drones are increasingly deployed to detect and deter Illegal, Unreported, and Unregulated (IUU) fishing, a major threat to marine sustainability *UNCTD (2023).* Platforms like Global Fishing Watch effectively leverage machine learning (ML) to analyze vast amounts of vessel movement data, pinpointing suspicious activities that might indicate illicit fishing operations. Beyond enforcement, AI also plays a crucial role in predictive analytics for fish stock assessment. By incorporating diverse environmental variables and historical catch data, sophisticated AI models can forecast fish stock dynamics with greater accuracy. This data-driven approach enables fisheries managers to implement more effective quota allocations and strategically plan seasonal closures, ensuring the long-term health of fish populations. Furthermore, AI is transforming smart aquaculture (Zaeri, 2025). Here, AI-enabled sensors and Internet of Things (IoT) devices continuously monitor critical parameters like water quality, feeding patterns, and fish health within aquaculture systems (Singh, 2024). This real-time data allows for optimized environmental control, improved productivity, and a significant reduction in waste, making aquaculture more efficient and sustainable.

Enhancing Marine Conservation with AI

Artificial Intelligence (AI) is rapidly emerging as a transformative force in marine conservation, offering powerful solutions to pressing threats facing our oceans, from biodiversity loss to pollution and climate change impacts. The sheer scale and complexity of marine ecosystems have historically made comprehensive monitoring and effective intervention challenging. However, AI's ability to process and analyze vast datasets, identify intricate patterns, and automate complex tasks is providing unprecedented capabilities for safeguarding ocean health. One of the most impactful applications of AI is in marine biodiversity monitoring and species protection. AI-powered computer vision systems can analyze underwater imagery and video footage from autonomous underwater vehicles (AUVs), remotely operated vehicles (ROVs), and diver-operated cameras (Sharma & Singh, 2025). These systems can automatically identify and classify marine species, track their movements, assess population sizes, and even detect individual animals with remarkable accuracy (Pandey & Joseph, 2025; Supina & Singh, 2025; Singh, 2024). This accelerates crucial biodiversity surveys, identifies endangered species, and helps understand migratory patterns, providing vital data for targeted conservation strategies. Similarly, AI-driven acoustic monitoring leverages hydrophones to listen to the ocean's soundscape (Ghoneim et al., 2025). AI algorithms can differentiate between natural sounds (like whale vocalizations) and anthropogenic noise (from shipping or sonar), allowing scientists to monitor marine mammal populations, identify critical habitats, and assess the impact of noise pollution, which is particularly harmful to species relying on sound for communication and navigation. Furthermore, AI is proving invaluable in the fight against plastic pollution. Machine learning models can analyze satellite imagery and drone footage to detect, map, and even predict the movement of plastic debris in the ocean. This intelligence guides cleanup efforts, allowing organizations like The Ocean Cleanup to deploy their systems more effectively. AI can also help identify the sources of pollution, leading to more targeted preventative measures on land. In the context of climate change impacts, AI provides sophisticated predictive modeling capabilities. By analyzing oceanographic data such as temperature, salinity, and oxygen levels, AI models can forecast phenomena like marine heatwaves, ocean acidification, and sea-level rise. This allows conservationists to anticipate threats to vulnerable ecosystems like coral reefs and coastal communities, enabling the development of adaptation and resilience strategies (Pires et al., 2023). AI can also analyze the health of coral reefs by processing underwater images to detect early signs of bleaching or disease, facilitating timely interventions to protect these vital habitats (Singh, 2024). Finally, AI enhances the effectiveness of Marine Protected Areas (MPAs). By integrating diverse data streams from satellite surveillance to vessel tracking and environmental sensors AI platforms can monitor human activities within MPAs, detect illegal fishing, and track compliance with regulations. This real-time intelligence empowers authorities to enforce protections more efficiently, ensuring that these designated sanctuaries effectively conserve biodiversity. While challenges remain, including data availability and ethical considerations, the accelerating integration of AI into marine conservation represents a significant leap forward, offering hope for a healthier and more resilient future for our oceans.

AI in Maritime and Coastal Infrastructure Resilience

The resilience of maritime and coastal infrastructure is paramount in an era of escalating climate change impacts, from rising sea levels and intensifying storms to coastal erosion. Artificial Intelligence (AI) is emerging as a critical enabler in bolstering this resilience, offering advanced capabilities for monitoring, prediction, and optimized management of these vital assets. One of the primary contributions of AI lies in real-time monitoring and predictive maintenance for port and maritime infrastructure. Ports, as critical nodes in global supply chains, are highly vulnerable to disruptions. AI-powered sensor networks deployed on cranes, wharves, bridges, and other vital equipment can continuously collect data on their structural integrity, operational performance, and environmental stressors (Sarangi, 2023; Kaur & Singh, 2025). Machine learning algorithms analyze this vast stream of data to detect subtle anomalies and predict potential failures before they occur. This allows port authorities to schedule proactive maintenance, minimizing costly downtime, extending the lifespan of infrastructure, and preventing catastrophic failures that could cripple operations during an extreme weather event or other crisis. For instance, AI can predict the weakening of a pier due to sustained wave action or forecast the corrosion rate of steel structures, enabling timely repairs (Chen et al., 2020; Kaur & Singh, 2025). Beyond individual assets, AI contributes to the overall resilience of port operations through optimization and automation. AI-driven traffic management systems can optimize vessel berthing and movement within a port, reducing congestion and improving throughput, which is critical for swift recovery after a disruption. In the face of impending storms, AI can facilitate more efficient evacuation of vessels and cargo, minimizing damage. Furthermore, the development of autonomous port operations, powered by AI for tasks like cargo handling and logistics, enhances efficiency and reduces reliance on human intervention in hazardous conditions, increasing the inherent resilience of these facilities. Digital twins virtual replicas of physical ports are also becoming increasingly prevalent (Sharma & Singh, 2025). These AI-enabled models allow port managers to simulate various disaster scenarios, test different response strategies, and optimize infrastructure layouts for maximum resilience against future shocks. For broader coastal infrastructure, AI offers unprecedented capabilities in monitoring and predicting environmental changes. Coastal erosion, a slow but relentless threat, can be effectively monitored using AI analysis of satellite imagery, drone footage, and LiDAR data. Machine learning models can track shoreline changes, assess the health of natural coastal defenses like mangroves and sand dunes, and predict erosion hotspots. This allows for data-driven decisions on where to implement protective measures, such as beach nourishment or the construction of seawalls and breakwaters. Moreover, AI is crucial for disaster preparedness and early warning systems in coastal communities. By integrating meteorological data, tidal gauges, and historical flood patterns, AI models can provide highly accurate forecasts of storm surges, king tides, and flooding events with increased lead time (Techera, 2018). This enables more effective evacuation planning, optimized allocation of emergency resources, and targeted communication to at-risk populations (Supina & Singh, 2025). The ability of AI to analyze complex environmental interactions and forecast future scenarios provides coastal communities and infrastructure planners with the foresight needed to design and implement adaptation strategies that can withstand the evolving challenges of climate change, ultimately fostering more resilient maritime and coastal regions. Rising seas and extreme weather threaten coastal infrastructure. AI contributes by:

Offshore Renewable Energy and AI

The expansion of offshore renewable energy, particularly offshore wind, wave, and tidal power, is crucial for global decarbonization efforts. However, the harsh marine environment and the intermittent nature of these resources present significant operational and integration challenges. Artificial Intelligence (AI) is proving to be an indispensable tool in overcoming these hurdles, enhancing the efficiency, reliability, and economic viability of offshore renewable energy projects (Sarangi, 2023; Techera, 2018). One of the most impactful applications of AI in this sector is predictive maintenance and operational optimization for offshore wind turbines. These massive structures are exposed to extreme weather, salt corrosion, and constant mechanical stress. Traditional scheduled maintenance is often inefficient and costly, requiring specialized vessels and personnel. AI-powered sensor networks integrated with machine learning algorithms analyze real-time data from accelerometers, temperature sensors, vibration monitors, and other critical components (Supina & Singh, 2025; Singh, 2024). This allows operators to detect subtle anomalies and predict potential equipment failures before they occur, such as cracks in blades, gearbox wear, or generator faults. By enabling proactive maintenance, AI significantly reduces unplanned downtime, extends the lifespan of turbines, and minimizes costly emergency repairs, thereby maximizing energy output and reducing operational expenses. AI can also dynamically adjust turbine settings, like blade pitch and yaw angles, in response to shifting wind conditions to maximize energy capture. Beyond individual assets, AI plays a vital role in optimizing the design and layout of entire offshore renewable energy farms. For offshore wind, AI-driven wind-flow simulations leverage vast datasets to predict long-term wind performance, ensuring turbines are strategically placed to capture the strongest and most consistent winds while minimizing wake effects between turbines. Similarly, for wave and tidal energy projects, AI models can analyze complex oceanographic data to identify optimal locations for energy converters, maximizing energy absorption and minimizing environmental impact. This also extends to the design of the collection systems, ensuring efficient power transmission back to shore. The integration of intermittent offshore renewable energy into national grids is another critical area where AI excels. The variability of wind and wave power can pose challenges to grid stability (Kaur & Singh, 2025; Sharma & Singh, 2025). AI-powered energy forecasting models analyze historical meteorological data, real-time weather inputs, and historical energy production records to provide highly accurate predictions of energy output from offshore farms (Voyer et al., 2018). This enhanced forecasting allows grid operators to better balance supply and demand, reducing the need for fossil fuel-based backup power and minimizing energy curtailment (wasted renewable energy). AI also optimizes energy storage solutions, managing battery charging and discharging cycles to efficiently store excess power during high production periods and release it during peak demand or low generation, further stabilizing the grid. Furthermore, AI facilitates autonomous monitoring and inspection using drones and robotics, reducing the need for dangerous and costly human interventions in harsh offshore environments (Supina & Singh, 2025; Singh, 2024). These AI-equipped systems can conduct detailed visual and thermal inspections of turbines and subsea infrastructure, automatically identifying defects. While challenges exist, such as data quality and integration, the transformative potential of AI in making offshore renewable energy more reliable, efficient, and cost-effective is undeniable, accelerating the transition to a sustainable energy future. Transitioning to low-carbon energy systems is vital for climate resilience. Oceans offer immense potential via wind, wave, and tidal energy. AI supports this sector through:

GOVERNANCE, EQUITY, AND ETHICAL CONSIDERATIONS

The burgeoning concept of the Blue Economy, while promising sustainable development from ocean resources, is inextricably linked with complex challenges in governance, equity, and ethical considerations. Without robust frameworks addressing these aspects, the pursuit of economic gain could inadvertently exacerbate existing inequalities and environmental degradation. Effective governance is the bedrock of a successful and truly sustainable Blue Economy. This entails establishing clear, comprehensive, and enforceable legal and regulatory frameworks at local, national, and international levels (Sharma & Singh, 2025). A key challenge lies in the fragmented nature of ocean governance, with overlapping jurisdictions and varying capacities among states. Therefore, international cooperation and strong regional agreements are paramount to managing transboundary marine resources, combating illegal activities like IUU fishing, and addressing global issues such as plastic pollution and climate change. Transparency and accountability in decision-making processes are crucial to prevent corruption and ensure that development initiatives genuinely serve public interest rather than private gain (Zaeri, 2025). Furthermore, adaptive governance models are needed to respond to the dynamic nature of marine ecosystems and the rapid pace of technological advancements in ocean industries. Equity is another critical dimension, ensuring that the benefits derived from the Blue Economy are shared fairly among all stakeholders, particularly coastal communities and developing nations who often depend most directly on ocean resources. Historically, the exploitation of marine resources has often disproportionately benefited powerful external actors, leading to the marginalization of local populations and the degradation of their traditional livelihoods (Singh, 2024). A truly equitable Blue Economy must prioritize social inclusion, empowering indigenous communities and small-scale fishers through secure tenure rights, access to fair markets, and participation in decision-making processes. This also involves investing in capacity building and technology transfer to enable developing countries to harness their marine resources sustainably, rather than merely becoming sources of raw materials for wealthier nations. Addressing existing power imbalances and ensuring equitable access to "blue jobs" and economic opportunities is essential to prevent a "blue grab" where powerful entities monopolize ocean space and resources (Voyer et al., 2018; Zaeri, 2025). Finally, the ethical considerations surrounding the Blue Economy are profound and multifaceted. At its core, there's a moral imperative to protect the inherent value of marine biodiversity and ecosystems, beyond their direct economic utility. This includes acknowledging the rights of nature and future generations to healthy oceans. The rapid advancement of technologies like deep-sea mining, marine genetic resource exploitation, and large-scale offshore energy projects raises significant ethical questions regarding potential irreversible environmental damage and the fair distribution of benefits from resources in areas beyond national jurisdiction. There's also an ethical responsibility to apply the precautionary principle, especially when dealing with novel technologies and activities in poorly understood marine environments. This means taking preventative action even in the absence of full scientific certainty when there's a risk of serious or irreversible harm (Sharma & Singh, 2025; Supina & Singh, 2025; Singh, 2024). Furthermore, ethical governance demands addressing issues of data privacy and intellectual property rights related to marine genetic resources and traditional knowledge of indigenous communities.

Ultimately, for the Blue Economy to be truly sustainable and just, it must be guided by principles of intergenerational equity, environmental stewardship, and respect for all life in our oceans. The deployment of AI in the blue economy must ensure fairness and transparency. Key aspects include:

Ocean Governance, Ethics, and Equity with AI

Ethical, moral, and involvement concerns are becoming more significant as the trend in the field of artificial intelligence (AI) advances into ocean observation, conservation, and governance. Although AI provides strong means of monitoring and decision-making, the reduction of advantages is not evenly distributed (Kaur & Singh, 2025). The AI-based ocean management can lead to the enhancement of existing disparities, the disadvantaged groups, and lack of confidence without comprehensive governance structures. To make sure that technological innovation will contribute to equitable and sustainable futures of the oceans, it is therefore critical to address data sovereignty, ethical use of AI, and participatory governance.

Data Sovereignty and Access

The principles of inclusive AI-enabled ocean governance include the data sovereignty and equal access. One of the most directly impacted regions of ocean change is the small island developing states (SIDS) and coastal communities, such as sea-level rise, extreme weather events, and dwindling fisheries. However, such communities do not have access to quality ocean information, digital infrastructure and sophisticated analytical tools to shape policy and adaptation plans (Sharma & Singh, 2025). AI systems are dependent on massive datasets acquired through satellites, sensors, and research institutions, which are often managed by highly-technological countries or by corporate organizations (Supina & Singh, 2025; Singh, 2024). This imbalance may restrict the participation of SIDS and the local communities in the ocean decision-making process in a meaningful manner. To prevent digital inequities, it is thus important to make sure that marine data and AI platforms are open, transparent, and affordable. Capacity-building solutions, data-sharing contracts, and investments in local digital infrastructure can prepare the communities of the coast to apply AI tools to the early warning systems, managing resources, and responding to climate changes and retain the control over the locally produced data.

Ethical AI Use

The responsible use of AI in the management of the sea is an ethical issue. AI models could also be biased according to the training data, and therefore provide biased predictions that favor certain areas, species or communities at the expense of others. An example where this can happen is when the models are trained on primarily data of well-researched oceans and used in the under-researched tropical or small-island setting and lead to incorrect risk estimates or policy advice. Surveillance and unintended consequences are also an issue (Supina & Singh, 2025). The potential solutions, such as a higher use of satellite tracking, automated vessel monitoring, and acoustic sensors, can increase the concerns about privacy or sovereignty, especially through small-scale fishers and indigenous people. Use of AI ethically involves transparency in model design, explicit accountability procedures as well as frequent auditing to detect and eliminate bias. The control of AI by people should be at the forefront, as the technology should not substitute expertise, moral thinking, and social principles, but act as their assistant. Along with the minimization of risks, inclusive design processes can create additional trust in AI-enabled systems, which involve diverse stakeholders.

Participatory Governance

The legitimacy, effectiveness, and long-term sustainability of ocean management based on AI require participatory governance. The stakeholder involvement makes sure technological solutions correspond to the local realities, cultural values, and even the socio-economic priorities. Coastal populations, fishermen, indigenous people, scientist, policymakers and civil societal bodies are all able to offer useful insights, which can improve the quality of the decisions (Novaglio et al., 2022). Keeping both old and new ecological knowledge and integrating them with AI-assisted scientific knowledge may result in more holistic and context-aware governance. Local and indigenous knowledge systems tend to offer observations of marine ecosystems, species behavior and environmental change over longer periods of time, which augments data-driven models. The approaches that can guarantee the use of AI tools to facilitate collective stewardship instead of being used to impose control are co-management methods where authority and responsibility are distributed between the governments and local stakeholders. Finally, ethical, equitable, and governance do not belong to the peripheral concerns but the key pillars of AI-powered ocean management. AI can aid in finding ways to become an empowering, resilient, and sustainable ocean futures tool by focusing on data sovereignty, ethical protection, and participatory governance instead of a source of exclusion or imbalance.

CHALLENGES AND FUTURE DIRECTIONS

The rapid integration of Artificial Intelligence (AI) into the blue economy, spanning oceanic research, climate monitoring, fisheries management, and offshore renewable energy, brings with it a host of challenges and exciting future directions. While AI offers transformative potential for sustainable ocean management, its widespread adoption is not without hurdles. One of the foremost challenges is data quality and availability (Pace et al., 2023). AI models are only as good as the data they're trained on. The marine environment is vast, dynamic, and often inaccessible, leading to gaps, inconsistencies, and biases in collected data. Ensuring a continuous flow of high-quality, diverse, and well-annotated datasets from various sensors satellites, AUVs, buoys, and underwater cameras remains a significant logistical and financial undertaking. Furthermore, interoperability and standardization of data across different platforms and national boundaries are crucial for building truly global AI solutions for the ocean (Mustafa et al., 2019; Kaur & Singh, 2025). Without common protocols, integrating data from disparate sources becomes a laborious task, limiting AI's analytical power. Understanding the Blue Economy involves recognizing the ocean as a critical space for sustainable development, balancing economic growth with social equity and environmental stewardship. It moves beyond traditional "ocean economy" activities, like fishing and shipping, to include emerging sectors such as renewable ocean energy, marine biotechnology, and sustainable aquaculture, all while emphasizing sustainable practices. The concept hinges on ensuring ocean health for future generations, fostering innovation, and promoting responsible resource management

CRITIQUES OF THE BLUE ECONOMY

The Blue Economy has a colossal possibility of generating sustainable economic development, improving food security, generating job opportunities, and conserving marine ecosystems. It aims at ensuring that the economic development is balanced with the environmental conservation and the social well-being by encouraging the responsible use of the ocean resources. Nevertheless, as promising as it may be, a fully-fledged sustainable and inclusive Blue Economy cannot be attained due to a series of interconnected problems that touch upon the environmental, technological, governance, and socio-economic levels. The over-exploitation of the marine resources is one of the most urgent problems (Supina & Singh, 2025). Most of the conventional ocean-based industries, especially capture fisheries are still working under unsustainable systems. Due to overfishing, damaging offers, and huge bycatch, fish stocks have been exhausted, and marine ecosystems deteriorated. In most places, economic incentives and livelihood pressure are of a short-term nature that prompt to overharvest at the expense of the long-term ecosystem and food security. The regenerative power of marine life is still threatened in the absence of good stock evaluation, enforcing, and alternative livelihood. Another threat that the Blue Economy faces is marine pollution. Marine environments are still polluted with plastic waste, agricultural runoff, industrial chemicals, oil spills and untreated sewage. These pollutants destroy marine organisms by ingesting them, entangling them, and degrading their habitats as well as reaching the human food chain through the seafood consumed by humans. The impact of pollution on coastal beaches, degraded coral reefs and contaminated waters on these coastal areas affect the economic activities of coastal tourism, aquaculture and fisheries by lowering the level of attractiveness and productivity of the area (Supina & Singh, 2025; Singh, 2024). Combining waste management systems, regulation, and production and consumption activities on land and at sea is a way to cope with marine pollution.

Climate change also enhances the fragility of any economy that predominantly depends on the sea. The increased sea level is threatening to the coastal structures, ports and the low-lying island states and the acidification of oceans makes shell forming organisms and coral reef systems ineffective. The consequences of marine heatwaves and altered ocean currents are disruption of fish migration patterns, decline in biodiversity and frequency of coral bleaching events. Cyclones and storm surges are extreme weather events that destroy coastal settlements and disturb the livelihoods that rely on oceans (Supina & Singh, 2025; Singh, 2024). These effects require major investments on climate adaptation, disaster risk reduction, and ecosystem-based strategies as a way of improving resiliency. The inadequate governance and low enforcement capacity also contribute to the weak sustainable development of Blue Economy. Maritime resources tend to cut across the borders of countries and thus close regional and international collaboration is essential. Weaknesses in marine management are caused by fragmented regulatory structures, institutional overlaps and poor monitoring mechanisms. Unreported, unregulated and illegal (IUU) fishing has become a significant menace, denying nations their earnings, jeopardizing conservation and putting the lives of honest fishers at risk. To overcome these challenges, strengthening the mechanisms of governance, increasing transparency, and increasing capacity to enforce are important. There are also technological gaps and shortage of infrastructure, which are also a hindrance especially to developing coastal and island countries. High-tech Blue Economy like offshore renewable energy, marine biotechnology, smart aquaculture, and ocean observation systems involve the large amount of money, highly trained human capital, and technology infrastructure. A number of nations do not get access to these resources and this restricts their involvement in the emerging ocean industries. Close these gaps with the help of technological transfer, capacity building, and cooperation at the international level are important to provide inclusive growth in the Blue Economy. The lack of data and its inconsistency is another thing that complicates the decision-making (Singh, 2024). Coherent, consistent and timely information on ocean health, fish stocks, biodiversity and economic activities are commonly dispersed or inaccessible particularly in developing countries. The absence of precise information means that policymakers will not be able to estimate risks, stipulate efficient regulations, or measure the effects of interventions. To facilitate evidence-based ocean governance, better data collection, standardization, and sharing must be improved by using digital technologies and scientific collaborations. Funding and investment continue to be a major challenge to the Blue Economy. As the demand of sustainable ocean investments increases, it is difficult to get enough and relevant funds to invest in long-term and environmentally friendly ventures. Most of the projects are characterized by high initial expenses, erratic returns and regulatory risks especially in the developing world. The mechanisms to be used in innovation financing, including blue bonds, blended finance, and public-private partnerships should be developed to attract investment and to match financial incentives and sustainability ambitions. Lastly, the problem of social equity and displacement should be addressed (Kaur & Singh, 2025). Sometimes the growth of new coastal industries, including coastal tourism, aquaculture, or offshore energy, can push out the traditional coastal populations, limit access to traditional fishing areas and increase social inequalities. When such benefits are not evenly shared, the local populations will have to take the costs of the environment and social life and might not be able to enjoy economic advantages. There is a need to ensure that the planning, community participation, and benefits sharing are inclusive, equitable, and just to avoid marginalization and encourage social justice in the Blue Economy. To sum up, although the Blue Economy provides a way to develop sustainable and resilient ocean-based development the success of this strategy requires a solution to these complex issues based on integrated, equitable, and forward-thinking approaches.

FUTURE DIRECTIONS FOR THE BLUE ECONOMY

Addressing these challenges requires a concerted and collaborative effort, driving the Blue Economy towards several key future directions:

- **Integrated Ocean Management and Spatial Planning:** Future efforts will focus on holistic, ecosystem-based management that considers all human activities and their cumulative impacts. This includes advanced *marine spatial planning* to allocate ocean space for various uses (conservation, energy, shipping, fisheries) in a way that minimizes conflict and maximizes sustainability.

- **Technological Innovation and AI Integration:** AI, robotics, big data analytics, and remote sensing will be central to improving monitoring, forecasting, resource management, and operational efficiency across all blue economy sectors. This includes AI for climate modeling, autonomous vessels for data collection, and smart aquaculture systems.
- **Circular Economy Principles:** Moving beyond linear "take-make-dispose" models, the future Blue Economy will increasingly adopt circular principles, focusing on waste reduction, resource efficiency, and the reuse and recycling of materials within ocean industries.
- **Nature-Based Solutions:** Greater emphasis will be placed on leveraging natural coastal ecosystems (mangroves, coral reefs, salt marshes) for coastal protection, carbon sequestration, and biodiversity conservation, integrating them into coastal development plans.
- **Sustainable Finance Mechanisms:** Developing innovative financial instruments, such as "blue bonds" and impact investment funds, will be crucial to channel capital towards sustainable ocean projects and incentivize responsible practices.
- **Strengthened Governance and International Cooperation:** Enhancing international legal frameworks (like the BBNJ Agreement for the High Seas), improving enforcement capabilities, and fostering multilateral partnerships will be vital to address transboundary issues and ensure global ocean governance.
- **Capacity Building and Knowledge Transfer:** Investing in education, training, and technology transfer, particularly for developing countries and small island developing states, will ensure inclusive participation and equitable benefits from the Blue Economy.
- **Community Engagement and Indigenous Knowledge:** Future initiatives will increasingly prioritize the active involvement of local and indigenous communities, integrating their traditional ecological knowledge and ensuring their rights and livelihoods are protected and enhanced.

By embracing these future directions, the Blue Economy can evolve from a theoretical concept into a practical framework for achieving genuine ocean sustainability and equitable prosperity.

CONCLUSION

The Oceanic Nexus represents a critical juncture where the burgeoning capabilities of AI converge with the urgent need for climate resilience within the blue economy.

As we navigate an era of unprecedented environmental challenges and economic opportunities, the synergy between these domains offers a powerful pathway towards a sustainable future. The integration of AI in monitoring, prediction, and adaptation strategies for our oceans is not merely an technological advancement; it is a fundamental shift in how humanity interacts with and manages its most vital natural resource. The journey ahead will undoubtedly present complexities, from ensuring equitable access to these technologies to addressing the ethical implications of AI deployment in sensitive marine environments. However, by fostering international collaboration, investing in robust research and development, and prioritizing inclusive governance frameworks, we can harness the full potential of this oceanic nexus. The future of the blue economy hinges on our collective ability to innovate responsibly, safeguarding marine ecosystems while unlocking new avenues for sustainable prosperity. The promise of the Oceanic Nexus is a testament to human ingenuity and our capacity to forge a resilient and thriving relationship with the global ocean. The convergence of AI, climate resilience, and the blue economy offers a powerful framework for addressing 21st-century ocean challenges. By enhancing our ability to monitor, manage, and sustainably develop marine resources, AI can empower societies to build a more resilient, inclusive, and regenerative oceanic future. Strategic investments, cross-sector partnerships, and ethical governance will be critical to harnessing AI's full potential in realizing the oceanic nexus.

REFERENCES

Ahmed, N., Ward, J. D., Thompson, S., Saint, C. P., & Diana, J. S. (2018). Blue-green water nexus in aquaculture for resilience to climate change. *Reviews in Fisheries Science & Aquaculture*, *26*(2), 139–154. DOI: 10.1080/23308249.2017.1373743

Blasiak, R., Jouffray, J. B., Amon, D. J., Moberg, F., Claudet, J., Søgaard Jørgensen, P., Pranindita, A., Wabnitz, C. C. C., & Österblom, H. (2022). A forgotten element of the blue economy: Marine biomimetics and inspiration from the deep sea. *PNAS Nexus*, *1*(4), pgac196. DOI: 10.1093/pnasnexus/pgac196 PMID: 36714844

Chen, S., De Bruyne, C., & Bollempalli, M. (2020). Blue economy: Community case studies addressing the poverty–environment nexus in ocean and coastal management. *Sustainability (Basel)*, *12*(11), 4654. DOI: 10.3390/su12114654

Elston, J., Pinto, H., & Nogueira, C. (2024). Tides of change for a sustainable blue economy: A systematic literature review of innovation in maritime activities. *Sustainability (Basel)*, *16*(24), 11141. DOI: 10.3390/su162411141

Gesami, B. K., & Nunoo, J. (2024). Artificial intelligence in marine ecosystem management: Addressing climate threats to Kenya's blue economy. *Frontiers in Marine Science*, *11*, 1404104. DOI: 10.3389/fmars.2024.1404104

Ghoneim, H., Amin, H. M., & Hassan, R. S. (2025). The Role of Technology in a Sustainable Blue Economy: Research Content and Future Directions. Handbook of Sustainable Blue Economy, 1-34.

Hoosain, M. S., Paul, B. S., Doorsamy, W., & Ramakrishna, S. (2023). The influence of circular economy and 4IR technologies on the climate–water–energy–food Nexus and the SDGs. *Water (Basel)*, *15*(4), 787. DOI: 10.3390/w15040787

Kaur, S., & Singh, A. (2025). India's Peaks Under Pressure: The Impact of Climate Change and Gender Issues on Mountain Tourism. In Balancing Mountain Tourism, Cultural Heritage, and Environmental Stability (pp. 309-320). IGI Global Scientific Publishing.

Mustafa, S., Estim, A., & Shapawi, R. (2019). Future-proofing oceans for food security and poverty alleviation. In *Decent Work and Economic Growth* (pp. 1–11). Springer.

Novaglio, C., Bax, N., Boschetti, F., Emad, G. R., Frusher, S., Fullbrook, L., Hemer, M., Jennings, S., van Putten, I., Robinson, L. M., Spain, E., Vince, J., Voyer, M., Wood, G., & Fulton, E. A. (2022). Deep aspirations: Towards a sustainable offshore blue economy. *Reviews in Fish Biology and Fisheries*, *32*(1), 209–230. DOI: 10.1007/s11160-020-09628-6 PMID: 33500602

Organisation for Economic Co-operation and Development. (2023). *Enhancing cybersecurity in the maritime transport sector.* OECD Publishing.

Pace, L. A., Saritas, O., & Deidun, A. (2023). Exploring future research and innovation directions for a sustainable blue economy. *Marine Policy, 148*, 105433. DOI: 10.1016/j.marpol.2022.105433

Pandey, A., & Joseph, J. (2025). Emerging Trends in Blue Economy: A Roadmap Through the Lens of Sustainable Development. Securing Sustainable Futures Through Blue and Green Economies, 17-36.

Pires Manso, J. R., Martínez Vázquez, R. M., Milán García, J., & de Pablo Valenciano, J. (2023). Renewable energies and blue economy: New trends in global research. *Energies, 16*(10), 4210. DOI: 10.3390/en16104210

Sarangi, U. (2023, February). Blue economy, blue finance and ocean governance for achieving sustainable development goals. [). Oxford, UK: Blackwell Publishing Ltd.]. *Natural Resources Forum, 47*(1), 3–21. DOI: 10.1111/1477-8947.12267

Sharma, R., & Singh, A. (2025). Integrating AI, Green Marketing, and Influencer Strategies: Shaping Sustainable Consumer Behavior in the Digital Age. In Strategic Brand Management in the Age of AI and Disruption (pp. 325-342). IGI Global Scientific Publishing.

Singh, A. (2024). Hospitality Sector: Ethical Challenges and Solutions. Responsible Implementations of Generative AI for Multidisciplinary Use, 367.

Supina, S., & Singh, A. (2025). Accentuating responsible travel in an ecotourism destination: Perspective and challenges from the stakeholders. In *Meaningful Tourism* (pp. 269–279). Emerald Publishing Limited. DOI: 10.1108/978-1-83797-837-320251035

Techera, E. J. (2018). Supporting blue economy agenda: Fisheries, food security and climate change in the Indian Ocean. *Journal of the Indian Ocean Region, 14*(1), 7–27. DOI: 10.1080/19480881.2017.1420579

United Nations Conference on Trade and Development. (2023). Review of maritime transport 2023. United Nations.

Voyer, M., Schofield, C., Azmi, K., Warner, R., McIlgorm, A., & Quirk, G. (2018). Maritime security and the Blue Economy: Intersections and interdependencies in the Indian Ocean. *Journal of the Indian Ocean Region, 14*(1), 28–48. DOI: 10.1080/19480881.2018.1418155

World Bank. (2023). *Cybersecurity and critical infrastructure protection in transport systems*. World Bank Group.

Zaeri, N. (2025). Securing Sustainability Through Artificial Intelligence and Blue Economy. In Securing Sustainable Futures Through Blue and Green Economies (pp. 37-82). IGI Global Scientific Publishing. DOI: 10.4018/979-8-3693-7893-9.ch003

Chapter 5

The Blue Economy in Türkiye: Potential, Policies, and Practices

Mehmet İnanç Özekmekçi
https://orcid.org/0000-0003-0996-6664

Faculty of Economics and Administrative Sciences, Department of Political Science and Public Administration, Erciyes University, Turkey

ABSTRACT

This chapter examines the development of the Blue Economy in Türkiye, emphasizing both its strategic potential and structural limitations. Positioned at the crossroads of global maritime trade, Türkiye's maritime sectors-including shipping, fisheries, aquaculture, tourism, and renewable energy-are presented as critical drivers of sustainable growth. Yet fragmented governance, the absence of a comprehensive national strategy, and weak integration of climate adaptation policies constrain this potential. Through case studies such as the Marmara Sea mucilage crisis, sustainable fisheries initiatives in the Black Sea, and conservation challenges in Kaş-Kekova, the chapter illustrates the tensions between ecological sustainability and economic development. By applying concepts such as governance fragmentation, wicked problems, and institutional misfit, the chapter offers a framework for understanding these challenges. It concludes with policy recommendations, highlighting the need for integrated strategies, stronger stakeholder participation, and alignment with global sustainability norms.

DOI: 10.4018/979-8-3373-5238-1.ch005

FRAMING THE BLUE ECONOMY IN TÜRKIYE: SECTORS AND SUSTAINABILITY

The concept of the blue economy has gained increasing visibility in recent years within global debates on sustainability, economic development, and ocean governance. Broadly defined as the sustainable use of marine and aquatic resources for economic growth, improved livelihoods, and ecosystem health, the framework of the blue economy emphasizes the delicate balance between utilization and conservation. The intellectual origins of the concept draw on different strands: on the one hand, Pauli's (2010) blue economy approach foregrounds innovative technologies and zero-waste production models; on the other hand, the usage advanced by institutions such as the World Bank (2017), the Food and Agriculture Organization of the United Nations (FAO, 2021), and the European Commission (EC, 2012) highlights the sustainable management of marine and coastal resources and the protection of ecosystem services. This diversity has led to the perception of the blue economy not as a singular definition but rather as a policy framework.

At the international level, the blue economy has been formalized through key policy frameworks such as the United Nations' Transforming our World: The 2030 Agenda for Sustainable Development (in particular SDG 14: Life Below Water) (United Nations, 2015), the Food and Agriculture Organization of the United Nations' (FAO) Blue Transformation vision (FAO, 2021), the African Union's 2050 Africa's Integrated Maritime Strategy (African Union, 2012), and the European Union's Blue Growth Strategy (European Commission, 2012). Within this framework, the blue economy is typically conceptualized as the intersection of sustainability, economic development, and ocean governance (World Bank, 2017). It extends beyond traditional maritime activities-such as shipping, fisheries, and port operations-to encompass emerging growth areas including renewable energy, marine biotechnology, smart ports, and digital maritime services. The underlying assumption is that marine and freshwater ecosystems are not merely economic resources but resilient ecological systems that must be preserved for future generations, highlighting that aquatic resources should be valued not only for their economic potential but also for their ecological resilience and intergenerational sustainability.

From the perspective of Türkiye, the blue economy represents both a strategic imperative and a significant opportunity. Geographically, Türkiye occupies a unique position, surrounded by four seas (the Mediterranean, Black Sea, Aegean, and Marmara). These seas not only shape the country's ecological diversity but also underpin its economic and geopolitical significance. Türkiye's maritime domains serve as critical corridors for global trade, sources of energy and food security, and hubs of tourism and cultural exchange. At the same time, the country faces serious challenges: marine pollution, overfishing, habitat degradation, climate change, and coastal erosion rank among the most pressing concerns. These issues are not solely ecological but are closely linked to governance capacity and institutional coordination gaps.

The core sectors of Türkiye's blue economy include maritime transport, fisheries and aquaculture, coastal and marine tourism, and offshore renewable energy and resources. The country holds a geopolitically strategic position in global maritime trade. The Bosporus and Dardanelles Straits constitute two of the world's busiest waterways, linking the Black Sea to the Mediterranean and positioning Türkiye as a central actor in regional supply chains. In recent years, investments in port infrastructure have increased container and bulk cargo capacity, strengthening Türkiye's ambition to become a logistics hub between Europe, Asia, and Africa. However, Turkish maritime transport is under mounting pressure to align with global sustainability standards. The International Maritime Organization's (IMO) target of achieving carbon neutrality by 2050 compels the sector to pursue alternative fuels, electrification, and energy efficiency measures (IMO, 2018). While Turkish ports and shipping companies have begun obtaining "green port" certifications and adopting digital logistics solutions, the transformation remains fragmented. Empirical findings suggest that Türkiye's shipping fleet possesses substantial emission-reduction potential through technical upgrades (Kanberoğlu, Turan, & Kökkülünk, 2023). Yet institutional preparedness remains limited, with a significant gap persisting between individual awareness and organizational implementation (Zincir et al., 2022). For Türkiye to fully leverage its geopolitical advantages, it must place sustainability and digitalization at the center of its maritime strategy.

A similar pattern is observed in fisheries and aquaculture. Fisheries have long been a cornerstone of coastal economies in the Black Sea, where species such as anchovy hold a defining role in both cultural and economic life. However, overfishing, illegal practices, and insufficient stock management have led to declining fish populations and the degradation of marine ecosystems. Climate change further exacerbates these challenges, with studies emphasizing potential shifts in migratory routes, increased transboundary mobility of fish stocks, and reduced resilience of species. Data from FAO and GFCM confirm that climatic and environmental change in the Black Sea significantly affect feeding, reproduction, and migratory behavior of fish, making transboundary cooperation in stock management indispensable (FAO & GFCM, 2024).

By contrast, aquaculture has become one of the fastest-growing sectors. Türkiye has risen to the forefront of European production of sea bream and sea bass, contributing substantially to foreign trade through exports. Yet aquaculture activities also bring environmental pressures, including habitat degradation, nutrient pollution, and conflicts between coastal use, tourism, and conservation (Demirak, Balcı, & Tüfekçi, 2006).

One of the most visible faces of Türkiye's blue economy is marine and coastal tourism. Concentrated along the Mediterranean and Aegean coasts, the tourism industry encompasses a wide variety of activities, ranging from mass tourism to yachting, diving, and cruise tourism, contributing both to national economic revenues and to the country's international profile. However, the rapid expansion of tourism infrastructure has increased pressures on coastal ecosystems, leading to habitat loss, pollution, and heightened vulnerability in the face of climate change (Barakazı, 2023). Against this backdrop, eco-tourism initiatives, certification schemes, and community-based management approaches have been gaining increasing importance.

Energy transition stands out as one of the key domains opening new horizons for the blue economy. Türkiye's geographical position offers significant advantages in this regard. In particular, favorable wind conditions in the Aegean and Marmara Seas present notable potential for offshore wind energy; accordingly, feasibility studies and pilot projects have been initiated in recent years (Düzcan & Kara, 2022). This trend demonstrates that the blue economy is not confined to traditional sectors such as maritime transport or fisheries, but also encompasses the renewable energy dimension. Similarly, wave and tidal energy represent as-yet underexplored opportunities within Türkiye's marine domains. Nevertheless, it is also clear that the energy transition agenda is not limited to renewable sources. The discovery of natural gas reserves in the Black Sea has transformed Türkiye's energy security discourse, creating a new fault line between the aim of reducing external dependence and the commitments to decarbonization. This dynamic illustrates that the vision of the blue economy must confront inherent tensions between fossil fuel extraction and climate change objectives (Presidency of the Republic of Türkiye, 2021).

The future of all these sectors is directly dependent on the health of Türkiye's marine and coastal ecosystems. Pressures such as pollution, overexploitation, habitat loss, and climate change threaten not only biodiversity but also the sustainability of economic domains such as fisheries, tourism, and energy. Marine pollution is particularly acute in semi-enclosed seas such as the Marmara and the Mediterranean, where untreated wastewater, industrial discharges, agricultural runoff, and plastic waste place severe stress on ecosystems. The mucilage crisis of 2021 in the Marmara Sea starkly revealed the ecological consequences of pollution and governance deficiencies (Ayaz, 2022). Triggered by a combination of nutrient loading, untreated wastewater, and rising sea surface temperatures, the crisis disrupted marine life, paralyzed fisheries, and posed risks to public health. Although large-scale clean-up operations and scientific support sought to mitigate its impacts, the case underscored the absence of preventive strategies and long-term ecological monitoring. In the context of the blue economy, the mucilage crisis is not only an environmental disaster but also a catalyst for governance reform.

The effects of climate change further intensify this picture. Rising sea surface temperatures alter migratory patterns, accelerate coastal erosion, and render ecosystems increasingly fragile. In addition, invasive species (particularly in the Mediterranean)are spreading rapidly; species such as lionfish and pufferfish threaten native ecosystems and negatively affect fisheries (Özvarol et al., 2021).

The future of Türkiye's blue economy cannot be adequately understood through sectoral dynamics alone. Issues such as carbon reduction in maritime transport, stock conservation in fisheries, or the management of ecological pressures in tourism are not merely economic or ecological matters but are directly tied to policymaking processes and governance capacity. Put differently, the underlying determinants of sectoral success or failure lie in fragmented institutional structures, the coherence of regulatory frameworks, and the degree of alignment with international norms.

POLICY AND GOVERNANCE LANDSCAPE

The effectiveness of Türkiye's blue economy depends not only on sectoral growth but also on the strength and coherence of its policy and governance frameworks. Because maritime and coastal issues span diverse domains (including trade, fisheries, energy, environment, and tourism) fragmented governance and regulatory gaps can significantly undermine sustainability efforts (Satiroğlu, 2024; Sesli et al., 2009). As emphasized in the literature, the governance of marine resources entails "wicked problems" that demand not only technical solutions but also strong institutional coordination and multi-stakeholder governance models (Jentoft & Chuenpagdee, 2009). Building a robust institutional and legal architecture that harmonizes economic development with ecological protection is therefore essential.

In this regard, Türkiye is party to a number of international and regional agreements shaping marine governance. The Barcelona Convention, which seeks to protect Mediterranean ecosystems; the Convention on Biological Diversity, which underlines the sustainability of species and habitat diversity; and two fundamental instruments adopted under the International Maritime Organization (IMO) constitute the most prominent elements of this framework. Among these, the International Convention for the Prevention of Pollution from Ships (MARPOL) provides the most comprehensive regulation against ship-sourced pollution, while the International Convention for the Safety of Life at Sea (SOLAS) establishes the primary legal standards for safety in maritime transport.

Nevertheless, Türkiye has not ratified the United Nations Convention on the Law of the Sea (UNCLOS), which remains central to the delimitation of maritime jurisdictional areas and global ocean governance. This legal position provides Türkiye with short-term strategic flexibility in its disputes with Greece and the Greek

Cypriot Administration of Southern Cyprus over Exclusive Economic Zone (EEZ) boundaries in the Eastern Mediterranean, yet it also risks deepening long-term diplomatic tensions (Acer, 2021, pp. 101-103). Although Türkiye has not acceded to UNCLOS, it recognizes many of its provisions on continental shelf and EEZ delimitation as customary international law, grounding its position particularly in the "principle of equity" and the requirement of an "equitable result" (Acer, 2021, pp. 104-106). However, due to limited access to arbitration and dispute settlement mechanisms envisaged by international law, Türkiye can benefit only partially from global governance mechanisms in areas such as seabed mining, the sharing of biological resources, and offshore energy exploration. This situation restricts Türkiye's participation in United Nations forums and constrains its influence in global debates on sustainable ocean management (Acer, 2021, pp. 108–109).

At this point, the European Union's Blue Growth Strategy provides Türkiye with a crucial benchmark in terms of integrating sustainability into maritime sectors while advancing economic development goals. The strategy highlights three central pillars: the adoption of innovative technologies, the enhancement of resource efficiency, and the protection of marine ecosystems (European Commission, 2012, pp. 2-3). This approach has been operationalized most prominently through the Common Fisheries Policy (CFP), which institutionalizes science-based stock management, establishes quotas to curb overfishing, and standardizes conditions of access to the common market, thereby ensuring regulatory coherence (EPRS, 2016, pp. 3-7). In contrast, fisheries governance in Türkiye remains fragmented across ministries and regulations, with stock management and enforcement mechanisms lacking the degree of institutionalization observed in the EU.

Similarly, maritime spatial planning (MSP) became mandatory for EU member states under the 2014 Directive, requiring the integrated use of marine areas for energy, transport, fisheries, and conservation objectives (Directive 2014/89/EU, 2014). In Türkiye, however, MSP remains confined to limited pilot projects, with weak integration between coastal planning and marine area management. Türkiye has yet to adopt a comprehensive national MSP law; the national planning framework is still at the approval stage, and existing planning practices operate under a "guiding" rather than binding status. Nevertheless, the Türkiye MSP Platform coordinates public and private stakeholders in alignment with coastal and marine sectors. The planning process incorporates land–sea interactions and integrated coastal–marine planning, covering a range of sectors such as fisheries, coastal tourism, renewable energy, port and transport, and marine protected areas (MSP Global 2030, n.d.).

With respect to the green transition, the EU has integrated maritime transport into its Emissions Trading System (ETS), thereby subjecting the sector to carbon pricing and steering its transformation toward the 2050 climate neutrality targets (European Commission, 2023). Türkiye, due to its status as an EU candidate country and its deep regional trade integration, is directly affected by these standards and must accelerate alignment in order to maintain competitiveness in shipping and port operations (Turkish Shipowners' Association, 2025, pp. 2-3, 9-10). Thus, Türkiye's formal non-alignment with EU policies does not diminish the importance of convergence; on the contrary, in the context of external trade and environmental sustainability objectives, adaptation to EU standards constitutes a strategic necessity.

At the national level, Türkiye's marine governance is characterized by a highly fragmented and multi-actor structure. Key responsibilities are divided among the Ministry of Transport and Infrastructure, the Ministry of Agriculture and Forestry, the Ministry of Energy and Natural Resources, and the Ministry of Environment, Urbanization, and Climate Change, among others. This dispersed structure, however, is not supported by a strong coordination mechanism, resulting in overlaps and gaps in practice. The Fisheries Law (Law No. 1380) regulates stock management and fishing practices; the Coastal Law (Law No. 3621) sets the framework for the use and planning of the coastal zone; and the Environmental Law (Law No. 2872) provides provisions on marine pollution and ecosystem protection. While these pieces of legislation constitute the cornerstones of marine governance, implementation and enforcement frequently remain inadequate (Karlı, 2024).

The mucilage crisis in the Marmara Sea in 2021 revealed the practical consequences of such fragmented governance. Jurisdictional overlaps among different agencies hindered rapid and effective intervention, illustrating how institutional fragmentation can exacerbate ecological crises (Ayaz, 2022, pp. 198-200). This challenge is further aggravated by the absence of a comprehensive National Blue Economy Strategy. Although the Ministry of Environment, Urbanization, and Climate Change has initiated preparations for a Sustainable Blue Economy Action Plan (Mavi Plan), a finalized strategy has yet to be adopted (Ministry of Environment, Urbanization, and Climate Change, 2025). The lack of such a strategic framework limits the creation of inter-sectoral synergies and the institutionalization of long-term planning, resulting in maritime, energy, tourism, and environmental policies advancing in largely disconnected ways.

The climate change dimension becomes a critical area when combined with governance deficiencies. Although Türkiye has formally adopted its 2053 net-zero target (Presidency of the Republic of Türkiye, Directorate of Communications, 2021), the integration of this goal into maritime policies remains limited. The Marmara Sea mucilage crisis of 2021 was not only an ecological disaster but also a textbook manifestation of "fragmented governance," as overlapping institutional jurisdictions impeded an effective response (Ayaz, 2022, pp. 198-200). Similarly, risks such as the rapid spread of invasive species in the Mediterranean (Özvarol et al., 2021) or the acceleration of coastal erosion are often addressed in a scattered fashion across national policy documents, without being systematically incorporated into climate adaptation strategies. This situation exemplifies what the literature defines as institutional misfit-a mismatch between policy objectives and the instruments used to achieve them (Young, 2017).

These challenges render governance problems more visible. Weak coordination among ministries, the lack of reliable data on fish stocks, pollution levels, and maritime activities, limited stakeholder participation, and the centralization of decision-making processes all hinder the development of blue economy policies on an inclusive and transparent basis. The limited institutional capacity and financial resources of local administrations create serious difficulties in implementing national directives, while insufficient involvement of actors such as civil society organizations and fishery cooperatives restricts the potential for community-based solutions. Taken together, these factors narrow the scope for evidence-based policymaking and weaken the application of sustainability principles.

In conclusion, Türkiye's policy and governance landscape presents a complex picture of both strengths and weaknesses. Active engagement in regional agreements and the presence of sectoral regulations constitute relative strengths, whereas institutional fragmentation, lack of strategic integration, and capacity deficits at the local level expose the system's vulnerabilities. Addressing these gaps requires the development of a comprehensive integration model. In this regard, the European Union's Integrated Maritime Policy (European Commission, 2007) and its practices of Marine Spatial Planning (Directive 2014/89/EU, 2014) represent instructive international examples, offering a framework through which Türkiye may harmonize economic development objectives with ecological sustainability.

On the other hand, innovation and digitalization are increasingly regarded as key enabling factors for the development of the blue economy. According to the OECD, innovation will be a critical driver of growth in the ocean economy. Emerging ocean-based industries are expected to gain momentum particularly through technological advances, expanded data collection capacities, and improvements in monitoring systems. Moreover, digitalization and big data hold the potential to enhance ocean governance by strengthening monitoring, transparency, and decision-making processes (OECD, 2021, pp. 19-21, 26). This trend parallels the European Union's "Digital Ocean" initiatives and support mechanisms such as the European Maritime, Fisheries and Aquaculture Fund (EMFAF) (European Commission, 2021), making convergence a strategic necessity for Türkiye given its EU candidate country status.

The digital transformation of maritime logistics is profoundly reshaping the functioning of global trade. Blockchain-based cargo tracking has become an essential tool for improving security and verification in delivery processes. In addition, port-wide deployments of IoT sensors and AI-driven systems, supported by real-time data flows, reinforce both operational safety and transparency. These technological innovations play a critical role in building the sustainable logistics infrastructure of the blue economy, particularly by enhancing container tracking and supply chain integration (Heikkilä, Saarni, & Saurama, 2022, pp. 10-12). In Türkiye, pilot initiatives toward "green port" and "smart port" applications have emerged in ports such as İzmir and Mersin, but overall investments in technological infrastructure remain limited. Most ports still rely on basic monitoring and identification systems, whereas achieving competitiveness requires more advanced digital solutions based on data connectivity, sensor technologies, and information sharing (Özkanlı & Denizhan, 2020, pp. 358-362). This makes comprehensive investment in automation, data analytics, and electrification indispensable. Furthermore, in light of the EU's smart port standards and emission monitoring regulations, Türkiye's current port infrastructure falls significantly short of compliance. The absence of updated regulatory frameworks and the limitations of public investment constitute further obstacles to scaling digitalization.

Artificial intelligence (AI) and big data analytics can also generate real-time insights into vessel traffic, pollution levels, and ecosystem health. For instance, satellite-based monitoring and AI-powered predictive models can track fish stock movements and thus contribute to the prevention of illegal, unreported, and unregulated (IUU) fishing (FAO, 2021). Similarly, digital platforms for vessel tracking can enhance maritime safety in congested waterways such as the Bosporus. Integrating such technologies into governance systems would enable more proactive and evidence-based decision-making. While the EU has established comprehensive data infrastructures through programs such as the Copernicus Marine Environment Monitoring Service (European Commission, 2020), data-sharing mechanisms in Türkiye remain constrained by institutional fragmentation and the lack of standardized protocols. This not only complicates convergence with EU practices but also generates weaknesses within Türkiye's own domestic policy-making processes.

Technological innovation in fisheries and aquaculture plays a critical role in maintaining sustainable production capacity. Remote sensing, underwater drones, smart feeding systems, and automated monitoring solutions help optimize resource use by reducing feed and energy waste, while simultaneously limiting environmental impacts and lowering the carbon footprint (Dikel, 2023, pp. 8-16). In Türkiye, some enterprises have begun using centrally managed automatic feeders and water quality monitoring systems, which not only reduce feed costs but also enable higher stocking densities and enhance overall efficiency. These developments align with the Food and Agriculture Organization's (FAO, 2021a) emphasis on traceability and digitalization as key elements in strengthening governance capacity at the global level.

The European Union's requirement for traceability in imported seafood has also pushed Türkiye to advance its monitoring systems. The Fisheries Sector Policy Document 2019-2023 prioritized the improvement of traceability standards, and the introduction of the Fisheries Information System (SÜBİS) allowed the recording of the entire production and marketing process, thereby enhancing transparency (Ministry of Agriculture and Forestry, 2019; Sağlam & Soyer, 2017, pp. 58–61). Since 2007, Türkiye has implemented computer-based traceability mechanisms such as SÜBİS and the Fisheries Observation and Monitoring System (BAGİS), enabling online monitoring of production and capture activities (Tolon, 2017). As of 2023, BAGİS tracked the activities of approximately 1,650 vessels using GSM and satellite technology (Fisheries Special Expertise Commission Report, 2023). While these infrastructures have strengthened Türkiye's adaptation to EU market requirements, challenges remain in extending their effectiveness across the entire sector.

Within the framework of the EU's Common Fisheries Policy, the European Maritime, Fisheries and Aquaculture Fund (EMFAF) has been supporting innovative projects during the 2021–2027 period, facilitating green and digital transitions in fisheries and aquaculture (European Parliamentary Research Service [EPRS], 2024, pp. 7, 16). In Türkiye, the Eleventh Development Plan (2019-2023) introduced the establishment of the "Technical Committee on Marine Technologies and Industry" and promoted the development of R&D culture within firms, aiming to contribute to high value-added production and export targets (Presidency of Strategy and Budget, 2019). Yet, despite these steps, Türkiye has not developed a comprehensive roadmap specifically dedicated to maritime and marine technologies. Nevertheless, as an associated country, Türkiye participates in Horizon Europe programs, which provide opportunities for collaborative projects in remote sensing, Internet of Things (IoT), artificial intelligence, and automation (European Commission, 2021). In this way, Türkiye -despite gaps in its domestic policy frameworks- has been able to capture strategic opportunities in marine technologies and digitalization through engagement with the European research area.

In conclusion, innovation and digitalization are not optional but central elements for advancing Türkiye's blue economy. The adoption of technology across all sectors will enhance Türkiye's competitiveness, safeguard sustainability, and strengthen resilience against both environmental and economic risks. Yet realizing this potential requires the development of a digitalization strategy aligned with EU standards, the removal of institutional barriers, and the inclusion of small-scale actors in the digital transition.

BLUE ECONOMY IN PRACTICE

Practical examples illustrate how the principles of the blue economy are either applied or left unfulfilled in Türkiye. These cases not only highlight sector-specific successes and failures but also expose the limits of governance mechanisms, the strengths and weaknesses of institutional capacity, and the structural challenges of policy design. The Marmara Sea mucilage crisis, sustainable fisheries projects in the Black Sea, and the Kaş–Kekova Specially Protected Area reveal, respectively, the dimensions of crisis management, sustainability in resource use, and the balance between conservation and economic activity in Türkiye's blue economy experience.

The Marmara Sea Mucilage Crisis and Response

The mucilage outbreak in the Marmara Sea in 2021 stands as one of the most severe ecological crises in Türkiye's recent history. A thick layer of organic matter covered the sea surface, partially or entirely halting photosynthesis, reducing dissolved oxygen, and causing mass die-offs and ecosystem degradation (Yümün, Kam, & Önce, 2023). Marine life came under acute stress, fisheries nearly collapsed, tourism was disrupted, and serious public health concerns arose.

The crisis was rooted in multiple factors: untreated wastewater discharges, industrial pollutants, unplanned coastal development, and rising sea surface temperatures linked to climate change. The Turkish Academy of Sciences (TÜBA) report emphasized the contribution of pollutants such as phosphorus, nitrogen, pesticides, microplastics, and heavy metals to mucilage formation (Aydın, 2021). Meteorological conditions also played a critical role: in spring 2021, above-average sea surface temperatures combined with unusually low wind speeds prevented dispersion and facilitated accumulation of the mucilage.

In response, the government announced the 22-Article Marmara Sea Action Plan within weeks of the outbreak. The plan sought not only to mitigate immediate pollution impacts but also to initiate structural transformations aimed at the long-term protection of the Marmara ecosystem. Measures included the conversion of all existing wastewater treatment facilities in the region into advanced biological treatment plants, a ban on untreated wastewater discharge into the sea, and stricter monitoring of industrial pollutants. To enhance scientific monitoring capacity, the number of observation stations in the Marmara Sea was slated to increase from 91 to 150, with the introduction of online, real-time data systems as well as satellite, radar, and drone-based basin-level surveillance. Ecosystem-based solutions also featured prominently, including the rehabilitation of seagrass meadows, the removal of ghost nets, and the establishment of coastal buffer zones (Ministry of Environment and Urbanization, 2021).

Furthermore, Presidential Circular No. 2021/10 formally established the "Marmara Sea Action Plan Coordination Board," thereby acknowledging the necessity of multi-actor coordination in tackling the crisis. The mucilage event thus underscored both the ecological vulnerabilities of semi-enclosed seas and the institutional shortcomings of fragmented governance-while also serving as a catalyst for policy reform in Türkiye's blue economy.

In conclusion, the 2021 Marmara mucilage crisis was not merely an ecological disaster but a multi-layered crisis with profound economic repercussions for the fisheries and tourism sectors, as well as social implications linked to public health concerns. While Türkiye's response capacity allowed certain immediate steps to be taken, fragmented institutional responsibilities and the lack of strategic foresight hindered the possibility of a lasting solution (Heinrich Böll Stiftung, 2022, pp. 12-15, 20-21). This case reveals the "reactive" logic prevailing in Türkiye's environmental policy: problems are typically addressed after they manifest as crises, rather than through pre-emptive risk-mitigation strategies. Thus, the Marmara experience strongly underlines that environmental governance within the framework of the blue economy must rest not only on emergency responses but also on long-term monitoring, inter-institutional coordination, and proactive policy design.

Sustainable Fisheries Projects in the Black Sea

The Black Sea is both economically vital and ecologically fragile. Türkiye derives the majority of its marine capture production from this basin; for instance, according to the TR90 fisheries sector report, approximately 78% of Türkiye's marine catch originates from the Black Sea (Su Ürünleri Merkez Araştırma Enstitüsü [SÜMAE], 2012). Yet overfishing, illegal and unreported practices, and insufficient enforcement have long placed fish stocks under severe pressure. This strain threatens not only ecosystems but also the livelihoods of coastal communities.

In recent years, Türkiye has introduced a range of management tools aimed at rendering fisheries more sustainable, including stock assessment programs, gear restrictions, seasonal bans, and spatial regulations. Within this framework, the Instruction on Quotas and Production Planning in Anchovy Fishing (Republic of Türkiye Ministry of Agriculture and Forestry, 2024) mandates electronic notifications at landing points, enforcement teams, and quota arrangements throughout the fishing season, thereby strengthening both legal compliance and traceability. These national initiatives are supported by regional projects conducted under the auspices of the Food and Agriculture Organization of the United Nations (FAO) and the General Fisheries Commission for the Mediterranean (GFCM). For example, the BlackSea4Fish program has developed joint acoustic survey protocols and data-sharing mechanisms, enabling scientific stock assessments of anchovy and providing evidence-based foundations for quota decisions (FAO, 2021). Similarly, the FishEBM-BS project has fostered ecosystem-based fisheries management among Black Sea states, enhancing regional capacity to mitigate overfishing pressures (FAO, 2022).

Taken together, Türkiye's multi-dimensional measures in fisheries extend beyond the mere conservation of stocks. They contribute to the ecological, economic, and social sustainability of marine resources in line with the blue economy vision based on governance, community participation, and international cooperation.

Nevertheless, the effectiveness of these initiatives remains fragile due to institutional capacity limitations. In fisheries enforcement, the limited use of technologies such as Vessel Monitoring Systems (VMS) and Automatic Identification Systems (AIS) hampers efforts to combat illegal, unreported, and unregulated (IUU) fishing (GFCM, 2021). Furthermore, in the absence of sustainable financing, many of these projects risk remaining confined to pilot-scale applications rather than evolving into permanent institutional mechanisms as required by the blue economy. The BRIDGE-BS project, for example, adopted a "pilot sites" approach in pursuit of blue economy objectives in the Black Sea, yet stressed that lasting outcomes would only be possible through strengthened long-term financing and governance capacity (Bat, 2022). In this respect, sustainable fisheries in the Black Sea are not limited to stock conservation alone but also depend on supporting ecosystem services, economic opportunities, and social well-being in a holistic manner, as envisioned by the blue economy.

The Kaş–Kekova Specially Protected Area

The Kaş–Kekova Specially Protected Area (SPA) constitutes one of the pioneering cases testing the applicability of blue economy principles in Türkiye. The designation of the region as an SPA aimed to establish a management framework balancing conservation with sustainable use. The management plan was developed through participatory processes involving the World Wide Fund for Nature (WWF) Türkiye, relevant public institutions, and local stakeholders such as fishers and diving operators (WWF Türkiye, 2012). The plan introduced mechanisms including zoning systems, no-take areas, restrictions on anchoring, and regulations on diving and tourism activities, all designed to alleviate tourism pressures and ensure the protection of marine ecosystems (WWF Türkiye, 2012).

In this context, Kaş–Kekova exemplifies the multiple dimensions of blue economy principles. First, eco-tourism -particularly diving activities- has contributed to the local economy, while tourism-related pressures have been managed through zoning and activity restrictions (Başak, 2012). Second, in the field of fisheries, the establishment of no-take zones has led to signs of recovery in grouper populations, yet declines have continued in species such as the white grouper and goldblotch grouper. This finding demonstrates that legal designation alone is insufficient; effective enforcement, monitoring, and reporting mechanisms are indispensable. Third, the participatory governance model was implemented through approximately forty consultation meetings and local stakeholder dialogues conducted under the coordination of WWF. By incorporating fishers and diving operators into decision-making processes, this approach fostered local ownership of conservation measures. Fourth, in terms of infrastructure innovation, a mooring buoy system was piloted to protect seagrass meadows, thereby sustaining tourism activities while simultaneously safeguarding ecosystem services (WWF-Türkiye, 2012).

Yet the Kaş-Kekova experience is not limited to success stories; it also highlights significant challenges in practice. These include financing constraints, limited enforcement capacity, the lack of strategic coherence within the national Specially Protected Areas (SPA) network, and tensions between conservation and development objectives. Indeed, conflicts of interest -such as the growth of tourism versus the restrictions on fisheries -reflect a typical area of difficulty in the application of blue economy principles (Başak, 2012).

In this respect, the Kaş-Kekova experience makes visible both the promises of the blue economy-eco-tourism, sustainability in fisheries, and participatory governance-and the institutional and structural barriers that constrain their realization. For Türkiye, the lesson is clear: rather than relying on temporary, project-based approaches, the full potential of the blue economy can only be achieved through permanent institutional mechanisms, cross-sectoral integration, and long-term strategic planning.

When the three case studies are considered together, it becomes evident that Türkiye's blue economy experience, while holding strong potential, is overshadowed by structural limitations. First, a reactive policy orientation emerges across all cases: in the Marmara Sea, the mucilage crisis was addressed only after the disaster unfolded; in the Black Sea, sustainable fisheries initiatives have largely remained confined to short-term pilot projects; and in Kaş-Kekova, despite the preparation of a management plan, enforcement and continuity challenges persist. Second, problems of institutional capacity and coordination constitute a common theme. Fragmented responsibilities among agencies in the mucilage crisis, insufficient monitoring technologies in the Black Sea, and the lack of strategic coherence in the national Specially Protected Areas (SPA) network in Kaş-Kekova have all hindered the consolidation of durable institutional structures for the blue economy. Third, issues of financing and continuity represent critical constraints across the cases: while resources could be mobilized for immediate interventions in Marmara, the sustainability of long-term investments remains uncertain; projects in the Black Sea are often dependent on international funding; and in Kaş-Kekova, local conservation efforts continue within a narrow scope due to inadequate financial support. Fourth, cross-sectoral conflicts of interest further limit the application of blue economy principles in Türkiye. Tensions between industrial pollution and ecological protection in Marmara, between fishers' short-term livelihood concerns and stock conservation in the Black Sea, and between tourism and fisheries in Kaş–Kekova all underscore the need for a more integrated strategy.

Ultimately, these three cases demonstrate that the blue economy requires a multi-layered approach encompassing not only environmental protection but also social welfare, economic sustainability, and participatory governance. The key lesson for Türkiye is that moving beyond project-based, short-term, and crisis-driven approaches toward a holistic blue economy vision (anchored in long-term strategic planning, permanent institutional mechanisms, technological capacity-building, and cross-sectoral integration) is indispensable.

CONCLUSION: FUTURE PERSPECTIVES AND POLICY RECOMMENDATIONS

Building a sustainable and competitive blue economy in Türkiye requires a forward-looking vision that integrates ecological protection, technological innovation, and institutional reform. While various sectoral initiatives have been implemented in recent years, the absence of a comprehensive national strategy hampers coordination and limits the potential synergies that could emerge. Future perspectives must therefore focus on developing an integrated framework that not only positions Türkiye as a blue economy actor at the regional and global levels but also safeguards environmental sustainability.

The most critical component of such a framework is the formulation of a National Blue Economy Strategy. This strategy should not merely define clear objectives, indicators, and monitoring mechanisms but also establish an institutional coordination body capable of overcoming the fragmented ministerial landscape. Regular reporting of indicators -such as carbon intensity, levels of marine pollution, biomass measures of fish stocks, the ecological impacts of coastal tourism, and the capacity of renewable energy production -would make the strategy measurable and accountable. Moreover, the strategy is expected to enhance Türkiye's alignment not only with domestic policy priorities but also with global and regional initiatives such as the United Nations Sustainable Development Goals and the European Union's Blue Growth Strategy. Indeed, the Ministry of Environment, Urbanization, and Climate Change's Institutional Financial Status and Expectations Report of 31 July 2025 explicitly states that preparations for a "Sustainable Blue Economy Action Plan (Blue Plan)," aligned with the 2053 Presidential targets, are still underway. This confirms that Türkiye has yet to adopt or implement a comprehensive national strategy (ÇŞİDB, 2025, p. 36).

A second pillar of the strategy should be the strengthening of the triple helix model of collaboration between the public sector, private sector, and academia. Joint projects in areas such as renewable energy, aquaculture technologies, smart logistics, and marine spatial planning would both enhance economic competitiveness and generate broader societal benefits.

At the same time, addressing persistent problems of traceability and sustainability in fisheries and aquaculture is imperative. The Fisheries Information System (SÜBİS) should be harmonized with EU standards, and mechanisms for combating illegal, unreported, and unregulated (IUU) fishing must be reinforced. Considering the impacts of climate change on Türkiye's seas -including mucilage outbreaks, invasive species, and coastal erosion -the 2053 net-zero target must be integrated into maritime, fisheries, and coastal management policies. Based on research demonstrating the emission-reduction potential of Türkiye's fleet, vessel renewal programs and green financing instruments should form part of this integration. In the field of marine conservation, as shown by the Kaş-Kekova experience, a strategic restructuring of the Specially Protected Areas (SEPA) network is necessary, with mechanisms to overcome financing and enforcement deficits. Strengthening local stakeholder participation and balancing tourism–fisheries tensions should be prioritized in this process.

Finally, the sustainability of the blue economy depends on its integration with green and circular economy models. Policies such as the recycling of marine plastics, powering port operations with renewable energy, and channeling marine waste into circular production chains can directly enhance resilience to climate change. Furthermore, to overcome the fragmented progress observed in digitalization and green port practices, a national roadmap should promote digital twin technologies, data-sharing platforms, and green logistics solutions. Such integrated approaches will not only strengthen environmental sustainability but also create new opportunities for Türkiye within global value chains.

REFERENCES

Acer, Y. (2021). Türkiye's legal approach to maritime boundary delimitation in the Eastern Mediterranean Sea. *Insight Türkiye, 23*(1), 99–114. DOI: 10.25253/99.2021231.8

African Union. (2012). *2050 Africa's Integrated Maritime Strategy (2050 AIM Strategy).* African Union.

Ayaz, Ç. E. (2022). Musilaj sorununun idari boyutu: İstanbul Büyükşehir Belediyesi ile İstanbul Su ve Kanalizasyon İdaresi özelinde bir inceleme. *Çankırı Karatekin Üniversitesi İktisadi ve İdari Bilimler Fakültesi Dergisi, 12*(1), 176–204. https://doi.org/DOI: 10.18074/ckuiibfd.989631

Aydın, M. E. (2021). Marmara Denizi'nde müsilaj oluşumu: Muhtemel sebepleri ve öneriler. In İ. Öztürk & M. Şeker (Eds.), *Marmara Denizi'nin ekolojisi: Deniz salyası oluşumu, etkileşimleri ve çözüm önerileri* (pp. 50–68). Ankara: Türkiye Bilimler Akademisi Yayınları (TÜBA).

Barakazı, M. (2023). Unsustainable tourism approaches in touristic destinations: A case study in Türkiye. *Sustainability (Basel), 15*(6), 4744. DOI: 10.3390/su15064744

Başak, E. (2012). Assessing the effects of the Kaş–Kekova Specially Environmental Protected Area (SEPA) on the sustainable development of the surrounding area. In Bleu, P. (Ed.), *The effects of preserving the coastal and marine environment for the local sustainable development: The case of Mediterranean Marine Protected Areas* (pp. 1–40). Plan Bleu.

Bat, L. (2022). Sustainable fisheries in the Black Sea within the framework of the BRIDGE-BS project [Conference presentation]. *Black Sea Sustainable Blue Economy Conference*, Trabzon, Türkiye.

Demirak, A., Balci, A., & Tüfekçi, M. (2006). Environmental impact of the marine aquaculture in Güllük Bay, Türkiye. *Environmental Monitoring and Assessment, 123*(1–3), 1–12. DOI: 10.1007/s10661-005-9063-y PMID: 17082906

Dikel, S. (2023). Endüstriyel akvakültürde mekanizasyonun vazgeçilmez gerekliliği. In Dikel, S., & Demirkale, İ. (Eds.), *Su ürünleri yetiştiriciliğinde teknoloji kullanımı ve çevresel etkileri* (pp. 5–22). İksad Publishing House.

Directive 2014/89/EU of the European Parliament and of the Council of 23 July 2014 establishing a framework for maritime spatial planning. *Official Journal of the European Union, L 257*, 135–145. https://eur-lex.europa.eu/legal-content/EN/TXT/?uri=CELEX:32014L0089

Duzcan, A., & Kara, Y. A. (2022). Offshore wind energy potential analysis of Turkish Marmara and Aegean seas. *International Journal of Environmental Science and Technology, 19*(5), 5571–5584. DOI: 10.1007/s13762-022-04577-y

European Commission. (2007). *An integrated maritime policy for the European Union* (COM(2007) 575 final). Publications Office.

European Commission. (2012). *Blue growth: Opportunities for marine and maritime sustainable growth (COM(2012) 494 final).* Luxembourg: Publications Office of the European Union. https://eur-lex.europa.eu/legal-content/EN/TXT/?uri=CELEX:52012DC0494

European Commission. (2023). EU Emissions Trading System (EU ETS) for maritime transport. *Questions & Answers.*

European Parliamentary Research Service. (2016). The Common Fisheries Policy. Publications Office of the European Union., PE 589.780.

European Parliamentary Research Service. (2024). *The European Maritime, Fisheries and Aquaculture Fund (EMFAF): 2021–2027.* Publications Office of the European Union.

FAO & General Fisheries Commission for the Mediterranean (GFCM). (2024, October 31). Karadeniz'de sürdürülebilir balıkçılık ve su ürünleri yetiştiriciliği için bölgesel iş birliği güçlendiriliyor. *Food and Agriculture Organization of the United Nations.* https://www.fao.org/gfcm/news/2024/31-oct/tr (Accessed February 10, 2025)

Food and Agriculture Organization of the United Nations (FAO). (2021). *GFCM 2030 strategy for sustainable fisheries and aquaculture in the Mediterranean and the Black Sea.* FAO.

Food and Agriculture Organization of the United Nations (FAO). (2021). *BlackSea4Fish project: Towards sustainable small pelagic fisheries in the Black Sea.* Rome: FAO. https://www.fao.org/gfcm/activities/fisheries/blacksea4fish

Food and Agriculture Organization of the United Nations (FAO). (2022). *FishEBM-BS: Fisheries ecosystem-based management in the Black Sea (FAO–GEF Project Summary).* Rome: FAO. https://www.fao.org/in-action/fishebm-bs

Food and Agriculture Organization of the United Nations (FAO). (2022). *BlackSea4Fish project: Enhancing cooperation for sustainable fisheries management in the Black Sea.* FAO.

Food and Agriculture Organization of the United Nations (FAO a). (2021). *Blue transformation: Roadmap for FAO fisheries and aquaculture.* FAO.

Global, M. S. P. 2030. (n.d.). *Türkiye – MSPglobal2030.* MSP Global Roadmap. https://www.mspglobal2030.org/msp-roadmap/msp-around-the-world/turkiye/ (Accessed February 10, 2025)

Heikkilä, M., Saarni, J., & Saurama, A. (2022). Innovation in smart ports: Future directions of digitalization in container ports. *Journal of Marine Science and Engineering, 10*(12), 1925. DOI: 10.3390/jmse10121925

Heinrich-Böll-Stiftung Türkiye. (2022). *The Sea of Marmara 2022: State of the ecosystem.* Heinrich-Böll-Stiftung Türkiye. https://tr.boell.org/en/2022/06/08/sea-marmara-2022-state-ecosystem

International Maritime Organization (IMO). (2018). *Initial IMO strategy on reduction of GHG emissions from ships*. IMO.

Jentoft, S., & Chuenpagdee, R. (2009). Fisheries and coastal governance as a wicked problem. *Marine Policy, 33*(4), 553–560. DOI: 10.1016/j.marpol.2008.12.002

Kanberoğlu, B., Turan, E., & Kökkülünk, G. (2023). Decarbonization of maritime transportation: A case study for Turkish ship fleet. *Journal of Marine Science and Application, 22*(3), 716–727. DOI: 10.1007/s11804-023-00370-6

Karlı, A. (2024). Türkiye'deki sürdürülebilir mavi ekonomi ve deniz yönetişimi imkânlarının değerlendirilmesi: Bakanlıklar düzeyinde bir inceleme. *Journal of Environmental and Natural Studies, 6*(3), 240–251. DOI: 10.53472/jenas.1576035

OECD. (2021). A new era of digitalisation for ocean sustainability? *Prospects, benefits, challenges* (OECD Science, Technology and Industry Policy Papers, No. 111). OECD Publishing.

Özkanlı, A., & Denizhan, B. (2020). Digitalization roadmap for Turkish seaports. *Avrupa Bilim ve Teknoloji Dergisi (European Journal of Science and Technology), (Özel Sayı)*, 358–363. https://doi.org/DOI: 10.31590/ejosat.araconf46

Pauli, G. (2010). *The blue economy: 10 years, 100 innovations, 100 million jobs.* Paradigm Publications.

Presidency of the Republic of Türkiye, Directorate of Communications. (2021, September 27). Türkiye announces net-zero emission target by 2053. Presidency of the Republic of Türkiye. https://www.iletisim.gov.tr/english/haberler/detay/turkiye-announces-net-zero-emission-target-by-2053 (Accessed February 10, 2025)

Sağlam, N. E., & Soyer, S. (2017). Türkiye'de Su Ürünleri Bilgi Sistemi (SÜBİS)'ne genel bir bakış. *Ordu Üniversitesi Bilim ve Teknoloji Dergisi, 7*(1), 57–82.

Satiroğlu, M. (2024). Integrated coastal zone management studies in Türkiye. *Sustainability, 16*(16), 7028. DOI: 10.3390/su16167028

Sesli, F. A., Şişman, A., & Aydınoglu, A. C. (2009). Coastal legislation and administrative structure in Türkiye. *Scientific Research and Essays, 4*(12), 1445–1453.

Strateji ve Bütçe Başkanlığı. (2019). *On Birinci Kalkınma Planı (2019–2023).* T.C. Cumhurbaşkanlığı Strateji ve Bütçe Başkanlığı.

Strateji ve Bütçe Başkanlığı. (2023). *Su ürünleri özel ihtisas komisyonu raporu (On İkinci Kalkınma Planı 2024–2028).* T.C. Cumhurbaşkanlığı Strateji ve Bütçe Başkanlığı.

Su Ürünleri Merkez Araştırma Enstitüsü (SÜMAE). (2012). *TR90 Doğu Karadeniz bölgesi su ürünleri sektörü raporu.* T.C. Doğu Karadeniz Kalkınma Ajansı.

T.C. Çevre ve Şehircilik Bakanlığı. (2021). *Marmara Denizi eylem planı.* T.C. Çevre ve Şehircilik Bakanlığı.

T.C. Tarım ve Orman Bakanlığı. (2019). *Su ürünleri sektör politika belgesi 2019–2023.* T.C. Tarım ve Orman Bakanlığı.

T.C. Tarım ve Orman Bakanlığı. (2024a). *Hamsi avcılığında kota ve üretim planlamasının uygulanmasına ilişkin talimat.* T.C. İstanbul İl Tarım ve Orman Müdürlüğü.

Tolon, M. T. (2017). Overview of fisheries traceability infrastructure in Türkiye. [AJER]. *American Journal of Engineering Research, 6*(9), 281–287.

Turkish Shipowners' Association. (2025, July). *Emission analysis of the Turkish shipping sector under the EU emissions trading system (EU ETS) – 2024: General analysis based on THETIS MRV data and allowance obligations.* Istanbul: Turkish Shipowners' Association.

United Nations. (2015). *Transforming our world: The 2030 Agenda for Sustainable Development.* United Nations. https://sdgs.un.org/2030agenda

World Bank. (2017). *The potential of the blue economy: Increasing long-term benefits of the sustainable use of marine resources for small island developing states and coastal least developed countries.* World Bank.

WWF-Türkiye. (2012). *Kaş–Kekova specially protected area marine management plan and its implementation: MedPAN South pilot project Türkiye, September 2012.* https://d2hawiim0tjbd8.cloudfront.net/downloads/wwf_ka_kekova_2012_en.pdf (Accessed February 10, 2025)

Young, O. R. (2017). *Governing complex systems: Social capital for the Anthropocene*. MIT Press. DOI: 10.7551/mitpress/9780262035934.001.0001

Yümün, Z. Ü., Kam, E., & Önce, M. (2023). Marmara Denizi'nde deniz salyası (müsilaj) kompozisyonu bulguları ışığında salya oluşma nedenleri ve çözüm önerileri. *Çevre, Şehir ve İklim Dergisi, 2*(3), 98–115.

Zincir, B., Inal, O. B., Dere, C., & Deniz, C. (2022). An overview on the readiness level of the Turkish maritime industry for decarbonization in shipping. *Marine Science and Technology Bulletin, 11*(1), 63–75. DOI: 10.33714/masteb.1061972

ADDITIONAL READING

Boswell, R. (Eds.). (2022). *The Palgrave handbook of blue heritage*. Palgrave Macmillan. DOI: 10.1007/978-3-030-99347-4

Heidkamp, C. P., Morrissey, K., & Duret, D. (Eds.). (2022). *Blue economy: People and regions in transitions*. Routledge.

Islam, S. N., & Bartell, S. M. (Eds.). (2024). *Global blue economy: Analysis, developments, and challenges*. Routledge.

Johnson, K., Masters, I., & Dalton, G. (Eds.). (2018). *Building industries at sea: Blue growth and the new maritime economy*. River Publishers.

Standing, G. (2022). *The blue commons: Rescuing the economy of the sea*. Penguin.

KEY TERMS AND DEFINITIONS

Digitalization in Maritime Sectors: Digitalization in maritime sectors refers to the integration of digital Technologies, artificial intelligence, blockchain, the Internet of Things (IoT), and automation systems (into ports, vessels, and fisheries). This transformation extends beyond enhancing operational efficiency: it also supports sustainability objectives such as monitoring fuel consumption and emissions, ensuring transparency in supply chains, and improving traceability in fisheries. Through digitalization, port and vessel operations can be optimized in real time, enabling energy savings and more effective control of environmental impacts.

Ecosystem-Based Management (EBM): Ecosystem-based management (EBM) emphasizes the management of natural resources by considering the integrity of ecosystems as a whole, rather than focusing on single species or single sectors. This approach takes into account the cumulative impacts of human activities on ecosystems and promotes precautionary and adaptive principles for long-term sustainability. In fisheries and the governance of marine protected areas, EBM is increasingly becoming a widely accepted standard.

Governance Fragmentation: Governance fragmentation refers to the dispersion of authority and responsibility across multiple institutions in the realms of policy and implementation, without the establishment of effective coordination mechanisms. Such fragmentation slows down decision-making processes, hinders the efficient use of resources, and results in policies that lack strategic coherence.

Green Ports: Green ports are port management models that prioritize the reduction of environmental impacts and the improvement of energy efficiency. Key practices include the adoption of low-emission cranes, the deployment of electric vehicles, the use of shore power systems, and the implementation of advanced waste management technologies.

Institutional Misfit: Institutional misfit denotes the mismatch between the structural design of existing institutions and the ecological or social systems they are tasked with managing. For instance, the static character of marine policies in the face of rapidly changing impacts of climate change, or the insufficient inclusion of local actors in decision-making processes, exemplifies this misalignment. As a result, persistent gaps emerge between policy objectives and the tools of implementation.

Marine Mucilage: Marine mucilage is a dense and sticky organic accumulation in marine ecosystems, triggered by the combined effects of nutrient overload, rising temperatures, and stagnant water conditions. Ecologically, it causes oxygen depletion, mass mortality of marine organisms, and loss of biodiversity; economically, it inflicts severe damage on fisheries, tourism, and maritime transport.

Marine Spatial Planning (MSP): Marine Spatial Planning (MSP) is a management tool that organizes diverse activities in coastal and marine areas in spatial and temporal terms. Its primary objective is to reduce conflicts among sectors such as fisheries, transport, tourism, and energy, while ensuring the protection of ecosystems. Although MSP in Türkiye remains largely limited to pilot initiatives, compared with practices in European Union countries, it is increasingly recognized as a domain expected to be integrated into national strategies.

Sustainability Transition: Sustainability transition refers to the long-term transformation of existing economic and technological systems into forms compatible with environmental limits. This process involves shifting energy production toward renewable sources, reducing carbon emissions, and mainstreaming ecological sensitivity across all sectors. Within the blue economy, such a transition entails not only the adaptation of sectors to technological innovations but also the restructuring of institutional capacities.

Traceability in Fisheries: Traceability in fisheries is the systematic recording and monitoring of species, quantities, time and location of capture, and the movements of products along the supply chain. This approach enhances consumer safety while supporting the fight against illegal, unreported, and unregulated (IUU) fishing and overexploitation.

Wicked Problems: The term "wicked problems" designates issues that are multidimensional, lack definitive solutions, and embody value conflicts among different stakeholders. In fisheries, the simultaneous need to ensure economic profitability, conserve fish stocks, and secure the livelihoods of small-scale fishers constitutes a paradigmatic example. Addressing such challenges requires participatory and adaptive governance models rather than simplistic technical solutions.

Chapter 6

Exploring the Impact of Artificial Intelligence on Blue Economy Strategies: A Primary Data Analysis of Stakeholder Perspectives and Sustainable Practice

Harmanpreet Kaur
https://orcid.org/0000-0003-3974-363X
University School of Business, Mohali, India

Dilpreet Kaur
https://orcid.org/0000-0003-0517-1688
Sri Balaji University, Pune, India

Pawan Kumar
https://orcid.org/0000-0003-4892-6374
Graphic Era University, Dehradun, India

Ercan Özen
Usak University, Turkey

ABSTRACT

This research investigates the transformative impact of Artificial Intelligence (AI) on the blue economy, focusing on how AI adoption influences sustainability outcomes. The purpose of the study is to assess the relationship between AI, government policy support, and technological infrastructure and their collective effect on promoting sustainable practices within the blue economy. Utilizing a quantitative research

DOI: 10.4018/979-8-3373-5238-1.ch006

design, a sample of 289 respondents from various marine sectors was surveyed, with data analyzed using regression analysis, ANOVA, and correlation techniques. The findings reveal significant positive relationships between AI adoption and sustainability outcomes, highlighting the importance of supportive policy frameworks and robust technological infrastructure in driving eco-efficient practices. The implications underscore the need for collaborative efforts between stakeholders to optimize AI technologies for resource management and environmental conservation.

1. INTRODUCTION:

The advent of Artificial Intelligence (AI) has triggered transformative shifts across multiple sectors, including those traditionally considered immune to rapid technological changes. Among the myriad industries undergoing significant disruption, the blue economy stands at the forefront of this revolution, harnessing AI to drive sustainability, optimize resource management, and ensure the long-term viability of marine ecosystems. The blue economy, encompassing the sustainable use of ocean resources for economic growth, livelihood development, and environmental protection, is increasingly integrating AI tools and technologies into its core strategies. This integration is not merely a trend but a critical adaptation, ensuring that the blue economy meets the growing demand for sustainable practices in the face of escalating environmental challenges (Sanusi et al., 2024). The evolution of AI provides the bedrock upon which these advancements are built. Over the past few decades, AI has transitioned from a conceptual notion in the realms of theoretical computer science to a powerful, tangible tool with applications spanning from automation to data analysis and predictive modeling. AI has become synonymous with innovation, efficiency, and precision. Its evolution has been marked by breakthroughs in machine learning, deep learning, and natural language processing—technologies that have empowered industries to unlock previously unattainable insights. As AI continues to mature, it increasingly supports complex decision-making processes, enhancing operational effectiveness in various sectors, including those integral to the blue economy (P. Kaur et al., 2022).

1.1 The Evolution of Artificial Intelligence

Artificial Intelligence, once considered a niche field in computer science, has expanded its reach into nearly every domain of human activity, driving innovations in healthcare, finance, manufacturing, and many other fields (Ortiz et al., 2025). The initial concept of AI, introduced in the mid-20th century, aimed to replicate human cognitive functions, including learning, problem-solving, and decision-making.

Early AI systems were primarily rule-based, operating under fixed instructions and limited adaptability. However, the field experienced exponential growth in the late 20th and early 21st centuries, thanks to the advent of more sophisticated algorithms, the availability of vast datasets, and advancements in computing power. In the 21st century, the development of machine learning and deep learning has propelled AI into the spotlight. Machine learning algorithms, which allow computers to learn from data and improve their performance over time, have become the backbone of many AI applications. This shift has been further accelerated by the rise of big data—the vast quantities of structured and unstructured data generated by modern technologies such as the Internet of Things (IoT) and sensor networks. AI's ability to process and derive insights from these datasets has led to its widespread adoption across industries. Deep learning, a subset of machine learning, has also revolutionized the capabilities of AI (Crosby et al., 2023). By mimicking the neural networks of the human brain, deep learning enables AI systems to perform complex tasks such as image and speech recognition with unprecedented accuracy. These advancements have had profound implications for various industries, ranging from autonomous driving to healthcare diagnostics. In the context of the blue economy, AI's role has expanded to include optimizing maritime logistics, monitoring ocean health, and managing marine resources, providing stakeholders with real-time insights and predictive tools essential for sustainable operations (Gesami & Nunoo, 2024).

1.2. Blue Economy

The blue economy refers to the sustainable use and management of ocean resources for economic growth, job creation, and improved livelihoods while also ensuring the preservation and protection of the marine environment. This concept extends beyond traditional industries, such as fishing and shipping, to encompass a diverse range of sectors, including aquaculture, marine biotechnology, renewable ocean energy, and ecotourism. The blue economy is intrinsically linked to the health of the oceans, aiming to strike a balance between economic development and environmental stewardship (Sasilatha et al., 2024). As the global population grows and the demand for natural resources increases, the blue economy has become an increasingly crucial component of the broader sustainable development agenda. The United Nations' Sustainable Development Goal 14, which emphasizes the conservation and sustainable use of oceans, seas, and marine resources, underscores the need for innovative solutions to address the dual challenges of economic growth and environmental protection. Achieving these goals requires the application of cutting-edge technologies, which is where AI plays a pivotal role. The ocean is a vital resource for humanity, providing food, energy, and raw materials while also playing a critical role in regulating the global climate. However, the rising pressures from overfishing, climate change, pollution, and habitat degradation threaten the viability of the blue economy. These challenges necessitate a paradigm shift towards more sustainable practices, ensuring that marine resources are not exploited at the expense of long-term environmental health. Here, AI can offer innovative solutions by enabling more efficient resource management, enhancing environmental monitoring, and supporting decision-making processes that balance economic activity with ecological sustainability (Dow et al., 2019).

1.3. AI and the Blue Economy: Synergies and Innovations

The convergence of AI with the blue economy presents a dynamic synergy where AI technologies can drive significant improvements in how marine resources are used and conserved. One of the most transformative applications of AI in the blue economy is in ocean health monitoring. Using AI-powered predictive models, real-time sensor data, and satellite imagery, stakeholders can track changes in ocean conditions, such as temperature shifts, pollution levels, and biodiversity loss. AI's ability to analyze large datasets with incredible speed and accuracy enables early detection of harmful environmental trends, facilitating timely interventions and mitigating the impact of adverse events on marine ecosystems (Ha, 2025).

In the field of fisheries management, AI algorithms are being increasingly employed to predict fish populations and their migratory patterns. By leveraging data from satellite observations, sonar systems, and autonomous underwater vehicles, AI systems can predict the optimal locations for fishing, reducing the risk of overfishing and ensuring the sustainable management of marine life (Chen et al., 2024; Foster & Rhoden, 2020; Voronkova et al., 2024). Additionally, AI technologies can optimize aquaculture operations by monitoring water quality, managing feed schedules, and improving efficiency in farm management. These innovations contribute to the long-term viability of the blue economy, ensuring that the use of marine resources is sustainable and does not compromise future generations' ability to benefit from them. Moreover, AI plays a crucial role in the development of renewable ocean energy, including tidal and wave power. By enhancing the efficiency of energy capture, transmission, and storage, AI can make these energy sources more competitive with traditional forms of power generation. In this way, AI facilitates the transition toward a cleaner, more sustainable energy system, contributing to both climate change mitigation and the economic growth of coastal regions that rely on marine-based industries.

1.4. Sustainable Practices and the Adoption of AI in the Blue Economy

The integration of AI into the blue economy offers unprecedented opportunities to foster sustainable practices. Sustainability in the context of the blue economy refers to managing marine resources in a way that meets present needs without compromising the ability of future generations to meet their own. Achieving sustainability requires not only optimizing resource use but also minimizing environmental degradation. AI enables these objectives through data-driven decision-making and predictive analytics. The adoption of AI in the blue economy can lead to more efficient resource utilization, thereby reducing waste and ensuring that marine resources are used judiciously (Spalding, 2024). For example, in marine agriculture, AI-powered systems can forecast the optimal times to plant and harvest, thereby optimizing yields while minimizing the environmental impact of excessive input use, such as fertilizers and pesticides. In shipping, AI can optimize routing and fuel consumption, thereby reducing emissions and lowering the carbon footprint of global trade. The ability of AI to enhance transparency and traceability in the blue economy is another significant driver of sustainable practices. AI tools, combined with blockchain technology, enable the monitoring of supply chains, ensuring that marine resources are sourced sustainably and are in compliance with environmental standards. This can help mitigate the risk of illegal, unreported, and unregulated (IUU) fishing, which continues to pose a significant challenge to marine conservation. Furthermore, AI can facilitate the development of circular economy models within the blue economy. By using AI to track waste products and identify opportunities for reuse or recycling, businesses can reduce their reliance on virgin materials, minimizing waste generation and promoting environmental sustainability. As the blue economy continues to evolve, the integration of AI offers transformative potential for achieving sustainable development goals (Emad et al., 2023). AI's capacity for data analysis, predictive modeling, and real-time decision-making can revolutionize industries within the blue economy, ensuring that marine resources are managed sustainably and efficiently. As technological advancements continue to unfold, the collaborative efforts between stakeholders in the blue economy, including policymakers, industry leaders, and environmental advocates, will be crucial in shaping a future where economic growth and environmental preservation coexist harmoniously. The adoption of AI in the blue economy is not just a technological advancement; it represents a paradigm shift toward more responsible and sustainable practices, ensuring the continued prosperity of ocean-based industries for generations to come (Er-Rousse & Qafas, 2024).

2. LITERATURE REVIEW

The confluence of Artificial Intelligence (AI) and the blue economy has garnered increasing attention in recent years, as both sectors play pivotal roles in the contemporary pursuit of sustainability. AI, with its capacity to analyze vast datasets, automate complex processes, and optimize decision-making, holds transformative potential for industries that rely on marine resources (Gambin et al., 2021). Concurrently, the blue economy, which refers to the sustainable use of ocean resources for economic growth, livelihood development, and environmental protection, has emerged as a critical area of focus for achieving global sustainability goals. The literature reveals a growing body of research exploring how AI can enhance the sustainability of the blue economy, providing insights into the dynamic interactions between AI technologies, marine ecosystems, and resource management (Bhatnagar & Rajaram, 2024; Kukreti et al., 2024b).

2.1. The Evolution of Artificial Intelligence

AI has evolved from a theoretical concept in computer science into a robust and ubiquitous technology that is reshaping industries worldwide. Initially, AI was limited to rule-based systems designed for specific tasks, such as playing chess or sorting data. However, with the advent of machine learning (ML) and deep learning (DL), AI systems have become capable of making predictions, detecting patterns, and performing complex analyses previously beyond human capacity. These advancements are underpinned by the increasing availability of big data, powerful computing resources, and sophisticated algorithms, all of which have dramatically expanded AI's capabilities (Kukreti et al., 2024c, 2024a; Sehajpal & Sidhu, 2024). In particular, AI's ability to process unstructured data—such as satellite imagery, sensor data, and environmental readings—has positioned it as a key tool for improving operational efficiency, optimizing resource management, and addressing global challenges like climate change and environmental degradation. The literature on AI in various industries has primarily focused on its applications in automation, predictive analytics, and operational optimization. In sectors like healthcare, automotive, and finance, AI has revolutionized diagnostic capabilities, autonomous vehicles, and risk assessments, respectively. The advancement of AI in these fields has demonstrated its potential for enhancing efficiency and accuracy while also creating new business models that capitalize on data-driven insights. This burgeoning field, however, is not confined to land-based industries but has found substantial application in maritime and oceanic domains, which form the core of the blue economy (Reshma et al., 2024; Reshma & Singh Attwal, 2024; Sharma et al., 2022).

2.2. The Blue Economy and Its Expanding Scope

The blue economy concept emphasizes the sustainable exploitation of marine resources for economic growth, poverty alleviation, and environmental protection. It is an umbrella term encompassing a wide array of activities that rely on the ocean, including fisheries, aquaculture, maritime transport, marine biotechnology, ecotourism, and the exploration of renewable energy resources from the sea, such as tidal and wave power. The blue economy is inherently linked to the health of marine ecosystems, which provide invaluable services, including climate regulation, food security, and biodiversity preservation. The traditional exploitation of marine resources has often been characterized by overfishing, habitat destruction, and pollution. As a result, the blue economy faces a unique set of challenges in striking a balance between economic growth and environmental sustainability (Singh et al., 2024; Tanwar, Aggarwal, et al., 2024; Tanwar, Malhan, et al., 2024). Scholars have long argued that a new paradigm is necessary—one that fosters environmentally sustainable development while enabling economic activities to flourish. This shift has been further catalyzed by increasing global demand for marine resources and the escalating threat of climate change. As climate change continues to impact ocean ecosystems through rising sea levels, ocean acidification, and disrupted marine biodiversity, the urgency of adopting sustainable practices within the blue economy has never been more apparent. Despite its promise, the blue economy's growth is constrained by these environmental pressures, which demand more innovative and efficient management approaches. The application of AI within the blue economy is increasingly viewed as a viable means of addressing these challenges, providing tools for improved decision-making, real-time monitoring, and resource optimization (Gupta et al., 2025; A. Kaur, Bhatnagar, et al., 2024; A. Kaur & Sharma, 2024; Yadav et al., 2024).

2.3. AI and the Blue Economy: Convergence for Sustainability

The integration of AI into the blue economy has gained traction as a key strategy for sustainable development. By harnessing AI technologies, industries in the blue economy can better monitor environmental conditions, predict ocean trends, and optimize operations to minimize resource depletion and waste. A significant body of literature now focuses on the synergies between AI and the blue economy, offering insight into the potential impacts of AI adoption on marine conservation, fisheries management, renewable ocean energy, and shipping industries.

In the fisheries sector, AI's capabilities are being harnessed to monitor fish stocks, predict migration patterns, and detect overfishing. Satellite imagery, combined with AI-driven machine learning algorithms, enables the real-time tracking of fishing activities, ensuring that regulations are enforced and providing invaluable data to inform sustainable management practices. Similarly, AI-driven predictive analytics can forecast fish populations, providing a proactive approach to resource management rather than relying solely on reactive methods. This transition from reactive to predictive management marks a significant advancement in ensuring the sustainability of marine ecosystems (A. Kaur, Kaur, et al., 2024). Moreover, aquaculture, a vital component of the blue economy, is undergoing a digital transformation driven by AI. AI technologies are optimizing operations by monitoring water quality, predicting growth cycles, and managing feed inputs with greater precision. These advancements have profound implications for resource efficiency, as they allow for the reduction of water and feed wastage, thereby improving the overall environmental footprint of aquaculture. The application of AI is also gaining momentum in the field of marine renewable energy, where it is used to optimize the operation of tidal and wave energy farms. By integrating AI into the monitoring and control systems of ocean-based renewable energy installations, operators can predict energy output, reduce maintenance costs, and enhance operational efficiency. In addition, AI can be utilized to assess the environmental impact of these installations on marine life, ensuring that renewable energy development aligns with environmental protection goals. In the shipping industry, AI technologies are revolutionizing logistical operations, from optimizing shipping routes to reducing fuel consumption and improving efficiency. Machine learning algorithms can predict weather patterns and traffic conditions, allowing ships to chart the most efficient course. This not only enhances operational efficiency but also significantly reduces emissions, contributing to the broader goal of reducing the environmental impact of maritime trade (Kumar et al., 2025).

2.4. Sustainable Practices with the Adoption of AI in the Blue Economy

The adoption of AI in the blue economy has profound implications for sustainable practices. As the need for sustainable resource management intensifies, AI offers tools for precision management that reduce waste, optimize resource use, and enhance the resilience of marine ecosystems. These AI-enabled strategies are particularly critical in mitigating the effects of climate change, overexploitation, and pollution on marine environments.

One of the key advantages of AI is its capacity to integrate diverse data sources into actionable insights. For example, the combination of remote sensing technologies, such as satellites, and AI-based analytics has led to the development of comprehensive ocean health monitoring systems. These systems can track temperature fluctuations, changes in sea level, and the movement of pollutants, providing stakeholders with valuable real-time information for informed decision-making.

With this data, industries can adapt their strategies, adjust operational processes, and engage in proactive conservation efforts. In the context of marine conservation, AI facilitates more effective management of protected areas and critical habitats. By utilizing AI-powered systems to analyze trends in biodiversity, researchers can more accurately predict threats to endangered species and marine habitats, enabling more effective intervention. Furthermore, AI can support ecotourism by providing data on sustainable tourism practices, thereby ensuring that human activities do not unduly harm marine ecosystems. AI also plays a vital role in the circular economy within the blue economy. By monitoring waste streams and optimizing material flows, AI can help industries reduce their reliance on raw materials, minimize waste production, and encourage the recycling and repurposing of marine by-products. This reduces the overall environmental footprint of marine industries and ensures that resource extraction does not outpace the ecosystem's capacity for regeneration. As AI technologies continue to evolve, their application within the blue economy will undoubtedly expand, further enabling the realization of sustainable practices that align economic growth with environmental stewardship. From resource management to conservation, AI is emerging as a powerful enabler of sustainable development in the blue economy, providing the tools necessary to ensure the long-term viability of marine ecosystems and the industries that depend on them. The intersection of AI and the blue economy holds immense potential for driving sustainability in marine industries. As AI technologies evolve and are increasingly integrated into the blue economy, they offer unprecedented opportunities to optimize resource management, mitigate environmental degradation, and promote sustainable practices across various sectors, including fisheries, aquaculture, marine renewable energy, and shipping. The growing body of literature highlights the pivotal role that AI plays in fostering the future of the blue economy, facilitating the transition towards more sustainable and efficient practices, and ensuring the long-term health of marine ecosystems. As the adoption of AI continues to expand, it will undoubtedly shape the trajectory of the blue economy, contributing to the realization of global sustainability goals and the sustainable use of ocean resources.

3. RESEARCH METHODOLOGY

This research employs a quantitative approach to investigate the impact of Artificial Intelligence (AI) on the blue economy, with a specific focus on how the adoption of AI affects sustainability outcomes. The primary aim is to assess the relationships between AI adoption, government policy support, and technological infrastructure and their collective impact on sustainable practices within the blue economy. The study employs a descriptive correlational research design to measure and analyze the interconnections among these variables systematically. The research is based on primary data collected through structured surveys distributed to 289 respondents, including stakeholders across multiple sectors of the blue economy. These sectors include marine resource management, fisheries, aquaculture, renewable energy, and marine biotechnology.

The research employs advanced statistical techniques, including regression analysis, ANOVA, and correlation analysis, to examine the relationships between the independent variables (AI adoption, policy support, and technological infrastructure) and the dependent variable (sustainability outcomes). This methodological approach enables a robust analysis of how technological and policy factors collectively impact the sustainability of the blue economy. The results reveal significant relationships between the predictors and sustainability outcomes, demonstrating that AI adoption, enhanced by supportive policy frameworks and robust technological infrastructure, contributes substantially to the improvement of sustainability practices across marine industries. Overall, the methodology integrates data-driven insights with empirical analysis, providing a comprehensive understanding of the potential synergies between AI and the blue economy while offering actionable recommendations for policymakers and industry leaders.

3.1 Purpose of the Research

The purpose of this research is to investigate the transformative role of Artificial Intelligence in shaping sustainable practices within the blue economy. By examining the relationship between AI adoption, governmental support, and technological infrastructure, the study aims to determine how these factors contribute to improved resource management and environmental conservation. Furthermore, the research aims to provide empirical insights into the strategic integration of AI technologies across marine-based industries, ensuring the sustainability of oceanic resources and promoting a balanced approach to economic growth and environmental protection.

4. DATA ANALYSIS

Independent Variables:

1. Artificial Intelligence Adoption
2. Government and Policy Support
3. Technological Infrastructure

Dependent Variable:

1. Sustainability Outcomes

Figure 1.

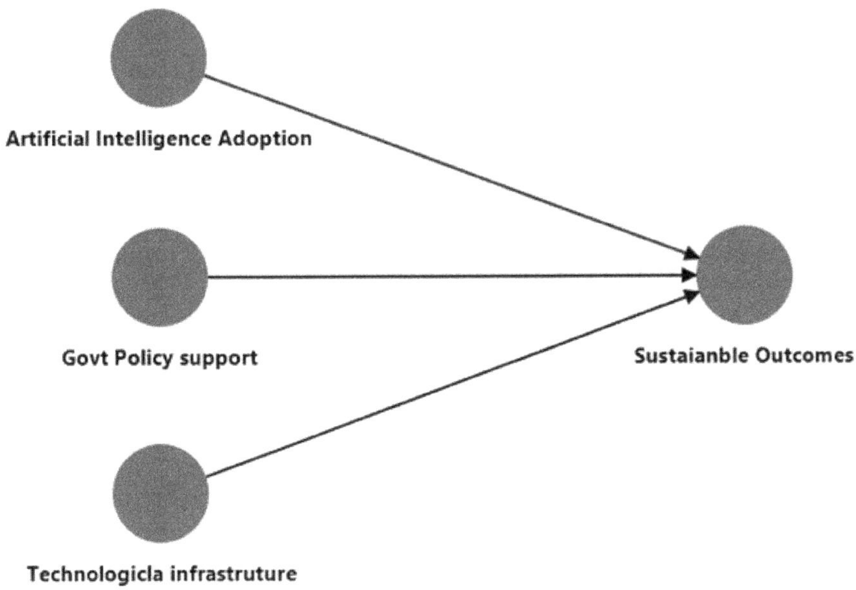

Table 1. Descriptive Statistics

	N	Minimum	Maximum	Mean	Std. Deviation	Skewness		Kurtosis	
	Statistic	Statistic	Statistic	Statistic	Statistic	Statistic	Std. Error	Statistic	Std. Error
AI_Adoption	280	1.00	7.00	4.6893	1.36254	-.485	.146	.040	.290
Policy_support	280	1.00	7.00	4.5179	1.53094	-.638	.146	.046	.290
Technological_ infrastruture	280	1.00	7.00	4.6071	1.48201	-.400	.146	-.167	.290
Sustainability_outcomes	280	1.00	7.00	4.7893	1.32059	-.509	.146	.203	.290
Valid N (listwise)	280								

Table 1 presents the descriptive statistics for the four variables: AI Adoption, Policy Support, Technological Infrastructure, and Sustainability Outcomes. The data reveal that the variables exhibit relatively high mean scores (ranging from 4.51 to

4.79), indicating moderate to high levels of adoption, support, and sustainability outcomes within the sample. The standard deviations (ranging from 1.32 to 1.53) indicate moderate variability in responses, suggesting that while most stakeholders rate these variables highly, there is still considerable diversity in their perspectives. The skewness values are all negative, indicating that the data is slightly skewed to the left, suggesting a tendency towards higher ratings. Additionally, the kurtosis values are generally low, implying a relatively flat distribution for these variables. This suggests that while most respondents are optimistic about AI adoption, policy support, and technological Infrastructure, a few respondents rate them lower.

Table 2. Model Summary

Model	R	R Square	Adjusted R Square	Std. Error of the Estimate	Change Statistics					Durbin-Watson
					R Square Change	F Change	df1	df2	Sig. F Change	
1	.876[a]	.767	.761	.83715	.602	139.428	3	276	.000	1.371

a. Predictors: (Constant), Technological_infrastruture, AI_Adoption, Policy_support
b. Dependent Variable: Sustainability_outcomes

Table 2 reports the model summary, which indicates that the regression model explains a substantial portion of the variance in Sustainability Outcomes (R^2 = 0.767). This suggests that AI Adoption, Policy Support, and Technological Infrastructure collectively account for approximately 77% of the variability in sustainability outcomes, which is a strong indication of the relevance of these predictors in determining the success of sustainability initiatives. The Durbin-Watson statistic of 1.37 indicates no significant autocorrelation in the residuals, suggesting that the model's assumptions are met. The F-change statistic (139.428) with a significance level of 0.000 confirms that the overall regression model is highly significant, indicating that the independent variables have a robust effect on the dependent variable, Sustainability Outcomes.

Table 3. ANOVAa

Model		Sum of Squares	df	Mean Square	F	Sig.
1	Regression	293.142	3	97.714	139.428	.000[b]
	Residual	193.426	276	.701		
	Total	486.568	279			

a. Dependent Variable: Sustainability_outcomes
b. Predictors: (Constant), Technological_infrastruture, AI_Adoption, Policy_support

Table 3 displays the results of the ANOVA test, which assesses the overall significance of the regression model. The F-statistic of 139.428, with a corresponding p-value of 0.000, strongly indicates that the regression model is statistically significant, meaning that the independent variables—AI Adoption, Policy Support, and Technological Infrastructure—significantly contribute to explaining variations in Sustainability Outcomes. The Sum of Squares for the regression (293.142) and residuals (193.426) reflect the proportion of variation explained by the model versus the unexplained variation, further validating the model's predictive power.

Table 4.Coefficientsa

Model		Unstandardized Coefficients		Standardized Coefficients	t	Sig.
		B	Std. Error	Beta		
1	(Constant)	1.136	.195		5.836	.000
	AI_Adoption	.206	.050	.213	4.090	.000
	Policy_support	.392	.047	.454	8.374	.000
	Technological_ infrastruture	.199	.044	.223	4.524	.000

a. Dependent Variable: Sustainability_outcomes

Table 4 presents the coefficients for each predictor variable in the regression model. The unstandardized coefficients indicate the expected change in Sustainability Outcomes for each one-unit change in the independent variables. Policy Support (B = 0.392) has the most significant positive effect on sustainability outcomes, followed by AI Adoption (B = 0.206) and Technological Infrastructure (B = 0.199). All coefficients are statistically significant, with p-values less than 0.001, indicating that each predictor makes a substantial contribution to explaining sustainability outcomes. The standardized coefficients (B) show that Policy Support has the most substantial standardized effect (B = 0.454), followed by AI Adoption (B = 0.213) and Technological Infrastructure (B = 0.223), emphasizing the relative importance of policy support in driving sustainability outcomes.

Table 5.Coefficient Correlationsa

Model			Technological_ infrastruture	AI_ Adoption	Policy_ support
1	Correlations	Technological_ infrastruture	1.000	-.264	-.384
		AI_Adoption	-.264	1.000	-.480
		Policy_support	-.384	-.480	1.000
	Covariances	Technological_ infrastruture	.002	-.001	-.001
		AI_Adoption	-.001	.003	-.001
		Policy_support	-.001	-.001	.002

a. Dependent Variable: Sustainability_outcomes

Table 5 displays the correlations among the predictor variables. AI Adoption and Policy Support have a moderate negative correlation (r = -0.480), suggesting that as one increases, the other tends to decrease. This may indicate that stakeholders perceiving higher policy support might see less need for aggressive AI adoption or vice versa. The technological Infrastructure exhibits weaker negative correlations with both AI Adoption and Policy Support, suggesting that these other factors may not influence Infrastructure as strongly. The correlations, although statistically significant, are not so strong as to suggest multicollinearity, meaning that each predictor variable contributes uniquely to the regression model.

Table 6.Casewise Diagnosticsa

Case Number	Std. Residual	Sustainability_outcomes	Predicted Value	Residual
16	-3.953	1.00	4.3095	-3.30946
34	3.669	6.00	2.9288	3.07116
48	3.426	7.00	4.1317	2.86827
50	3.204	7.00	4.3177	2.68228
159	4.626	7.00	3.1277	3.87225

a. Dependent Variable: Sustainability_outcomes

Table 6 reports the casewise diagnostics, highlighting extreme residuals or outliers in the data. For instance, case number 16 has a standard residual of -3.953, indicating a significantly low predicted Sustainability Outcome relative to its actual value. Similarly, case number 34 shows a standard residual of 3.669, indicating an unusually high predicted sustainability outcome. These outliers can affect model accuracy, and their presence warrants further investigation, as they may represent unique or atypical responses that diverge from the general trends observed in the data.

Table 7.Residuals Statisticsa

	Minimum	Maximum	Mean	Std. Deviation	N
Predicted Value	1.9331	6.7140	4.7893	1.02503	280
Residual	-3.30946	3.87225	.00000	.83264	280
Std. Predicted Value	-2.786	1.878	.000	1.000	280
Std. Residual	-3.953	4.626	.000	.995	280

a. Dependent Variable: Sustainability_outcomes

Table 7 presents the residual statistics, which measure the difference between the observed and predicted values of Sustainability Outcomes. The mean residual is 0, indicating that the model is unbiased in its predictions. The standard deviation of the residuals (0.83264) suggests that the model has a reasonable level of accuracy, with the residuals mostly concentrated around zero. The predicted values range from 1.93 to 6.71, and the standardized residuals range from -3.953 to 4.626, indicating that the model fits the data relatively well, with a few extreme residuals that may require further investigation for potential outliers or influential data points.

5. RESULTS AND FINDINGS

The study's results reveal significant insights into the interplay between the adoption of Artificial Intelligence (AI), government policy support, technological infrastructure, and sustainability outcomes within the context of the blue economy. The analysis, utilizing regression, ANOVA, and correlation techniques, highlights the profound impact of these independent variables on the dependent variable's sustainability outcomes. From the descriptive statistics, it is evident that respondents exhibit moderate to high levels of engagement with the variables under study. The mean scores for AI adoption (M = 4.6893), policy support (M = 4.5179), and technological infrastructure (M = 4.6071) indicate that, while there is substantial adoption of AI technologies and policy frameworks, significant variance remains in perceptions and experiences among stakeholders. This variability is further highlighted by the standard deviations (ranging from 1.32 to 1.53), indicating a moderate spread in the responses across the sample. Notably, the negative skewness values across all variables reflect a slight tendency toward higher ratings, suggesting that most stakeholders perceive a favorable impact of these factors on sustainability initiatives within the blue economy. The model summary reveals that the independent variables collectively explain a substantial 76.7% of the variance in sustainability outcomes (R^2 = 0.767), affirming the robustness of the predictors. The F-statistic of 139.428, with a significance value of 0.000, indicates that the regression model is highly significant, reinforcing the pivotal role of AI, policy support, and technological infrastructure in shaping sustainable practices. The Durbin-Watson statistic of 1.37 further suggests that there is no significant autocorrelation in the residuals, indicating that the model's assumptions are met mainly. In terms of coefficients, policy support ($\beta = 0.454$) emerges as the most influential predictor, followed by AI adoption ($\beta = 0.213$) and technological infrastructure ($\beta = 0.223$). This hierarchy underscores the importance of strong policy frameworks in driving sustainability initiatives while also highlighting the complementary roles that AI and infrastructure play in facilitating sustainable blue economy strategies. These findings suggest that policymakers and stakeholders should prioritize enhancing policy support and fostering the adoption of AI to optimize sustainability outcomes. The correlations between the predictors reveal moderate negative relationships, particularly between AI adoption and policy support ($r = -0.480$), suggesting that an overreliance on one factor may potentially mitigate the effectiveness of the other. The moderate correlation between technological infrastructure and the other variables suggests that infrastructure development is essential but not entirely dependent on AI or policy measures alone. In summary, the adoption of AI, combined with robust policy support and advanced technological infrastructure, plays a crucial role in driving sustainable outcomes in the blue economy.

6. PRACTICAL, SOCIAL AND MANAGERIAL IMPLICATIONS

The findings of this study carry significant practical, social, and managerial implications for stakeholders involved in the blue economy, especially in sectors related to marine conservation, renewable energy, and sustainable fisheries. From a managerial perspective, the most crucial implication is the need for businesses and policymakers to recognize the integrated role of AI adoption, policy frameworks, and technological infrastructure in achieving sustainability goals. AI technologies offer the potential to revolutionize sustainability practices by optimizing resource management, enhancing predictive capabilities, and facilitating data-driven decision-making. However, the policy environment plays a crucial role in enabling or hindering AI adoption. Managers should advocate for comprehensive policy frameworks that incentivize technological innovation while ensuring equitable access to these advancements. The technological infrastructure required to support AI adoption in the blue economy must be robust, secure, and scalable. Managers in industries like renewable energy and marine resource management must invest in cutting-edge infrastructure, including data centers, cloud computing, and AI-enabled systems, to capitalize on the benefits of AI. Furthermore, businesses and governments must address the digital divide, ensuring that smaller entities within the blue economy also have access to the tools and technologies necessary to drive sustainable practices. Socially, the findings suggest that increasing policy support and fostering AI adoption can have far-reaching benefits for communities, especially in coastal and marine-adjacent areas. By leveraging AI to improve sustainability outcomes, businesses and governments can contribute to the creation of sustainable jobs in emerging fields such as marine data science, renewable energy technologies, and conservation analytics. Additionally, AI's potential to enhance environmental monitoring and resource management can help mitigate the adverse effects of climate change on vulnerable populations, fostering a more inclusive and resilient society. From a strategic standpoint, managers in the blue economy should prioritize the development of cross-sector collaborations to integrate AI effectively. This could include partnerships between private enterprises, government agencies, and academic institutions to jointly develop AI-driven solutions for climate change mitigation and sustainable resource use. Furthermore, managerial emphasis on green finance and investment in AI solutions could help create long-term value, not only for businesses but also for the broader societal goal of environmental sustainability.

7. DISCUSSION

This study highlights the transformative intersection of Artificial Intelligence (AI), governmental policy support, and technological infrastructure in promoting sustainability within the blue economy. The findings reveal that the integration of AI technologies into the marine sector holds immense promise, not as an isolated innovation but as a strategic instrument that must be embedded within an enabling ecosystem of policy and infrastructure. AI facilitates the optimization of marine resource management, enhances environmental monitoring, and enables real-time, data-driven decision-making all of which are pivotal for ensuring ecological resilience and sustainable economic growth. However, the influence of AI does not function in a vacuum. Policy frameworks emerge as the cornerstone that legitimizes, incentivizes, and guides the ethical and effective deployment of AI systems. Without institutional coherence and regulatory clarity, the adoption of AI risks being fragmented and ineffectual. Moreover, technological infrastructure serves as the foundational backbone upon which AI solutions are operationalized. Robust digital networks, high-performance computing capabilities, and accessible platforms are essential prerequisites for the successful integration of intelligent systems into the marine industry. The interplay among these factors underscores a broader systemic dynamic. Practical sustainability in the blue economy necessitates not only advanced technologies but also strategic alignment between innovation and governance. However, the study reveals latent tensions—where overdependence on technology may be counterproductive if not matched by sufficient policy guidance or, conversely, where regulatory overreach might stifle innovation. This delicate balance highlights the importance of holistic frameworks that synchronize technological potential with institutional intent. Furthermore, while the theoretical benefits of AI are widely acknowledged, practical implementation varies widely across sectors and regions. This disparity reflects uneven levels of preparedness, resource allocation, and stakeholder engagement. Thus, capacity building, education, and inclusive policy dialogues become essential to democratize the benefits of AI across diverse marine economies. Ultimately, the findings affirm that AI is a powerful enabler rather than a singular solution. Its impact on sustainability is magnified when embedded within a comprehensive strategy that includes supportive policies and resilient infrastructure. In the face of mounting environmental challenges and resource pressures, the blue economy must evolve through a concerted effort that merges technological ingenuity with responsible governance. This integrated approach not only enhances operational efficacy but also ensures that marine ecosystems are preserved for future generations, affirming sustainability as both a developmental imperative and a moral obligation.

8. CONCLUSIONS

In conclusion, this study highlights the transformative potential of Artificial Intelligence in advancing the blue economy and promoting sustainable practices across key sectors, including marine conservation, renewable energy, and sustainable fisheries. The findings suggest that AI adoption, coupled with strong policy support and a robust technological infrastructure, significantly contributes to achieving sustainability outcomes in the blue economy. By leveraging AI, stakeholders can optimize resource management, enhance operational efficiency, and promote environmentally responsible practices that align with global sustainability objectives. The results also highlight the interdependent relationship between policy support, technological infrastructure, and AI adoption, underscoring the necessity of a multi-faceted approach for successful implementation. Managers in blue economy sectors should focus on creating an ecosystem that integrates cutting-edge technologies with supportive policy frameworks to foster sustainability. Moreover, technological infrastructure must be strengthened to ensure that AI solutions are accessible and scalable across diverse actors within the blue economy, particularly in developing regions where access to such technologies may be limited. The study's policy implications suggest that governments must prioritize the development of policies that support the integration of AI while incentivizing research and development in green technologies. This would require a collaborative effort between policymakers, industries, and technological innovators to ensure that sustainability becomes a core principle driving blue economy initiatives. From a social perspective, the adoption of AI can lead to significant positive changes, particularly in coastal communities that heavily rely on marine resources. By enhancing sustainability outcomes, AI can help mitigate the impacts of climate change, safeguard marine ecosystems, and promote economic resilience in the face of environmental challenges.

Furthermore, AI can create new opportunities for green jobs, contributing to social equity and broader societal well-being. In summary, AI presents profound opportunities for the blue economy to evolve sustainably; however, its full potential can only be realized when accompanied by adequate policy frameworks, robust technological infrastructure, and collaborative efforts across various sectors. Future research should delve deeper into the specific mechanisms by which AI impacts economic performance and environmental outcomes, offering a more granular understanding of its transformative potential in the blue economy.

REFERENCES

Bhatnagar, M., & Rajaram, R. (2024). Correlating investor patterns and ESG performance with neuro marketing insights: Behavioural finance meets sustainability. In *Neuromarketing's Role in Sustainable Finance*. IGI Global., DOI: 10.4018/979-8-3693-9117-4.ch001

Chen, V. Y., Wu, Y.-W., Hu, C.-W., & Han, Y.-S. (2024). Enhancing green sea turtle (Chelonia mydas) conservation for tourists at Little Liuqiu island, Taiwan: Application of deep learning algorithms. *Ocean and Coastal Management*, *252*, 107111. Advance online publication. DOI: 10.1016/j.ocecoaman.2024.107111

Crosby, A., Orenstein, E. C., Poulton, S. E., Bell, K. L. C., Woodward, B., Ruhl, H., Katija, K., & Forbes, A. G. (2023). Designing Ocean Vision AI: An Investigation of Community Needs for Imaging-based Ocean Conservation. *Conference on Human Factors in Computing Systems - Proceedings*. DOI: 10.1145/3544548.3580886

Dow, R. M., Ferretti, S., & Vernile, A. (2019). The Atlantic ocean and space applications. *Proceedings of the International Astronautical Congress, IAC, 2019-October*.

Emad, G. R., Shahbakhsh, M., Abdussamie, N., & Rabiul Islam, T. M. (2023). Implementation of industry 4.0 and its effect on human element and sustainable development in the blue economy. *Proceedings of the International Conference on Offshore Mechanics and Arctic Engineering - OMAE, 10*. DOI: 10.1115/OMAE2023-101806

Er-Rousse, O., & Qafas, A. (2024). Artificial intelligence for the optimization of marine aquaculture. In S. K., G. O., & K. M.L. (Eds.), *E3S Web of Conferences* (Vol. 477). EDP Sciences. DOI: 10.1051/e3sconf/202447700102

Foster, M. N., & Rhoden, S. L. N. H. (2020). The integration of automation and artificial intelligence into the logistics sector: A Caribbean perspective. *Worldwide Hospitality and Tourism Themes*, *12*(1), 56–68. DOI: 10.1108/WHATT-10-2019-0070

Gambin, A. F., Angelats, E., Gonzalez, J. S., Miozzo, M., & Dini, P. (2021). Sustainable Marine Ecosystems: Deep Learning for Water Quality Assessment and Forecasting. *IEEE Access : Practical Innovations, Open Solutions*, *9*, 121344–121365. DOI: 10.1109/ACCESS.2021.3109216

Gesami, B. K., & Nunoo, J. (2024). Artificial intelligence in marine ecosystem management: Addressing climate threats to Kenya's blue economy. *Frontiers in Marine Science*, *11*, 1404104. Advance online publication. DOI: 10.3389/fmars.2024.1404104

Gupta, S. T., Joshi, P., Kaur, A., & Kumar, P. (2025). Digital tools and technologies: Enabling knowledge sharing and collaborative business culture. In *Knowledge Sharing and Fostering Collaborative Business Culture*. IGI Global., DOI: 10.4018/979-8-3373-0710-7.ch016

Ha, L. T. (2025). Wavelet connectedness between FinTech innovation, artificial intelligence development, and blue economy performance during the COVID-19 pandemic and the Ukraine–Russia tension. *Discover Sustainability*, *6*(1), 426. Advance online publication. DOI: 10.1007/s43621-025-01100-4

Kaur, A., Bhatnagar, D., & Singh, G. (2024). Exploring the machine learning-based decision tree technique in assessing credit risk likelihood of P2P Fintech lending for effective marketing. In *Neuromarketing's Role in Sustainable Finance*. IGI Global. DOI: 10.4018/979-8-3693-9117-4.ch019

Kaur, A., Kaur, H., & Tanwar, A. (2024). Are robo-advisors revolutionizing portfolio management for enhanced returns over human advisors? In *Risks and Challenges of AI-Driven Finance: Bias, Ethics, and Security*. IGI Global., DOI: 10.4018/979-8-3693-2185-0.ch008

Kaur, A., & Sharma, V. (2024). Nexus between climate change and mitigation approaches for sustainable development: A bibliometric review. *Global Knowledge, Memory and Communication*. Advance online publication. DOI: 10.1108/GKMC-11-2023-0418

Kaur, P., Aziz, A., Jain, D., Patel, H., Hirokawa, J., Townsend, L., Reimers, C., & Hua, F. (2022). Sea Situational Awareness (SeaSAw) Dataset. *IEEE Computer Society Conference on Computer Vision and Pattern Recognition Workshops, 2022-June*, 2578 – 2586. DOI: 10.1109/CVPRW56347.2022.00290

Kukreti, M., Sehajpal, S., Tiwari, R., & Sood, K. (2024a). Driving global health and sustainable development goals with smart technology. In *Driving Global Health and Sustainable Development Goals With Smart Technology*. IGI Global., DOI: 10.4018/979-8-3373-0240-9

Kukreti, M., Sehajpal, S., Tiwari, R., & Sood, K. (2024b). Introduction. In *Driving Global Health and Sustainable Development Goals With Smart Technology*. IGI Global. DOI: 10.4018/979-8-3373-0240-9

Kukreti, M., Sehajpal, S., Tiwari, R., & Sood, K. (2024c). Preface. In *Driving Global Health and Sustainable Development Goals With Smart Technology*. IGI Global. DOI: 10.4018/979-8-3373-0240-9

Kumar, A., Khan, R., & Deepika. (2025). Heart disease prediction using machine learning techniques. In *Sustainable Materials, Structures and IoT [SMSI-2024]*. CRC Press. DOI: 10.1201/9781003596776-30

Ortiz, J. A., Ortiz-Montes, A., & Valdés-Acosta, M. T. (2025). Applications of artificial intelligence for the sustainable management of fishing activities. In *Machine and Deep Learning Solutions for Achieving the Sustainable Development Goals*. IGI Global., DOI: 10.4018/979-8-3693-8161-8.ch013

Reshma, S., T., Pareek, S., & Kuhar, S. (2024). House Price Prediction using Machine Learning Approaches. *2024 OPJU International Technology Conference on Smart Computing for Innovation and Advancement in Industry 4.0, OTCON 2024*. DOI: 10.1109/OTCON60325.2024.10687534

Reshma, & Singh Attwal, K. P. (2024). A Review on Medicinal Plant identification through Leaf Image. In M. H. (Ed.), *Proceedings - 2024 6th International Conference on Computational Intelligence and Communication Technologies, CCICT 2024* (pp. 27 – 33). Institute of Electrical and Electronics Engineers Inc. DOI: 10.1109/CCICT62777.2024.00015

Sanusi, S. M., Paul, S. I., & Makarfi, A. M. (2024). Internet of fisheries things (IOFT) for blue economy & ecosystem. In *Data Science for Agricultural Innovation and Productivity*. Bentham Science Publishers., DOI: 10.2174/9789815196177124010010

Sasilatha, T., Karthickmanoj, R., Padmapriya, J., & Thiruvasagam, G. (2024). Unveiling the Potential of AI-Enhanced Resource Management for Sustainable Fisheries and Aquaculture in the Indian Ocean. *Proceedings of the International Association of Maritime Universities Conference*, 23 – 29.

Sehajpal, S., & Sidhu, N. K. (2024). Neuroscience, nursing, and financial innovation: A synergistic approach to optimized patient care and economic well-being. In *Neuromarketing's Role in Sustainable Finance*. IGI Global., DOI: 10.4018/979-8-3693-9117-4.ch008

Sharma, K. K., Mmdu, A. B., & Reshma. (2022). Biological Impacts of MRI Examinations of Different Ratings on Human Tissue. *Proceedings - 2022 International Conference on Computing, Communication and Power Technology, IC3P 2022*, 172 – 175. DOI: 10.1109/IC3P52835.2022.00043

Singh, V., & Kaushik, H. V., & Reshma. (2024). Social Media Sentiment Analysis Using Twitter Dataset. *2nd IEEE International Conference on Data Science and Information System, ICDSIS 2024*. DOI: 10.1109/ICDSIS61070.2024.10594648

Spalding, M. J. (2024). AI and Improving the Blue Economy. *Journal of Ocean Technology*, *19*(2), 27–35.

Tanwar, A., Aggarwal, R., & Kaur, A. (2024). Explainability of AI in Improving the Quality of Service in the Actuarial Industry. In *Artificial Intelligence and Actuarial Science: Applications and Case Studies from Finance and Insurance*. CRC Press., DOI: 10.1201/9781003472551-7

Tanwar, A., Malhan, S., Punj, N., & Kaur, A. (2024). Integration of Artificial Intelligence and Internet of Things in the Banking Sector: A Convergence of Frontiers. In *Integration of Cloud Computing and IoT: Trends, Case Studies and Applications*. CRC Press., DOI: 10.1201/9781032656694-25

Voronkova, V., Nikitenko, V., Sobol, Y., Oleksenko, R., Zelenin, Y., Kravchenko, O., Levchenko, L., & Yukhymenko, N. (2024). The Digital Transformation Of The Most Dynamic Region In The World (China, Malaysia, India) As A Key Factor In Economic Development Within The Context Of INDUSTRY 5G: Trends, Challenges, And Strategies. *Pakistan Journal of Life and Social Sciences*, *22*(1), 1522–1531. DOI: 10.57239/PJLSS-2024-22.1.00105

Yadav, B. K., Keyal, P. K., Khan, R., Chaube, N., Chaudhary, P., & Chaurasiya, P. (2024). Advanced Threat Detection: Leveraging Data Analytics for Cybersecurity Enhancement. *Proceedings of International Conference on Emerging Technologies and Innovation for Sustainability,* EmergIN 2024, 62 – 67. DOI: 10.1109/EmergIN63207.2024.10961019

Chapter 7
Integrating Artificial Intelligence in the Blue Economy:
A Case-Based Study of Smart Marine Ecosystems for Climate Change Mitigation and Circular Economy

K. Balaji
https://orcid.org/0000-0002-3065-3294
Christ University, Bengaluru, India

ABSTRACT

The Blue Economy represents a critical pathway toward sustainable development, balancing economic growth with the conservation of marine ecosystems. This chapter explores the transformative role of Artificial Intelligence (AI) in advancing sustainable practices within the Blue Economy through a series of global case studies. It examines how leading companies and initiatives—such as Sinay, Saildrone, Sea.AI, Buffalo Automation, IBM, Microsoft, Ocean Cleanup, and Veolia Group—leverage AI-driven solutions for real-time environmental monitoring, predictive climate modeling, autonomous navigation, and circular waste management. Key insights reveal that AI enables data-driven decision-making, reduces carbon emissions, enhances operational efficiency, and supports closed-loop recycling systems, while confronting challenges related to data integration, regulatory frameworks, and economic viability. Policy implications highlight the urgent need for adaptive regulations, open data ecosystems, and public-private partnerships to scale smart marine ecosystems globally.

DOI: 10.4018/979-8-3373-5238-1.ch007

1. INTRODUCTION

The Blue Economy, as the concept was first defined in the Rio+20 United Nations Conference on Sustainable Development in 2012, is a concept of sustainable use of ocean resources to achieve economic growth, better livelihoods, and ocean ecosystem health. It covers all areas extensively and they include fisheries, sea transport, tourism along the coastal areas, marine life biotechnology, and in offshore renewable energy. By 2023, there are approximately over 3 billion people who rely on marine and coastal biodiversity to sustain their livelihoods and provide economic opportunity and the global Blue Economy is estimated to be valued at over 3-4 trillion annually, with this figure expected to grow. The growing dangers of climate change that have been increasing in the last decade, including acidification of ocean, rising sea level, and reduction in biodiversity, have highlighted the necessity to find smarter, technology-based solutions. Machine Learning (ML) and Artificial Intelligence (AI) has become a ground-breakin resource in this field. ML models allow forecasting the marine ecosystem, real-time control of environmental parameters, and efficient use of resources in the marine industry(Awuor et al., 2008).

The intersection of AI and the Blue Economy has a transformative potential: research has shown that digital technologies have the potential to help reduce carbon emissions in shipping by up to 20% and increase the efficiency of marine conservation by 30-40% through real-time detection and decision-support systems. Moreover, AI-driven analytics have greatly increased the pace of incorporating the concepts of a circular economy, including the transformation of marine waste into renewable assets. Even though these aspects are encouraging, the implementation of ML in marine systems around the world is still not even and not well recorded especially in policy frameworks and ecosystem-based adaptation. This paper intends to fill that gap by providing three case studies in three regions where ML is in use with success in the Blue Economy to achieve sustainable, climate-resilient, and circular transformation of marine ecosystems(Awuor et al., 2008).

1.1. Blue Economy and Its International Importance.

Blue Economy is the notion of utilizing the ocean resources in the most sustainable way in order to achieve economic progress, better living standards, and the healthier ecosystem. Due to the ever-increasing population of the world, oceans are becoming more important in terms of supplying food, energy, transportation and creating economic opportunities. The Blue Economy has proven to be a trillion-dollar yearly contributor to the global economy and a provider of millions of jobs, especially in the fisheries, maritime transport, renewable energy and coastal tourism sectors. The Blue Economy has been acknowledged by the United Nations as the key to the

Sustainable Development Goals (SDGs), and in particular SDG 14 - Life Below Water. The issue is however on balancing economic growth against protecting delicate sea life since most of the activities carried out at sea could drain natural resources, pollute, and cause degradation of biodiversity. Thus, it is necessary to incorporate sustainable processes and new technologies to achieve the maximum potential of the Blue Economy without the ecological integrity decline(Bag et al., 2023).

1.2. The Future of the Artificial Intelligence in the Industry.

Artificial Intelligence (AI) is a disruptive technology that is changing the contemporary industries in all fields including manufacturing, healthcare, finance, agriculture, and logistics. AI contributes to increased efficiency, accuracy, and autonomous workplace by allowing machines to handle high amounts of data, identify trends, and make predictions. The AI can be used in factories to perform predictive maintenance, optimization of the supply chain, and real-time decision-making, thus saving operational expenses and improving efficiency. In addition, AI-driven platforms, including machine learning, computer vision, and natural language processing, aid in the intelligent data analysis and allow companies to act in advance of market trends and customer demands. The rising global problems like climate change and scarcity of resources have made AI solutions more important. In particular, AI aids in more intelligent monitoring, optimized resource use, and predictive forecasting in the context of environmental sustainability, which will bring more resilient and adaptive industrial ecosystems.

1.3. The Problems of Climate Change in the Marine Life.

The marine ecosystem is one of the environments that are the most susceptible to negative effects of climate change. The increase in sea temperatures, the acidification of oceans, sea rise and weather extremes have greatly changed the dynamics of marine biodiversity and ecosystems. Bleaching of coral reefs, which sustains about 25 percent of all marine life, is widespread as a result of high water temperature and biodiversity and livelihoods of millions rely on the reef ecosystems are under threat. Furthermore, ocean acidification, which is caused by the absorption of CO2 in large quantities, affects the organisms that form the food web such as shellfish and plankton, and consequentially, the entire ocean food web. Both natural habitat and human settlements are affected by the coastal erosion and saltwater intrusion, whereas fisheries and food security are directly threatened by the changing marine currents, affecting the way fish migrate. Moreover, its build-up of plastic waste and contaminants aggravates the destruction of marine environments. These issues need sophisticated real-time monitoring tools, predictive modeling tools and adaptive management tools that are able to react to the evolving environmental conditions and achieve conservation and sustainable usage(Barbier, 2023).

1.4. Significance of switching to the Circular Economy in the Marine Industry.

It is essential to shift to Circular Economy in the marine sector as the idea encourages sustainable development and decreases environmental degradation. In contrast to the classical linear model of economy which is based on the take-make-dispose paradigm, the Circular Economy focuses on the efficiency of resources, their minimization, and their endless utilization in form of recycling, re-manufacturing, and reuse. Within the framework of the Blue Economy, this shift contributes to the minimization of marine pollution, in particular, plastic waste, and facilitates sustainable processes in such industries as fisheries, aquaculture, shipbuilding, and offshore energy. To illustrate, in aquaculture, the closed-loop systems can be turned on, and water and nutrients can be used again, which will minimize the release of wastes and pollution of the environment. Likewise, waste management systems based on AI allow tracking of marine waste and its recycling and can greatly enhance the efficiency of the waste collection program. The models of the Circular Economy bring about sustainability in the natural resources besides generating economic opportunities through innovation and new business models(Betin & Titova, 2023).

The main aim of the research is to discuss and examine how machine learning (ML) as the essential element of artificial intelligence (AI) is being applied to practical usage in the Blue Economy to make it more sustainable, resilient, and circular. The study will focus on assessing the efficacy of AI systems that are driven by ML to transform marine environments into smart and data-driven systems that would help overcome the issue of the changing climate, biodiversity decline, and unsustainable utilization of resources..

2. THEORETICAL FRAMEWORK

The Blue Economy, the principle of the Climate Change Mitigation, the principles of the Circular Economy, and the technologies of Artificial Intelligence (AI) are interconnected based on the theoretical framework. It incorporates the major concepts, models, and technological paradigms that inform the sustainable development of the marine ecosystem and handle the environmental and economic issues. The framework can be used to analyze the role that AI-solutions play in enabling smarter decision-making, the efficient use of resources, and ecosystem sustainability in the framework of the Blue Economy(Betin & Titova, 2023).(Brönner et al., 2023).

The United Nations and the World Bank are examples of the international bodies that support the implementation of the Blue Economy as a solution to Sustainable Development Goals (SDGs), specifically SDG 14: Life Below Water. It advocates a systemic perspective whereby economic growth is in tandem with social justice and ecological sustainability with the view of long term sustainability of marine environments.

2.1. Mitigation of Climate Change in the Marine Ecosystem.

The marine ecosystems are the first to experience the effects of climate change through increasing ocean temperatures, acidification, rise in sea-level, and increasing extreme weather patterns. The mitigation of climate change in the marine ecosystem is aimed at lowering greenhouse gas (GHG) emissions, increasing carbon sequestration, and building the resilience of the marine ecosystem. Some of the key measures are the promotion of sustainable fisheries management, protection and restoration of coastal ecosystems like mangroves, seagrasses and salt marshes that act as sinks of carbon. Also, the use of renewable energy sources such as offshore wind farms will minimise the use of fossil fuels which will help in reducing the carbon emission. The state-of-the-art surveillance and predictive modeling helps to provide early warning systems of climate related risks like coral bleaching events or sea level rise effects on coastal populations. The combination of these strategies and technology-based ones allows making an informed decision and responding flexibly to the situation, preserving the ecosystem services and maintaining sustainable economic activities(Chemeli et al., 2021).

2.2. Circular Economy Principles.

The Circular Economy is a paradigm shift with the traditional linear approach of take, make, dispose to a regenerative system with focus on resource efficiency, reduction of waste, and reuse. The application of Circular Economy principles in the marine industry implies the development of products and processes that have the least harmful consequences on the environment and at the same time offer the best recovery of resources. This will involve the design of closed loop designs in aquaculture where nutrients and water are recycled, marine plastic waste collection and recycling technologies and the re-manufacturing of ship parts (Chen et al., 2021).

2.3. *Summary of Artificial Intelligence Technology in use in marine industries.*

Artificial Intelligence (AI) is a complex of modern technologies that allow machines and software systems to replicate the intelligence of a person based on the use of data, identification of patterns, and automation. Machine Learning (ML), Big Data Analytics, and the Internet of Things (IoT) are the AI technologies that have a transformative effect in the marine industry, allowing to solve the main challenges and open up new opportunities. Machine Learning algorithms can handle a large amount of data gathered by satellites, underwater sensors, and autonomous drones to forecast fish stocks, track marine biodiversity, and identify illegal fishing in the real-time. The Big Data Analytics can be used to integrate and analyze heterogeneous data such as oceanographic data, weather models, and economic indicators to aid predictive modeling and decision-making systems to be climate resilient and manage resources in a sustainable manner. Smart sensors based on IoT can constantly measure the essential parameters in the environment (temperature, salinity, pollutants, etc.), which allows to form feedback and provide automatic intervention. All these AI-driven technologies assist in making smarter, data-driven efforts in optimizing resources, pollution reduction, and tracking the ecosystem, which means more efficiency, sustainability, and resilience of the Blue Economy(Doerr et al., 2023).

3. REVIEW OF LITERATURE

In the last ten years, the use of Machine Learning (ML) in marine and oceanic systems has become the focus of considerable attention, especially in the framework of the implementation of Artificial Intelligence (AI) to promote sustainability in the Blue Economy. Such a change has been brought about by the increasing pressure to take measures on the effects of climate change on marine environment and at the same time to utilize the economic potential of the ocean in terms of economic value. The initial pioneering research on the topic was by (Fava et al., 2023)who established the efficacy of the supervised learning models, such as Random Forest and Support Vector Machines (SVM), in streamlining operations in aquaculture by forecasting water quality, fish production, and disease incidences and patterns. Their paper formed the foundation of applying data-based models in small-scale marine operations, showing that despite the small datasets, even ML may provide high efficiency and environmental returns. This was later followed by contributions by where(Doney et al., 2020) applied ML in areas outside of aquaculture into large marine monitoring systems. They aimed at measuring the potential of ML to handle and examine satellite and sensor-based data streams to track oceanic pollutants, sea surface temperature shifts and erosion along the coasts.

One of the major developments was made by (Gaines et al., 2019), who performed an extensive survey of AI-based automated surveillance devices on marine conservation. This resulted in their work pinpointing the inefficiencies of manual ecological surveys and focusing on the transformative nature of ML in cost-cutting and enhancing the real-time nature of monitoring work. These models have facilitated real time monitoring of marine endowment in both picture and sound data and have contributed immensely in the conservation of the coral reef and fish populations. Simultaneously, (Gangisetty & Rai, 2022) emphasized that deep learning (DL) models (convolutional neural networks (CNNs) and Long Short-Term Memory (LSTMs) networks) have been increasingly used in ecological models in the last five years. They claimed that traditional statistical models are effective in linear ecological trends, but the DL models have better performance in explaining complicated and nonlinear interactions in the ocean ecologies. They have also noted in their analysis that there is a need of transparency in ML systems, particularly when used to make environmental policy and managing resources.

Another more specific use of ML has appeared in the waste management and marine pollution detection. In a 2024 bibliometric review published in (Government of Kenya, 2021), the authors described the application of AI in the detection of oil spills, algal blooms, microplastics, and the discharge of industrial waste in oceans. The researchers observed that the conventional techniques of monitoring pollution like manual sampling and laboratory monitoring were time consuming and spatial constrained. Meanwhile, ML algorithms that were trained based on remote sensing and UAV images provided more accurate and fast-scaling assessment. The review emphasized the fact that classical ML models, such as decision trees and SVMs, were predominant before but deep learning models, especially CNNs and vision transformers (ViTs) have since surpassed them in terms of performance particularly in image classification tasks of marine debris and underwater surveillance.

The contribution of ML towards facilitating the practice of a circular economy of the marine industries has also been growing rapidly. In the years 2023 and 2024, engineering research probed how marine waste (e.g. plastic and biomass) can be transformed into usable energy sources using AI-optimized processes. As an example, image recognition models were used to classify the various kinds of plastic waste floating in the ocean to allow special collection and recycling. CNN models like VGG16 and YOLO have shown to be accurate in identifying microplastics through drone images to facilitate the development of localized processes to collect waste. These instruments did not only aid in circularity, but also helped to cut down the carbon emissions through the reduction of the incineration of waste and the promotion of sustainable production cycles.

In climatic prediction and modeling of ecosystem, there is a great advancement. In the study by(Haigh et al., 2015), a state-of-the-art deep learning model (graph-based) was created to predict Marine Heatwaves (MHWs) in various ocean areas around the world. They used spatial-temporal data inputs such as sea currents, ocean surface temperature and wind patterns to provide predictions that would have lead times of six months. This innovation is vital in the resource managers and policymakers that require a warning of climate-induced marine anomalies. Likewise, (Hamza et al., 2022) proposed a platform, the Marine Digital Twin, to Spain's Mar Menor lagoon, a modular and artificial intelligence-driven tool that simulates the dynamics of the marine ecosystem. The system brings into the system the real-time environmental information and applies the reinforcement learning to optimize intervention strategies. Their results highlighted the strength of the digital twins in conducting scenario-based simulations thus helping to develop adaptive and evidence-based marine policies.

On the border between community resilience and technical implementation, (Hughes et al., 2017) were interested in the application of AI in the Blue Economy of the coastal area of Kenya. Their paper analyzed AI tools to monitor coral bleaching, collapse of fish stocks and coastline erosion. According to them, ML-aided data platforms especially those intended to be used by the local population enhanced early warning and informed sustainable harvesting. They also, however, observed that localization of ML models and the presence of indigenous knowledge in the training data is the key to long-term success. The study confirmed that AI-adapted decision-making tools would significantly empower climate resilience of small island and developing states (SIDS).

The initiatives of the private sector have also started to take advantage of AI to create sustainable maritime operations. In 2024 the Sinay Paris-based company received attention regarding the application of ML to underwater noise pollution, ship emissions and marine biodiversity in port areas. Their platforms integrated both environmental information and logistics and shipping times to formulate sustainable operating models (IPCC et al., 2019). Although their technologies won more than the EUR20 billion of climate-tech investments, experts also expressed worries of data bias, model transparency and the environmental cost of AI infrastructure, in particular, data center energy consumption. However, it is possible to see how the practical potential of AI to decarbonize logistics-intensive Blue Economy areas is demonstrated in the work by Sinay(Jacobs et al., 2021).

The other key area of AI impact is operational efficiency and reduction of emissions in shipping. Research conducted by (Jacobs et al., 2021) demonstrated that AI could optimize the vessel routes and maintenance times and save up to 20% of fuel. Their development of AI into such key performance indicators as the Energy Efficiency Design Index (EEDI) and the Energy Efficiency Operational Indicator

(EEOI) had measurable positive results. Such observations confirm the assertion that intelligent logistics systems can play a major role in the decarbonisation pathways, particularly when the services are backed by predictive modelling of the ocean weather and current information(Jain et al., 2023).

The use of ML is another state-of-the-art application in the assessment of biodiversity and quantification of blue carbon. In 2024, a study aimed at investigating Greenlands deep-sea ecosystems, conducted with the help of AI-based image recognition and automated the process of classifying benthic species of which many are carbon-sequestrating species(Karanja & Saito, 2018). It was a model that was fed with ten years of underwater video footage and generated accurate estimates of biodiversity and biomass. This method allowed better calculation of the storage potential of carbon which is essential to national climate inventories and global carbon credit systems (Kulp & Strauss, 2018).

The data infrastructure is also changing to meet AI and ML implementation in ocean systems. In a study published in Frontiers in Marine Science in 2025, the authors examined the idea of the application of AI to develop open-access oceanographic data through global initiatives, such as the Digital Ocean Hubs by UNEP. Ontologies were also aligned using AI and vocabularies harmonized with AI, in addition to metadata tagging being automated to make marine data more FAIR (Findable, Accessible, Interoperable, and Reusable). The researchers emphasized that the interoperability is the basis of the scaled AI applications across geographies and across disciplines, particularly in the joint fisheries management and the marine biodiversity network(Lam et al., 2019).

4. RESEARCH METHODOLOGY

The study employs multiple case study methodology with the aid of content analysis of secondary data and is qualitative in nature. The four international companies, Sinay (France) and Saildrone (USA), Sea.AI (France) and Buffalo Automation (USA) were chosen due to their demonstrated ML-based marine solutions and cross-sectoral applicability as well as their contribution to sustainability and operational efficiency. Peer-reviewed academic sources, official company publications, industry reports, technical briefs, and interviews when available were used as the source of data collection. (Mihiretu et al., 2023).

4.1. Case studies of companies that adopt Machine Learning and AI in the Blue Economy.

4.1.1. *Case Study 1: AI-controlled Marine pollution surveillance - Watershed Clean Goods Project.*

Among the most ambitious and well known projects which are targeted at addressing the problem of marine pollution in the form of plastics, it should be noted the Ocean Cleanup Project which was created in 2013 by Dutch inventor Boyan Slat. The organization has its headquarters in the Netherlands and operates across the world with the objective of designing sophisticated technologies to clean up plastic pollution in the oceans of the world as well as ensuring that more waste gets no way into water bodies. As the problem of marine pollution increasingly becomes a staple of ecological and socio-economic concern, Ocean Clean-up incorporates Artificial Intelligence (AI) to further advance in its capacities to identify and monitor as well as eliminate plastic debris in marine environments. The fundamental technology applied to the Ocean Clean-up Project is a combination of satellite images, autonomous surface vessels (interceptors), and AI-controlled sensors. Satellite remote sensing is used to find out the areas that contain a lot of plastic waste, especially the notorious Great Pacific Garbage Patch. The system can be trained to predict pollution hotspots accurately using machine learning algorithms on big datasets of data covering oceanography and waste distribution. The information is incorporated into geospatial models, which direct autonomous vessels to such areas on a real-time basis. The decision-support system is based on data processing and predictive modelling. The Ocean Clean-up uses platforms of Big Data Analytics to process terabytes of oceanic condition data, such as currents, weather patterns and historical trends of pollution. The AI models apply predictive analytics to determine the direction of waste in order to make the process of cleaning up timely and accurate. These models enhance efficiency in operations by reducing the use of fuels and maximising on waste collection(Musingi et al., 2021).

The problem that the AI-based solution of the Ocean Clean-up entails is far-reaching. It enhances the accuracy of detecting pollution as well as it allows early warning mechanisms of emerging areas of plastic accumulation to preventive measures. Interceptor vessels that are controlled by the organization work in rivers and harbours, and they intercept plastic before it gets to the open ocean. With the use of the AI-assisted detection and predictive modelling, Ocean Clean-up has been able to capture millions of kilograms of plastic wastes hence playing a major role in curbing ocean pollution. The international integration of the solutions of Ocean Clean-up is gradually rising, and alliances with governments, educational institutions, and individual companies are formed. The technology provides a global solution to the issue of marine plastic pollution on a large scale. The project also delivers useful information to policymakers by developing comprehensive data reports, which would further the investigation of the cause of pollution and the regulation. Value creation of the project is not restricted to environmental conservation, but also to sustainable business practices, corporate social responsibility and preservation of the ecosystem(Musingi et al., 2021).

4.1.2. Case Study 2: IBM Watson Decision Platform of Agriculture: AI in Sustainable Fisheries Management.

IBM (International Business Machines Corporation) is one of the oldest and most influential technological companies in the world having been formed in 1911. During the recent years, IBM has expanded its advanced Artificial Intelligence (AI) products to agriculture and fisheries industries with the Watson Decision Platform of Agriculture. Even though the platform was originally intended to serve land-based agriculture, it has been modified to manage sustainable fisheries, which represents an important technological breakthrough in the Blue Economy. The history of the project lies in the fact that there is an urgent necessity to reconcile the rising demand of seafood with sustainability. Overfishing endangers the marine life and the livelihoods of millions of people on fisheries. The IBM solution uses AI-based predictive analytics to help governments and trade organizations implement sustainable fisheries, which include how to better estimate fish stocks, predict the dynamics of ecosystems, and improve the timing of harvesting. In this case study, AI applications comprise machine learning models that process enormous amounts of data gathered with the help of satellite images, sonar sensors, environmental sensors, and past fishing history. With the help of such models, it is possible to predict the movements of fish stocks, patterns of its growth, and optimal harvesting times. IBM incorporates the implementation of smart sensors on fishing vessels and buoys with IoT capabilities that continuously monitor the real-time data of the sea temperature, salinity, and plankton density. Such a set of data points enables the platform to create a dynamic model of fish behaviour and population dynamics.

Outcomes attained in this manner have been great. IBM Watson Decision Platform has been used to lower the levels of overfishing in pilot programs established in partnership with fisheries in Southeast Asia and Northern Europe, allowing them to offer actionable insights into the availability of stocks. Fishermen would be able to change their the timing and location of fishes according to real-time predictions and come up with more sustainable yields without putting fish at risk(Nost & Colven, 2022). Moreover, the system enhanced the traceability of food products of the sea, thereby increasing regulatory compliance and confidence to consumers. The solution is also being implemented globally by fisheries and governments in an attempt to achieve sustainability goals as part of the UN SDG framework, and particularly SDG 14 (Life Below Water). Combination of AI in fisheries facilitates the sustainability of the economy by lowering the costs of the activities, minimizing fuel use, and enhancing the quality of yields, whereas the ecological sustainability is ensured through retention of marine biodiversity. In addition, the system is connected to real-time solutions that evolve with the changing conditions of the environment, avoiding abrupt stock crashes or ecosystem upsets. The example of IBM solution is one of how the conventional technology giants in the world are transforming the old industries to become greener in their future. The scalable architecture of the platform allows both small-scale and large commercial fisheries to gain and improve the data-driven decision-making of the Blue Economy(D. O. Obura, 2001).

4.1.3. Case Study 3: Circular Economy Models in Marine Plastic Waste Management – Veolia Group

The Veolia Group is a French construction company established in 1853 and is known internationally as one of the best resource managers. As a firm, Veolia has been on the forefront in promoting the Circular Economy models in many areas, such as waste management, water treatment, and energy services. Veolia, in its context of the Blue Economy, has introduced various projects utilizing Artificial Intelligence (AI) to optimize the analysis of plastic waste in the ocean and is in line with the overall global trends to decrease marine pollution and maximize resource use. The history behind the project of Veolia is the increasing issue of marine plastic waste that impacts the environment, ocean life, and tourists on the shoreline. Veolia has created AI based waste tracking and recycling system that tracks the waste flows in coastal and marine areas that would allow smarter intervention strategies. Veolia uses smart waste management systems that deploy AI algorithms, IoT sensors, and sophisticated recycling to deliver waste management services in strategic collaborations with local municipalities and port authorities.

Waste tracking is an AI-driven application based on machine learning models that were developed based on environmental sensors and satellite imagery and waste collection points. These algorithms are used to examine space and time data and identify the areas of accumulation and the most efficient routes and schedules of waste collection. Sensors installed in garbage cans and on marine platforms are IoT-enabled and constantly update in real-time information about trash volume, its type and composition, and send it to a centralized data platform(D. Obura et al., 2017).

A major feature of the model of Veolia is the implementation of closed-loop systems. The marine plastics collected are not disposed but rather sorted and turned to raw materials that are reused in the production cycle. The innovative sorting technologies developed by Veolia are based on image recognition technology using AI to recognize the type of plastic and separate it to recycle or use again. The strategy will play a major role in minimising the release of plastics to the ocean and the establishment of sustainable value chains. The effect of the model of Veolia can be evaluated and is substantial. Using AI, Veolia has increased the efficiency of marine waste collection by more than 30, decreasing operational expenses, and emissions of greenhouse gases by the collection vehicles. In addition, recycling has also increased the rate of recovery of the materials by 25 percent as a result of accurate sorting methods. In addition to the positive impact on the environment, the system increases transparency and traceability of waste flows, which enhances regulatory compliance and reporting to the population.

The circular economy model has been adopted worldwide and Veolia has implemented the model in Europe, Asia, and North America. The implementation of the solutions in port cities, coastal areas, and islands communities is done by local governments and private stakeholders in collaboration with Veolia. The value created entails not only the enhanced environmental health but also the economic benefits in terms of new industries to be developed around the recycled marine plastics, employment opportunities in the waste management process and the reduction of costs to the municipalities. The case study demonstrates how combining the principles of the Circular Economy and AI-based technologies has a transformative potential to make the industry of Blue Economy sustainable and scalable(Okafor-Yarwood et al., 2020).

4.1.4. Case Study 4: AI-Enhanced Climate Change Impact Forecasting- Microsoft AI for Earth Program.

The Microsoft Artificial Intelligence Program created in 2017 is a worldwide-known program of the company, the Microsoft Artificial Intelligence (AI) program, but in this case it is aimed at assisting organizations in their endeavour's to resolve environmental issues through the application of Artificial Intelligence (AI). Bringing the support of one of the largest technology companies of the world, the program offers cloud-based solutions, data, and AI to address acute concerns in biodiversity, climate change, water management, and agriculture. Within the marine sphere, the program is very important in climate change impact prediction making it easy to better predict sea level rise, coral bleaching and marine biodiversity loss. The AI for Earth initiative is a partnership with research organizations, governments, and non-governmental organizations to use machine learning and data analytics in the analysis of complicated marine systems. The history of this effort is the international desire to have more accurate, data-rich models of how climate change will impact marine ecosystems and in order to have evidence-based conservation plans.

Large-scale predictive modelling based on the Microsoft Azure cloud computing infrastructure uses AI applications. These models utilize terabytes of data that are collected through satellite, in-situ marine sensors, oceanographic databases and climate models. Machine learning algorithms are used to examine patterns in the past and give predictions to future trends, including the severity of coral bleaching due to temperature change, acidification of oceans, and pollutants. The granularity and accuracy of the forecasts are improved, as AI will learn new data constantly. The findings have been the pioneers. The program was conducted with the Australian Institute of Marine Science (AIMS) and was able to formulate predictive models that have proven to be effective in predicting coral bleaching events at least months before it occur and this provides scientists and policymakers with essential lead time to take necessary actions to mitigate against it. In the Arctic, AI models have enhanced the study of the process of sea ice melting to facilitate dynamic management of shipping routes and fishing areas(Okafor-Yarwood et al., 2020).

Microsoft AI for Earth initiative has facilitated the provision of real-time solutions through the provision of friendly dashboards with interactive data visualisation. Cloud platforms and APIs can provide decision-makers with a predictive insight and incorporate the findings in their conservation and resource management strategies. The open data sharing is also supported by the program which promotes global cooperation and speeds up the creation of new solutions.

The program has been implemented in over 500 organizations in 100+ countries around the globe supporting the United Nations Sustainable Development Goals, especially SDG 13 (Climate Action) and SDG 14 (Life Below Water). Its scalable architecture facilitates the use of it by small research institutions and big governmental organizations. The AI on Earth program has a multidimensional value creation, which includes improvement of scientific knowledge, resilience of the policy, minimization of the environmental impact of human behaviour, and allowing sustainable economic development of vulnerable coastal communities. To reduce the risk of climate change in marine life, the program can be a prime example of the importance of AI in converting large, complicated data into meaningful actions(OECD, 2016).

4.1.5. Case study 5: Smart Marine Environmental Monitoring and Sustainable Port Operations: Sinay (France)

Sinay is a climate-tech innovator that was founded in 2017 by French entrepreneur Yanis Souami, and with headquarters located in Caen, France. By using machine learning and data analysis based on AI, Sinay allows maritime actors, including ports and shipping companies as well as marine energy platforms, to track the environmental effects and improve operations in the most sustainable way. The Sinay Hub of the company is a hub that unites various data streams, such as vessel tracks of AIS, weather, air and water quality, underwater noise and satellite imagery all displayed on a single dashboard.

It has one flagship deployment at Port La Nouvelle, which is a large French Mediterranean terminal that is under expansion. In this case, the Hub by Sinay has three combined modules of air, water, and acoustic noise, which process sensor data in real-time. Machine learning limits raise alarms when the environmental variables are beyond the set limits and this ensures that the disturbances are reduced as soon as they are detected by the port management algorithms in place. Outside the monitoring, the Estimated Time of Arrival (ETA) module by Sinay operates an ML on the AIS data to enhance the vessel scheduling and minimize energy wastage, which will directly reduce carbon emissions and traffic jams.

The influence of Sinay is two-fold, that is, environmental stewardship and value creation. To the port authorities, the system provides environmental reporting that is ready to comply, which promotes trust by stakeholders and prevents penalties by the regulatory body. Precise ETA predictions reduce the period of idle time and fuel usage, which results in tangible emissions reductions to shipping customers. Olympus Analytics pointed out that Sinay uses AI in open-source, commercial AIS and proprietary IoT sensor data, integrating machine-learned indicators specific to individual clients.

Sinay has enhanced strategic growth by acquiring Open Ocean in 2022 which has added advanced metocean forecasting and long-term analytics to the Sinay Hub and has been seamlessly integrated into the Sinay Hub. The company has alliances in Europe and expanding to other markets such as marine renewables, underwater cabling and environmental protection areas(Overbeeke et al., 2022).

Nowadays Sinay is present in key world ports and firms at sea. Carbon footprint reduction, efficient logistics, as well as a strong circular strategy on environmental data stewardship are a part of its value proposition. With the development of AI, Sinay is constantly incorporating the latest methods of ML like deep learning and predictive modelling to increase real-time ecological understanding and support sustainable Blue Economy initiatives(Pauly & Christensen, 1995).

4.1.6. Case study 6: Autonomous Ocean Observation and Data Collection Climate Research- Saildrone (USA)

In 2012, engineer Richard Jenkins created Saildrone Inc. that became the first company to transform ocean observation, with wind- and solar-powered unmanned surface vehicles (USVs). The fleet, which consists of 23-ft Explorers up to 65-ft Surveyors, is based in Alameda, California and uses AI, ML and onboard sensors to autonomously navigate, do environmental monitoring, bathymetric mapping and analyze the ecosystem.

Saildrone had a cooperative agreement with NOAA in 2014 to add metocean sensors and acoustic fish-detection systems to USVs as a way of integrating with technology needed to support its mission. Having raised Series A (~$14 million in 2016) and Series C (100m) in 2021, the platform launched missions in fisheries, mapping, climate research and maritime defence.

Acoustic, oceanographic, and imaging measurements are used to identify fish, plastic debris, underwater noise, and climate-related parameters in real-time onboard ML algorithms. One of the missions received coverage in 2019 was the autonomous circumnavigation of Antarctica (SD 1020), which is a 13,670 nautical mile-long journey occurring over 196 days, in which USVs also collected hourly CO2 readings to enhance global carbon budget models. This proved the way AI-motivated USVs gather constant and high-resolution ocean information- particularly of distant or climate- delicate areas(Perrault et al., 2019).

Saildrone initiated its 65 ft Surveyor in 2021, which has multibeam echo-sounders and acoustic Doppler profilers to map the seabed in high-resolution - needed in offshore renewable or habitat-conservation studies. Surveyors detect bathymetric changes and biodiversity-rich areas autonomously by applying real-time ML-based anomaly detection algorithms to detect these changes.

The company has introduced world-wide adoption: NOAA, NASA, US Navy, CSIRO, ECMWF and many research institutions use Saildrone missions. The industry applications are to assist in offshore wind planning and mapping of carbon sinks, assessment of fisheries, detection of marine debris, and defence. The 2022 Carbon Impact Report of Saildrone estimated that USVs reduce the emission of greenhouse gases relative to the conventional survey by the use of traditional vessels(Pielke, 2021).

This method produces value environment intelligence with reduced cost and carbon footprint. The model promotes circular science: on-going data collection contributes to cloud ML systems to constantly improve on models to do surveys, species detection, climate predictions, and compliance. Saildrone is also working to spread digital infrastructure to make ocean monitoring more democratic to climate and conservation use with 20 Surveyors identifying high-resolution bathymetry around the world by 2030(Pendy, 2023).

4.1.7. Case Study 7: AI-ready Real-time Obstacle Detection and Maritime Safety Systems- Sea.AI (France).

Introduced in 2019 by Raphael Biancale, Sea.AI addresses one of the most important issues in the maritime domain: the real-time detection of obstacles by the ships. The start-up is the creation of the need of French ocean racers, and it can incorporate thermal cameras and ML-driven object recognition to identify debris, buoys, mammals, and small crafts- objects that are often invisible to the traditional radar.

Sea.AI predicts and creates proximity alerts using ML models based on a proprietary dataset of more than 10 million manually labelled objects. Such warnings could be shown on the multi-function display of the vessel and provoke audible alarms-often these warnings could save some important seconds to avoid collisions. Sea.AI has already shown that it can be used with high-speed and high-stakes conditions, albeit in professional racing fleets, such as with Team Malizia (Will Harris has sailed 100,000+ miles using the system).

Garmin has partnered with Sea.AI to incorporate the alerts of its marine displays, which makes it easy to be adopted. Multi-layers at the top of a variety of camera systems make the platform compatible with the different types of vessels, including racing yachts and commercial ferries. Performance indicators will involve a 33% decrease in close-encounter situations and enhanced situational awareness to the crew in difficult circumstances like night or busy surroundings-areas of minimizing accidents, wastage of fuel and marine pollution risks. As a commercial venture, it has already been adopted by 500 vessels, in the ocean racers, search and rescue craft and recreational boats market in Europe and is moving into the UK and the German rescue services. The next product in the portfolio of the new Brain by Sea.AI the software only version which is compatible with third-party thermal cameras will expand the market reach by reducing hardware dependence and deploying onboard in seconds.

Steering integration and automatic avoidance features are already under the trial and will provide a way of autonomous reactions to an emergency in the future. The platform, with sensors that can identify objects as small as floating plastics and marine mammals, is an important AI-controlled device in the maritime safety system and is in line with the overarching decarbonization policy, which minimizes unpredicted slowdowns or reroutes(Roxy et al., 2020).

The business model of Sea.AI is a hardware and SaaS with recurring upgrades in the model of the ML. Some of the adoption metrics are cross sector deployment in racing, commercial and rescue fleets. Value is derived from enhanced safety, reduced downtime, lower insurance liability, and environmental protection by avoiding spills. Sea.AI's real-time vision and ML-powered detection underpin smarter, greener navigation aligned with Blue Economy principles.

4.1.8. Case study 8: Autonomous and Smart Marine Transportation- Buffalo Automation (USA)

Buffalo Automation, founded in 2015 by University at Buffalo engineers, offers the AutoMate platform: an AI-driven neural network system integrating vessel sensor data (cameras, sonar, radar, LiDAR, AIS, weather, GPS) to enable autonomous marine navigation and collision avoidance . With initial seed funding of $900k, the start-up focused on marine autonomy, targeting both recreational and commercial vessels.

AutoMate utilizes deep-learning object detection, sensor fusion, and real-time decision models to classify navigational hazards—from swimmers to small boats—and execute compliance-based manoeuvring per COLREGS regulations. Supported by edge computing, the platform performs continuous inference within seconds of sensor input .

In 2021, Buffalo launched an autonomous, solar-electric water taxi demo in Knoxville, Tennessee. The vessel, equipped with AutoMate, successfully navigated urban waterways, detecting obstacles with ML and navigating autonomously—demonstrating scalable solutions for urban maritime services. This highlighted the platform's versatility from marine leisure to passenger transport(Ryther, 1969).

Buffalo has partnered with municipal authorities and boat manufacturers to deploy AI-enabled navigation systems. Its approach directly supports decarbonization: safety-enabled autonomy reduces crew needs and optimises vessel behaviour, reducing emissions and running costs. The engine-agile control systems and emergency overrides ensure regulatory compatibility and insurance compliance.

Commercial adoption includes workboats and small ferry operators, with pilot projects in Tennessee and planned rollouts in tourist, resort, and cargo sectors. Value generation stems from operational optimization, automated docking, reduced crew requirements, and potential carbon-credit eligibility from reduced emissions. Buffalo is currently advancing toward fully autonomous commercial ferries globally.

As the Blue Economy evolves, AutoMate represents a scalable edge-AI tool for harbour transit, marine logistics, and eco-cruise operations. Its real-time ML-controlled hazard detection and autonomous navigation address safety and emissions goals while offering operators cost savings and environmental credentials essential for climate-smart marine services(Rutenberg et al., 2021).

4.1.9. Cross-case insights/ comparative analysis of case narratives

The comparative analysis of how AI applications across the Blue Economy is shown in the table 1.

Table 1. Comparative analysis of case narratives

Case / Organization	Core Domain	AI Technologies Used	Primary Application Area	Sustainability Outcomes	Scale & Adoption	Key Value Creation
The Ocean Cleanup	Marine pollution surveillance	ML-based hotspot prediction, satellite analytics, autonomous vessels	Ocean & river plastic interception	Early detection of plastic accumulation; large-scale waste removal	Global (oceans, rivers, harbors)	Reduced marine plastic loads; policy-relevant datasets; CSR and ecosystem preservation
IBM – Watson Decision Platform	Sustainable fisheries	Predictive ML models, IoT sensors, satellite & sonar analytics	Fish stock estimation & harvest optimization	Reduced overfishing; biodiversity conservation	Pilot & global rollouts (EU, SE Asia)	Sustainable yields; regulatory compliance; improved seafood traceability
Veolia	Circular marine waste economy	AI-based waste tracking, image recognition, IoT sensors	Plastic waste collection & recycling	30% efficiency gains; higher recycling rates	Europe, Asia, North America	Cost reduction; circular value chains; job creation
Microsoft – AI for Earth	Climate impact forecasting	Cloud AI, predictive modeling, big data analytics	Sea-level rise, coral bleaching, ice melt prediction	Early-warning systems; ecosystem resilience	500+ orgs, 100+ countries	Evidence-based policymaking; open-data collaboration
Sinay	Smart ports & marine monitoring	ML-based ETA prediction, real-time sensor analytics	Port emissions, noise & water monitoring	Reduced emissions; real-time compliance	Deployed in major European ports	Operational efficiency; ESG compliance; reduced fuel use
Saildrone	Autonomous ocean observation	Onboard ML, acoustic sensing, anomaly detection	Climate research & fisheries assessment	Low-carbon data collection; ecosystem mapping	NOAA, NASA, global research fleets	High-resolution data; lower survey costs; climate intelligence
Sea.AI	Maritime safety & navigation	Computer vision, thermal imaging, deep learning	Obstacle detection & collision avoidance	Reduced accidents; lower spill risks	500+ vessels (EU, UK)	Enhanced safety; insurance savings; fuel efficiency
Buffalo Automation	Autonomous marine transport	Sensor fusion, edge AI, neural networks	Collision-free autonomous navigation	Emissions reduction; energy optimization	Pilot deployments (US waterways)	Cost savings; autonomous ferries; climate-smart transit

Source: Review of literature. Compiled by: Author.

This comparative analysis(Table 1) demonstrates how AI applications across the Blue Economy converge toward shared sustainability goals while differing in technological depth and operational focus. Monitoring-oriented solutions (Ocean Cleanup, Sinay, Microsoft) emphasize early detection and compliance, whereas autonomy-driven systems (Saildrone, Sea.AI, Buffalo Automation) prioritize safety, efficiency, and emissions reduction. Circular economy models (Veolia) translate environmental intelligence into closed-loop value creation. Collectively, the cases highlight AI's role as a scalable enabler of environmentally responsible, economically viable marine operations.

5. KEY FINDINGS AND INSIGHTS

The analysis of the case studies—Sinay (France), Saildrone (USA), Sea.AI (France), Buffalo Automation (USA), IBM's Watson Decision Platform for Agriculture, Microsoft's AI for Earth Program, Ocean Clean-up, and Veolia Group—reveals important insights into how Artificial Intelligence (AI) is reshaping the Blue Economy. AI plays a pivotal role in enabling data-driven decision-making across the maritime sector. For example, Sinay's Hub consolidates diverse data streams—including vessel AIS tracks, environmental sensors, and satellite imagery—into a unified dashboard, enabling real-time monitoring of air, water, and acoustic parameters to optimize port operations and reduce emissions. Similarly, IBM's Watson Decision Platform enhances sustainable fisheries management through predictive modelling of fish stock dynamics and environmental factors, reducing overfishing risks and improving traceability in seafood supply chains(Sirmacek & Vinuesa, 2022). Saildrone's fleet of autonomous surface vehicles (USVs) exemplifies the integration of AI for remote environmental monitoring, providing continuous data on oceanographic conditions, carbon flux, and bathymetric mapping. The platform enables researchers and industry stakeholders to monitor marine ecosystems efficiently and cost-effectively. In parallel, Microsoft's AI for Earth initiative uses cloud-based predictive models to forecast critical environmental changes, such as coral bleaching and sea-level rise, helping decision-makers implement conservation strategies proactively.

The Ocean Clean-up project uses AI-powered satellite imaging and predictive modelling to identify plastic pollution hotspots in oceans and rivers. This enables real-time detection of debris accumulation zones, while its autonomous interceptors collect plastics before they spread further. The combination of machine learning thresholds and data analytics helps reduce marine pollution, fulfilling a key objective of the circular economy by transforming waste into manageable, recyclable resources(Tuda & Omar, 2012). Similarly, Veolia Group's AI-driven waste tracking and recycling system plays a vital role in managing marine plastic waste. Machine learning algorithms analyse sensor data and satellite imagery to track the flow of waste from coastal regions into marine environments. Veolia's closed-loop recycling models transform captured plastic into reusable raw materials, improving material recovery efficiency and reducing the ecological footprint of waste management(Taherdoost, 2022).

On the safety front, Sea.AI's ML-powered obstacle detection system enhances vessel navigation by identifying debris, marine mammals, and small boats in real time, preventing collisions and environmental accidents. Complementing this, Buffalo Automation's AutoMate platform offers autonomous navigation and collision avoidance solutions for commercial ferries and urban water taxis, contributing to decarbonization by reducing fuel usage and human error. Despite these successes, the deployment of AI in the Blue Economy faces significant challenges. Technically, integrating diverse data sources under harsh marine conditions and developing reliable ML models remains complex. Regulatory frameworks have not yet fully adapted to accommodate autonomous AI-driven systems, creating legal and liability uncertainties. Economically, high development and implementation costs continue to limit widespread adoption, especially in developing regions.

Nonetheless, these case studies clearly demonstrate that AI contributes substantially to sustainable and circular Blue Economy practices. Solutions like Ocean Clean-up and Veolia's recycling systems directly reduce plastic pollution, while real-time environmental monitoring (Sinay, Saildrone, Microsoft AI for Earth) enables proactive conservation and efficient resource utilization. Overall, these innovations help create resilient ecosystems that balance ecological preservation with economic activity(UNFCCC, 2006).

6. POLICY IMPLICATIONS AND RECOMMENDATIONS

A major policy implication is the need for robust regulatory frameworks supporting the adoption of AI in the Blue Economy. Current maritime laws focus largely on safety and environmental compliance but do not adequately address the accountability, data integrity, and autonomous decision-making features of AI-powered systems such as those implemented by Ocean Clean-up, Sinay, and Sea.AI. Policymakers must develop adaptive standards for algorithm certification, operational safety in autonomous navigation, and environmental reporting protocols.

Another key recommendation concerns data-sharing policies. Many AI-driven solutions rely on diverse data inputs from proprietary AIS data (IBM, Sinay) to public environmental observations (Saildrone, Microsoft). Yet, data silos, proprietary restrictions, and privacy concerns hinder collaborative progress. Governments should create clear, enforceable policies encouraging standardized, interoperable data ecosystems. Open data platforms will accelerate innovation by improving the accuracy and transparency of machine learning models and promoting global scientific collaboration.

Public-private partnerships (PPPs) represent a strategic approach for scaling up smart marine ecosystems. Case studies show that collaborations between public institutions and technology innovators (e.g., Microsoft's collaboration with Australian Institute of Marine Science, Ocean Cleanup's work with governments and NGOs) are essential for advancing large-scale deployments. Policy incentives such as tax credits and R&D grants—should support these partnerships to reduce financial barriers, especially in developing countries where Blue Economy potential is high but resources are limited.

Practitioner-Oriented Implementation Prerequisites for AI Adoption in the Blue Economy

To enhance practical relevance for non-academic readers, it is important to recognize that the successful implementation of AI-driven solutions in the Blue Economy depends on three critical prerequisites. First, robust and interoperable data infrastructures are essential, including access to high-quality real-time environmental, operational, and satellite data. Second, organizations must invest in specialized human capital, combining domain expertise in marine systems with skills in data science, AI, and digital operations. Finally, strong governance and regulatory frameworks are required to ensure ethical AI use, data sharing, cybersecurity, and alignment with sustainability and compliance objectives, thereby enabling scalable and responsible adoption across sectors.

7. FUTURE RESEARCH DIRECTIONS

Despite the significant progress observed in these case studies, there remain numerous areas for further research. First, advanced AI techniques such as federated learning, explainable AI, and deep reinforcement learning offer potential to enhance adaptive and autonomous decision-making, particularly in unpredictable and remote marine environments. These methods could enable systems like Saildrone's USVs and Sea.AI's obstacle detection to continually improve based on new data without compromising data privacy.

There is also a strong need to investigate the socio-economic impacts of AI adoption in the Blue Economy. Additionally, research must address the ethical dimensions of AI use ensuring transparency, fairness, and accountability in autonomous decision-making. Emerging technologies that complement AI should also be researched in tandem. For example, blockchain technologies for traceability of marine resources can enhance trust in sustainable supply chains. Edge computing solutions allow AI models to perform inference directly on vessels or remote stations, minimizing latency and dependence on cloud connectivity in harsh marine conditions. Furthermore, the rollout of 5G networks promises to improve high-speed data transmission across maritime zones, enhancing real-time AI applications. Finally, longitudinal studies are essential to quantify the ecological and economic outcomes of AI-driven marine interventions over extended timeframes.

The study acknowledges methodological limitations, including reliance on secondary data sources, qualitative case-based analysis, and the absence of primary empirical validation, which may constrain generalizability.

8. CONCLUSION

This chapter has explored how Artificial Intelligence is emerging as a powerful enabler of sustainable development within the Blue Economy. The case studies of Sinay, Saildrone, Sea.AI, Buffalo Automation, IBM, Microsoft, Ocean Clean-up, and Veolia Group demonstrate the wide-ranging applications of AI—from real-time environmental monitoring and predictive modelling to autonomous navigation and plastic waste management.

AI's primary strength lies in its ability to process massive, heterogeneous datasets into actionable insights, enabling smarter decision-making, reducing waste, and enhancing operational efficiency in marine industries. Despite challenges such as technical integration, regulatory gaps, and economic costs, these innovations showcase significant strides toward a circular and sustainable marine economy.

Policymakers must establish forward-looking frameworks, encourage open data ecosystems, and support public-private partnerships to scale up smart marine ecosystems globally. Further research into advanced AI techniques, socio-economic impacts, and complementary technologies will strengthen the resilience and adaptability of marine industries in the face of climate change and resource constraints.

In closing, the transformative role of AI in the Blue Economy is not merely technological—it represents a paradigm shift toward responsible, efficient, and sustainable ocean governance. Achieving this future requires coordinated action by industry leaders, academic researchers, and policymakers to foster innovation, environmental stewardship, and economic viability.

REFERENCES

Awuor, C. B., Orindi, V. A., & Adwera, A. O. (2008). Climate change and coastal cities: The case of Mombasa, Kenya. *Environment and Urbanization*, *20*(1), 231–242. DOI: 10.1177/0956247808089158

Bag, S., Rahman, M. S., Rogers, H., Srivastava, G., & Pretorius, J. H. C. (2023). Climate change adaptation and disaster risk reduction in the garment industry supply chain network. *Transportation Research Part E, Logistics and Transportation Review*, *171*, 103031. DOI: 10.1016/j.tre.2023.103031

Barbier, E. B. (2023). Greening the ocean economy. *Frontiers in Environmental Economics*, *2*, 1096303. Advance online publication. DOI: 10.3389/frevc.2023.1096303

Betin, O., & Titova, G. (2023). The formation of the economy of the world ocean. *Rybnoe Hozâjstvo*, 18–23. DOI: 10.37663/0131-6184-2023-2-18-23

Brönner, U., Sonnewald, M., & Visbeck, M. (2023). Digital Twins of the Ocean can foster a sustainable blue economy in a protected marine environment. *The International Hydrographic Review*, *29*(1), 26–40. DOI: 10.58440/ihr-29-a04

Chemeli, A., Njoroge, J. M., & Agufana, P. B. (2021). Climate change and immovable cultural heritage in Kenya: Impact and response strategies. In Leal Filho, W., Luetz, J., & Ayal, D. (Eds.), *Handbook of Climate Change Management*. Springer., DOI: 10.1007/978-3-030-22759-3_91-1

Chen, Y., Zou, X., Li, K., Yang, X., & Chen, C. (2021). Multiple local 3DCNNs for region-based prediction in smart cities. *Information Sciences*, *542*, 476–491. DOI: 10.1016/j.ins.2020.06.026

Doerr, A. N., Anderson, L., & Scorse, J. (2023). Marine entrepreneurship. In *Routledge eBooks* (pp. 244–259). Routledge., DOI: 10.4324/9781003058151-18

Doney, S. C., Busch, D. S., Cooley, S. R., & Kroeker, K. J. (2020). The impacts of ocean acidification on marine ecosystems and reliant human communities. *Annual Review of Environment and Resources*, *45*(1), 83–112. DOI: 10.1146/annurev-environ-012320-083019

Fava, I., Thijsse, P., Sipos, G., & Schaap, D. (2023). Using AI and ML to support marine science research. *EGU General Assembly, 2023*. Advance online publication. DOI: 10.5194/egusphere-egu23-4929

Gaines, S., Cabral, R., Free, C. M., & Golbuu, Y. (2019). *The expected impacts of climate change on the ocean economy*. World Resources Institute. DOI: 10.69902/757f40cd

Gangisetty, S., & Rai, R. R. (2022). FloodNet: Underwater image restoration based on residual dense learning. *Signal Processing Image Communication, 104*, 116647. DOI: 10.1016/j.image.2022.116647

Government of Kenya. (2021). *Rising Water Levels in Kenya's Rift Valley Lakes, Turkwel Gorge Dam, and Lake Victoria: A Scoping Report*. Government of Kenya and UNDP.

Haigh, R., Ianson, D., Holt, C. A., Neate, H. E., & Edwards, A. M. (2015). Effects of ocean acidification on temperate coastal marine ecosystems and fisheries in the Northeast Pacific. *PLoS One, 10*(2), e0117533. DOI: 10.1371/journal.pone.0117533 PMID: 25671596

Hamza, A. J., Esteves, L. S., & Cvitanović, M. (2022). Changes in mangrove cover and exposure to coastal hazards in Kenya. *Land (Basel), 11*(10), 1714. DOI: 10.3390/land11101714

Hughes, T. P., Barnes, M. L., Bellwood, D. R., Cinner, J. E., Cumming, G. S., Jackson, J. B. C., Kleypas, J., van de Leemput, I. A., Lough, J. M., Morrison, T. H., Palumbi, S. R., van Nes, E. H., & Scheffer, M. (2017). Coral reefs in the anthropocene. *Nature, 546*(7656), 82–90. DOI: 10.1038/nature22901 PMID: 28569801

IPCC, Pörtner, H. O., Roberts, D. C., Masson-Delmotte, V., Zhai, P., Poloczanska, E., Mintenbeck, K., Tignor, M., Alegría, A., Nicolai, M., Okem, A., Petzold, J., Rama, B., & Weyer, N. M. (2019). *Technical Summary: IPCC Special Report on the Ocean and Cryosphere in a Changing Climate*. IPCC.

Jacobs, Z., Yool, A., Jebri, F., Srokosz, M., van Gennip, S., Kelly, S., Roberts, M., Sauer, W., Queirós, A. M., Osuka, K. E., Samoilys, M., Becker, A. E., & Popova, E. (2021). Key climate change stressors of marine ecosystems along the path of the East African coastal current. *Ocean and Coastal Management, 208*, 105627. DOI: 10.1016/j.ocecoaman.2021.105627

Jain, H., Dhupper, R., Shrivastava, A., Kumar, D., & Kumari, M. (2023). AI-enabled strategies for climate change adaptation: Protecting communities, infrastructure, and businesses from the impacts of climate change. *Computational Urban Science, 3*(1), 25. DOI: 10.1007/s43762-023-00100-2

Karanja, J. M., & Saito, O. (2018). Cost-benefit analysis of mangrove ecosystems in flood risk reduction: A case study of the Tana Delta, Kenya. *Sustainability Science, 13*, 503–516. DOI: 10.1007/s11625-017-0427-3

Kulp, S. A., & Strauss, B. H. (2018). CoastalDEM: A global coastal digital elevation model improved from SRTM using a neural network. *Remote Sensing of Environment, 20*, 231–239. DOI: 10.1016/j.rse.2017.12.026

Lam, V. W. Y., Chavanich, S., Djoundourian, S., Dupont, S., Gaill, F., Holzer, G., Isensee, K., Katua, S., Mars, F., Metian, M., & Hall-Spencer, J. M. (2019). Dealing with the effects of ocean acidification on coral reefs in the Indian Ocean and Asia. *Regional Studies in Marine Science, 28*, 100560. DOI: 10.1016/j.rsma.2019.100560

Mihiretu, A., Okoyo, E. N., & Lemma, T. (2023). Climate variability trends community perspective-livelihood adaptation strategy nexus in the arid-tropics, Ethiopia. *Journal of Arid Environments, 210*, 104929. DOI: 10.1016/j.jaridenv.2022.104929

Musingi, D. M., Munywoki, M. M., & Musingi, B. M. (2021). Evaluation of seawater intrusion in coastal aquifers using both numerical methods and causal research: A case study of Mombasa, Kenya. *Engineering, 13*(7), 388–398. DOI: 10.4236/eng.2021.137028

Nost, E., & Colven, E. (2022). Earth for AI: A political ecology of data-driven climate initiatives. *Geoforum, 130*, 23–34. DOI: 10.1016/j.geoforum.2022.01.016

Obura, D., Gudka, M., Rabi, F. A., Gian, S. B., Bijoux, J., Freed, S., & Associates. (2017). *Coral Reef Status Report for the Western Indian Ocean.* http://hdl.handle.net/20.500.11822/28371

Obura, D. O. (2001). Kenya. *Marine Pollution Bulletin, 42*(12), 1264–1278. DOI: 10.1016/S0025-326X(01)00241-7 PMID: 11827110

OECD. (2016). *The Ocean Economy in 2030*. OECD Publishing., DOI: 10.1787/9789264251724-

Okafor-Yarwood, I., Kadagi, N. I., Miranda, N. A. F., Uku, J., Elegbede, I. O., & Adewumi, I. J. (2020). The blue economy–cultural livelihood–ecosystem conservation triangle: The African experience. *Frontiers in Marine Science, 7*, 586. Advance online publication. DOI: 10.3389/fmars.2020.00586

Overbeeke, F., Shepherd, L., Canac, S., & Grosskopf, A. (2022). *Blue entrepreneurship scoping study for Kenya: Unlocking business solutions that benefit people, the ocean and climate*. IUCN.

Pauly, D., & Christensen, V. (1995). Primary production required to sustain global fisheries. *Nature*, *374*(6519), 255–257. DOI: 10.1038/374255a0

Pendy, B. (2023). Role of AI in business management. *Brilliance*, *3*(1), 48–55. DOI: 10.47709/brilliance.v3i1.2191

Perrault, R., Shoham, Y., & Associates. (2019). *The AI Index 2019 Annual Report*. AI Index Steering Committee, Human-Centered AI Institute, Stanford University. https://hai.stanford.edu/sites/xg/files/sbiybj10986/f/ai_index_2019_report.pdf

Pielke, R. A. (2021). The role of models in prediction for decision. In *Princeton University Press eBooks* (pp. 111–136). Princeton University Press., DOI: 10.2307/j.ctv1dwq0tq.11

Roxy, M. K., Gnanaseelan, C., Parekh, A., Chowdary, J. S., Singh, S., & Modi, A. (2020). Indian ocean warming. In *Assessment of Climate Change Over the Indian Region* (pp. 191–206). Springer., DOI: 10.1007/978-981-15-4327-2_10

Rutenberg, I., Gwagwa, A., & Omino, M. (2021). Use and impact of artificial intelligence on climate change adaptation in Africa. In Oguge, N., Ayal, D., Adeleke, L., & da Silva, I. (Eds.), *African Handbook of Climate Change Adaptation*. Springer International Publishing.

Ryther, J. H. (1969). Photosynthesis and fish production in the sea. *Science*, *166*(3901), 72–76. DOI: 10.1126/science.166.3901.72 PMID: 5817762

Sirmacek, B., & Vinuesa, R. (2022). Remote sensing and AI for building climate adaptation applications. *Results in Engineering*, *15*, 100524. DOI: 10.1016/j.rineng.2022.100524

Taherdoost, H. (2022). Machine Learning Algorithms: Features and Applications. In Wang, J. (Ed.), *Encyclopedia of Data Science and Machine Learning* (pp. 938–960). IGI Global., DOI: 10.4018/978-1-7998-9220-5.ch054

Tuda, A., & Omar, M. (2012). Protection of marine areas in Kenya. *The George Wright Forum*, *29*(1), 43–50.

UNFCCC. (2006). *Technologies for adaptation to climate change*. UNON Publishing Service.

Chapter 8

Artificial Intelligence to Predict Climate Change-Related Risks in Marine Ecosystems

Silvio Andrae
https://orcid.org/0000-0002-8586-7812
Independent Researcher, Germany

ABSTRACT

Artificial intelligence (AI) is increasingly utilized to predict climate change–related risks in marine systems. This analysis establishes a typology of four model classes: deep learning, probabilistic and ensemble models, reinforcement learning, and hybrid physics-informed approaches. These classes are systematically mapped to key marine-climate applications, including extreme events, ecosystem change, and adaptive resource management. The results indicate that AI can significantly improve short-term risk detection and scenario exploration, whereas long-term projections benefit from hybrid models that incorporate physical constraints and uncertainty quantification. Nevertheless, structural data gaps, regional biases, and limited interpretability restrict the generalizability and policy relevance of many AI models. The chapter contends that the value of AI for marine climate risk assessment depends not only on predictive accuracy but also on transparent model design, robust data foundations, and alignment with governance requirements.

DOI: 10.4018/979-8-3373-5238-1.ch008

INTRODUCTION

Oceans are integral to the Earth system, regulating climate, storing heat and carbon, and supporting marine ecosystems that are essential for global food security and coastal livelihoods. They absorb over 90 percent of the excess heat produced by anthropogenic greenhouse gas emissions and approximately one quarter of emitted carbon dioxide, thereby moderating the rate of atmospheric warming (IPCC, 2019; IPCC, 2023). However, this buffering capacity incurs high costs, including ocean warming, acidification, deoxygenation, and altered circulation patterns, which increasingly destabilize marine ecosystems. Documented impacts include coral bleaching, shifts in fish distributions, changes in primary productivity, and a higher frequency of extreme events such as marine heatwaves and harmful algal blooms, resulting in cascading ecological and socio-economic effects (Hoegh-Guldberg et al., 2018; Smale et al., 2019).

Accurate prediction of climate change–related risks in marine systems represents a significant scientific and policy challenge. Traditional Earth system and ocean models, which are based on physical and biogeochemical equations, have advanced the representation of large-scale dynamics. Nevertheless, these models are limited by computational constraints, uncertainties in subgrid-scale processes, sparse observational coverage in many ocean regions, and challenges in capturing nonlinear cross-scale interactions (Palmer, 2017; IPCC, 2021). Consequently, critical phenomena such as abrupt ecosystem shifts, compound extremes, and regionally specific impacts are often inadequately resolved or characterized by high uncertainty, reducing their utility for operational risk assessment and decision support.

In this context, artificial intelligence (AI) and machine learning (ML) have emerged as valuable complements to established climate modeling techniques. Advances in deep learning, probabilistic modeling, and hybrid physics–data approaches have already transformed fields such as image recognition and numerical weather prediction (Goodfellow et al., 2016; Rolnick et al., 2019). In climate science, AI-based models are increasingly used to accelerate numerical simulations, improve spatial resolution through statistical downscaling, identify complex patterns in large observational datasets, and develop early warning systems for extreme events (Reichstein et al., 2019). Recent developments, particularly transformer-based global weather models, indicate that data-driven methods can, for specific prediction tasks, equal or exceed the performance of state-of-the-art numerical models while requiring significantly less computational resources (Lam et al., 2023).

Despite these advancements, the application of AI to marine climate change remains inconsistent and conceptually fragmented. Current research often focuses on either atmospheric prediction or isolated ecological applications, such as species distribution modeling and image-based habitat mapping, and typically lacks a unified analytical framework (Robinson et al., 2017; Villon et al., 2022). Several fundamental questions remain inadequately addressed: Which AI model types are most appropriate for various categories of marine climate risks? How do data availability, preprocessing decisions, and observational biases influence model performance and uncertainty? To what extent can AI-based methods generalize to future climate regimes that differ markedly from historical conditions? Furthermore, how can these models be integrated into governance and policy frameworks that demand transparency, robustness, and accountability?

This chapter addresses these issues by offering a structured and critical analysis of AI-based methods for predicting climate change–related risks in marine systems. Rather than conceptualizing AI as a single technology, it introduces a typology that differentiates four principal model classes: (i) neural networks and deep learning architectures, (ii) probabilistic and ensemble-based methods, (iii) reinforcement learning for adaptive scenario exploration, and (iv) hybrid models that combine physical constraints with data-driven learning (Willard et al., 2020). This typology provides an analytical framework for evaluating both empirical performance and structural limitations across diverse marine climate applications.

Methodologically, this chapter synthesizes insights from three key areas of literature: AI-driven climate and weather modeling, machine learning applications in marine ecology, and emerging research on AI governance and climate risk assessment (Reichstein et al., 2019; Rolnick et al., 2019). Empirical examples are selected from recent case studies on marine heatwaves, biogeochemical changes, biodiversity shifts, and atmospheric tracer transport, with a focus on model validation, uncertainty quantification, and robustness under distributional change. Instead of presenting new empirical findings, the chapter integrates and evaluates existing evidence in a manner directly relevant to scientific practice and policy-oriented decision-making.

The central argument of this chapter is twofold. First, AI has the potential to improve the prediction of marine climate risks significantly, provided that models are appropriately aligned with the problem context, data characteristics, and decision-making horizon. Second, an uncritical reliance on exclusively data-driven approaches may reinforce existing biases, obscure uncertainty, and erode trust in climate information, especially in governance settings (Gilpin et al., 2018). Consequently, hybrid approaches that incorporate physical constraints, probabilistic reasoning, and domain expertise into AI architectures represent a particularly promising strategy for achieving both predictive accuracy and scientific credibility (Willard et al., 2020; Iglesias-Suarez et al., 2024).

The structure of the chapter is as follows. Section 2 reviews the relevant literature at the intersection of AI, climate modeling, and marine ecology. Section 3 introduces the typology of AI approaches and examines their methodological characteristics. Section 4 relates these model classes to empirical applications in marine climate change. Section 5 addresses data sources, preprocessing challenges, and structural biases in ocean observations. Section 6 explores technical, ethical, and governance considerations, and Section 7 concludes by identifying priorities for future research and policy-oriented implementation of AI in marine climate risk assessment.

LITERATURE REVIEW

AI in Climate and Earth System Modeling

Over the past decade, the application of artificial intelligence to climate and Earth system modeling has grown substantially. Initial research primarily addressed statistical emulation and post-processing of numerical model outputs. More recent developments have integrated deep learning architectures into core modeling workflows. Reviews by Reichstein et al. (2019) and Rolnick et al. (2019) demonstrate that advances in machine learning improve pattern detection, model acceleration, and uncertainty characterization in both atmospheric and oceanic contexts.

A significant advancement has occurred in numerical weather prediction, with deep learning models, particularly transformer-based architectures, demonstrating performance comparable to or surpassing state-of-the-art numerical systems for medium-range forecasts (Lam et al., 2023). Concurrently, related developments in climate modeling address statistical downscaling, bias correction, and the emulation of computationally intensive components such as subgrid processes (Willard et al., 2020). Despite these advances, the literature remains predominantly focused on atmospheric dynamics, with marine applications often treated as secondary or addressed only through coupled models.

Machine Learning in Marine Ecology and Ocean Observation

Within marine science, machine learning is predominantly utilized for ecological and observational tasks rather than comprehensive dynamical prediction. A substantial body of research employs supervised learning methods, including Random Forests, Support Vector Machines, and neural networks, for species distribution modeling, habitat suitability analysis, and biodiversity monitoring (Robinson et al., 2017). These techniques are prevalent in fisheries science, where machine learning models link environmental variables to observed changes in fish stocks and ecosystem structure.

Recently, deep learning techniques have been applied to image- and video-based monitoring of marine ecosystems. Notable applications include automated coral reef mapping, plankton classification, and underwater video analysis, frequently utilizing convolutional neural networks (Villon et al., 2022). Although these methods have enhanced spatial and temporal coverage, the literature consistently highlights ongoing challenges, including data imbalance, limited generalization, and the open-world complexity of marine ecosystems.

Ecological machine learning studies are typically conducted at regional or local scales and maintain only limited connections to climate modeling frameworks. This separation constrains their capacity to address long-term climate risks, compound extremes, or large-scale teleconnections. These gaps underscore the need for stronger integration between marine ecology and climate-focused AI research.

From Prediction to Decision Support: Governance-Relevant AI

A third body of literature investigates artificial intelligence in the context of decision support, risk assessment, and governance. In this domain, the emphasis extends beyond predictive accuracy to encompass interpretability, uncertainty communication, and institutional usability. Gilpin et al. (2018) offer a foundational analysis of explainability challenges in machine learning, which are especially relevant in policy settings that demand accountability and transparency.

In climate-related applications, probabilistic and ensemble-based approaches are increasingly prioritized to support decision-making under uncertainty (Rolnick et al., 2019). Instead of generating single deterministic forecasts, these models produce distributions and scenario ranges that align with risk-based governance frameworks. However, multiple authors caution that AI systems trained on historically biased or incomplete data may reinforce existing inequities, particularly in regions with limited observational coverage, such as much of the Global South.

Synthesis and Research Gaps

Collectively, the literature reveals a fragmented landscape. While AI-driven climate modeling has progressed rapidly, it remains predominantly focused on atmospheric processes. Marine ecological applications of machine learning are methodologically advanced but frequently lack integration with climate risk frameworks. Governance-oriented discussions prioritize transparency and uncertainty, yet seldom address the technical complexities of ocean modeling and data limitations.

This chapter advances the field by explicitly connecting model typology, marine climate applications, and governance relevance. Through synthesizing insights across these domains, it addresses a central gap in the literature: the absence of an integrated framework for evaluating how various classes of AI models contribute to predicting and managing climate change–related risks in marine systems.

TYPOLOGY OF AI APPROACHES ON MARINE CLIMATE RESEARCH

Conceptual Rationale

Artificial intelligence in climate research encompasses a diverse array of methodological approaches, each characterized by distinct epistemic assumptions, data requirements, and appropriateness for specific decision contexts. Many existing studies address AI applications primarily from an application-driven perspective, often neglecting systematic differentiation between underlying model classes. This lack of distinction impedes comparability, obscures methodological trade-offs, and complicates the translation of model outputs into policy-relevant knowledge.

To address this gap, a functional typology is adopted that categorizes AI approaches based on their learning mechanisms, methods of uncertainty representation, integration with physical knowledge, and capacity to support decision-making. Drawing on prior syntheses in climate science and marine ecology (Reichstein et al., 2019; Rolnick et al., 2019; Robinson et al., 2017), four broad classes are identified:

(1) neural networks and deep learning,
(2) probabilistic and ensemble-based models,
(3) reinforcement learning for adaptive scenarios, and
(4) hybrid models combining physics-based constraints with data-driven learning.

This typology does not establish a hierarchy among methods. Instead, it offers an analytical framework for evaluating the suitability of different approaches for specific marine climate risks, data environments, and governance requirements.

Neural Networks and Deep Learning

Neural networks form the foundation of contemporary AI applications in climate science. In marine contexts, convolutional neural networks (CNNs) are particularly effective for analyzing high-dimensional spatial data, such as satellite imagery, facilitating the detection of marine heatwaves, harmful algal blooms, and coral reef degradation (Goodfellow et al., 2016; Villon et al., 2022). For temporally structured data, such as sea surface temperature, oxygen concentration, or pH time series, recurrent neural networks (RNNs) and long short-term memory networks (LSTMs) are commonly used.

Recently, transformer-based architectures have introduced a qualitative advancement in climate-related prediction tasks. Models such as GraphCast demonstrate that global weather forecasts can be produced within minutes, achieving accuracy comparable to leading numerical models across multiple metrics (Lam et al., 2023). These developments indicate that similar architectures may apply to oceanic and coupled climate systems, especially in contexts where long-range dependencies and multivariate interactions are critical.

Despite their predictive capabilities, deep learning models generally require large datasets and are susceptible to overfitting in areas with limited observations. Their restricted interpretability further limits their direct application in governance contexts, where transparency and explainability are essential (Gilpin et al., 2018). Consequently, purely data-driven neural networks are most suitable for well-observed domains and for short- to medium-term prediction horizons.

Probabilistic Models and Ensemble Learning

A second class of AI approaches emphasizes uncertainty representation rather than point prediction. Probabilistic models—including Bayesian networks, Random Forest ensembles, and boosting methods—are widely used in climate and marine research to estimate the likelihood of extreme events and ecosystem shifts. By generating distributions rather than single forecasts, these models support risk-based decision-making under uncertainty.

In marine applications, ensemble approaches are particularly valuable for assessing the probability of marine heatwaves, El Niño–related impacts, and harmful algal blooms (Robinson et al., 2017). Their capacity to integrate heterogeneous data sources—satellite observations, in situ measurements, and model outputs—makes them well suited for environments characterized by incomplete and noisy data.

However, probabilistic models face their own limitations. Ensemble performance depends critically on the representativeness of training data, and uncertainty intervals may be misinterpreted or ignored in policy contexts. Moreover, probabilistic methods often rely on simplified representations of underlying dynamics, limiting their ability to capture nonlinear feedbacks central to long-term climate change.

Reinforcement Learning and Adaptive Scenarios

Reinforcement learning (RL) constitutes a distinct paradigm in which agents acquire optimal strategies through interaction with simulated environments. Rather than predicting future states, RL emphasizes identifying adaptive policies that balance competing objectives over time. In the context of marine climate change, RL is particularly relevant for resource management under uncertainty.

Applications encompass dynamic adjustment of fishing quotas, design of adaptive marine protected areas, and analysis of trade-offs between ecological sustainability and economic outcomes (Rolnick et al., 2019). By embedding agents within simulated socio-ecological systems, RL enables the examination of the long-term consequences of alternative management strategies.

However, RL models are susceptible to the specification of reward functions and assumptions regarding system dynamics. Transferring strategies from simulated environments to real-world ocean systems remains a significant challenge, particularly in the presence of regime shifts and non-stationarity associated with climate change. Therefore, RL is best regarded as an exploratory tool for scenario analysis rather than as a direct predictive model.

Hybrid Models: Integrating Physics and Data

Hybrid models integrate machine learning techniques with physical constraints derived from established climate and ocean theory. In contrast to purely data-driven approaches, these models incorporate conservation laws, dynamical equations, or causal structures into learning architectures, thereby enhancing stability and generalization (Willard et al., 2020).

Notable examples include NeuralGCM, which integrates neural networks into general circulation models; mass-conserving tracer transport models such as CarbonBench; and deep-learning-based subgrid parameterizations that replace traditional schemes for clouds or convection. Recent research on causally informed neural networks further demonstrates that constraining input–output relationships can enhance physical plausibility and long-term stability.

Hybrid approaches are especially promising for marine climate applications, where observational data are limited and long-term consistency is critical. The primary challenges include computational complexity, model validation, and ensuring transparency in the interactions between learned and physical components. Nonetheless, these approaches represent a crucial pathway toward the development of robust, policy-relevant AI systems.

Synthesis: Matching Models to Marine Climate Risks

In addition to offering a descriptive classification, the typology developed in this section serves analytical and governance-oriented functions. Rather than organizing AI approaches exclusively by algorithmic families, it distinguishes model classes based on their relationship to physical knowledge, their handling of uncertainty, and their ability to support decision-relevant interpretation. This perspective highlights that the appropriateness of AI methods in marine climate research is determined not only by predictive skill, but also by underlying epistemic assumptions and institutional usability.

From an epistemic perspective, the four model classes serve distinct roles in generating climate-relevant knowledge. Deep learning approaches are primarily designed for pattern recognition and short-term inference, excelling at detecting anomalies and emergent signals in large, high-dimensional datasets. In contrast, probabilistic and ensemble-based models focus on risk characterization by explicitly representing uncertainty and enabling scenario comparisons. Reinforcement learning further shifts the focus from prediction to policy exploration, allowing examination of the long-term implications of adaptive strategies under uncertain future conditions. Hybrid models combine explanation and prediction by embedding physical constraints into data-driven learning, thus preserving causal structure while maintaining statistical flexibility.

This differentiation helps clarify why methodological debates in climate-related AI extend beyond performance benchmarks. Deep learning models, probabilistic approaches, reinforcement learning, and hybrid architectures each involve specific trade-offs among flexibility, robustness, interpretability, and generalizability. Explicitly articulating these trade-offs is essential for determining which approaches are suitable for particular marine climate risks, data environments, and temporal horizons. In marine contexts, where observational data are limited, system dynamics are strongly coupled across scales, and non-stationarity due to climate change is significant, these epistemic choices are especially consequential. Different institutional settings imply different tolerances for uncertainty, opacity, and error. Early warning systems may prioritize speed and sensitivity, whereas long-term planning, regulatory decisions, and resource allocation require transparency, robustness, and credible bounds on uncertainty. By articulating these distinctions, the typology provides a structured basis for aligning AI model choice with specific decision contexts.

The typology thus establishes a conceptual link between methodological diversity and empirical application. It enables systematic mapping of model classes to specific use cases, as discussed in the following section, and provides a foundation for evaluating AI models as both technical tools and integral elements of broader socio-technical systems for marine climate risk assessment. This framing is especially relevant for marine applications, where long-term uncertainty, data scarcity, and high decision stakes require careful alignment of modeling choices with institutional expectations.

EMPIRICAL APPLICATIONS OF AI IN THE CONTEXT OF MARINE CLIMATE CHANGE

Rationale and Scope

Empirical applications of artificial intelligence in marine climate research encompass a wide range of topics, from short-term forecasting of extreme events to long-term assessments of ecosystem change and resource management. Instead of providing a comprehensive catalogue of case studies, this section adopts a comparative perspective by organizing AI applications according to the model typology introduced in Section 3 and mapping them to key classes of marine climate risks.

This section pursues two primary objectives. The first is to identify which AI approaches have demonstrated empirical value for specific marine climate challenges. The second is to clarify the conditions under which these applications are robust, transferable, and policy-relevant, with particular attention to data availability, prediction horizon, and uncertainty.

Extreme Events and Short-Term Marine Climate Risks

AI-based models have achieved notable success in predicting and detecting short-term marine climate extremes, particularly in contexts with dense observational data. Deep learning architectures, such as convolutional neural networks (CNNs) and transformer-based models, are extensively applied to satellite and reanalysis data to detect marine heatwaves, harmful algal blooms, and sea-ice anomalies. These methods are effective at recognizing spatial patterns and temporal precursors of extreme events and can provide near-real-time outputs to support early warning systems.

Probabilistic and ensemble-based methods complement these applications by providing uncertainty estimates essential for operational risk management. Ensemble forecasts of marine heatwaves and El Niño–related anomalies enable decision-makers to assess likelihoods rather than relying solely on deterministic thresholds. However, these models generally operate on short to medium time horizons and may have limited capacity to extrapolate to unprecedented conditions.

Marine Heatwaves – Deep Learning and Probabilistic Early Warning Systems

Marine heatwaves are among the most thoroughly documented short-term climate risks in aquatic systems, with substantial impacts on coral reefs, fish stocks, and coastal ecosystems. Empirical studies demonstrate that convolutional neural networks (CNNs) and ensemble models effectively identify spatiotemporal patterns in satellite data and generate early warning signals. CNN-based methods facilitate high-resolution detection of sea surface temperature anomalies, while probabilistic ensembles yield probability distributions for heatwave intensity and duration. This combination significantly enhances risk communication compared to deterministic predictions by explicitly quantifying uncertainties. Nevertheless, generalizability is limited: models perform best in data-rich regions, whereas uncertainties increase in tropical and polar marine areas. This case illustrates the complementary contributions of deep learning and ensemble methods to short-term risk assessment, though they are less effective for mapping long-term ecological tipping points.

Ecosystem Change, Biodiversity, and Fisheries

A second primary application domain involves marine ecosystems and biodiversity, where AI is primarily utilized to analyze species distributions, habitat suitability, and ecosystem responses to environmental change. Machine learning models, including Random Forests and neural networks, are used to link ecological variables such as temperature, oxygen, and chlorophyll to observed shifts in fish stocks and community composition.

These models have demonstrated empirical value for fisheries management, particularly in identifying climate-driven redistribution of commercially important species. Deep learning also enables large-scale analysis of underwater imagery and video, thereby improving the spatial coverage of coral reefs and benthic habitats. However, most ecological applications remain region-specific and are highly dependent on data quality and representativeness, which limits their capacity to inform global or long-term climate risk assessments.

Adaptive Management and Scenario Exploration

Reinforcement learning has emerged as a promising tool for adaptive management under climate uncertainty. Unlike predictive models, reinforcement learning focuses on learning policies that balance competing objectives such as ecological sustainability and economic yield over time. In marine contexts, reinforcement learning has been applied to explore dynamic fishing strategies, adaptive spatial management, and long-term trade-offs in socio-ecological systems.

Empirically, these applications remain largely exploratory and are primarily based on simulations. Their main contribution is not precise prediction but rather the stress-testing of governance strategies under uncertain future scenarios. Therefore, reinforcement learning should be viewed as a complement to predictive modeling, particularly for evaluating the robustness of management options under deep uncertainty.

Long-Term Projections and Hybrid Modeling Approaches

Long-term marine climate risks, including ocean deoxygenation, acidification, and large-scale circulation changes, present significant challenges for AI applications. Purely data-driven models often lack robustness to distributional changes, whereas traditional physical models are computationally intensive and uncertain at fine spatial scales.

Hybrid approaches address these challenges by integrating machine learning into physically constrained modeling frameworks. Examples include neural general circulation models, mass-conserving tracer transport emulators, and deep-learning-based subgrid parameterizations. Empirical evidence indicates that these models can reproduce key climate statistics and variability while substantially reducing computational costs. However, their primary limitations are complexity, validation requirements, and restricted accessibility outside specialized research environments.

Long-term oceanic changes – hybrid models for tracer and deoxygenation processes

Long-term changes, such as ocean deoxygenation and the large-scale distribution of greenhouse gas tracers, impose specific demands on AI models. Purely data-driven approaches are limited in these contexts because future states often fall outside the historical training range. Hybrid models that integrate machine learning with physical conservation laws present a promising solution. Examples include mass-conserving emulators for atmospheric and oceanic tracer transport and neurally augmented circulation models that efficiently approximate biogeochemical processes. Empirical results demonstrate that these models can reproduce key statistical properties and variability patterns while significantly reducing computational effort compared to classical models. Their primary strength lies in the robust exploration of long-term scenarios under physical constraints, rather than in precise prediction. This case demonstrates that hybrid approaches add value, particularly when long-term consistency, explainability, and governance relevance are prioritized over short-term forecast accuracy.

Comparative Overview of Empirical Applications

Table 1 synthesizes the empirical landscape by linking AI model classes to marine climate applications, data requirements, time horizons, and governance relevance. Instead of ranking approaches by performance, the table emphasizes functional complementarities and trade-offs.

Table 1. Empirical applications of AI approaches in marine climate research

AI Model Class	Typical Methods	Marine Climate Applications	Time Horizon	Key Strengths	Key Limitations / Risks
Neural Networks & Deep Learning	CNNs, RNNs/ LSTMs, Transformers (e.g. GraphCast)	Marine heatwaves, harmful algal blooms, SST anomalies, coral bleaching detection	Short to medium term	High predictive accuracy, real-time capability, scalability	Data-intensive, limited interpretability, weak extrapolation
Probabilistic & Ensemble Models	Random Forests, Boosting, Bayesian methods	Extreme-event risk, species distribution, ecosystem sensitivity	Short to medium term	Explicit uncertainty quantification, decision support	Dependence on data quality, simplified dynamics
Reinforcement Learning	Model-free and model-based RL	Adaptive fisheries management, scenario exploration	Long-term (exploratory)	Policy stress-testing, adaptive strategies	Simulation dependence, reward design sensitivity
Hybrid Models (Physics + ML)	NeuralGCM, CarbonBench, DL subgrid schemes	Long-term climate projections, tracer transport, circulation impacts	Medium to long term	Physical consistency, improved generalization	Computational complexity, validation effort

Source: own representation.

Synthesis

Empirical evidence demonstrates that no single AI approach is universally optimal for marine climate risk assessment. Effectiveness depends on aligning model class, data environment, prediction horizon, and governance context. Deep learning and ensemble methods are most effective for short-term risks, while hybrid approaches are necessary for long-term and structurally uncertain processes. Reinforcement learning contributes by expanding the analytical space of possible responses rather than by enhancing prediction. This synthesis informs the subsequent discussion of data constraints, bias, and governance challenges in Section 5. It reinforces the central argument of this chapter: AI most effectively enhances marine climate risk assessment when applied in a pluralistic and problem-oriented manner.

DATA FOUNDATIONS, PREPROCESSING, AND STRUCTURAL BIASES

Data Sources for AI-Based Marine Climate Research

The quality, coverage, and consistency of the underlying data fundamentally determine the performance and reliability of AI-based climate models. In marine environments, data sources are inherently heterogeneous and unevenly distributed across spatial, vertical, and variable dimensions. Three primary categories of data sources are especially pertinent.

Satellite-based observations offer near-global coverage of surface-related variables, including sea surface temperature, chlorophyll a, sea level anomalies, surface currents, and sea ice extent. Missions such as Sentinel (ESA), MODIS (NASA), and VIIRS are essential for monitoring large-scale marine processes and short-term extremes. The primary advantage of these systems is their spatial continuity and temporal frequency. However, most variables are limited to surface layers and are susceptible to atmospheric interference.

In-situ observing systems provide vertical structure and biogeochemical detail that remote sensing cannot capture. The global Argo program, comprising over 4,000 autonomous profiling floats, delivers temperature and salinity profiles to depths of 2,000 meters. Extensions such as BGC-Argo and Deep Argo increasingly measure oxygen, nutrients, pH, and deep-ocean processes (Roemmich et al., 2019). Additional in-situ sources include moored buoys, ship-based measurements, and autonomous underwater vehicles such as gliders. These datasets are critical for training and validating AI models that seek to represent subsurface and ecosystem-relevant dynamics.

Model-based re-analyses, such as ERA5 and other ocean reanalysis products, integrate observations with physical models using data assimilation techniques. These products are widely employed as training targets or reference datasets in AI applications, especially in regions with persistent observational gaps. Although re-analyses provide internally consistent, gridded datasets, they also incorporate the assumptions and biases inherent in the underlying physical models.

Preprocessing and Data Integration

Transforming diverse data sources into inputs suitable for AI models requires extensive preprocessing. In marine applications, preprocessing is a significant determinant of model behavior, bias, and interpretability rather than a simple technical preparatory step. Decisions made during this stage fundamentally influence the patterns that AI models can learn, the uncertainties that are amplified or suppressed, and the interpretability of results in scientific and policy contexts.

Key preprocessing steps include quality control and noise reduction, particularly for satellite data affected by cloud cover, atmospheric interference, or sensor artefacts. Gap filling and interpolation are often necessary to generate continuous spatio-temporal fields, especially in sparsely observed ocean regions. Bias correction and harmonization are critical when integrating datasets from different sensors, platforms, or time periods, as uncorrected inconsistencies may otherwise dominate the patterns learned by AI models. Data assimilation, which combines observations into physically consistent model fields, is closely related. For many AI applications, assimilation products provide training data that balance empirical observations with dynamical coherence. However, these products also embed the structural assumptions of the underlying physical models, which can be propagated and potentially reinforced by AI systems.

Feature engineering converts raw variables into physically meaningful indicators, such as ocean heat content, anomaly indices, or biogeochemical proxies, thereby improving learning efficiency and interpretability. Aggregation choices regarding spatial resolution, temporal averaging, and anomaly definitions directly affect model sensitivity to extremes, trends, and regime shifts. Excessive smoothing can dampen early warning signals, whereas overly fine-grained representations may amplify noise and reduce robustness. These trade-offs are especially significant in marine systems, where localized extremes such as heatwaves or hypoxic events occur within the context of large-scale circulation patterns.

Multimodal data integration is critical in marine research. Many ecologically relevant variables, such as oxygen concentration, nutrient availability, or carbonate chemistry, are sparsely observed and cannot be inferred from remote sensing alone. Practical AI applications require combining satellite-derived surface indicators with in-situ measurements and model-based reanalyses. AI models trained on a single data modality are at risk of systematic blind spots, particularly when extrapolated across regions or time periods.

Recognizing preprocessing as an epistemic bottleneck has significant implications. Preprocessing pipelines implicitly encode assumptions about spatial continuity, temporal stability, and variable relevance. If these assumptions are undocumented, AI models may internalize preprocessing artefacts as genuine climate signals. Transparent documentation of preprocessing choices, sensitivity analyses regarding key assumptions, and the use of ensemble preprocessing strategies are central to responsible AI design. Thus, preprocessing is a core component of scientifically robust, governance-relevant AI-based marine climate research rather than an ancillary technical step.

Structural Biases and Data Inequality

In addition to technical preprocessing challenges, AI-based marine climate models are influenced by structural biases inherent in the global observation system. Data availability is geographically uneven; the North Atlantic and North Pacific are comparatively well observed, whereas tropical regions, marginal seas, polar areas, and much of the Global South remain under-sampled (Tanhua et al., 2019). This imbalance introduces geographic bias into AI models, which typically perform best in data-rich regions and display greater uncertainty in under-observed areas.

Further biases result from sampling and temporal heterogeneity. Historical datasets frequently reflect changes in measurement technologies, mission priorities, and geopolitical interests, leading to discontinuities and regime-specific blind spots. AI models trained on these data risk encoding historical contingencies as structural features.

Distributional shift under climate change presents an additional challenge. As ocean conditions extend beyond the historical range represented in training data, model generalization becomes increasingly uncertain. This limitation is especially significant for long-term projections of deoxygenation, acidification, and ecosystem tipping points.

Implications for Model Robustness

These data constraints directly affect the robustness and relevance of AI-based marine climate predictions to governance. Models trained on biased or incomplete datasets may exhibit high apparent accuracy while systematically misrepresenting risks in under-observed regions. Such distortions are particularly problematic in policy contexts that depend on AI-supported risk assessments for adaptation planning or resource management.

The literature discusses several mitigation strategies. Hybrid modeling approaches that incorporate physical constraints, such as mass or energy conservation, help reduce overfitting and improve extrapolation under distributional change (Willard et al., 2020). Probabilistic and ensemble methods explicitly represent uncertainty, thereby making model limitations more transparent to decision-makers. At a systemic level, expanding global observation infrastructures through initiatives such as GOOS, OneArgo, and BGC-Argo is essential to prevent the reinforcement of existing epistemic inequalities.

Synthesis

Data foundations and preprocessing are central determinants of model performance, bias, and legitimacy in AI-based marine climate research. The aquatic environment intensifies these challenges due to observational sparsity, depth-dependent processes, and rapid non-stationarity associated with climate change. Addressing these constraints requires both technical solutions and institutional investment in global data infrastructures, as well as transparent communication of uncertainty.

These issues provide the foundation for the discussion in Section 6, which examines the ethical, governance, and policy implications of deploying AI-based climate models in decision-making contexts.

ETHICAL AND GOVERNANCE-RELATED IMPLICATIONS ETHICAL OF AI-BASED MARINE CLIMATE MODELS

From Predictive Skill to Institutional Use

The growing application of AI in marine climate research introduces ethical and governance challenges that surpass considerations of technical performance. Although predictive accuracy is essential for scientific credibility, it alone does not ensure institutional adoption within policy and management settings. Decision-makers require models that demonstrate not only predictive skill but also transparency, robustness, and accountability, especially when outputs inform critical decisions regarding climate adaptation, fisheries management, or coastal protection.

A primary governance challenge involves translating AI outputs into actionable knowledge. Many advanced AI models, particularly deep learning architectures, function as black boxes, which makes it difficult to determine how specific inputs affect predictions. This lack of transparency complicates the attribution of responsibility and erodes trust, especially when model outputs diverge from established assessments or yield regionally inconsistent results (Gilpin et al., 2018).

Transparency, Explainability, and Trust

Explainability is now widely regarded as essential for the legitimate application of AI in climate governance. In marine contexts, the need for explainability is heightened by data sparsity and uncertainty, which increase the risk of spurious correlations and overconfident predictions. In the absence of interpretable mechanisms, stakeholders may find it difficult to determine whether AI-derived signals represent genuine climate risks or are artefacts of data limitations and model design.

Multiple strategies have been proposed to address this challenge. Post-hoc interpretability methods, simplified surrogate models, and physics-informed architectures can enhance transparency while maintaining predictive performance (Willard et al., 2020). However, explainability entails trade-offs: more interpretable models may lack flexibility, whereas highly complex models may defy meaningful explanation. Governance frameworks must therefore define acceptable levels of opacity in relation to the significance of the decisions being supported.

Equity, Data Access, and Epistemic Imbalances

Ethical concerns also stem from the uneven distribution of data and computational resources worldwide. As noted in Section 5, AI models generally achieve higher performance in data-rich regions, which are primarily situated in the Global North. This disparity increases the risk that AI-based climate services will systematically underrepresent risks in data-poor regions, thereby perpetuating existing epistemic and political inequalities.

From a governance standpoint, this imbalance prompts concerns regarding fairness and inclusion. When AI-based assessments inform international climate finance, adaptation priorities, or conservation planning, biased models may unintentionally disadvantage already vulnerable regions. Mitigating these risks necessitates both technical interventions, such as uncertainty-aware modeling, and institutional commitments to open data, capacity building, and inclusive model development (Tanhua et al., 2019; UNESCO-IOC, 2020).

Accountability and Model Responsibility

Another unresolved governance issue involves accountability for decisions supported by AI. In contrast to traditional physical models, AI systems typically involve multiple actors, including data providers, model developers, platform operators, and end users. This distribution of responsibility complicates accountability when predictions are inaccurate or lead to adverse outcomes.

Hybrid modeling approaches partially address this issue by grounding AI components in established physical principles, which clarifies the domain of model validity. Probabilistic and ensemble methods further enhance accountability by explicitly conveying uncertainty instead of providing single-point predictions. However, institutional frameworks remain insufficiently developed for the validation, documentation, and auditability of AI-based climate models.

Governance-Oriented Design Principles

Collectively, these considerations indicate that ethical and governance challenges should not be regarded as external constraints applied after model development. Instead, they necessitate governance-oriented design principles that incorporate transparency, uncertainty representation, and equity considerations into AI systems from the beginning.

Instead of pursuing universal solutions, governance requirements should be tailored to the specific decision context and associated risk level. For example, early warning systems for marine heatwaves may permit greater opacity than long-term planning tools that affect resource allocation or regulatory decisions. Thus, ethical AI in marine climate research emphasizes aligning model capabilities with institutional expectations and societal values rather than achieving technical perfection.

Synthesis

AI-based models offer significant potential to improve marine climate risk assessment, yet their legitimacy for governance depends on factors beyond predictive performance. Transparency, equity, and accountability are central criteria for responsible deployment. While hybrid modeling strategies, probabilistic outputs, and expanded observation infrastructures address some challenges, institutional frameworks must also advance to support these developments.

These governance challenges inform the concluding discussion in Section 7, which synthesizes the chapter's findings and identifies priorities for future research and the policy-oriented integration of AI in marine climate risk assessment.

DISCUSSION AND CONCLUSION

Synthesis of Findings

This chapter critically examines the contributions of artificial intelligence to the prediction of climate change–related risks in marine systems. Rather than presenting AI as a single solution, the analysis demonstrates that different AI model classes fulfill distinct analytical roles, rely on specific data environments, and differ in their appropriateness for governance and decision-making.

The typology developed in Section 3 offers a comprehensive framework for organizing this diversity. Deep learning approaches are practical for pattern recognition and short-term prediction in data-rich contexts. Probabilistic and ensemble-based models facilitate risk assessment under uncertainty, while reinforcement learning supports adaptive scenario exploration. Hybrid models integrate physical consistency with data-driven flexibility. Empirical applications reviewed in Section 4 indicate that no single model class is universally optimal for all marine climate challenges. Instead, predictive value depends on aligning model selection with the specific risk, available data, and decision horizon.

Implications for Marine Climate Risk Assessment

The findings indicate that AI most effectively enhances marine climate risk assessment when applied selectively and with contextual awareness, rather than as a complete substitute for physical models. Short-term risks, such as marine heat-waves or harmful algal blooms, benefit from deep learning and ensemble methods. In contrast, long-term, structurally uncertain processes, such as deoxygenation or circulation changes, require hybrid approaches that maintain physical constraints.

The analysis further demonstrates that data foundations and preprocessing choices are central determinants of model behavior, bias, and legitimacy, rather than secondary technical considerations. Structural data inequalities, observational gaps, and distributional changes resulting from climate change impose significant constraints on model generalization. Without explicit representation of uncertainty and transparent communication of limitations, AI-based predictions may overstate confidence and mislead decision-makers.

Governance-Relevant Insights

From a governance perspective, the chapter identifies a fundamental shift: the value of AI in marine climate research is determined not only by predictive accuracy but also by institutional usability. Explainability, uncertainty quantification, and equitable data access are essential criteria for responsible deployment. While hybrid modeling strategies and probabilistic outputs provide partial solutions, they must be supported by institutional frameworks that establish standards for validation, documentation, and accountability.

The analysis also suggests that ethical considerations and governance challenges are closely linked to technical design choices. Models optimized exclusively for performance may be unsuitable for policy contexts. In contrast, governance-oriented design principles, such as calibrated transparency and risk-sensitive model selection, can enhance trust and legitimacy while maintaining scientific rigor.

Contribution of the Chapter

The primary contribution of this chapter is the development of an integrated analytical framework that connects AI model typology, empirical marine climate applications, data constraints, and governance implications. By synthesizing insights from climate modeling, marine ecology, and AI governance, the chapter advances beyond application-specific case studies to offer a coherent perspective on AI-enabled marine climate risk assessment. In this way, the chapter addresses a significant gap in the literature: the absence of structured guidance for evaluating, comparing, and deploying different AI approaches in relation to specific marine climate risks and decision contexts.

Limitations and Future Research Directions

Several limitations warrant acknowledgment. First, the chapter is based on published empirical studies and thus reflects existing observational and methodological biases, especially the predominance of data-rich regions and short-term applications. Second, the rapid evolution of AI architectures may change performance benchmarks, requiring ongoing reassessment of best practices.

Future research should prioritize three areas. First, expanding global ocean observation systems to address structural data inequalities, second, developing standardized benchmarks and evaluation protocols for AI-based marine climate models, particularly for long-term and out-of-distribution scenarios, and third, enhancing interdisciplinary collaboration among climate scientists, ecologists, data scientists, and governance experts to ensure that AI tools are both scientifically robust and institutionally relevant.

Concluding Remarks

Artificial intelligence offers significant potential for enhancing the prediction and management of climate change–related risks in marine systems. However, this potential depends on careful model selection, transparent uncertainty management, and alignment with governance requirements. AI should be regarded not as a universal solution, but as a complementary element within a diverse modeling ecosystem.

When integrated into physically informed frameworks, supported by robust data infrastructures, and guided by governance-oriented design principles, AI can contribute meaningfully to marine climate risk assessment. The primary challenge is not the development of increasingly complex models, but the responsible, equitable, and transparent deployment of existing tools with a clear understanding of their limitations.

REFERENCES

Gilpin, L. H., Bau, D., Yuan, B. Z., Bajwa, A., Specter, M., & Kagal, L. (2018). Explaining explanations: An overview of interpretability of machine learning. 2018 IEEE 5th International Conference on Data Science and Advanced Analytics (DSAA), 80–89. https://doi.org/DOI: 10.1109/DSAA.2018.00018

Goodfellow, I., Bengio, Y., & Courville, A. (2016). *Deep Learning*. MIT Press.

Hoegh-Guldberg, O., Jacob, D., Bindi, M., Brown, S., Camilloni, I., Diedhiou, A., . . . Zougmoré, R. B. (2018). Impacts of 1.5 C global warming on natural and human systems. In *Global warming of 1.5 C.: An IPCC Special Report* (pp. 175-311). IPCC Secretariat.

Iglesias-Suarez, F., López-Santiago, J., Arias, P. A., Palmer, T. N., & Wehrli, K. (2024). Causally-informed deep learning to improve climate models and projections. *Journal of Geophysical Research: Atmospheres, 129*(18), e2024JD040073. https://doi.org/DOI: 10.1029/2024JD040073

Intergovernmental Oceanographic Commission of UNESCO (IOC-UNESCO). (2020). *Global Ocean Science Report 2020: Charting capacity for ocean sustainability*. UNESCO Publishing., DOI: 10.18356/9789210054527

IPCC. (2021). *Climate Change 2021: The Physical Science Basis. Contribution of Working Group I to the Sixth Assessment Report of the Intergovernmental Panel on Climate Change*. Cambridge University Press., DOI: 10.1017/9781009157896

Lam, R., Sanchez-Gonzalez, A., Willson, M., Wirnsberger, P., Fortunato, M., Alet, F., Ravuri, S., Ewalds, T., Eaton-Rosen, Z., Hu, W., Merose, A., Hoyer, S., Holland, G., Vinyals, O., Stott, J., Pritzel, A., Mohamed, S., & Battaglia, P. (2023). Learning skillful medium-range global weather forecasting. *Science*, *382*(6677), 1416–1421. DOI: 10.1126/science.adi2336 PMID: 37962497

Palmer, T. N. (2017). The primacy of doubt: Evolution of numerical weather prediction from determinism to probability. *Journal of Advances in Modeling Earth Systems*, *9*(2), 730–734. DOI: 10.1002/2017MS000999

Reichstein, M., Camps-Valls, G., Stevens, B., Jung, M., Denzler, J., Carvalhais, N., & Prabhat, . (2019). Deep learning and process understanding for data-driven Earth system science. *Nature*, *566*(7743), 195–204. DOI: 10.1038/s41586-019-0912-1 PMID: 30760912

Robinson, N. M., Nelson, W. A., Costello, M. J., Sutherland, J. E., & Lundquist, C. J. (2017). A systematic review of marine-based species distribution models (SDMs) with recommendations for best practice. *Frontiers in Marine Science*, *4*, 421. Advance online publication. DOI: 10.3389/fmars.2017.00421

Roemmich, D., Alford, M. H., Claustre, H., Johnson, K. S., King, B., Moum, J., Oke, P., Owens, W. B., Pouliquen, S., Purkey, S., Scanderbeg, M., Suga, T., Wijffels, S., Zilberman, N., Bakker, D., Baringer, M., Belbeoch, M., Bittig, H. C., Boss, E., & Yasuda, I. (2019). On the future of Argo: A global, full-depth, multi-disciplinary array. *Frontiers in Marine Science*, *6*, 439. DOI: 10.3389/fmars.2019.00439

Rolnick, D., Donti, P. L., Kaack, L. H., Kochanski, K., Lacoste, A., Sankaran, K., ... (2019). Tackling climate change with machine learning. *arXiv*. https://doi.org//arXiv.1906.05433DOI: 10.48550

Smale, D. A., Wernberg, T., Oliver, E. C., Thomsen, M., Harvey, B. P., Straub, S. C., Burrows, M. T., Alexander, L. V., Benthuysen, J. A., Donat, M. G., Feng, M., Hobday, A. J., Holbrook, N. J., Perkins-Kirkpatrick, S. E., Scannell, H. A., Sen Gupta, A., Payne, B. L., & Moore, P. J. (2019). Marine heatwaves threaten global biodiversity and the provision of ecosystem services. *Nature Climate Change*, *9*(4), 306–312. DOI: 10.1038/s41558-019-0412-1

Tanhua, T., McCurdy, A., Fischer, A., Appeltans, W., Bax, N., Currie, K., DeYoung, B., Dunn, D., Heslop, E., Glover, L. K., Gunn, J., Hill, K., Ishii, M., Legler, D., Lindstrom, E., Miloslavich, P., Moltmann, T., Nolan, G., Palacz, A., & Wilkin, J. (2019). What we have learned from the framework for ocean observing: Evolution of the global ocean observing system. *Frontiers in Marine Science*, *6*, 471. Advance online publication. DOI: 10.3389/fmars.2019.00471

Villon, S., Iovan, C., Mangeas, M., & Vigliola, L. (2022). Confronting deep-learning and biodiversity challenges for automatic video-monitoring of marine ecosystems. *Sensors (Basel)*, *22*(2), 497. DOI: 10.3390/s22020497 PMID: 35062457

Willard, J., Jia, X., Xu, S., Steinbach, M., & Kumar, V. (2020). Integrating scientific knowledge with machine learning for engineering and environmental systems. *ACM Computing Surveys*, *53*(4), 1–37. DOI: 10.1145/3514228

Chapter 9

Looking at the Problem of Access to Freshwater Through the Lens of Climate Change and Artificial Intelligence

Erkin Cihangir Karataş
https://orcid.org/0000-0002-4906-9696
Independent Researcher, Turkey

ABSTRACT

This study investigates the global challenge of accessing safe freshwater through the perspectives of climate change and artificial intelligence (AI). Freshwater, only 2.5% of global water, is vital for life, yet over 2.1 billion people lack safe access. Climate change disrupts water cycles via rising temperatures, altered precipitation, glacial melt, and extreme weather, worsening scarcity and quality issues across regions like South Asia, Africa, and Europe. AI offers transformative solutions in real-time monitoring, efficient treatment, optimized irrigation, and disaster management, with successful cases in Singapore, India, and Africa. However, high costs, data dependency, ethical concerns, and limited local capacity hinder AI adoption. The study debates water as a human right versus a commodity, advocating a balanced approach to ensure equitable access and environmental sustainability. Inclusive, ethical AI policies and global cooperation are essential to address this crisis, aligning with Sustainable Development Goals for social justice and sustainable water management.

DOI: 10.4018/979-8-3373-5238-1.ch009

INTRODCUTION

Water is a vital resource for all known life forms on Earth; all life on the planet depends on it. Consequently, in searches for extraterrestrial life, the presence of water is the primary indicator sought, as its detection provides necessary conditions for many potential life forms. On Earth, water exists as fresh or salt water; although salt water supports numerous species, it is not suitable for all life. Many terrestrial species, including humans, depend directly on fresh water. Humans have historically settled near fresh water sources, shaping the civilizations they built around water availability. Water is essential not only for the continuation of natural life but also for meeting basic human needs such as drinking and domestic use. Beyond these, it serves as a fundamental input across all economic sectors: agricultural and industrial activities, critical to modern life, cannot function without it—used for irrigation and product washing in agriculture, and for cooling and cleaning in industry. Thus, water acts as a direct or indirect input in nearly all production processes and must meet specific quality requirements. Not all water types are suitable for every purpose; sources must be inherently appropriate or treated accordingly, with fresh water being the most suitable for human needs. Fresh water sources are classified as surface or groundwater, where surface waters are generally more accessible and preferred. While salt water dependence exists, humans rely far more directly on fresh water to sustain life. Therefore, this study focuses exclusively on fresh water, with the term "water" hereafter referring to fresh water.

Globally, only 2.5% of water is fresh, the vast majority being oceanic salt water. Of this 2.5%, 68.7% is locked in glaciers, 30.1% in groundwater, and merely 1.2% in surface sources—the most accessible fraction for human use. Groundwater, though usable, is less accessible and costlier to extract, while glacial water is largely unavailable due to high costs and natural barriers. Beyond these distributional constraints, water resources face a critical limitation: availability. Although the hydrologic cycle makes water renewable in principle, contrary to common perception, it is a finite resource subject to depletion. Availability defines the safe withdrawal rate; excessive extraction creates pressure known as water stress, manifesting in quantity and quality forms. Quantity pressure occurs when withdrawal depletes the resource, leaving insufficient volume for the hydrologic cycle and dependent ecosystems. Quality pressure arises when pollution renders water unsuitable for natural life or human use, even if volume is adequate. Both types can reach critical levels. Today, water resources face mounting pressure from human activities and anthropogenic climate change; a fully circular water economy remains unachieved, with most activities exerting pressure on resources. Access to fresh water has become a global challenge due to population growth, unsustainable practices, declining availability, climate change, and economic disparities. Severity varies regionally, influenced by development, education, population density, and geography. Additionally, safety is critical: water from quality-stressed sources may be unsafe, adversely affecting large populations worldwide.

The world is undergoing irreversible global transformations: climate change appears poised to permanently alter climatic patterns, while artificial intelligence technology promises to reshape socio-economic structures, akin to the Industrial Revolution. Against these pivotal variables, how might the fresh water access problem evolve? This constitutes the central research question. The study examines climate change's impact on the issue while exploring whether artificial intelligence can serve as a tool to alleviate it. Employing a qualitative research method, the study comprises four parts. The first reviews literature to discuss whether fresh water constitutes a fundamental human right or a commercial commodity, offering insight into the problem's core nature. The second assesses global fresh water access status, addressing availability and safety dimensions while examining economic aspects and potential parallels between access issues and economic development levels. The third evaluates climate change's effects on the problem. The fourth explores potential contributions of artificial intelligence to improving safe fresh water access and distribution.

Research Problem, Gap, and Objectives

The global challenge of access to safe freshwater is exacerbated by two major contemporary forces: irreversible climate change and the rapid advancement of artificial intelligence (AI). While climate change intensifies water scarcity through altered hydrological cycles, extreme weather events, and quality degradation, AI emerges as a potential transformative tool for monitoring, optimization, and equitable distribution. However, existing literature often examines these dimensions in isolation—focusing either on climate impacts (IPCC, 2021; UNESCO, 2020), water governance and rights (Akel Ünal, 2020; Barlow, 2016), or AI applications in discrete water management tasks (Zhang & Ng, 2024). There remains a significant research gap in integrative analyses that simultaneously address how climate change aggravates freshwater access problems and how AI can serve as a mitigation instrument within governance and equity constraints. This chapter fills this gap by providing a qualitative synthesis that links climate-induced water insecurity with AI-enabled solutions, emphasizing ethical, social, and institutional dimensions. The primary research question guiding this study is:

How can artificial intelligence technologies contribute to alleviating the problem of safe freshwater access in the context of climate change?Specific objectives are:

- To examine the debate on freshwater as a human right versus a commodity and its implications for access.
- To assess the current global state of safe freshwater access, including regional and socioeconomic disparities.
- To analyze the multifaceted impacts of climate change on freshwater quantity, quality, and accessibility.
- To evaluate the potential contributions of AI in monitoring, treatment, prediction, distribution, and policy development for sustainable water management.
- To identify ethical challenges, social equity issues, and governance barriers associated with AI deployment in water security.

Methodology

This study employs a qualitative synthesis approach, drawing on an extensive review of peer-reviewed academic literature, international reports (e.g., IPCC, UNESCO, WHO, FAO), and policy documents published primarily between 2015 and 2025. Sources were selected based on relevance to the intersection of climate change, freshwater access, and AI applications, prioritizing high-impact journals, authoritative institutional reports, and empirical case studies with global coverage. Evidence was analyzed thematically to identify patterns, contradictions, and opportunities across governance, technical, and socio-ethical dimensions. No primary data collection was conducted; the synthesis integrates existing knowledge to provide a holistic framework.

Key Terms and Definitions

To ensure clarity, the following key terms are defined as used in this chapter:

- Freshwater: Water with salinity less than 0.5 parts per thousand, encompassing surface and groundwater sources suitable for human consumption and ecosystems (UNESCO, 2020).
- Safely managed drinking water: Drinking water from an improved source that is accessible on premises, available when needed, and free from faecal and priority chemical contamination (WHO/UNICEF JMP, 2025).
- Water scarcity: The condition where demand for water exceeds sustainable supply, encompassing physical scarcity (quantity) and economic scarcity (access/infrastructure) (Falkenmark & Lindh, 1974; Mekonnen & Hoekstra, 2016).
- Integrated Water Resources Management (IWRM): A process that promotes coordinated development and management of water, land, and related re-

sources to maximize economic and social welfare equitably without compromising sustainability (UN-Water, 2019).

- Artificial Intelligence (AI) in water management: The application of machine learning, deep learning, IoT integration, and data analytics to optimize water monitoring, treatment, distribution, forecasting, and decision-making (Zhang & Ng, 2024).
- Digital divide: Unequal access to digital technologies and data, potentially exacerbating social inequalities in AI-driven water solutions (Olawade et al., 2024).

Conceptual Framework

This chapter adopts an integrated conceptual framework that links climate change as a primary driver of freshwater insecurity with governance and socio-economic factors as mediating variables, and artificial intelligence as a potential enabling intervention (Figure 1).

The framework rests on three core pillars derived from the literature on water security and sustainable development:

- Climate Change as Driver: Rising temperatures, altered precipitation patterns, glacial melt, and extreme events directly reduce freshwater quantity and quality while increasing demand (IPCC, 2021; Trenberth, 2011; Huss & Hock, 2018).
- Mediating Factors:
    - Governance and institutional structures (e.g., water rights regimes, integrated water resources management – IWRM) determine allocation efficiency and equity (Akel Ünal, 2020; Hodgson, 2006).
    - Socio-economic dimensions including poverty, gender disparities, and digital divide influence vulnerability and access (UN Women, 2018; Olawade et al., 2024).
- AI as Enabling Tool: AI applications (monitoring, forecasting, optimization, equitable distribution) can mitigate impacts and support adaptive governance, but only if ethical risks (bias, privacy, exclusion) and implementation barriers are addressed (Zhang & Ng, 2024; Kae, 2023).

The framework illustrates feedback loops: unmitigated climate impacts exacerbate scarcity, which strains governance and deepens social inequalities; conversely, effective AI-supported IWRM can reduce pressure and enhance resilience.

Figure 1. Conceptual Framework Linking Climate Change, Freshwater Access, Governance, and AI Interventions

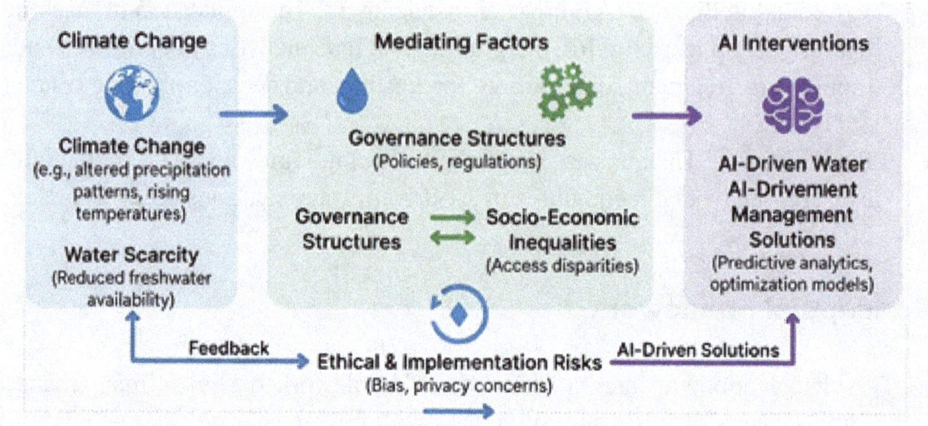

This framework provides analytical coherence to the chapter, guiding the sequential discussion from problem characterization, climate impacts, to AI-enabled solutions and limitations.

IS WATER A FUNDEMENTAL RIGHT OR A COMMODITY?

Before addressing global freshwater access challenges and the current state of water access, it is useful to consider whether water is a fundamental right or a commodity. If viewed as a fundamental right, everyone should have access to it, though the scope and limits require clarification—issues beyond this study's scope. Conversely, treating water as a commodity implies unequal access. Both extremes (unlimited right or fully tradable commodity) pose challenges. Thus, establishing a balanced perspective by examining the pros and cons of each approach is essential before evaluating global access.

Legally, the "right to water" entails sufficient, safe, physically and economically accessible water for human needs. Due to its vital role, it is deemed inalienable. This right may be enshrined constitutionally, legislatively, or through court decisions, influenced by international law, and recognized as a social right intertwined with rights to a healthy environment, health, and life. It functions as both individual and collective, with some countries emphasizing public ownership and sustainable development as state duties. Jurists regard it as a human right at individual and societal levels, guaranteeing basic needs (Akel Ünal, 2020).

Water ownership concerns resources or extracted water, but many systems prioritize use rights (for drinking, irrigation, energy, industry, etc.), independent of ownership or property rights. In Roman law, perpetual resources (seas, lakes) were res communes omnium (common to all), non-privatizable, while finite ones could be private; everyone could take needed amounts from commons. Continental European systems, influenced by this, often place permanent resources under state sovereignty (e.g., France, Spain, Germany, Italy), distinct from private ownership: the state holds usus and fructus but not abusus. Private rights may apply to intermittent sources, but use of state resources requires permits, licenses, or fees; rights are temporary, conditional, non-fully transferable, and revocable (Akel Ünal, 2020). In Anglo-Saxon systems, use rights dominate, with less state sovereignty and more market influence, based on riparianism (landowners bordering water have reasonable use rights without impairing flow/quality; roots in Roman law) (LII, 2024) and prior appropriation ("first in time, first in right"), prevalent in scarce regions like the American West. Here, rights depend on historical priority for beneficial use, stemming from settlers' distrust of government and self-developed customs (U.S. Department of the Interior, 2017). Barlow (2016) argues this doctrine enabled separating water from land rights, forming the basis for commodification.

Both Continental (state-dominant) and Anglo-Saxon (market-oriented) systems have strengths and weaknesses in resource protection and access. Criticism of the state-based system centers on inefficiency from excluding market pricing, which proponents claim would promote efficient allocation, vigilant protection, and rational consumption by redirecting to high-value uses and signaling scarcity. Examples illustrate these issues: In Vietnam, state control with private processing aligns with market efficiency in costs/investments (Nguyen, 2019). Conversely, full state control caused disasters like the USSR's Aral Sea exploitation for cotton quotas, drying the basin and affecting 3.5 million people across 690,000 km^2 (Grabish, 2017). In India, subsidized electricity for farmers led to excessive irrigation and resource pressure (Foley, 2020). Criticism of the market-based system is primarily humanitarian: commodification risks undermining the human right to safe water (UN Digital Library, 2010), encouraging profit-driven exploitation and unchecked price rises for a vital, finite resource. Recent cases confirm this: Over 30 years, water rights trading has boomed, sold to agribusiness, bottlers, and investors; in the US, trillions of liters traded to billionaires or black markets; in Spain, government-encouraged trades diverted water to tourism/unsuitable crops; in Australia (early 2000s), private ownership created billion-dollar markets with oversold rights, price hikes, and risky agreements (Barlow, 2016).

These flaws necessitate a modern integrated approach incorporating elements from both systems for flexible adaptation. Disagreement arises from water's economic value (Hodgson, 2006): jurists emphasize human rights, economists public good status—implying equitable access but not free. Pricing debates focus on cost recovery without exceeding affordability (UNDP: max 3% household income; OECD/WB: 5%), avoiding full investment burden, compensating environmental pressure, and using progressive tariffs, subsidies, or quotas (Akel Ünal, 2020). In modern adopters, resources remain state-owned, with conditional temporary transfers prioritizing social benefit; private involvement aids efficiency (e.g., Germany: public ownership, state/private supply) (Umweltbundesamt, 2017). This hybrid recognizes water as a limited human right with economic value, requiring priced access for societal/environmental gains.

ACCESS TO FRESHWATER IN THE WORLD

Despite extensive global literature, reports, and initiatives like the Sustainable Development Goals (SDGs), access to safe freshwater remains a persistent challenge. Progress is uneven worldwide, influenced by countless variables and externalities, including negative impacts from the Industrial Revolution (Malik et al., 2024; Whelan et al., 2022; Wolf et al., 2021) and earlier (Longman et al., 2018), still evident in Europe's water resources today.

Figure 2. Access to Freshwater in the World

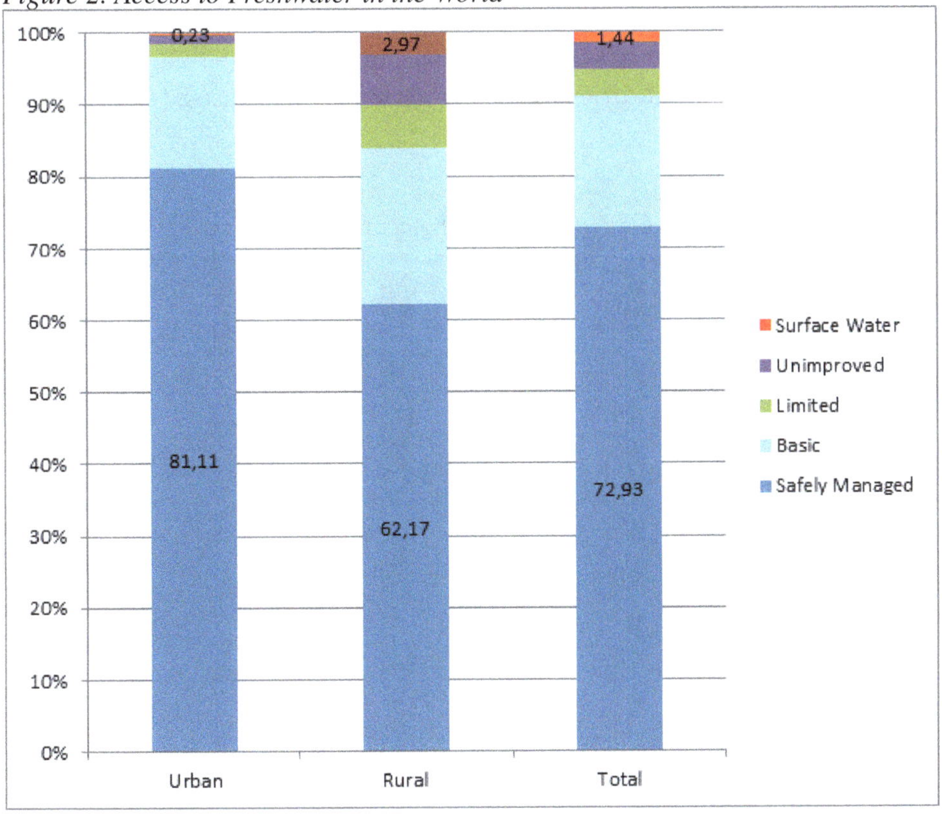

Safely managed: Drinking water from an improved water source that is accessible on premises, available when needed and free from faecal and priority chemical contamination.

Basic: Drinking water from an improved source, provided collection time is not more than 30 minutes for a roundtrip including queuing.

Limited: Drinking water from an improved source for which collection time exceeds 30 minutes for a roundtrip including queuing.

Unimproved: Drinking water from an unprotected dug well or unprotected spring.

Surface water: Drinking water directly from a river, dam, lake, pond, stream, canal or irrigation canal(WASH, 2025).

Source: WASH Data Base

Globally, over 2.1 billion people lack access to safe water: approximately 835 million in urban areas (19% of 4.4 billion urban population) and 1.3 billion in rural areas (38% of 3.4 billion rural population) (World Bank, 2023). Additionally, 111 million rely on unmanaged surface water (0.23% urban: 10.1 million; 2.97% rural:

100.9 million), exposing them to chemical waste, fecal contamination, health risks, and epidemics. Regionally (Chart 2), Europe and North America have the highest safely managed access, while Oceania leads in unmanaged use (~17%, >2.3 million people), and Sub-Saharan Africa has the most people on unmanaged sources (>73 million at >6%).

Figure 3. Distribution of Global Freshwater Access by SDG Regions

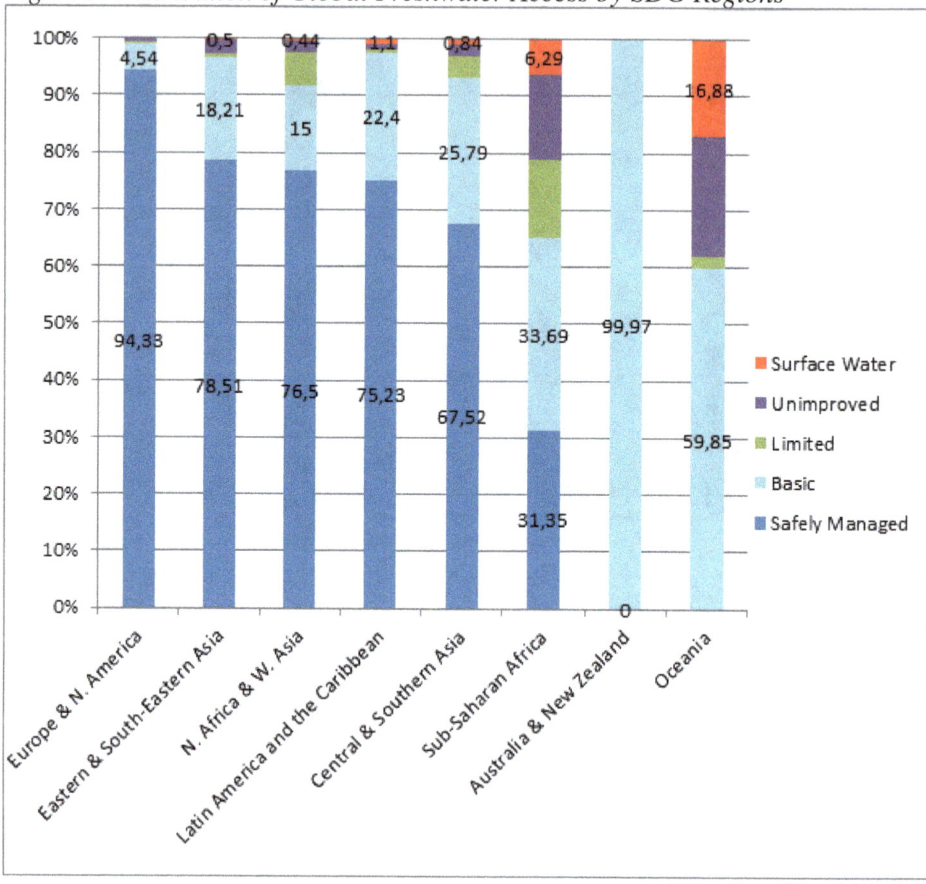

Source: WASH Data Base

Figure 4. Access to Freshwater by Income

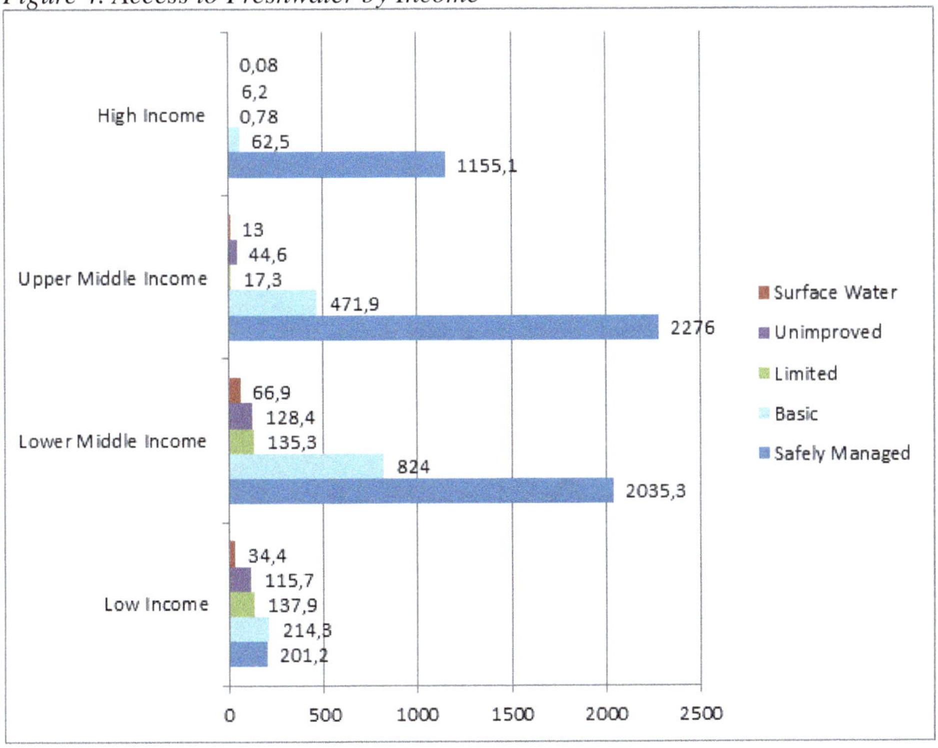

Source: WASH Data Base

Figure 5. Access to Freshwater by Income (%)

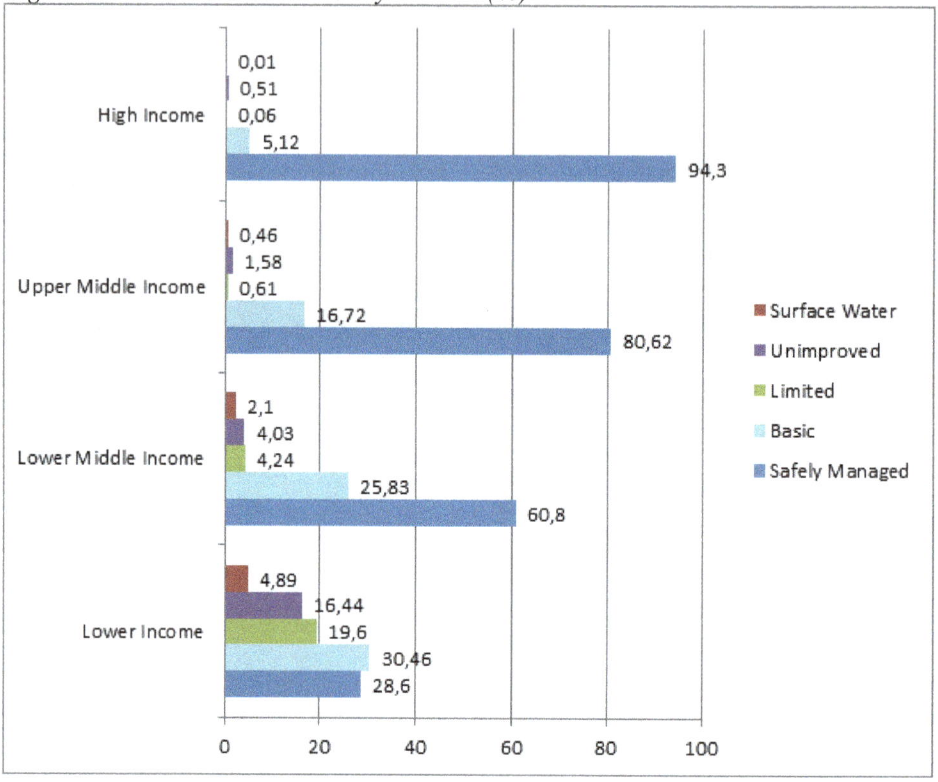

Source: WASH Data Base

Income levels strongly correlate with access (Charts 3 and 4): High-income groups (>94% of 1.32 billion safely managed); upper-middle (>80% safely managed, 2.6% problematic); lower-middle (60% safely managed, >10% lacking); low-income (only ~30% basic management, 41% insecure/unmanaged). This suggests a prima facie link between economic well-being and safe water access, though causal modeling is needed.

HOW CLIMATE CHANGE AFFECTS FRESHWATER ACCESS

Climate change, primarily through global warming, disrupts freshwater quantity, quality, accessibility, and equity. This section synthesizes major impacts based on IPCC (2021) projections and empirical evidence, emphasizing vulnerabilities in low-income and climate-sensitive regions (IPCC, 2021; UNESCO, 2020; WMO, 2024).

Changes in the Water Cycle and Regional Effects

Rising temperatures and altered precipitation intensify hydrological extremes and accelerate glacial retreat, creating short-term surpluses but long-term scarcity (IPCC, 2021; Huss & Hock, 2018; Trenberth, 2011). For instance, in glaciated regions, initial melt increases flow but eventually leads to river depletion (Immerzeel et al., 2020; Sorg et al., 2012).

Table 1. Key Regional Impacts of Climate Change on Freshwater Resources

Region	Primary Impacts	Consequences	References
South Asia	Himalayan glacial melt, irregular monsoons	Unsustainable river flows by 2050s (1.6B affected)	Immerzeel et al., 2020; Mall et al., 2007; Seth et al., 2012; Eyler, 2019; Jayaraman & Murari, 2014; Mohapatra, 2025; Dash & Abraham, 2024
Sub-Saharan Africa	Prolonged droughts, erratic rainfall	Food insecurity, reduced crop yields	Kim et al., 2021; Niang et al., 2014; Lobell et al., 2011; Ngarakana-Gwasira et al., 2016; Akter, 2024
Latin America	Andean glacier loss, Amazon drying	Reduced river flow, salinization	Vuille et al., 2018; Marengo et al., 2017; Marengo et al., 2021; Marengo et al., 2016; Junk et al., 2014; Rignot et al., 2019
MENA Region	Groundwater depletion, chronic scarcity	Absolute scarcity levels	Mekonnen & Hoekstra, 2016; Ghanim, 2019; Choukr-Allah et al., 2016; Baba & Tayfur, 2011
Mediterranean Europe	Increasing droughts	Agricultural and ecosystem stress	García-Ruiz et al., 2011; Gobiet et al., 2014; CREA, 2025; FRI, 2025; Soana et al., 2024; HELCOM, 2018; Uehlinger et al., 2009
North America	Snowpack reduction	Colorado River flow decline	Barnett et al., 2005; Diffenbaugh et al., 2015; O'Reilly et al., 2015; Smith et al., 2019; Hobbs, 2014
Australia	Basin-wide droughts	Groundwater decline, production losses	Pittock & Finlayson, 2011; DWER, 2022; Wong et al., 2013
Other (e.g., Arctic, Pacific Islands)	Ice melt, sea level rise	Local surpluses but global risks	Rignot et al., 2019; Sharan et al., 2023; WMO, 2024; Raj et al., 2025

These regional variations exacerbate governance weaknesses in developing countries, creating a global trend (Karthe et al., 2017).

Threats to Water Quality

Elevated temperatures and extremes degrade quality through algal blooms, salinization, and pollutant concentration (Paerl & Huisman, 2008; Werner et al., 2013;

Mosley, 2015). For example, increased algal blooms in lakes increase treatment costs (Özbayram et al., 2020; İleri, 2013).

Notable cases include:

- Algal bloom crises (e.g., Lake Erie 2014, affecting 500,000 people; Michalak et al., 2013; O'Reilly et al., 2015; Smith et al., 2019).
- Coastal groundwater salinization (Bangladesh: Mahmuduzzaman et al., 2014; Mediterranean: Grima-Olmedo et al., 2023; Australia: DWER, 2022; Maldives: WMO, 2024; Nurse et al., 2014).
- Flood-related contamination in major rivers (Rhine: Uehlinger et al., 2009; Yamuna: Sharma et al., 2024; Büyük Menderes: Zeynalova et al., 2024; Amazon: Junk et al., 2014).

These threats increase health risks and make access more difficult, especially in low-income areas (WHO, 2018).

Sectoral and Socio-Economic Impacts

Agriculture (~70% of withdrawals) is most vulnerable, followed by energy and industry (FAO, 2018; Opperman et al., 2022). Socioeconomic consequences disproportionately affect low-income, rural, female, and indigenous populations, increasing displacement and conflict risks (UN Women, 2018; UNHCR, 2021; Haque et al., 2019; Cooley et al., 2019).

Table 2. Selected Sectoral Impacts

Sector	Key Impact	Examples	References
Agriculture	Irrigation shortages	20–30% yield losses (Sahel, India, Australia)	Lobell et al., 2011; Misra, 2014; Karthe et al., 2017; Diffenbaugh et al., 2015; Pittock & Finlayson, 2011; Restiani et al., 2016; Mohapatra, 2025
Energy	Reduced hydroelectric output	Up to 61% global decline by 2050	Opperman et al., 2022; Schneider & Froggatt, 2019; Beilfuss & Brown, 2010; Eyler, 2019; Mueller et al., 2021
Health	Waterborne disease proliferation	Cholera, malaria expansion	WHO, 2018; Federspiel & Ali, 2018; Ngarakana-Gwasira et al., 2016; Zain et al., 2024; Şeker et al., 2020
Industrial	Production constraints	Increased costs in textile/ chemical sectors	Zhou et al., 2020; Restiani et al., 2016
Tourism	Resource strain	Mediterranean, Maldives risks	Gössling et al., 2012; Nurse et al., 2014; Koç et al., 2023; Moore, 2010
Education	Indirect access barriers	School absences due to water fetching	UN Women, 2018; Hossein et al., 2022

These effects deepen poverty and gender inequality, increasing the potential for social conflict (Karthe et al., 2017; Ayferam, 2023).

Future Projections and Adaptation Challenges

Projections indicate over 50% of the global population at risk of scarcity by 2050 (UNESCO, 2020). Adaptation via IWRM, recycling, and desalination is promising but constrained by financing, capacity, and cooperation gaps (UN-Water, 2019; AfDB, 2020; Tal, 2018; Puppala et al., 2025; Topak & Acar, 2010; Rahaman, 2024; Marengo et al., 2016).In summary, climate change amplifies freshwater insecurity through interconnected hydrological, quality, and equity pressures, necessitating integrated, technology-aided responses (Eyler, 2019; Karthe et al., 2017).

POSSIBLE CONTRIBUTIONS OF ARTIFICIAL INTELLIGENCE

Artificial intelligence (AI) technologies offer promising tools to mitigate climate-induced freshwater scarcity by enhancing efficiency, prediction, and equity in water management. While traditional approaches struggle with increasing pressures, AI enables data-driven, adaptive solutions across monitoring, optimization, and governance (Zhang & Ng, 2024). This section categorizes AI applications thematically, analyzes their contributions and limitations, and presents in-depth case studies to illustrate real-world impact.

Real-Time Monitoring and Forecasting

AI integrates satellite, IoT, and big data to provide accurate, real-time insights into water resources, enabling proactive responses to scarcity and extremes. This category addresses a core gap in conventional monitoring, where delayed data hinders timely intervention.

Key applications include high-resolution global mapping (Pekel et al., 2016) and regional groundwater tracking (Noui, 2025; Abdullaev & Rakhmatullaev, 2016). In vulnerable areas, AI models predict glacial melt (Rignot et al., 2019), seawater intrusion (Yang et al., 2023), and surface quality (Rana et al., 2023; Ma et al., 2019). Advanced forecasting tools, such as deep learning for drought and flood risks, outperform traditional methods in accuracy and speed (Huntingford et al., 2019; Mosavi et al., 2018; Rudra & Sarkar, 2023; Hofmann et al., 2025; Aganda & Homaifar, 2017; Karaca & Yürekli, 2025; Raj et al., 2025; Dash & Abraham, 2024; Imazon, 2022; Silva et al., 2022). Analytically, these tools reduce uncertainty in climate-adapted planning, particularly in data-scarce regions, but require robust data infrastructure for reliability.

Optimization of Treatment, Distribution, and Agricultural Efficiency

AI optimizes resource use in treatment and irrigation, targeting high-loss areas like agriculture (70% of global withdrawals; FAO, 2024). This category promotes sustainability by minimizing waste and energy consumption. Treatment and distribution benefits include predictive maintenance, leak detection, and chemical optimization (PUB, 2020; Yin et al., 2025; Nagpal et al., 2024; Skoczko, 2025; Alenezi & Alabaiadly, 2025; Al-Addous et al., 2024; Liu et al., 2024; Sönmez, 2021; Joseph et al., 2022; Sharma et al., 2024). In agriculture, precision systems achieve significant savings: 27% in general optimization (Preite & Vignali, 2024), 20-50% in drip irrigation (Izhar, 2025), and up to 99% efficiency in targeted regions (Audu, 2025; IDB, 2021; AWS, 2025; Gaitan et al., 2025; Lawal & Adebisi, 2025; Tiankan, 2025; Jumaev, 2024; DSI, 2024). These applications demonstrate economic and environmental gains, freeing water for domestic use, though scalability depends on infrastructure investment.

Water Quality Analysis and Pollution Control

AI excels in detecting contaminants and modeling pollution dynamics, supporting preventive quality management essential for safe access. Techniques include machine learning for groundwater assessment (Mohammadpour et al., 2025; Rana et al., 2023), microplastic tracking (Jin et al., 2024), and river interaction analysis (Şengörür et al., 2015; Pérez-Beltrán et al., 2024; Zhu et al., 2022; Sharma et al., 2024; Gilbert, 2025; Durgun, 2024). By processing large datasets, AI identifies threats faster than traditional methods, enabling targeted interventions and reducing health risks in polluted basins.

Disaster Response

AI strengthens resilience against climate-amplified disasters by providing early warnings and recovery support. Examples include flood prediction and infrastructure protection (Liu et al., 2025; AWS, 2025; Mosavi et al., 2018; Dash & Abraham, 2024; Ghaffarian et al., 2023; Albahri et al., 2024; Sun et al., 2020), drought monitoring (Funk et al., 2019), and wildfire prevention to protect watersheds (Giannakidou et al., 2024; USC, 2022). These studies highlight the role of artificial intelligence in minimizing access disruptions during times of crisis, revealing its potential to save lives and save resources in vulnerable regions.

Reducing Access Disparities and Innovative Integration

AI can address inequities by prioritizing vulnerable groups and integrating with emerging technologies. Equity-focused tools map disparities and optimize allocation in slums, rural areas, and refugee camps (Stuart & Setty, 2022; Grundfos, 2025; PSD, 2025; UNICEF, 2024; Gu et al., 2023; Gökçek et al., 2024; Nakib, 2024; Daniel, 2024). Hybrid systems combine AI with blockchain for transparency (Patil, 2024), robotics for surveillance (Ou et al., 2024), and renewable energy for sustainable pumping (Grundfos, 2025; Moyo, 2021; Mamodiya et al., 2025). These integrations promote inclusive management, though digital divides risk widening gaps without targeted policies.

Policy Development, Cooperation, and Capacity Building

AI facilitates governance by simulating scenarios and building capacity.Transboundary tools support conflict resolution (Kae, 2023; EMM, 2022; Wilkins, 2023; Jumaev, 2024; Sarkar & Jha, 2025; Serrao, 2025; Quereshi, 2024). Training platforms foster adoption (UNESCO, 2025; LSBA, 2025; TONEX, 2025; Kishore, 2025; Sajja et al., 2025; Orn et al., 2020; Farooq et al., 2025). Analytically, AI shifts policy from reactive to predictive, enhancing international cooperation.

In-Depth Case Studies

- Singapore: AI-optimized recycling (NEWater) meets 40% of demand with reduced energy, demonstrating urban resilience but highlighting cost barriers (PUB, 2020; Zhang & Ng, 2024). Outcome: High scalability in water-stressed cities.
- Israel: Precision irrigation saves 20-50% water, boosting yields in arid zones, yet infrastructure dependency limits transferability (Izhar, 2025; Tal, 2018). Outcome: Model for agriculture-dominant economies.
- India (Yamuna and Monsoon Systems): AI pollution detection and forecasting improve safety for millions, but data gaps challenge rural application (Sharma et al., 2024; Dash & Abraham, 2024). Outcome: Potential for large-scale equity gains with better integration.

Risks, Limitations, and Policy Recommendations

AI's benefits are tempered by risks: high costs and digital divide (Olawade et al., 2024), data bias (Kae, 2023), energy/water footprint (Ding et al., 2025; Karataş, 2025), ethical issues (Pons-Ausina, 2025), and exclusion of local communities.

Recommendations include inclusive development, data democratization, ethical frameworks, capacity building, and public-private partnerships to ensure AI supports equitable, sustainable access. In summary, AI provides integrated tools for freshwater management, with strong evidence from categorized applications and cases, but requires balanced governance to maximize benefits amid climate pressures.

Ethical Challenges, Social Equity, and Governance Barriers in AI Deployment

While AI offers transformative potential for freshwater management, its deployment raises significant ethical, social, and governance concerns that could exacerbate existing inequalities if not addressed (Olawade et al., 2024; Kae, 2023). These challenges are particularly acute in low-income and climate-vulnerable regions, where water insecurity intersects with poverty, gender disparities, and marginalization of indigenous communities.

Algorithmic bias and data quality issues pose ethical risks: AI models trained on incomplete or skewed datasets may perpetuate inequities in resource allocation, favoring urban or affluent areas (Kae, 2023; Pons-Ausina, 2025). For example, in developing countries, lack of localized data can lead to inaccurate predictions, disproportionately affecting rural and indigenous populations reliant on traditional water sources.

Social equity dimensions are critical: Women and girls in low-income households often bear the burden of water collection, spending hours daily – a task exacerbated by scarcity (UN Women, 2018). AI solutions that ignore gender-specific needs (e.g., time-saving tools for remote communities) risk widening gaps. Similarly, poverty limits access to AI-enabled infrastructure, creating a digital divide that excludes the most vulnerable (Grundfos, 2025; PSD, 2025). Indigenous communities and displaced populations (e.g., refugees in camps) face additional barriers, as AI systems may overlook cultural water practices or transient needs (Daniel, 2024; UNHCR, 2021).

Governance barriers further complicate deployment: Weak institutional frameworks in many countries hinder data sharing and regulation, while private sector dominance risks commodifying water through proprietary AI tools (Barlow, 2016; Serrao, 2025). Transboundary waters amplify conflicts, where AI simulations could help but require trust and equitable data access (Sarkar & Jha, 2025; Quereshi, 2024).

Energy and environmental footprints add irony: AI's high computational demands consume significant water and electricity, potentially counteracting sustainability goals in water-stressed areas (Ding et al., 2025; Karataş, 2025).

Policy Recommendations for Inclusive AI Adoption

To mitigate these risks, an inclusive approach is essential:

- Prioritize ethical frameworks that protect water as a human right, incorporating bias audits and community participation (UN Digital Library, 2010).
- Promote open-source AI and data democratization to reduce digital divide (Kae, 2023).
- Integrate gender and social equity in design, targeting women, poor, and indigenous groups (UN Women, 2018).
- Strengthen governance through public-private partnerships and international cooperation for transboundary issues.
- Invest in capacity building and low-energy AI to ensure sustainability (Strubell et al., 2019; AfDB, 2020).

These measures can transform AI from a potential divider to an equalizer in freshwater access.

CONCLUSION

This study has examined the complex problem of access to safe freshwater through the dual lenses of climate change and artificial intelligence, adopting a qualitative synthesis approach. Freshwater, comprising only 2.5% of global water resources, is under increasing pressure from quantity and quality constraints, exacerbated by population growth, unsustainable practices, and socioeconomic disparities.

The analysis reaffirmed water as both a fundamental human right and an economic asset, with modern governance models seeking balance between equity and efficiency (Akel Ünal, 2020; Hodgson, 2006). Global access remains uneven, with over 2.1 billion people lacking safely managed drinking water, particularly in rural and low-income regions (WASH, 2025).

Climate change intensifies these challenges through disrupted hydrological cycles, glacial melt, quality degradation, and sectoral impacts, disproportionately affecting vulnerable populations (IPCC, 2021; UNESCO, 2020). Regional variations and future projections underscore the urgency for adaptive strategies.

Artificial intelligence emerges as a powerful enabler, offering solutions in monitoring, optimization, forecasting, quality control, disaster response, and policy simulation. Categorized applications and case studies (Singapore, Israel, India) demonstrate practical benefits in efficiency and resilience, yet ethical risks, social inequities (gender, poverty, digital divide), and governance barriers must be addressed to prevent exacerbation of disparities (Olawade et al., 2024; UN Women, 2018; Kae, 2023).

Key Trends and Insights:

- Climate impacts are accelerating, shifting many regions toward chronic scarcity.
- AI's data-driven approach can bridge traditional management gaps, but success depends on inclusive design.
- Governance failures and social dimensions remain central bottlenecks.

Implications for Policy and Practice:

- Adopt integrated frameworks prioritizing water as a human right, with AI tools supporting equitable allocation and transboundary cooperation.
- Invest in ethical AI development, open data, and capacity building in low-income countries.
- Promote public-private partnerships for scalable, low-energy solutions.

Implications for Future Research

- Empirical studies on AI's long-term equity impacts in developing contexts.
- Comparative analyses of governance models integrating AI.
- Interdisciplinary work on ethical frameworks for digital water technologies.

Ultimately, safe freshwater access is a matter of social justice, environmental sustainability, and global cooperation. Leveraging AI within ethical and inclusive governance offers a pathway to resilience amid irreversible climate change, aligning with Sustainable Development Goals.

REFERENCES

Abdullaev, I., & Rakhmatullaev, S. (2016). Data management for integrated water resources management in Central Asia. *Journal of Hydroinformatics, 18*(6), 990–1007. DOI: 10.2166/hydro.2016.006

AfDB. (2020). *Africa's Water Sector: Financing and Investment Needs*. African Development Bank.

Akel Ünal, Y. (2020). *Türkiye'de Su Hakkı*, Published Doctoral Thesis, Galatasaray University, Institute of Social Sciences, Istanbul.

Akter, S. (2024). Impacts of flooding on rural communities in Bangladesh. *Journal of Fisheries, 12*(2), 122206. Advance online publication. DOI: 10.17017/j.fish.731

Al-Addous, M., Bdour, M., Rabaiah, S., Boubakri, A., Schweimanns, N., Barbana, N., & Wellmann, J. (2024). Innovations in Solar-Powered Desalination: A Comprehensive Review of Sustainable Solutions for Water Scarcity in the Middle East and North Africa (MENA) Region. *Water (Basel), 16*(13), 1877. DOI: 10.3390/w16131877

Albahri, A.S., Khaleel, Y.L., Habeeb, M.A., Ismael, R.D., Hameed, Q.A., Deveci, M., Homod, R.Z., Albahri, O.S., Alamoodi, A.H. & Alzubaidi, L. (2024). A systematic review of trustworthy artificial intelligence applications in natural disasters, *Computers and Electrical Engineering*, 118(B):109409. https://doi.org/DOI: 10.1016/j.compeleceng.2024.109409

Alenezi, A., & Alabaiadly, Y. (2025). Artificial Intelligence Applications in Water Treatment and Desalination: A Comprehensive Review. *Water (Basel), 17*(8), 1169. DOI: 10.3390/w17081169

Audu, J. (2025). A Meta-Review on the Use of Artificial Intelligence in Irrigation Practices in Nigeria. *6th African Regional Conference on Irrigation and Drainage*, At: Abuja, Nigeria.

AWS. (March 2025). *AWS AI powers new water projects in Spain*, Retrieved August 16, 2025, from https://www.aboutamazon.eu/news/sustainability/aws-ai-powers-new-water-projects-in-spain

Ayferam, G. (2023). *The Nile Dispute: Beyond Water Security*, Retrieved August 06, 2025, from https://carnegieendowment.org/sada/2023/01/the-nile-dispute-beyond-water-security?lang=en

Baba, A., & Tayfur, G. (2011). Groundwater contamination and its effect on health in Turkey. *Environmental Monitoring and Assessment, 183*(1-4), 77–94. DOI: 10.1007/s10661-011-1907-z PMID: 21336483

Barlow, M. (2016). Su Hakkı, Yeni İnsan Yayınevi, 2016, İstanbul.

Barnett, T. P., Adam, J. C., & Lettenmaier, D. P. (2005). Potential Impacts of a Warming Climate on Water Availability in Snow-Dominated Regions. *Nature, 438*(7066), 303–309. DOI: 10.1038/nature04141 PMID: 16292301

Beilfuss, R., & Brown, C. (2010). Assessing environmental flow requirements and trade-offs for the Lower Zambezi River and Delta, Mozambique. *International Journal of River Basin Management, 8*(2), 127–138. DOI: 10.1080/15715121003714837

Choukr-Allah, R., Nghira, A., Hirich, A., & Bouchaou, L. (2016). Water Resources Master Plan for Sustainable Development of the Souss-Massa River Basin (1-26). In: Choukr-Allah, Redouane & Ragab, R. & Bouchaou, Lhoussaine & Barcelo, Damia. (2016). *The Souss-Massa River Basin*, Morocco.

Cooley, H., Phurisamban, R., & Gleick, P. (2019). The cost of alternative urban water supply and efficiency options in California. *Environmental Research Communications, 1*(4), 042001. DOI: 10.1088/2515-7620/ab22ca

CREA. (2025). *Climate Change and Its Impacts in the Alps,* Retrieved July 27, 2025, from https://creamontblanc.org/en/climate-change-and-its-impacts-alps

Daniel, C. (2024). Circular Economy 2.0: How Artificial Intelligence is Transforming Sustainability in Water Use. *Journal of Research and Development (Srinagar), 12*(2), 261.

Dash, Y., & Abraham, A. (2024). Integrating empirical analysis and deep learning for accurate monsoon prediction in Kerala, India, *Applied. Computers & Geosciences, 24*, 100211. DOI: 10.1016/j.acags.2024.100211

Diffenbaugh, N. S., Swain, D. J., & Touma, D. (2015). Anthropogenic Warming Has Increased Drought Risk in California. *Proceedings of the National Academy of Sciences of the United States of America, 112*(13), 3931–3936. DOI: 10.1073/pnas.1422385112 PMID: 25733875

Ding, Z., Wang, J., Song, Y., Zheng, X., He, G., Chen, X., Zhang, T., Lee, W., & Song, J. (2025). Tracking the carbon footprint of global generative artificial intelligence. *Innovation (Cambridge (Mass.)), 6*(5), 100866. DOI: 10.1016/j.xinn.2025.100866 PMID: 40432776

DSI. (June 2024). *Sulamada Yapay Destekli Otomasyon Çalışmaları Devam Ediyor*, Retrieved August 16, 2025, from https://www.dsi.gov.tr/Haber/Detay/12868

Durgun, Y. (2024). Real-time water quality monitoring using AI-enabled sensors: Detection of contaminants and UV disinfection analysis in smart urban water systems. *Journal of King Saud University. Science*, *36*(9), 103409. DOI: 10.1016/j.jksus.2024.103409

DWER. (2022). *Gnangara groundwater allocation plan 2022*. Department of Water and Environmental Regulation.

EMM. (2022). *How can artificial intelligence tackle the water crisis*, Retrieved August 20, 2025, from https://www.energymanagermagazine.co.uk/how-can-artificial-intelligence-tackle-the-water-crisis/

Eyler, B. (2019). *Last Days of the Mighty Mekong*. Zed Books. DOI: 10.5040/9781350221031

Falkenmark, M., & Lindh, G. (1974). *Water for a Starving World*. Westview Press.

FAO. (2018). *The State of Food Security and Nutrition in the World 2018*. Food and Agriculture Organization.

FAO. (2024). *World Water Day, 22 March 2024: Water for Peace*, Food and Agriculture Organization, Retrieved August 16, 2025, from https://www.fao.org/land-water/events/world-water-day-celebrations/wwd-2024/en/

Farooq, M. U., Murtaza, G., Mahmood, M. H., Rai, S. I., Tanveer, H., Haq, E. U., Muhammad, H., & Sabir, R. M. (2025). AI-Powered Solutions for Sustainable Water Conservation. In Garwi, J., & Masengu, R. (Eds.), *AI and Ecological Change for Sustainable Development* (pp. 473–512). IGI Global Scientific Publishing., DOI: 10.4018/979-8-3373-0680-3.ch016

Federspiel, F., & Ali, M. (2018). The cholera outbreak in Yemen: Lessons learned and way forward. *BMC Public Health*, *18*(1), 1338. DOI: 10.1186/s12889-018-6227-6 PMID: 30514336

Foley, M. (June 2020). *Unshackling India's Energy-Water-Agriculture Nexus: Is Solar Power the Panacea it Promises to be?* Retrieved January 29, 2025, from https://www.worldbank.org/en/news/feature/2020/06/05/unshackling-india-energy-water-agriculture-nexus

Funk, C., Harrison, L., Alexander, L., Peterson, P., Behrangi, A., & Husak, G. (2019). Exploring trends in wet-season precipitation and drought indices in wet, humid and dry regions. *Environmental Research Letters*, *14*(11), 115002. DOI: 10.1088/1748-9326/ab4a6c

Gaitan, N. C., Batinas, B. I., Ursu, C., & Crainiciuc, F. N. (2025). Integrating Artificial Intelligence into an Automated Irrigation System. *Sensors (Basel)*, *25*(4), 1199. DOI: 10.3390/s25041199 PMID: 40006428

García-Ruiz, J. M., López-Moreno, J. I., Vicente-Serrano, S. M., Lasanta–Martínez, T., & Beguería, S. (2011). Mediterranean Water Resources in a Changing Climate. *Earth-Science Reviews*, *105*(3-4), 85–100. DOI: 10.1016/j.earscirev.2011.01.006

Ghaffarian, S., Taghikhah, F. R., & Maier, H. R. (2023). Explainable artificial intelligence in disaster risk management: Achievements and prospective futures. *International Journal of Disaster Risk Reduction*, *98*, 104123. DOI: 10.1016/j.ijdrr.2023.104123

Ghanim, A. (2019). Water Resources Crisis in Saudi Arabia Challenges and Possible Management Options: An Analytic Review. *World Academy of Science, Engineering and Technology*, *13*(2), 51–56. DOI: 10.5281/zenodo.2571928

Giannakidou, S., Radoglou-Grammatikis, P., Lagkas, T., Argyriou, V., Goudos, S., Markakis, E. K., & Sarigiannidis, P. (2024). Leveraging the power of internet of things and artificial intelligence in forest fire prevention, detection, and restoration: A comprehensive survey. *Internet of Things : Engineering Cyber Physical Human Systems*, *26*, 101171. DOI: 10.1016/j.iot.2024.101171

Gilbert, L. (2025). *First-of-its-Kind Monitoring Tool Uses AI to Forecast Water Contamination*, Retrieved August 19, 2025, from https://www.usu.edu/today/story/first-of-its-kind-monitoring-tool-uses-ai-to-forecast-water-contamination/

Gobiet, A., Kotlarski, S., Beniston, M., Heinrich, G., Rajczak, J., & Stoffel, M. (2014). 21st century climate change in the European Alps—A review. *The Science of the Total Environment*, *493*, 1138–1151. DOI: 10.1016/j.scitotenv.2013.07.050 PMID: 23953405

Gökçek, A. E., Yılmaz, M., Sekmen, U., & Toptaş, M. (2024). Su ve Kanalizasyon İdarelerinde Veri Tabanlı İzleme ve Kontrol Sistemleri (SCADA) Uygulaması: Büyükşehir Belediyeleri. *AJEAS*, *2*(2), 33–51.

Gossling, S., Peeters, P., Hall, C. M., Ceron, J., Dubois, G., Lehmann, L. V., & Scott, D. (2012). Tourism and Water Use: Supply, Demand, and Security. An international review. *Tourism Management*, *33*(1), 1–15. DOI: 10.1016/j.tourman.2011.03.015

Grabish, B. (2017). *Dry Tears of the Aral*, Retrieved January 29, 2025, from https://www.un.org/en/chronicle/article/dry-tears-aral

Grima-Olmedo, J., Ballesteros-Navarro, B., Pulido-Velazquez, D., Renau-Pruñonosa, A., Alcalá, F. J., Llopis-Albert, C., Jiménez-Gavilán, P., Milkov-Ivanov, N., Baena-Ruiz, L., & Grima-Olmedo, C. (2023). A Framework to Assess Natural Chloride Background in Coastal Aquifers Affected by Seawater Intrusion in Eastern Spain. *Water (Basel)*, *15*(15), 2728. DOI: 10.3390/w15152728

Grundfos. (2025). *How to provide sustainable water through automated water kiosks in off-grid communities*, Retrieved August 21, 2025, from https://www.grundfos.com/solutions/learn/ecademy/all-courses/community-water-supply-systems/how-to-provide-sustainable-water-through-automated-water-kiosks-in-off-grid-communities

Gu, Z., Li, W., Hanemann, M., Tsai, Y., Wutich, A., Westerhoff, P., Landes, L., Roque, A. D., Zheng, M., Velasco, C. A., & Porter, S. (2023). Applying machine learning to understand water security and water access inequality in underserved colonia communities. *Computers, Environment and Urban Systems*, *102*, 101969. DOI: 10.1016/j.compenvurbsys.2023.101969

Haque, S. F., Hope, R., Arif, S. T., Akhter, T., Naz, M., & Salehin, M. (2019). A social-ecological analysis of drinking water risks in coastal Bangladesh. *The Science of the Total Environment*, *679*, 23–34. DOI: 10.1016/j.scitotenv.2019.04.359 PMID: 31078772

HELCOM. (2018). *State of the Baltic Sea*. Helsinki Commission.

Hobbs, K. (2014). Toledo's Drinking Water Crisis, Retrieved July 28, 2025, from https://www.nrdc.org/bio/karen-hobbs/toledos-drinking-water-crisis

Hodgson, S. (2006). *Modern water rights - Theory and practice*. FAO.

Hofmann, J., Holt, A., & Welten, S. (2025). *AI for Rhine Low-Flow Forecasting: Strategic Insights & The Path to Surrogate Models*, FloodWaive, Retrieved August 14, 2025, from https://www.digishape.nl/wp-content/uploads/2025/06/Floodwaive-Presentation-AI-for-Rhine-Low-Flow-Forecasting.pdf

Huntingford, C., Jeffers, E., Bonsall, M., Christensen, H., Lees, T., & Yang, H. (2019). Machine learning and artificial intelligence to aid climate change research and preparedness. *Environmental Research Letters*, *14*(12), 124007. DOI: 10.1088/1748-9326/ab4e55

Huss, M., & Hock, R. (2018). Global-Scale Hydrological Response to Future Glacier Mass Loss. *Nature Climate Change*, *8*(2), 135–140. DOI: 10.1038/s41558-017-0049-x

IDB. (June 2021). *Use of AI in Precision Agriculture for a watershed in the Chihuahua Desert, Mexic*, Retrieved August 16, 2025, from https://fairlac.iadb.org/en/use-ai-precision-agriculture-watershed-chihuahua-desert-mexico

Imazon (2022). *Artificial Intelligence Platform Estimates Deforestation Risk of 15 thousand km² in the Brazilian Amazon in 2022*, Retrieved August 14, 2025, from https://imazon.org.br/en/imprensa/artificial-intelligence-platform-estimates-deforestation-risk-of-15-thousand-km%C2%B2-in-the-brazilian-amazon-in-2022/

Immerzeel, W. W., Lutz, A. F., Andrade, M., Bahl, A., Biemans, H., Bolch, T., Hyde, S., Brumby, S., Davies, B. J., Elmore, A. C., Emmer, A., Feng, M., Fernández, A., Haritashya, U., Kargel, J. S., Koppes, M., Kraaijenbrink, P. D. A., Kulkarni, A. V., Mayewski, P. A., & Baillie, J. E. M. (2020). Importance and Vulnerability of the World's Water Towers. *Nature*, *577*(7790), 364–369. DOI: 10.1038/s41586-019-1822-y PMID: 31816624

IPCC. (2021). Sixth Assessment Report [Intergovernmental Panel on Climate Change, Geneva.]. *Climatic Change*, •••, 2021. PMID: 34866716

Jayaraman, T., & Murari, K. (2014). Climate Change and Agriculture: Current and Future Trends, and Implications for India. *Review of Agrarian Studies*, *4*(1), 1–49. DOI: 10.25003/RAS.04.01.0001

Jin, H., Kong, F., Li, X., & Shen, J. (2024). Artificial intelligence in microplastic detection and pollution control. *Environmental Research*, *262*(1), 119812. DOI: 10.1016/j.envres.2024.119812 PMID: 39155042

Joseph, K., Sharma, A. K., & van Staden, R. (2022). Development of an Intelligent Urban Water Network System. *Water (Basel)*, *14*(9), 1320. DOI: 10.3390/w14091320

Jumaev, M. (2024). AI for Water Management: Smart Solutions for Conservation in Uzbekistan and Central Asia. *American Journal of Library And Information Science Innovation*, *2*(8), 15–18.

Junk, W. J., Piedade, M. T. F., Lourival, R., Wittmann, F., Kandus, P., Lacerda, L. D., Bozelli, R. L., Esteves, F. A., Nunes da Cunha, C., Maltchik, L., Schöngart, J., Schaeffer-Novelli, Y., & Agostinho, A. A. (2014). Brazilian wetlands: Their definition, delineation, and classification for research, sustainable management, and protection. *Aquatic Conservation*, *24*(1), 5–22. DOI: 10.1002/aqc.2386

Kae, J. (April 2023). *Can AI Help Solve the Global Water Crisis?* Retrieved August 20, 2025, from https://www.hidropolitikakademi.org/tr/article/31233/can-ai-help-solve-the-global-water-crisis

Karaca, Ö.F. & Yürekli, K. (2025). Determination of the most suitable artificial neural network method for sediment estimation in the Euphrates-Tigris Basin, *International Journal of Sediment Research*, In Press, Corrected Proof, https://doi.org/DOI: 10.1016/j.ijsrc.2025.07.011

Karataş, E. C. (2025). Will the Creation Be Able to Overwhelm the Destruction: Water, and AI Nexus? In Günar, A. (Ed.), *Economic and Political Consequences of AI: Managing Creative Destruction* (pp. 437–458). IGI Global Scientific Publishing., DOI: 10.4018/979-8-3693-7036-0.ch019

Karthe, D., Abdullaev, I., Boldgiv, B., Borchardt, D., Chalov, S., Jarsjö, J., Li, L., & Nittrouer, J. A. (2017). Water in Central Asia: An integrated assessment for science-based management. *Environmental Earth Sciences, 76*(20), 690. https://doi.org/. DOI: 10.1007/s12665-017-6994-x

Kim, D. G., Grieco, E., Bombelli, A., Hickman, J. E., & Sanz-Cobena, A. (2021). Challenges and opportunities for enhancing food security and greenhouse gas mitigation in smallholder farming in sub-Saharan Africa. A review. *Food Security, 13*(2), 457–476. DOI: 10.1007/s12571-021-01149-9

Kishore, K. (July 2025). *How can we harness AI to tackle the complexity of disaster risk?* Retrieved August 21, 2025, from https://www.undrr.org/news/how-can-we-harness-ai-tackle-complexity-disaster-risk

Koç, C., Bayazıt, Y., & Özgül, S. (2023). Impact of tourists on urban water needs in Marmaris, Türkiye. *Environment, Development and Sustainability, 25*(8), 8837–8855. DOI: 10.1007/s10668-023-03461-0

Lawal, Y., & Adebisi, O. (2025). AI and IoT-Based Smart Irrigation: A Review of Challenges and Future Trends, *IV. International Congress of the Turkish Journal of Agriculture - Food Science and Technology, Niğde, Türkiye,* (936–942).

LII. (April 2024). *riparian doctrine*, Retrieved January 27, 2025, from https://www.law.cornell.edu/wex/riparian_doctrine

Liu, F., Yu, X., & Tang, J. (2024). Water-saving control system based on multiple intelligent algorithms. *Autonomous Intelligent Systems, 4*(13), 13. Advance online publication. DOI: 10.1007/s43684-024-00068-8

Liu, Z., Coleman, N., Patrascu, F. I., Yin, K., Li, X., & Mostafavi, A. (2025). Artificial intelligence for flood risk management: A comprehensive state-of-the-art review and future directions. *International Journal of Disaster Risk Reduction, 117*, 105110. DOI: 10.1016/j.ijdrr.2024.105110

Longman, J., Veres, D., Finsinger, W., & Ersek, V. (2018). Exceptionally high levels of lead pollution in the Balkans from the Early Bronze Age to the Industrial Revolution. *Proceedings of the National Academy of Sciences of the United States of America, 115*(25). Advance online publication. DOI: 10.1073/pnas.1721546115 PMID: 29844161

LSBA. (August, 2025). *Professional Certificate in Water Management with AI,* Retrieved August 20, 2025, from https://www.lsba.org.uk/Home/CourseDetail?courseId=26234

Ma, Y., Matta, E., Meissner, D., Schellenberg, H., & Hinkelmann, R. (2019). Can Machine Learning Improve the Accuracy of Water Level Forecasts for Inland Navigation? Case Study: Rhine River Basin, Germany, *38th IAHR World Congress.* DOI: 10.3850/38WC092019-0274

Mahmuduzzaman, M., Uddin, Z., Ahmed, S., Nuruzzaman, A., & Rabbi, F. (2014). Causes of Salinity Intrusion in Coastal Belt of Bangladesh. *International Journal of Plant Research, 4*(4A), 8–13. DOI: 10.5923/s.plant.201401.02

Malik, A., Sharma, S., Batra, I., Sharma, C., Kaswan, M. S., & Garza-Reyes, J. A. (2024). Industrial revolution and environmental sustainability: An analytical interpretation of research constituents in Industry 4.0. *International Journal of Lean Six Sigma, 15*(1), 22–49. DOI: 10.1108/IJLSS-02-2023-0030

Mall, R. K., Singh, R., Gupta, A., Srinivasan, G., & Rathore, L. S. (2007). Impact of Climate Change on Indian Agriculture: A review. *Climatic Change, 82*(1-2), 225–231. DOI: 10.1007/s10584-006-9236-x

Mamodiya, U., Kishor, I., Garine, R., Ganguly, P., & Naik, N. (2025). Artificial intelligence based hybrid solar energy systems with smart materials and adaptive photovoltaics for sustainable power generation. *Scientific Reports, 17370*(15), 17370. DOI: 10.1038/s41598-025-01788-4 PMID: 40389492

Marengo, J. A., Galdos, M.V., Challinor, A., Cunha, A.P., Marin, F.R., Vianna, M.S., Alvala, R.C.S., Alves, L.M., Moraes, O.L. & Bender, F., (2021). Drought in Northeast Brazil: A review of agricultural and policy adaptation options for food security, *Climate Resilience and Sustainability*, 1(1) Special Issue: Climate Science for Service Partnership Brazil: Collaborative Research towards climate solutions in Brazil, (1-20). https://doi.org/DOI: 10.1002/cli2.17

Marengo, J. A., Tomasella, J. & Nobre, C.A., (2016). Climate Change and Water Resources, In book: *Waters of Brazil* (171-186). https://doi.org/DOI: 10.1007/978-3-319-41372-3_12

Mekonnen, M. M., & Hoekstra, A. Y. (2016). Four Billion People Facing Severe Water Scarcity. *Science Advances*, *2*(2), 1500323. DOI: 10.1126/sciadv.1500323 PMID: 26933676

Michalak, A. M., Anderson, E. J., Beletsky, D., Boland, S., Bosch, N. S., Bridgeman, T. B., Chaffin, J. D., Cho, K., Confesor, R., Daloglu, I., Depinto, J. V., Evans, M. A., Fahnenstiel, G. L., He, L., Ho, J. C., Jenkins, L., Johengen, T. H., Kuo, K. C., Laporte, E., & Zagorski, M. A. (2013). Record-setting algal bloom in Lake Erie caused by agricultural and meteorological trends consistent with expected future conditions. *Proceedings of the National Academy of Sciences of the United States of America*, *110*(16), 6448–6452. DOI: 10.1073/pnas.1216006110 PMID: 23576718

Misra, A. K. (2014). Climate Change and Challenges of Water and Food Security. *International Journal of Sustainable Built Environment*, *3*(1), 153–165. DOI: 10.1016/j.ijsbe.2014.04.006

Mohammadpour, A., Gharehchahi, E., Golaki, M., Gharaghani, M. A., Ahmadian, F., Abolfathi, S., Samaei, M. R., Uddin, M. G., Olbert, A. I., & Mousavi Khaneghah, A. (2025). Advanced water quality assessment using machine learning: Source identification and probabilistic health risk analysis. *Results in Engineering*, *27*, 105421. DOI: 10.1016/j.rineng.2025.105421

Mohapatra, S. (2025). *Impact of Climate Change on Indian Agriculture*, University of Auckland Research Repository. http://dx.doi.org/DOI: 10.2139/ssrn.5198832

Moore, W. (2010). The impact of climate change on Caribbean tourism demand. *Current Issues in Tourism*, *13*(5), 495–505. DOI: 10.1080/13683500903576045

Mosavi, A., Özturk, P., & Chau, K. (2018). Flood Prediction Using Machine Learning Models: Literature Review. *Water (Basel)*, *10*(11), 1536. DOI: 10.3390/w10111536

Mosley, L. M. (2015). Drought impacts on the water quality of freshwater systems; review and integration. *Earth-Science Reviews*, *140*, 203–214. DOI: 10.1016/j.earscirev.2014.11.010

Moyo, R. T. (2021). *Artificial intelligence based solar/diesel hybrid water pumping system*, Unpublished master's thesis, Department of Mechanical Engineering, Durban University of Technology. https://doi.org/DOI: 10.51415/10321/3722

Mueller, A., Detges, A., Pohl, B., Reuter, M.H., Rochowski, L., Volkholz., J. & Woertz, E., (2021). *Climate change, water and future cooperation and development in the Euphrates-Tigris basin*, Cascades Project Report 2021.

Nagpal, M., Siddique, M. A., Sharma, K., Sharma, N., & Mittal, A. (2024). Optimizing wastewater treatment through artificial intelligence: Recent advances and future prospects. *Water Science and Technology*, *90*(3), 731–757. DOI: 10.2166/wst.2024.259 PMID: 39141032

Nakib, A. M. (2024). Automated Water Management System with AI-based Demand Prediction, *Informatyka, Automatyka, Pomiary W Gospodarce I Ochronie Środowiska*, 14(4), (136–140). https://doi.org/DOI: 10.35784/iapgos.6106

Ngarakana-Gwasira, E. T., Bhunu, C. P., Masocha, M., & Mashonjowa, E., (2016). Assessing the Role of Climate Change in Malaria Transmission in Africa, *Malaria Research and Treatment*, 2016(1-7):7104291. https://doi.org/DOI: 10.1155/2016/7104291

Nguyen, T. (2019). *The Dynamics of Water Privatization, Decentralized Power, and the Tragedy of the Commons in Water Management in Vietnam, US-Vietnam Review*, Retrieved January 28, 2025, from https://usvietnam.uoregon.edu/en/dynamics-water-privatization-decentralized-power-tragedy-commons-water-management-vietnam/

Niang, I., Boote, K. J., Kimball, B. A., Ziska, L. H., Izaurralde, R. C., Thomson, D., Ort, A. M., & Wolfe, D. (2014) Africa. In: Barros, V.R., et al., Eds., *Impacts, Adaptation, and Vulnerability. Part B: Regional Aspects.* Contribution of Working Group II to the Fifth Assessment Report of the Intergovernmental Panel on Climate Change, Cambridge University Press, Cambridge, UK, (1199-1265). https://doi.org/DOI: 10.2134/agronj2010.0303

Noui, A. (2025). Contribution of Artificial Intelligence, Geophysics, GIS, and ICT to the Fight Against Groundwater Drawdown in the Algerian, Arab, and African Arid Regions, Retrieved August 7, 2025, from https://sdgs.un.org/partnerships/contribution-artificial-intelligence-geophysics-gis-and-ict-fight-against-groundwater

Nurse, L. A., McLean, R. F., Agard, J., Briguglio, L. P., Duvat-Magnan, V., Pelesikoti, N., & Webb, A. (2014). Small islands. In *Climate Change 2014: Impacts, Adaptation, and Vulnerability. Part B: Regional Aspects. Contribution of Working Group II to the Fifth Assessment Report of the Intergovernmental Panel on Climate Change (1613-1654)*. Cambridge University Press., DOI: 10.1017/CBO9781107415386.029

O'Reilly, C. M., Sharma, S., Gray, D. K., Hampton, S. E., Read, J. S., Rowley, R. J., Schneider, P., Lenters, J. D., McIntyre, P. B., Kraemer, B. M., Weyhenmeyer, G. A., Straile, D., Dong, B., Adrian, R., Allan, M. G., Anneville, O., Arvola, L., Austin, J., Bailey, J. L., & Zhang, G. (2015). Rapid and highly variable warming of lake surface waters around the globe. *Geophysical Research Letters*, *42*(24), 773–781. DOI: 10.1002/2015GL066235

Olawade, D. B., Wada, O. Z., Ige, A. O., Egbewole, B. I., Olojo, A., & Oladapo, B. I. (2024). Artificial intelligence in environmental monitoring: Advancements, challenges, and future directions. *Hygiene and Environmental Health Advances, 12*, 100114. DOI: 10.1016/j.heha.2024.100114

Opperman, J. J., Camargo, R. R., Laporte-Bisquit, A., Zarfl, C., & Morgan, A. J. (2022). Using the WWF Water Risk Filter to Screen Existing and Projected Hydropower Projects for Climate and Biodiversity Risks. *Water (Basel), 14*(5), 721. DOI: 10.3390/w14050721

Orn, D., Duan, L., Liang, Y., Siy, H., & Subramaniam, M. (2020). Agro-AI Education: Artificial Intelligence for Future Farmers, In *Proceedings of the 21st Annual Conference on Information Technology Education (SIGITE '20)*, Association for Computing Machinery, New York, NY, USA, (54–57). https://doi.org/DOI: 10.1145/3368308.3415457

Özbayram, E. G., Köker, L., Akçaalan, R., İnce, O., & Albay, M. (2020). Bacterial community composition of sapanca lake during a cyanobacterial bloom. *Aquatic Sciences and Engineering, 35*(2), 52–56. DOI: 10.26650/ASE2020652073

Paerl, H. W., & Huisman, J. (2008). Blooms Like It Hot. *Science, 320*(5872), 57–58. DOI: 10.1126/science.1155398 PMID: 18388279

Patil, S.B., (2024). The Use of Blockchain for Water Rights in Irrigation Management. *International Water and Irrigation*, 44(1), (01-06). https://doi.org/DOI: 10.52783/iwi.v44i1.82

Pekel, J. F., Cottam, A., Gorelick, N., & Belward, A. D. (2016). High-resolution mapping of global surface water and its long-term changes. *Nature, 540*(418–422), 418–422. Advance online publication. DOI: 10.1038/nature20584 PMID: 27926733

Pérez-Beltrán, C. H., Robles, A. D., Rodriguez, N. A., Ortega-Gavilán, F., & Jiménez-Carvelo, A. M. (2024). Artificial intelligence and water quality: From drinking water to wastewater. *Trends in Analytical Chemistry, 172*, 117597. DOI: 10.1016/j.trac.2024.117597

Pittock, J., & Finlayson, C. M. (2011). Australia's Murray-Darling Basin: Freshwater ecosystem conservation options in an era of climate change. *Marine and Freshwater Research, 62*(3), 232–243. DOI: 10.1071/MF09319

Pons-Ausina, J. F., Hosseini, S. N., & Soriano Olivares, J. (2025). AI for Smart Water Solutions in Developing Areas: Case Study in Khelvachauri (Georgia). *Water (Basel), 17*(8), 1119. DOI: 10.3390/w17081119

Preite, L., & Vignali, G. (2024). Artificial intelligence to optimize water consumption in agriculture: A predictive algorithm-based irrigation management system. *Computers and Electronics in Agriculture, 223*, 109126. DOI: 10.1016/j.compag.2024.109126

PSD. (April 2025), *AI for Water Equity in Urban Slums*, Retrieved August 15, 2025, from https://prism.sustainability-directory.com/scenario/ai-for-water-equity-in-urban-slums/

PSD. (April 2025), *Community-Based AI for Water Management in Rural Areas*, Retrieved August 15, 2025, from https://prism.sustainability-directory.com/scenario/community-based-ai-for-water-management-in-rural-areas/

PUB. (2020). *Annual Report 2019/2020*. Singapore's National Water Agency.

Puppala, H., Arora, M. K., Peddinti, P. R. T., Tamvada, J. P., & Das, K. (2025). Enhancing access to rainwater harvesting in regions with saline groundwater. *Discover Sustainability, 6*(440), 440. Advance online publication. DOI: 10.1007/s43621-025-01250-5

Quereshi, S. (2024). *Treating East Africa's Water Conflicts With AI and Big Data*, Retrieved August 20, 2025, from https://www.thinkglobalhealth.org/article/treating-east-africas-water-conflicts-ai-and-big-data

Rahaman, M.Md. Mizanur Rahman. (2024). Water Sharing Geopolitics between India and Bangladesh: Recent Trends. *International Journal of Applied and Advanced Multidisciplinary Research, 2*(8), 569–592. DOI: 10.59890/ijaamr.v2i8.2418

Raj, A. A., Yang, M. H., & Yu, L. (2025). Artificial Intelligence in Combating Climate Change in the Pacific Island Countries: Challenges, Barriers, and Pathways Forward. *Journal of Island and Marine Studies, 3*(1), 110015. DOI: 10.59711/jims.12.110015

Rana, R., Kalia, A., Boora, A., Alfaisal, F. M., Alharbi, R. S., Berwal, P., Alam, S., Khan, M. A., & Qamar, O. (2023). Artificial Intelligence for Surface Water Quality Evaluation, Monitoring and Assessment. *Water (Basel), 15*(22), 3919. DOI: 10.3390/w15223919

Restiani, P., & Khandelwal, A. (2016). *Water Governance Mapping Report: Textile Industry Water Use in India*. Stockholm International Water Institute.

Rignot, E., Mouginot, J., Scheuchl, B., & Morlighem, M. (2019). Four Decades of Antarctic Ice Sheet Mass Balance. *Proceedings of the National Academy of Sciences of the United States of America, 116*(4), 1095–1103. DOI: 10.1073/pnas.1812883116 PMID: 30642972

Rudra, R. R., & Sarkar, S. K. (2023). Artificial neural network for flood susceptibility mapping in Bangladesh. *Heliyon, 9*(6), e16459. DOI: 10.1016/j.heliyon.2023.e16459 PMID: 37251459

Sajja, R., Pursnani, V., Sermet, Y., & Demir, I. (2025). AI-Assisted Educational Framework for Floodplain Manager Certification: Enhancing Vocational Education and Training Through Personalized Learning. *IEEE Access : Practical Innovations, Open Solutions*, *13*, 42401–42413. DOI: 10.1109/ACCESS.2025.3548591

Sarkar, S., & Jha, P. K. (2025). Application of AI/ML in Water Resource Management to Resolve Transboundary Water Conflict. In: Nanda, A., Gupta, P.K., Gupta, V., Jha, P.K., Dubey, S.K. (eds) *Navigating the Nexus*. Water Science and Technology Library, 102. Springer, Cham. https://doi.org/DOI: 10.1007/978-3-031-76532-2_18

Schneider, M., & Froggatt, A. (2019). *World Nuclear Industry Status Report 2019*. Mycle Schneider Consulting.

Şeker, M., Koyuncu, İ., & Öztürk, İ. (2020). Türkiye'de İklim Değişimi ve Halk Sağlığı Raporu. *Türkiye Bilimler Akademisi Yayınları TÜBA Raporları*, (37). Advance online publication. DOI: 10.53478/TUBA.2020.120

Şengörür, B., Köklü, R., & Ateş, A. (2015). Water Quality Assessment Using Artificial Intelligence Techniques: SOM and ANN—A Case Study of Melen River Turkey. *Water Quality, Exposure, and Health*, *7*(4), 469–490. DOI: 10.1007/s12403-015-0163-9

Serrao, M. M. (2025). *Flowing toward peace: How AI and ML models can transform water security and diplomacy*, Retrieved August 20, 2025, from https://smartwatermagazine.com/blogs/marcello-michael-serrao/flowing-toward-peace-how-ai-and-ml-models-can-transform-water-security

Sharan, A., Lal, A., & Datta, B. (2023). Evaluating the impacts of climate change and water over-abstraction on groundwater resources in Pacific island country of Tonga. *Groundwater for Sustainable Development*, *20*, 100890. DOI: 10.1016/j.gsd.2022.100890

Sharma, M., Rawat, S., Kumar, D., Awsati, A., Sarkar, A., Sidola, A., Choudhury, T., & Kotecha, K. (2024). The state of the Yamuna River: A detailed review of water quality assessment across the entire course in India. *Applied Water Science*, *14*(175), 175. Advance online publication. DOI: 10.1007/s13201-024-02227-x

Silva, C. A., Guerrisi, G., Del Frate, F., & Sano, E. E. (2022). Near-real time deforestation detection in the Brazilian Amazon with Sentinel-1 and neural networks. *European Journal of Remote Sensing*, *55*(1), 129–149. DOI: 10.1080/22797254.2021.2025154

Skoczko, I. (2025). Energy Efficiency Analysis of Water Treatment Plants: Current Status and Future Trends. *Energies, 18*(5), 1086. DOI: 10.3390/en18051086

Smith, R., Bass, B., Sawyer, D., Depew, D., & Watson, S. (2019). Estimating the economic costs of algal blooms in the Canadian Lake Erie Basin. *Harmful Algae, 87*, 101624. DOI: 10.1016/j.hal.2019.101624 PMID: 31349881

Soana, E., Gervasio, M. P., Granata, T., Colombo, D., & Castaldelli, G. (2024). Climate change impacts on eutrophication in the Po River (Italy): Temperature-mediated reduction in nitrogen export but no effect on phosphorus. *Journal of Environmental Sciences (China), 143*, 148–163. DOI: 10.1016/j.jes.2023.07.008 PMID: 38644013

Sönmez, M. S. (2021). *İSKİ İzleme Sistemleri Akıllı Şebeke Uygulamaları*. İSKİ.

Sorg, A., Bolch, T., Stoffel, M., Solomina, O., & Beniston, M. (2012). Climate change impacts on glaciers and runoff in Tien Shan (Central Asia). *Nature Climate Change, 2*(10), 725–731. DOI: 10.1038/nclimate1592

Strubell, E., Ganesh, A., & McCallum, A. (2019). "Energy and Policy Considerations for Deep Learning in NLP, In *Proceedings of the 57th Annual Meeting of the Association for Computational Linguistics*, (3645–3650). DOI: 10.18653/v1/P19-1355

Stuart, K., & Setty, K. (January 2022), Applying AtlasAI data for pro-poor water programming, Retrieved August 15, 2025, from https://aquaya.org/applying-atlasai-data-for-pro-poor-water-programming/?gad_source=1&gad_campaignid=17242733466&gclid=CjwKCAjwtfvEBhAmEiwA-DsKjp_GicwwLgKlTnhIFSyvSf250GELk-NSSprN1qZzHVbS-_hDMMR35RoC1mMQAvD_BwE

Sun, W., Bocchini, P., & Davison, B. D. (2020). Applications of artificial intelligence for disaster management. *Natural Hazards, 103*(3), 2631–2689. DOI: 10.1007/s11069-020-04124-3

Tal, A. (2018). Addressing Desalination's Carbon Footprint: The Israeli Experience. *Water (Basel), 10*(2), 197. DOI: 10.3390/w10020197

Tiankan, Z. (May 2025). *Coding crops: How AI is sowing change in rural China*, Retrieved August 16, 2025, from https://www.thinkchina.sg/technology/coding-crops-how-ai-sowing-change-rural-china

TONEX. *AI for Environmental and Water Resource Management (AIEWRM)*, Retrieved August 21, 2025, from https://www.tonex.com/training-courses/ai-for-environmental-and-water-resource-management-aiewrm/

Topak, R., & Acar, B. (2010). Sustainable Irrigation and Importance of Technological Irrigation Systems for Konya Basin. *Tarım Bilimleri Araştırma Dergisi, 3*(2), 65–70.

Trenberth, K. E. (2011). Changes in Precipitation with Climate Change. *Climate Research*, *47*(1), 123–138. DOI: 10.3354/cr00953

Uehlinger, U., Wantzen, K., Leuven, R. S. E. W., & Arndt, H. (2009). The Rhine River Basin. In *Tockner, K. (u.a) Rivers of Europe* (pp. 199–245). Acad. Pr., DOI: 10.1016/B978-0-12-369449-2.00006-0

Umweltbundesamt, (October 2017), *Legal Framework*, Retrieved Feburary 29, 2025, from https://www.umweltbundesamt.de/sites/default/files/medien/1410/publikationen/wawi_flyer_eng_web_aktualisiert.pdf

UN Digital Library. (2010). *The human right to water and sanitation: resolution/adopted by the General Assembly*, Retrieved January 29, 2025, from https://digitallibrary.un.org/record/687002?v=pdf

UN-Water. (2019). *Annual Report – 2019*. United Nations.

UNDP. (2024). *AI Hub for Sustainable Development Strengthening Local AI Ecosystems through Collective Action*, Retrieved January 29, 2025, from https://policycommons.net/artifacts/13537479/ai-hub-for-sustainable-development/14435235/

UNESCO. (2020). *United Nations World Water Development Report 2020*. UNESCO.

UNESCO. (June, 2025). *Artificial intelligence applications for water management*, Retrieved August 21, 2025, from https://www.unesco.org/en/articles/artificial-intelligence-applications-water-management

UNHCR. (2021). *Climate Change and Displacement*. United Nations High Commissioner for Refugees.

UNICEF. (November 2024). *Mapping of Rural Water Supply Publications, using AI assistance,* Retrieved January 29, 2025, from https://knowledge.unicef.org/wash/resource/fact-sheet-mapping-rural-water-supply-publications-using-ai-assistance

U.S. Department of the Interior. (May, 2017). *Water Rights Act*, Retrieved January 26, 2025, from https://www.doi.gov/ocl/water-rights-act

USC. (June, 2022). *Water, Wildfire, and Climate Change*, Retrieved August 29, 2025, from https://www.ucs.org/resources/water-wildfire-and-climate-change

Vuille, M., Carey, M., Huggel, C., Buytaert, W., Rabatel, A., Jacobsen, D., Soruco, A., Villacis, M., Yarleque, C., Timm, O., Condom, T., Salzmann, N., & Sicart, J. (2018). Rapid decline of snow and ice in the tropical Andes-Impacts, uncertainties and challenges ahead. *Earth-Science Reviews*, *176*, 195–213. DOI: 10.1016/j.earscirev.2017.09.019

WASH. *Drinking water ladder*, Retrieved March 10, 2025, from https://washdata.org/monitoring/drinking-water

Werner, A. D., Bakker, M., Post, V. E. A., Vandenbohede, A., Lu, C., Ataie-Ashtiani, B., Simmons, C. T., & Barry, D. A. (2013). Seawater Intrusion Processes, Investigation and Management: Recent advances and future challenges. *Advances in Water Resources*, *51*, 3–26. DOI: 10.1016/j.advwatres.2012.03.004

Whelan, M. J., Linstead, C., Worrall, F., Ormerod, S. J., Durance, I., Johnson, A. C., Johnson, D., Owen, M., Wiik, E., Howden, N. J. K., Burt, T. P., Boxall, A., Brown, C. D., Oliver, D. M., & Tickner, D. (2022). Is water quality in British rivers "better than at any time since the end of the Industrial Revolution? *The Science of the Total Environment*, *843*, 157014. DOI: 10.1016/j.scitotenv.2022.157014 PMID: 35772542

Wilkins, H. (2023). *Can AI Help Solve Diplomatic Dispute Over the Grand Ethiopian Renaissance Dam?* Retrieved August 20, 2025, from https://www.voanews.com/a/can-ai-help-solve-diplomatic-dispute-over-the-grand-ethiopian-renaissance-dam-/6985115.html

WMO. (2024). *Climate change transforms Pacific Islands*, Retrieved July 25, 2025, from https://wmo.int/news/media-centre/climate-change-transforms-pacific-islands

Wolf, S., Esser, V., Schüttrumpf, H., & Lehmkuhl, F. (2021). Influence of 200 years of water resource management on a typical central European river. Does industrialization straighten a river? *Environmental Sciences Europe*, *33*(15), 15. Advance online publication. DOI: 10.1186/s12302-021-00460-8

Women, U. N. (2018). *Gender and Water: Empowering Women for Sustainable Water Management*. United Nations.

Wong, T. H. F., Allen, R., Brown, R. R., Deletić, A., Gangadharan, L., Gernjak, W., Jakob, C., Johnstone, P., Reeder, M., Tapper, N., Vietz, G., & Walsh, C. J. (2013). *Stormwater Management in a Water Sensitive City: Blueprint 2013*. Cooperative Research Centre for Water Sensitive Cities Ltd.

World Bank. (2023). *Urban Development*, Retrieved March 04, 2025, from https://www.worldbank.org/en/topic/urbandevelopment/overview

Yang, H., Sun, H., Liu, T., Yang, X., Yang, F., & Jiao, J. (2023). Characterization of seawater intrusion based on machine learning and implications for offshore management under shared socioeconomic paths. *Journal of Hydrology (Amsterdam)*, *623*, 129862. DOI: 10.1016/j.jhydrol.2023.129862

Yin, R., Xu, H., Wei, J., & Varga, L. (2025). Addressing data sharing challenges for leakage management in water distribution networks – An MCDM-based assessment of barriers and solutions. *Journal of Environmental Management, 391*, 126481. DOI: 10.1016/j.jenvman.2025.126481 PMID: 40645083

Zain, A., Sadarangani, S. P., Shek, L. P., & Vasoo, S. (2024). Climate change and its impact on infectious diseases in Asia. *Singapore Medical Journal, 65*(4), 211–219. DOI: 10.4103/singaporemedj.SMJ-2023-180 PMID: 38650059

Zenalova, N., Demirak, A., Döndü, M., & Keskin, F. (2024). Assessment of Büyük Menderes River (West of the Türkiye) Surface Water Quality with Water Quality Index (WQI), Comprehensive Pollution Index (CPI) and Geographic Information Systems (GIS). *Acta Aquatica Turcica, 20*(2), 108–127. DOI: 10.22392/actaquatr.1308181

Zhang, H., & Ng, C. (2024). Applications of Artificial Intelligence, Machine Learning, and Data Analytics in Water Environments. *ACS ES&T Water, 4*(3), 761–763. DOI: 10.1021/acsestwater.4c00140

Zhou, F., Bo, Y., Ciais, P., Dumas, P., Tang, Q., Wang, X., Liu, J., Zheng, C., Polcher, J., Yin, Z., Guimberteau, M., Peng, S., Ottle, C., Zhao, X., Zhao, J., Tan, Q., Chen, L., Shen, H., Yang, H., & Wada, Y. (2020). Deceleration of China's human water use and its key drivers. *Proceedings of the National Academy of Sciences of the United States of America, 117*(14), 7702–7711. DOI: 10.1073/pnas.1909902117 PMID: 32209665

Zhu, M., Wang, J., Yang, X., Zhang, Yu., Zhang, L., Ren, H., Wu, B., & Ye, L. (2022). A review of the application of machine learning in water quality evaluation. *Eco-Environment & Health, 1*(2), 107–116. DOI: 10.1016/j.eehl.2022.06.001 PMID: 38075524

Chapter 10
Safeguarding the Seas and Promoting a Sustainable Blue Economy Through Maritime Security

Burak Şakir Şeker
https://orcid.org/0000-0002-8536-1790
Faculty of Economics and Administrative Sciences, Ankara Hacı Bayram Veli University, Turkey

ABSTRACT

This chapter explores the intersection of maritime security and the sustainable blue economy, emphasizing the critical role of secure seas in advancing ecological health, economic prosperity, and global stability. It highlights how maritime threats—ranging from piracy and smuggling to climate-driven risks and cyber vulnerabilities—undermine efforts to build resilient ocean economies. Drawing on regional case studies such as the Red Sea, Indo-Pacific, Arctic, and African maritime zones, the discussion underscores the interplay between security frameworks, governance mechanisms, and sustainable development. The chapter also examines the role of international law, regional cooperation, and technological innovation in mitigating risks. By linking security with sustainability, it argues that safeguarding maritime spaces is not only a matter of protecting trade routes and national interests but also a prerequisite for ensuring long-term ecological and economic resilience.

DOI: 10.4018/979-8-3373-5238-1.ch010

INTRODUCTION

The world's oceans sustain life on a scale that is both profound and underappreciated. They cover more than 70 percent of the Earth's surface, regulate the climate, produce half of the planet's oxygen, and provide a primary source of protein for billions of people (United Nations, 2022). They are also the arteries of globalization: more than 80 percent of international trade is transported by sea (UNCTAD, 2023). Yet, beneath this narrative of abundance lies a more fragile reality. The oceans are increasingly contested spaces, vulnerable to competing claims, ecological pressures, and security threats.

At the heart of today's maritime debate lies the concept of the "blue economy." Coined in the early 2010s (Pauli, 2010), the term captures the idea that oceans can fuel economic growth while sustaining ecosystems and ensuring social inclusion. The blue economy includes fisheries, aquaculture, shipping, renewable energy, seabed mining, and coastal tourism. In theory, it promises a sustainable future where economic prosperity and environmental stewardship are not opposing goals but mutually reinforcing ones. In practice, however, the realization of a sustainable blue economy depends heavily on one often overlooked prerequisite: maritime security.

Maritime security is the condition in which maritime spaces are free from threats such as piracy, terrorism, smuggling, illegal fishing, and environmental crimes. It is not limited to the deployment of naval forces; it encompasses legal frameworks, international cooperation, and the governance of maritime spaces. Without it, fishing communities cannot thrive, offshore wind farms cannot be protected, and commercial shipping cannot operate efficiently. As the European Union's (EU) maritime strategy emphasizes, security is not a separate pillar from sustainability but rather the foundation upon which a blue economy can grow (Council of the European Union, 2023).

This research argues that the preservation of marine ecosystems through effective protection measures of maritime security is important to ensuring sustainable blue economy. It examines the mechanisms whereby threats to the sea undermine ecological sustainability as well as economic growth, examines regional as well as global reactions, and highlights the role that technological innovation as well as governance processes play. Ultimately, the analysis recommends an inclusive approach whereby security integrates with sustainability to ensure that the sea becomes a storehouse of wealth rather than a battlefield for conflict.

THE BLUE ECONOMY: ECONOMIC POTENTIAL, ECOLOGICAL AND GOVERNANCE CHALLENGES, AND THE SECURITY LINK

The "blue economy" came to mainstream prominence during the United Nations Conference on Sustainable Development (Rio+20) in 2012, where it was floated as a plan to support the attainment of the Sustainable Development Goals (SDGs). Very broadly, the blue economy is the sustainable exploitation of marine resources to support economic growth, raise the standard of living, and maintain vibrant ecosystems (World Bank, 2017). It goes from classical sectors such as shipping and fishing, to newer ones such as marine biotechnology, mining the seabed, and marine renewables.

Worldwide, the blue economy value stands at USD 3 to 6 trillion every year, offering employments to over 350 million people (OECD, 2020). While coastal tourism alone offers 26 million employments, the fisheries enable the livelihoods of numerous communities that are very vulnerable. Moreover, the oceans also contain untapped potential; wind energy off the coast will offer 10 percent of the global electricity supply by the year 2040 (International Energy Agency, 2019). For small island developing states (SIDS), the blue economy is not just a development strategy but a survival imperative.

The EU continues to invest heavily in offshore wind to meet their climate targets (European Parliament, 2025), while the US significantly scaling back investments. In January 2025, the Trump Administration issued a presidential memorandum effectively closing new offshore wind lease auctions in all regions on the outer continental shelf (The White House, 2025a). This corresponds with a larger strategy of distancing from the Paris Agreement (United Nations, 2015), prioritizing fossil fuel production over renewable energy and effectively segregating the global blue economy into two different worlds: a European-Asian[1] vision emphasizing greenhouse gas reductions, and a US vision emphasizing oil and gas production prowess and security.

The potential of the blue economy is not evenly distributed, but its promise is undeniable. For example, in Southeast Asia, the South China Sea generates approximately USD 3.5 trillion in trade annually (Kaplan, 2021). In Africa, the African Union's 2050 Integrated Maritime Strategy envisions the blue economy as a driver of industrialization, food security, and regional integration. Similarly, the EU has invested in its "Blue Growth" strategy, linking the sustainability of fisheries and marine energy to regional prosperity (Council of the European Union, 2023).

Maritime trade remains the backbone of globalization. Container ships carry goods from Asia to Europe, crude oil tankers cross the Gulf of Aden, and liquefied natural gas (LNG) shipments from Qatar fuel industries worldwide. The COVID-19 pandemic underscored this dependency: disruptions in shipping routes created shortages of essential goods and highlighted vulnerabilities in supply chains (International Chamber of Shipping, 2021).

Despite its potential, the blue economy faces serious ecological and governance challenges. Overfishing is depleting stocks faster than they can regenerate. According to the Food and Agriculture Organization (FAO) (2022), more than one-third of global fish stocks are overexploited. Climate change exacerbates these pressures, warming seas, acidifying oceans, and intensifying extreme weather events. Rising sea levels also threaten coastal infrastructure, displacing populations and eroding the foundations of coastal economies.

Governance presents an equally complex challenge. Maritime spaces are vast, and jurisdictional disputes often complicate their management. The South China Sea disputes (Strating, 2020) illustrate how competing territorial claims can escalate tensions, undermining both security and economic opportunities. In regions with limited naval or coast guard capabilities, such as parts of West Africa, maritime crime flourishes (Ogunniyi & Akpu, 2019). The lack of enforcement mechanisms creates a vicious cycle: illegal exploitation weakens local economies, which in turn reduces the capacity of states to enforce maritime governance.

One of the large and less-regulated frontiers of blue economy is deep sea mining. As the focus on green energy accelerates with time, the demand for cobalt, nickel, and other strategic minerals means that people have turned their attention to extracting these minerals from the bottom of the ocean (Kasanawaqa, 2024). However, this sector highlights a major governance problem. Although a broad regulatory framework is established in UNCLOS, the specific regulation parameters are being developed by the International Seabed Authority (Lodge, 2017).

The planetary implications are massive, moreover, NGOs and scientists have indicated that deep-sea mining will have devastating effects on biodiversity and deep-sea environments which function at a geologically slow pace (Verbeek, 2020). While fishing grounds can theoretically regenerate, destroying seabed environments is irreversible damage. Here, again, the blue economy finds itself locked in a conflict between ambition and sustainability. The ocean will end up being simply another resource to mine if governance fails, thus destroying the very essence of a sustainable blue economy.

While ecological and governance challenges receive significant attention, the security dimension often remains underemphasized. Yet, without security, the blue economy cannot thrive. Illegal, unreported, and unregulated (IUU) fishing robs coastal states of billions in revenue annually (Crowell & Turvold, 2020). Piracy increases insurance premiums and disrupts shipping routes (Dixon, 2024). Smuggling networks use maritime routes for trafficking drugs, arms, and people, undermining governance and fueling instability.

In this sense, the blue economy and maritime security are two sides of the same coin. Security ensures that resources are protected, trade flows remain stable, and investments in renewable energy or tourism are viable. Sustainability, in turn, provides legitimacy and local support for security initiatives. Together, they form the foundation for a future where oceans can be harnessed for shared prosperity rather than contested for short-term gains.

MARITIME SECURITY: FROM NAVAL POWER TO OCEAN GOVERNANCE

Maritime security is not a new concern. For centuries, the world's oceans have been both conduits of prosperity and arenas of insecurity. The earliest recorded instances of maritime insecurity trace back to piracy in the Mediterranean during the Roman Empire. Roman fleets were forced to patrol and secure sea lanes against pirates who threatened commerce and grain supplies from North Africa (Horden & Purcell, 2000). The British Empire later invested heavily in naval power, not only to project force globally but also to secure trade routes from the Indian Ocean to the Caribbean.

What has changed in the 21st century is the range and complexity of threats. Maritime security today encompasses far more than traditional piracy. It includes terrorism, cyberattacks on ports, trafficking of narcotics and arms, human smuggling, and environmental crimes. These threats are transnational and interconnected, requiring responses that go beyond naval dominance. Maritime security has moved on beyond the hallowing association with naval defence to include a wider framework referred to as "ocean governance" (Silver et al., 2015).

The expanded comprehension of maritime security underscores its intrinsic relationship with governance. Naval forces by themselves are insufficient to address issues such as unlawful fishing, oil contamination, or human trafficking. Rather, security ought to be integrated within a governance structure that incorporates surveillance, law enforcement, judicial mechanisms, and international collaboration. This transition from naval supremacy to ocean governance signifies an acknowledgment that the oceans are not solely arenas of conflict but communal areas necessitating collective management.

Its law is essentially framed by the United Nations Convention on the Law of the Sea (UNCLOS) of 1982. It outlines several maritime areas, specifies corresponding rights and obligations, as well as mechanisms for the settling of controversies. However, it was conceived in a time where cyberattacks as well as environmental crimes were insignificant. While UNCLOS still constitute an invaluable foundation, it needs reexamination as well as complementing to address challenges of the modern era adequately (Kraska, 2011).

Organizations like the International Maritime Organization (IMO) and the International Tribunal for the Law of the Sea (ITLOS) also play significant roles; their enforcement capability, nonetheless, usually rests on the political will as well as capabilities of the states. Regional initiatives—the Djibouti Code of Conduct within the East African subregion (IMO, 2017) or the regional ASEAN maritime security initiatives within the ranks of Southeast Asian states (ASEAN, 2022)—are, correspondingly, increasingly valuable complementary international governance.

Despite advancements in international cooperation, piracy remains a persistent challenge. The Somali piracy crisis of the late 2000s highlighted the fragility of maritime security in weak or failing states. At its peak in 2011, Somali pirates carried out 237 attacks, holding hundreds of seafarers hostage and extracting millions of dollars in ransom (Oceans Beyond Piracy, 2012). These attacks destabilized one of the world's most critical chokepoints—the Gulf of Aden, through which 10 percent of global trade passes (MSCIO, 2025).

International responses included the establishment of multinational naval task forces such as EU NAVFOR Atalanta, NATO's Operation Ocean Shield, and the Combined Maritime Forces (CMF) (CTF-151). By 2017, Somali piracy was largely suppressed, illustrating how coordinated patrols, armed vessel protection, and regional capacity-building could contain maritime insecurity (UNSC, 2021).

However, piracy did not disappear; it shifted geographically. Today, the Gulf of Guinea has become the global epicenter of piracy, accounting for over 23 percent of crew kidnappings worldwide. Unlike Somali piracy, which focused on ransom of ships, West African piracy often targets crew kidnappings for ransom, directly threatening human life (IMB, 2024). The Yaoundé Code of Conduct (IMO, 2013) has improved cooperation between Gulf of Guinea states, but resource constraints, corruption, and governance challenges persist. Incidents within Southeast Asian waters especially in the Singapore Straits has been rising to the number of 43 that 93% of the ships were boarded (IMB, 2024).

The 21st century has introduced hybrid threats that blur the lines between traditional naval warfare and irregular tactics. States and non-state actors alike have turned to gray-zone operations at sea, where attribution is difficult, and escalation is carefully managed. The South China Sea offers a striking example: rather than overt naval confrontation, China has deployed maritime militias and coast guard vessels to assert territorial claims, harass foreign fishermen, and intimidate rival claimants (Poling, 2020). These tactics exploit the ambiguity of international law while avoiding outright conflict.

Cybersecurity has also emerged as a critical maritime concern. Ports are increasingly automated, and global supply chains depend on digital logistics systems. In

2017, the Maersk shipping company was crippled by the NotPetya cyberattack, which paralyzed operations at 76 ports worldwide and cost the company USD 300 million in damages (Greenberg, 2018, 2019). Such attacks expose the vulnerability of global trade infrastructure to digital threats that transcend geography.

ILLEGAL, UNREPORTED, AND UNREGULATED (IUU) FISHING

Illegal, unreported, and unregulated (IUU) fishing is one of the most pressing threats to both maritime security and the blue economy. The FAO (2022) estimates that IUU fishing accounts for up to 26 million tons of fish annually, valued at USD 10–23 billion. These figures are not just statistics—they represent stolen livelihoods, depleted ecosystems, and weakened governance in some of the world's most vulnerable regions.

West Africa illustrates the severity of the issue. It also loses approximately USD 2.3 billion annually as a result of illicit, unreported, and unregulated fishing. It deteriorates the livelihoods of small-scale fishermen and complicates the ability of millions to access decent food (Doumbouya et al., 2017). In the Pacific, big boats from high-income countries seize opportunities presented by poor surveillance, claiming resources that small island states desperately need to survive (Canyon et al., 2021).

IUU fishing also has a very destructive impact on the environment. It damages sustainable fishing by taking young fish, damaging habitats, and exceeding catch limits. When populations of fish plummet, it takes years to recover, if they do so at all. The failure of the North Atlantic cod fishery back in the early 1990s (Myers et al., 1997) is a warning: decades afterwards, there has still not fully recovered, creating long-term social as well as economic issues among coastal communities throughout the Northwest Atlantic (Pedersen et al., 2017).

In addition, IUU fishing normally employs destructive practices such as bottom trawling, dynamite fishing, or unlawful equipment that destroys ecosystems. Coral reefs as well as seagrass habitats, valuable to numerous species, are particularly vulnerable (Shaver & Yozell, 2018). Climate change exacerbates these issues, as the added warmth as well as ocean acidification (Licker et al., 2019) place further strain on already stressed marine ecosystems that are already harmed through overfishing.

IUU fishing is both an environmental crime as well as a governance and security issue. In regions where enforcement is poor, IUU fishing is sometimes part of larger criminal organizations. Vessels that intercept unlawful fishing may also transport drugs, arms, or individuals who are trafficked. Corruption that allows unlawful fishing to occur the same that also undermines other aspects of law enforcement (Myers & Yozell, 2018).

In Somali waters, in the 1990s and early 2000s, foreign illegal fishing deprived local people of resources and contributed to the rise of piracy. Domestic fishermen had no competitive advantage over foreign vessels, so they were forced to resort to thievery on the high seas, which gave rise to recurrent insecurity (Jacobsen & Høy-Carrasco, 2018). This reveals the very close connection between resource management and high-seas crime.

IUU fishing has been addressed internationally through collaboration and tech. FAO's Port State Measures Agreement (PSMA) was agreed upon in 2009, aiming to prevent landing rights to those involved in IUU fishing. Closing ports to unlawful operators minimizes the economic gain for the practice (FAO, 2016). It has been ratified so far by 84 states up to 2025, although enforcement is uneven (FAO, 2025).

Regional fisheries management organisations (RFMOs) also step in, setting quotas and enforcing compliance. Good practice, as the case of the Western and Central Pacific Fisheries Commission shows, can rehabilitate depleted stocks of tuna (Pomeroy et al., 2020).

Technological innovations provide new tools: satellite tracking, ship monitoring systems (VMS), and automatic identification systems (AIS) help governments track dubious ships (United Nations, 2023). Institutions like Global Fishing Watch (2025) use publicly sourced satellite data to identify unlawful activities, bringing them more transparency and responsibility. These measures, however, work based on political will, the ability to enforce laws, as well as local cooperation.

It is the human face of IUU fishing that tends to be overlooked. Smaller-scale fishermen in low-income countries are most harmed by the illicit activities of the large fishing vessels. In West Africa, the fishermen report their hauls are diminishing, so they venture further off shore on precarious boats, which has the potential to cause fatal consequences (Belhabib et al., 2015). The erosion of livelihoods can fuel migration pressures, exacerbate poverty, and destabilize coastal communities. In this sense, combating IUU fishing is not just about protecting ecosystems but also about safeguarding human security. Besides, blue economy is often shortchanged when it comes to labor exploitation. As blue economies compete to harvest cheaper fish, IUU vessels may employ forced labor and human trafficking (Haenlein, 2017). A blue economy of sustainability can never operate under a regime of modern slavery, hence labor rights must become a matter of security for governance, not just a human rights matter. It requires empowering local communities, ensuring equitable access to resources, and integrating security measures with development strategies.

PIRACY, ARMED ROBBERY, AND OTHER MARITIME CRIMES

Piracy has haunted the seas for as long as humans have engaged in maritime trade. From the corsairs of the Mediterranean to the "Golden Age of Piracy" in the Caribbean, pirates have always thrived where governance is weak, sea lanes are lucrative, and enforcement is absent (Courdingly, 2006). In the 21st century, piracy has re-emerged not as an anachronism, but as a symptom of global inequalities, fragile states, and transnational crime. While the image of Somali pirates boarding oil tankers dominated headlines in the late 2000s, piracy today is more diverse and geographically dispersed. The International Maritime Bureau (IMB, 2025) shows that incidents are concentrated in three hotspots: the Gulf of Guinea in West Africa, the Strait of Malacca and surrounding waters in Southeast Asia, and residual cases in the Horn of Africa.

The Somali piracy crisis of 2005–2012 remains the most dramatic example of modern maritime insecurity. With Somalia's central government collapsed since 1991, coastal communities were left vulnerable to illegal fishing and toxic waste dumping by foreign vessels. What began as local self-defense evolved into a lucrative criminal enterprise (Elmi et al., 2015). Pirates hijacked vessels, held crews hostage, and extracted ransoms that sometimes exceeded USD 10 million (Faruk, 2024). The crisis disrupted global shipping lanes. The Gulf of Aden, through which 12 percent of global trade passes, became a high-risk zone. Shipping companies rerouted vessels around the Cape of Good Hope, increasing costs and insurance premiums. The international response was unprecedented: NATO, the EU (Operation Atalanta), and the U.S.-led Combined Maritime Forces deployed warships, while private security firms began placing armed guards on vessels (Larsen & Nissen, 2017). By 2017, Somali piracy had sharply declined. Several factors explain this success: sustained naval patrols, the use of armed vessel protection, and the establishment of regional courts to prosecute pirates. Crucially, Somali authorities and local communities were gradually integrated into international strategies. However, the root causes—poverty, state fragility, and weak governance—remain unresolved. Analysts warn that piracy could re-emerge if international attention wanes (Melvin, 2019).

Unlike Somali piracy, which focused on hijacking vessels for ransom, piracy in the Gulf of Guinea often targets crew kidnappings. Armed groups operating from Nigeria's Niger Delta have extended their reach offshore, abducting seafarers for ransom. Ransom payments, increased insurance premiums, and rerouting add billions to global shipping costs annually (Norbert, 2017). For West African states, the impact is not just economic but reputational, deterring foreign investment and undermining the promise of the blue economy. Regional cooperation has improved since the signing of the Yaoundé Code of Conduct (IMO, 2013), which established frameworks for information-sharing and joint patrols. Yet, challenges remain: many states lack sufficient naval assets, judicial systems are under-resourced, and corruption undermines enforcement. Unlike in Somalia, where international navies had a strong presence, external involvement in West Africa is more limited. Greater EU, U.S., and Chinese support will be necessary to stabilize the region (Wilson, 2021).

The Strait of Malacca, one of the busiest waterways in the world, has long been a hotspot for piracy. However, coordinated patrols by Indonesia, Malaysia, Singapore, and Thailand (the "Malacca Strait Patrols") have reduced incidents since the mid-2000s (Collin, 2025). Today, most attacks in Southeast Asia are small-scale armed robberies against ships at anchor rather than large-scale hijackings (Dixon, 2024). Still, the region remains vulnerable due to its geography. Thousands of small islands and narrow straits create hiding places for pirates and smugglers. Moreover, the overlap between piracy and other forms of maritime crime—such as fuel smuggling and human trafficking—complicates enforcement (Edwards, 2022).

Piracy does not exist in isolation; it intersects with other forms of maritime crime. Criminal networks use maritime routes to smuggle narcotics, weapons, and people. In Latin America, drug cartels use "narco-submarines" and fishing vessels to transport cocaine from the Andes to North America and Europe (Harithas, 2024). In Southeast Asia, trafficking networks exploit porous maritime borders to move migrants, often under dangerous and inhumane conditions (Missbach & Hoffstaedter, 2025). The overlap between IUU fishing, piracy, and trafficking highlights the need for integrated approaches. If authorities focus narrowly on piracy without addressing underlying criminal economies, efforts will be short-lived. Maritime security must be understood as a web of interrelated challenges rather than isolated threats.

ENVIRONMENTAL PROTECTION AND MARITIME SECURITY

The ecological health of the oceans is deteriorating rapidly. Overfishing, pollution, habitat destruction, and climate change are pushing marine ecosystems toward tipping points. The Intergovernmental Panel on Climate Change (IPCC, 2019) warns that ocean warming and acidification threaten coral reefs, fisheries, and coastal livelihoods worldwide.

Yet, these environmental challenges are often sidelined in security discussions, seen as "soft" issues compared to piracy or terrorism. This neglect is short-sighted because environmental degradation undermines the very foundations of the blue economy and can trigger instability. Collapsing fish stocks push communities into poverty, oil spills devastate coastal economies, and rising sea levels displace populations. In this sense, environmental protection is not peripheral but central to maritime security.

Accidental and deliberate pollution events highlight the vulnerability of marine ecosystems to human negligence and conflict. The 2010 Deepwater Horizon disaster in the Gulf of Mexico spilled nearly 5 million barrels of oil, causing one of the worst environmental catastrophes in U.S. history. Fisheries collapsed, tourism revenues plummeted, and ecosystems were devastated (Graham et al., 2011). In conflict zones, oil spills are often deliberate, for instance, during the Gulf War in 1991, Iraqi forces released millions of barrels of oil into the Persian Gulf, causing massive ecological damage (Toukan, 1991). More recently, attacks on tankers in the Red Sea during the Yemen conflict (Lloyd's List, 2025) have raised fears of environmental disasters. Maritime security strategies must therefore include not only prevention of such attacks but also rapid response mechanisms for environmental crises.

As an another dilemma, climate change is reshaping the maritime landscape. Rising seas threaten to submerge low-lying states such as the Maldives, Tuvalu, and Kiribati. Extreme weather events, such as cyclones and typhoons, are becoming more frequent and destructive, damaging ports, naval bases, and coastal infrastructure (Institute for Economics & Peace, 2024). Melting Arctic ice opens new shipping routes but also heightens geopolitical competition and environmental risk (McVey, 2022). For small island developing states, climate change is not a distant threat but an existential one. These states often lack resources for adaptation and rely heavily on fisheries and coastal tourism. Integrating climate resilience into maritime security strategies is thus critical for their survival.

Another overlooked threat is illegal dumping of waste at sea. Toxic waste, plastics, and other pollutants are often discharged illegally, with devastating consequences for marine ecosystems. Weak enforcement and corruption exacerbate the problem. For instance, in the 1990s, European companies were accused of dumping toxic waste off the coast of Somalia, taking advantage of the country's lawlessness (Rengelink, 2012). The fight against marine pollution requires stronger enforcement of international conventions such as the International Maritime Organization's MARPOL agreement. It also requires better monitoring technologies, including drones and satellites, to detect and deter illegal dumping.

Environmental degradation at sea ultimately translates into human insecurity. Communities dependent on fisheries face poverty when stocks collapse. Oil spills devastate coastal livelihoods. Rising seas force migration, creating climate refugees[2]. These insecurities can destabilize states, fueling grievances and conflict. Climate change and environmental degradation act as threat multipliers, exacerbating existing vulnerabilities and insecurities (Jayawardhan, 2017). Recognizing this, maritime security frameworks must integrate environmental protection not as an afterthought but as a core mission. Navies and coast guards can play roles as environmental stewards, supporting disaster relief, monitoring pollution, and enforcing conservation measures.

Another major, less-regulated sectors in the blue economy is deep-sea mining. As we look to a more green energy future, pressure for such blue economy resources as cobalt and nickel is leading people to look at seabed mining (McCarthy, 2020). However, this up-and-coming industry illustrates a very important governance problem. UNCLOS (United Nations, 1982) defines rules of law in resolving seabed mining resource conflicts, but if seabed mining continues, damage to biodiversity in our deep ocean environments could be devastating. Organizations such as Greenpeace highlight (Greenpeace, 2025) this catastrophic potential damage to deep ocean biodiversity and demand a moratorium on species removal. The blue economy can bring positive GDP gains in resource exploitation, but this 'blue'–or ecological—side of things can be badly damaged in achieving this. Such a critical problem clearly illustrates a major problem with ocean governance. While fishing can replenish over time, deep-sea environments exist on a geologically relevant time scale, not a short human time scale. Organizations such as the International Seabed Authority and a new High Seas Treaty (United Nations, 2023) have rules in place to regulate such activity and protect biodiversity, but in a gold-rush mentality towards these new blue economy resources, this regulation can't keep up.

A governance gap evolved into a full-blown diplomatic confrontation because the International Seabed Authority concluded its sessions without completing work on the Mining Code[3], which is a critical set of rules governing commercial mining, thus postponing any regulatory certainty into 2026 or beyond (International Seabed Authority, 2025b). Meanwhile, companies are racing to establish operations, for instance, The Metals Company (TMC) indicated that with support from Nauru, it aims to apply for an exploitation license with production later on (TMC, 2025). Such a pace creates a precarious precedent where mining can commence before a tangible framework is put in place.

However, on the scientific front, opposition has increased in intensity. A major study conducted in 2025, interestingly sponsored by industries, indicated a reduction in Pacific animal populations by 37% in test mining in the Pacific (Stewart et al.,2025) and this triggered demands for a precautionary pause. As a consequence, a network of countries, Deep Sea Conservation Coalition (DSCC)[4], advocating a moratorium broadened to over 40 nations, including the UK, France, and Canada (DSCC, 2025), creating an increasingly divided international perception of mining the deep ocean or not.

Then, in April 2025, a new development arose when, in an effort to end reliance on Chinese supply chains with respect to cobalt, nickel, and manganese, the President of the US issued an executive order. Through this order, US agencies were instructed to fast-track permits for mining on the continental shelf (The White House, 2025b). This puts America on a collision course with an increasingly large bloc of countries, including France and UK, which call for a moratorium on mining. In this way, by shifting the focus of seabed mining from an ecological problem to a matter of national security, America has politicized the seabed and turned a debate over preservation into a scramble for the seabed bottom.

SMART OCEAN GOVERNANCE: TECHNOLOGY, INNOVATION, AND MARITIME DOMAIN AWARENESS

Technology in the maritime domain is not a show to watch but is rather a means which underpins governance. Yet if prosecutions are not possible, then information from satellites does not matter. Furthermore, without being able to see exactly what is happening, good laws will not matter one bit. Therefore, true maritime security must have clever ocean governance, which is an efficient connection between Maritime Domain Awareness and enforcement laws.

The seas have always been spaces of uncertainty and from the perspective of ancient sailors, the ocean was a mysterious and often hostile expanse. Even today, with radar, satellites, and GPS, large areas of the oceans remain effectively invisible to authorities. Ships vanish, illicit cargoes move undetected, and accidents occur beyond the gaze of regulators. Maritime Domain Awareness (MDA) seeks to close these gaps by fusing technology, data, and intelligence to provide a real-time picture of activities at sea (Hicks & Metrick, 2018). Without such awareness, maritime security becomes guesswork, and through this, authorities cannot protect what they cannot see. This simple logic explains why technology—spanning satellites, drones, artificial intelligence, and big data—now sits at the heart of modern maritime strategies.

Satellite technology has revolutionized the monitoring of global seas. Systems such as Automatic Identification System (AIS) tracking allow ships to broadcast their identity, location, and course. Yet, AIS is far from foolproof: vessels can turn off transponders, falsify signals (spoofing), or simply sail without them. This makes satellites that detect radar or visual imagery indispensable (Goward, 2017). The EU's Copernicus program, for instance, provides Earth observation data used to track illegal fishing, maritime security challenges and monitor environmental hazards (EU, 2025a). Private firms such as Planet Labs (GAO, 2022) and Spire Global (NASA, 2025) now offer near-real-time imagery, allowing even smaller states to access satellite intelligence once reserved for major powers. These tools are increasingly used not only by navies but also by NGOs and journalists investigating illicit maritime activity.

Unmanned aerial vehicles (UAVs)[5], surface drones and unmanned surface vessel (USV) extend surveillance reach. These small, often autonomous boats and drones can patrol high-risk areas, conduct reconnaissance, or even perform interdiction missions. As costs fall, they may democratize surveillance, allowing resource-constrained states to achieve wider coverage of their waters. Unlike costly patrol ships, drones can loiter for hours, providing persistent coverage over remote areas. The U.S. Coast Guard has tested long-endurance UAVs and USVs to monitor drug trafficking routes in the Caribbean (Orgill & Moon, 2024), while Indonesia has used them to track illegal fishing vessels in its sprawling archipelago (Dirhamsyah et al., 2022).

Perhaps the most transformative development is the integration of artificial intelligence (AI) into maritime security. Vast amounts of data—AIS signals, satellite images, weather patterns, shipping manifests—are generated daily. Human analysts cannot process this information at scale. AI algorithms, however, can identify anomalies: a vessel broadcasting suspicious AIS patterns, or a ship lingering in unusual areas known for smuggling. For example, Global Fishing Watch, a partnership between Google, Oceana, and SkyTruth, uses AI to detect illegal fishing by analyzing vessel movements (Oceana, 2024). Similarly, customs authorities are experimenting with predictive analytics to identify high-risk cargo shipments before they arrive at ports (Transport Security International, 2023). The potential is enormous, but so are the risks. Overreliance on AI could lead to false positives, wasted enforcement resources, or even diplomatic disputes if accusations are based on flawed algorithms. Data sharing between states also raises sensitive issues of sovereignty and privacy. Basically, tech isn't a standalone solution and it is a part of a larger governance system which involves both trouble identification and trouble solving.

Maritime digitalization brings vulnerabilities. Modern ships are floating computers, reliant on GPS, automated navigation, and satellite communications. Ports operate through highly interconnected digital systems. This creates opportunities for hackers, whether criminals seeking ransom or state actors conducting sabotage. In 2017, the NotPetya cyberattack crippled Maersk, the world's largest shipping company, causing losses of over USD 300 million (Grobarcik et al., 2022). In 2021, South Africa's largest port operator was hit by a ransomware attack that disrupted trade flows for days (UNCTAD, 2021). The International Maritime Organization (IMO, 2025) has since mandated cybersecurity risk management for shipping companies, but implementation remains uneven. Cybersecurity is therefore not just a land-based issue—it is central to maritime resilience. Simply put, all this is a symptom of a failure in governance. Vessels are becoming floating data centers, and without common cyber laws, a small problem can snowball into a massive inconvenience in a way that undermines international trade. Navies, coast guards, and port authorities must now integrate cyber defense into their operational planning.

While technology offers solutions, it also creates dependencies and inequalities. Wealthier states and companies can afford sophisticated satellites and AI platforms, while poorer coastal states may remain blind to activities in their waters. This risks widening the maritime security gap, leaving vulnerable states at the mercy of external actors. Moreover, technology is not neutral. Surveillance tools can be misused by authoritarian governments to monitor political opponents under the guise of security (Greenwood, 2020). . Balancing efficiency with accountability is thus essential.

REGIONAL COOPERATION AND GLOBAL GOVERNANCE

The ocean does not respect borders, a fishing vessel that plunders stocks in one exclusive economic zone (EEZ) may sail into another the next day. Pirates rarely limit themselves to a single jurisdiction. Oil spills drift across boundaries, and climate impacts reverberate globally. No state, however powerful, can secure the seas alone. This reality underpins the importance of regional and global cooperation. Often described as the "constitution of the oceans," UNCLOS (1982) sets out the rights and responsibilities of states at sea. It establishes territorial seas, EEZs, and the principle of freedom of navigation. Crucially, it provides a legal framework for resolving disputes, whether over fisheries, seabed mining, or maritime boundaries. However, UNCLOS faces challenges, as an instance, The United States, despite relying heavily on its provisions, has never ratified it. China's expansive claims in the South China Sea openly defy the 2016 ruling by the Permanent Court of Arbitration, which invalidated its "nine-dash line" claims (Permanent Court of Arbitration, 2016). Enforcement thus remains the Achilles heel of global law of the sea. So, regional frameworks in action such as The EU's Operation Atalanta (Horn of Africa)[6], The 2013 Code of Conduct between West and Central African states[7], ASEAN in Southeast Asia[8] and Arctic Council[9] are crucial for the collective answer to the rising issues regarding the relation between maritime security and blue economy.

While cooperation is essential, the seas are also arenas of great power rivalry. China's Belt and Road Initiative has extended its influence through port construction (Hillman, 2020) and control across Asia, Africa, and Europe. The United States has responded by strengthening alliances and emphasizing freedom of navigation operations in contested waters (Jaehyon,2025). Russia, too, has reasserted its naval presence, particularly in the Black Sea and Arctic (van Loon et al., 2025). Meanwhile, middle and mediating powers such as India, Japan, and Türkiye are increasingly active in maritime diplomacy. These dynamics risk militarizing the oceans, where cooperation and rivalry coexist uneasily.

Global governance is not solely the domain of states while Shipping companies, insurers, and NGOs wield considerable influence. For instance, the shipping industry's decision to reroute vessels away from the Red Sea during Houthi attacks after October 2023 forced governments to respond militarily (Poudineh & Apostolopoulou, 2025). NGOs like Greenpeace highlight the environmental costs of deep-sea mining, pushing for moratoriums (Hurley et al., 2025). The complexity of actors underscores that maritime governance is a mosaic of formal treaties, regional frameworks, industry standards, and civil society campaigns.

Ultimately, safeguarding the seas requires integrating security, economic, and environmental priorities but fragmentation remains a key obstacle. Fisheries agreements may ignore security threats, while naval strategies often neglect ecological impacts. The challenge of the 21st century is to weave these threads together into a holistic governance framework. Some progress is visible as like the High Seas Treaty, agreed in 2023 after two decades of negotiation, aims to protect biodiversity in international waters by creating marine protected areas and regulating resource exploitation (United Nations, 2023).[10] Though implementation will be slow, it reflects growing recognition that the oceans are a shared heritage requiring collective stewardship.

CASE STUDIES IN MARITIME SECURITY AND THE BLUE ECONOMY

Case Study 1: The EU's Operation Aspides and the Red Sea

The Red Sea, stretching from the Suez Canal in the north to the Bab el-Mandeb Strait in the south, is one of the most strategic waterways in the world. Nearly 15 percent of global trade passes through it, including significant volumes of the world's liquefied natural gas (LNG) and oil (IMO, 2024). For Europe in particular, the Red Sea is a lifeline: goods from Asia, Middle Eastern energy supplies, and food shipments all transit its narrow corridors before reaching the Mediterranean. Yet, this artery of globalization is highly vulnerable. The Bab el-Mandeb is just 32 kilometers wide at its narrowest point, making it a chokepoint where disruptions can ripple across global supply chains. Recent attacks by Yemen's Houthi rebels, including drone and missile strikes on commercial vessels in 2023 and 2024, underscored this fragility. Insurance premiums soared, shipping companies rerouted vessels around the Cape of Good Hope, and energy markets reacted with volatility (Strain, 2025).

In response, the EU launched Operation Aspides in early 2024 as part of its Common Security and Defence Policy (CSDP). Aspides—named after the Greek word for "shield"—is a naval mission tasked with protecting merchant shipping in the Red Sea and Gulf of Aden. Its mandate includes escorting vessels, deterring attacks, and coordinating with international partners such as the U.S.-led Operation Prosperity Guardian (Council of the European Union, 2025). The mission builds on the EU's earlier experience with Operation Atalanta, which successfully curbed Somali piracy in the 2010s. However, Aspides is more complex. Whereas

Atalanta confronted non-state pirates seeking ransom, Aspides faces state-backed insurgents armed with advanced weaponry, including Iranian-supplied drones and missiles (Calcagno et al., 2025). This raises the stakes, blurring the line between counterterrorism and interstate confrontation.

The link between maritime security and the blue economy is evident in the Red Sea. Disruptions to shipping directly affect the cost of goods in European markets and undermine energy security. For coastal states such as Djibouti and Egypt, whose ports depend on transit traffic, instability threatens revenues and foreign investment (Mendez-Parra et al., 2025) Moreover, insecurity deters investment in blue economy initiatives such as offshore wind farms or sustainable fisheries. Investors will not commit capital to regions where infrastructure can be destroyed overnight.

Aspides, therefore, is not only a security mission but also an economic stabilization effort. Its effectiveness will determine whether the Red Sea can continue to serve as a corridor of prosperity rather than a flashpoint of conflict. However, Aspides, like many international missions, risks treating symptoms rather than root causes. The Houthis attack shipping not for piracy but as part of a broader political and military struggle in Yemen. Without addressing the underlying conflict, naval patrols may offer only temporary relief. There is also the question of European unity as some EU states, wary of escalation with Iran, prefer a limited defensive posture. Others advocate for robust engagement to assert European strategic autonomy (Wintour, 2025). Balancing these positions while coordinating with U.S. and regional partners will be a persistent challenge.

Case Study 2: The Indo-Pacific

If the Red Sea is a vulnerable artery, the Indo-Pacific is the beating heart of global trade. Home to half the world's population and two-thirds of global GDP, the region's sea lanes carry goods worth trillions annually. The Strait of Malacca alone handles about 40 percent of global trade and 80 percent of China's oil imports (Myers, 2023). The blue economy here is immense, ranging from fisheries that feed hundreds of millions to booming port industries and offshore energy exploration. Yet, the same waters are contested by overlapping territorial claims, militarization, and ecological pressures. The South China Sea epitomizes these tensions: rich in fish, oil, and gas, it is claimed almost entirely by China (Hirayama, 2025) but also by Vietnam, the Philippines, Malaysia, Brunei, and Taiwan.

India's approach to maritime security reflects its ambition to be a net security provider in the Indian Ocean since 2009 through US support (Shangri-La Dialogue, 2009). But then, it has expanded its naval presence, established coastal radar chains, and deepened partnerships with island states such as the Maldives, Mauritius, and Seychelles. Through initiatives like SAGAR[11] (Deb & Dutta, 2023), India frames maritime security as inseparable from blue economy development—helping states combat illegal fishing while offering aid and capacity-building.

For Japan, a resource-poor island nation, secure sea lanes are existential. Tokyo has invested heavily in coast guard diplomacy, providing patrol vessels and training to Southeast Asian partners. Its Free and Open Indo-Pacific strategy links maritime stability with democratic values, economic connectivity, and ecological sustainability (Bito, 2025). Japan also pushes for innovation in blue economy sectors, from offshore wind to hydrogen energy, but acknowledges that without secure sea lanes, such ventures cannot thrive.

ASEAN faces the delicate task of balancing sovereignty concerns with collective action. The Malacca Strait Patrols, launched in 2004 by Indonesia, Malaysia, Singapore, and later Thailand, succeeded in reducing piracy through coordinated patrols and information sharing (MINDEF Singapore, 2016). However, broader cooperation on the South China Sea remains elusive due to divergent national interests, such as the statue of India[12], and external pressures. Therefore, for Southeast Asian states, the blue economy is central to development strategies, yet insecurity—whether from maritime disputes or environmental crimes—remains a constant drag on potential.

Since the arising issues here, two new groupings have reshaped Indo-Pacific maritime security. First, The Quad[13], while often described as a diplomatic and security dialogue, it increasingly focuses on maritime security, joint exercises, and infrastructure financing. Its agenda includes promoting resilient supply chains and combating IUU fishing, linking security to blue economy sustainability (Baruah, 2024). Secondly, AUKUS[14] is best known for its nuclear-powered submarine deal, but AUKUS also includes cooperation on cyber, AI, and undersea technologies. Critics see it as fueling an arms race and proliferation of nuclear weapons (Dalton & Levite, 2023), while supporters argue it strengthens deterrence against coercion (Monaghan et al., 2025). Both initiatives illustrate how maritime security in the Indo-Pacific is inseparable from broader geopolitical competition, with implications for trade, energy, and ecological governance.

The unity of transnational trade corridors is being disrupted by another wave of protectionism. The US Executive Order titled "Restoring America's Maritime Dominance" in April 2025 imposed deterrent tariffs on Chinese-made ships and container cranes based on national security grounds (The White House, 2025c). As a consequence, shipping companies have had to rebalance their trade corridors to exclude US ports for a subset of their vessels, thus establishing a bifurcated maritime logistics system for the US market and a separate one for other markets in the world.

Case Study 3: The Arctic

The Arctic is warming nearly four times faster than the global average (Duffey et al., 2023). As sea ice retreats, new maritime routes such as the Northern Sea Route (NSR) (Townsend & Kendall-Taylor, 2021) and the Northwest Passage (Sodero, 2022) are becoming seasonally navigable. For shipping companies, this promises shorter transit times between Asia and Europe—potentially cutting journeys by two weeks compared to the Suez Canal route, albeit, for Arctic states, it represents both opportunity and risk.

The blue economy potential in the Arctic is significant: vast reserves of oil, gas, and minerals; expanding fisheries as stocks migrate northward; and tourism in fragile polar ecosystems (Steinberg & Kristoffersen, 2018). Yet these prospects are shadowed by enormous ecological costs, because accidents in icy, remote waters are difficult to contain. Increased shipping threatens marine biodiversity and indigenous livelihoods (Lynch et al., 2022). Moreover, competition over resources risks militarizing the region (Vakulenko, 2025).

The Arctic Council, established in 1996, provides the primary forum for cooperation among Arctic states[15] along with indigenous organizations. While it excludes military security from its mandate, it has advanced environmental protection and scientific cooperation. Agreements on oil spill preparedness and search-and-rescue demonstrate practical progress (Arctic Council, 2021). However, Russia's invasion of Ukraine in 2022 disrupted Arctic governance. Western members suspended most cooperation with Moscow (Cunningham, 2024), undermining a forum that had symbolized post–Cold War collaboration. This raises concerns about the future of shared stewardship in a rapidly changing region.

Russia has invested heavily in Arctic infrastructure, reopening Soviet-era bases and deploying advanced weaponry (Buchanan, 2023). China, though a non-Arctic state, describes itself as a near-Arctic power (Dams et al., 2020) and has launched scientific expeditions and invested in Arctic energy projects. NATO has increased exercises in northern waters, signaling its awareness of the Arctic's strategic significance (Brzezinski & Arick, 2025). For the blue economy, this militarization creates uncertainty. Shipping companies may hesitate to commit to polar routes if geopolitical tensions rise. Indigenous communities, often excluded from high-level security debates, face disruptions to their way of life without sufficient consultation.

Case Study 4: Africa

The Gulf of Guinea, stretching from Senegal to Angola, has become the global epicenter of piracy since Somali piracy declined in the mid-2010s. Unlike Somali pirates, who sought ransom for hijacked ships, West African pirates often kidnap crew members for ransom while targeting cargo theft (Gaibulloev & Sandler, 2016). The region's vast oil exports (EIA, 2023), coupled with weak law enforcement, make it particularly vulnerable. The insecurity in the Gulf of Guinea drains billions annually from West African economies—funds that could otherwise fuel blue economy initiatives.

The Yaoundé Architecture, launched in 2013, created a framework for regional cooperation, including coordination centers in West and Central Africa. Though progress is uneven due to resource shortages, the framework has improved information sharing and joint operations. International partners such as the EU, U.S., and Japan provide training and funding, but African ownership remains key (Mohammed, 2023). At the same time, West African states recognize the blue economy as a pillar of development. Nigeria's Deep Blue Project, for example, aims to build capacity against maritime crime while developing port infrastructure and fisheries (Imide & Osadume, 2025). Besides other West African countries such as Ghana, Senegal, and Côte d'Ivoire are investing in coastal tourism and aquaculture (Beaumont et al., 2025). Without security, however, these ambitions risk stalling.

In the East Africa, Somali piracy dominated global headlines in the late 2000s, peaking at 237 attacks in 2011 (IMB, 2012).[16] International naval coalitions, including the EU's Operation Atalanta, Combined Task Force-151 and NATO's Ocean Shield, combined with private security measures and Somali capacity-building, eventually reduced attacks to near zero by 2017[17].

The lessons from Somalia highlight both the possibilities and limits of international intervention. Naval patrols were effective in deterring attacks, but long-term stability required addressing conditions ashore: poverty, weak governance, and lack of livelihood alternatives. Many coastal communities had turned to piracy after foreign vessels depleted their fish stocks through illegal fishing (Peters & Paoli, 2025). Today, Somali piracy remains dormant but not eradicated. Meanwhile, the region faces new maritime challenges, from illegal fishing by foreign fleets to Houthi attacks spilling into the Gulf of Aden. For Somalia, Kenya, and Tanzania, developing blue economy sectors—such as fisheries, offshore energy, and shipping hubs—depends on sustaining maritime security and governance reforms.

CONCLUSION AND POLICY RECOMMENDATIONS

The twenty-first century challenges humanity to rethink the relationship between the seas, security, and economic development. Maritime security is no longer a narrow military concern—it is the foundation upon which the blue economy depends.

Protecting oceans, ensuring free and safe passage, and fostering governance structures are inseparable from efforts to create sustainable, inclusive economic growth.

The central message of this paper is that the seas are not empty spaces to be secured only when crises erupt. They are living ecosystems, highways of commerce, and sources of human identity and culture. Securing them is not simply about deploying warships or passing treaties—it is about weaving together security, sustainability, and governance in a coherent vision. The blue economy will fail if treated as a purely economic project divorced from ecological limits. Maritime security will falter if treated as a purely military endeavor divorced from human development. The future lies in integrating the two, acknowledging that prosperity and peace at sea are inseparable. However, all this is now threatened by a divergent reality in geopolitics. With the withdrawal of the US from climate agreements and the aggressive move for seabed minerals, a consensus on ocean governance is fraying. The next decade will therefore see a struggle not just against pirates and polluters but a struggle between two different visions – one emphasizing multilateral governance and another emphasizing unilateral security.

This study began with a simple but profound proposition: the blue economy—the sustainable use of ocean resources for growth, livelihoods, and ecosystem health—cannot exist without maritime security. Across the globe, from the narrow chokepoints of the Red Sea to the melting expanses of the Arctic and the turbulent coasts of West Africa, this interdependence is evident. Where insecurity prevails, investments wither, ecosystems suffer, and communities struggle. Where stability is established, opportunities for trade, energy, and sustainable development multiply.

Yet, maritime security itself is evolving. The threats are no longer confined to piracy or naval skirmishes. They now encompass illegal, unreported, and unregulated (IUU) fishing, environmental crime, cyberattacks on ports, and gray-zone coercion by powerful states. These threats blur traditional categories of security and economy, demanding new frameworks that integrate both. If the blue economy is to deliver on its promise—providing food security, renewable energy, decent work, and climate resilience—then maritime security must be reimagined not as a military afterthought but as a foundation of sustainability.

The case studies examined in this paper offer both cautionary tales and models of success. Within the Red Sea, the Operation Aspides undertaken by the European Union's naval mission emphasizes the vulnerability of global trade routes. It demonstrates that formidable naval capabilities can protect commerce; however, it also emphasizes the constraints of military responses in the lack of political resolutions. In order for the blue economy to prosper in the Red Sea area, the establishment of stability in Yemen and the promotion of inclusive governance among coastal nations are equally as vital as the provision of naval escorts.

At sea, security in the Indo-Pacific cannot be dissociated from great-power rivalry. India, Japan, ASEAN, and newer entities like the Quad and AUKUS reflect their pursuit of balance in a competitive setting. Their initiatives signal that sustainable growth, whether off the wind farms or the fishing grounds, is not strong enough to be detached from geopolitical rivalry. While pragmatic coordination, like the Malacca Strait Patrols, suggests that, even with rivalries, regional actors are capable of building entities that keep insecurity to a minimum.

The melting of ice presents both novel opportunities and significant risks in the Arctic region. The Arctic Council has historically served as a forum for collaborative efforts; however, prevailing geopolitical tensions currently undermine its effectiveness. For indigenous communities and vulnerable ecosystems, the increased militarization poses an immediate threat. The situation in the Arctic underscores that climate change transcends environmental concerns, acting as a catalyst for emerging maritime boundaries and possibly inciting new conflicts.

In Africa, The Gulf of Guinea and the Horn of Africa illustrate the developmental cost of insecurity. Piracy, IUU fishing, and weak governance deprive coastal states of billions annually. Yet regional initiatives such as the Yaoundé Architecture and the Deep Blue Project point to homegrown solutions. Africa's experience shows that external naval patrols may suppress symptoms, but sustainable security must come from capacity-building, governance reform, and community engagement.

The Arctic and Africa are opposing ends of the spectrum. While the Arctic is a borderland of possibility based on climate change as well as great-power competition, Africa is defined as a terrain where insecurity thwarts basic development within the blue economy. Still, the bottom line does not change: the actualization of the blue economy's potential depends on security, ecological sustainability, as well as inclusive governance. While the Arctic states face the need to prevent rising opportunities from causing ecological destruction or great-power conflict, African states face the need to build core security and governance standards that allow investability as well as livelihoods to persist. Regardless, the inseparable connection between security as well as sustainability within the maritime spaces comes into sharp focus.

In all these cases, there is a common factor: security and sustainability cannot be achieved individually. Effective blue economy strategies combine enforcement and ecological stewardship, regional involvement and local engagement, and technological advancement and governance. And, after all, if some policy suggestions need to be followed to meet these objectives, the following statements would be highest dimensions. National blue economy strategies as well as regional blue economy strategies need to go beyond economic planning to integrate security as a core tenet. Fisheries management plans need to include the measures to prevent IUU fishing. Offshore renewable energy projects must account for protection against sabotage and cyber threats. Development banks financing maritime infrastructure should assess security risks alongside environmental ones.

The African Union's 2050 Integrated Maritime Strategy (African Union, 2012) provides a model by explicitly linking maritime security with development. Similar integrated approaches should be replicated in Southeast Asia, Latin America, and the Pacific. The transboundary nature of maritime threats demands regional solutions. Frameworks like the Yaoundé Code of Conduct, the Malacca Strait Patrols, and EU CSDP missions demonstrate that cooperation works when political will aligns with shared interests. Yet many regions still lack effective mechanisms. Policies to tackle these challenges should prioritize: Building regional coordination centers that integrate information-sharing, like those in West Africa. Promoting joint patrols in high-risk areas, supported by cost-sharing arrangements. Establishing regional fisheries organizations with enforcement powers, not just advisory roles.

Satellite tracking, drones, and AI surveillance are changing how we understand maritime areas. However, not all countries have the same access, which could leave poorer coastal states unaware of what happens in their waters. International efforts—like the G7's partnership with Global Fishing Watch—should be increased to provide free tools for monitoring. At the same time, protections are needed to stop the abuse of surveillance technologies for controlling people at home. Openness, legal rules, and involvement from the community can help find a balance between security and responsibility.

It is not merely about capability, either—but about legitimacy. Weakly institutionalized states cannot enforce laws, while excluded communities may oppose security measures. Filling the governance gaps demands: Enhancing the laws conforming to the UNCLOS and the Treaty of the High Seas. Investing in judicial systems and coast guard to prosecute offenses on the high seas. Ensuring coastal communities gain access to blue economy initiatives, depriving parties of illicit activity incentives.

Climate change is a threat to security and can harm the economy. Higher sea levels put ports at risk, severe weather can disrupt shipping, and changing ecosystems affect fisheries. Because of this, maritime security plans need to include ways to adapt to climate change. This means building strong infrastructure, using low-carbon shipping technologies, and investing in natural solutions like restoring mangroves (UNESCO, 2025), which protect coastlines and help wildlife.

Shipping companies, insurers, and technology firms are indispensable stakeholders in maritime security. Their actions—rerouting vessels, adopting cybersecurity standards, or funding monitoring platforms—shape outcomes as much as government policies. Public-private partnerships should therefore be institutionalized. For example, information-sharing between navies and shipping companies can enhance threat detection. Insurers can incentivize compliance with sustainability standards through premium reductions. Technology companies can collaborate with governments to deploy AI responsibly in maritime monitoring.

The coming decades will test humanity's capacity to govern the oceans responsibly. By 2050, global population growth, rising consumption, and accelerating climate change will place unprecedented pressure on maritime spaces. Choices made now—about resource extraction, naval competition, and environmental protection—will shape whether the seas remain a source of shared prosperity or become arenas of ecological collapse and geopolitical conflict.

The oceans' vastness can give an illusion of inexhaustibility, but as this paper has shown, they are fragile. Safeguarding them requires foresight, cooperation, and humility. If nations can integrate maritime security with the sustainable blue economy, the seas can continue to nourish life, fuel trade, and inspire human imagination. If they cannot, the costs will be felt far beyond the shorelines.

REFERENCES

African Union. (2012). *2050 Africa's Integrated Maritime Strategy (2050 AIM Strategy)*. https://au.int/sites/default/files/documents/30929-doc-2050_aim_strategy_eng_0.pdf

Arctic Council. (2021). Arctic Council Strategic Plan 2021-2030. Reykjavik: Arctic Council Secretariat. https://oaarchive.arctic-council.org/server/api/core/bitstreams/99c66b04-b82e-4fe7-848c-ffafa111cf3e/content

ASEAN. (2022). *The ASEAN Senior Officials Meeting on Transnational Crime (SOMTC) Work Programme to implement the ASEAN plan of Action to Combat Transnational Crime 2022-2024*. https://asean.org/wp-content/uploads/2023/08/4-SOMTC-WP-TC-2022-2024.pdf

Baruah, D. M. (2024). *The Contest for the Indian Ocean: And the Making of a New World Order*. Yale University Press.

Beaumont, J., Hampton, S., & Willima, D. (2025). Africa's blue economy and climate change: key sectors. In *Africa's blue economy and climate change: Status and emerging pathways* (pp. 5–30). Institute for Security Studies.

Belhabib, D., Sumaila, U. R., Lam, V. W. Y., Zeller, D., Billon, P. L., Kane, E. A., & Pauly, D. (2015). Euros vs. Yuan: Comparing European and Chinese fishing access in West Africa. *PLoS One*, *10*(3), e0118351. DOI: 10.1371/journal.pone.0118351 PMID: 25793775

Bhaumik, A. (2025). India secures Singapore's support to its plan to join patrolling in Malacca Strait. *Deccan Herald*. https://www.deccanherald.com/india/india-secures-singapores-support-to-its-plan-to-join-patrolling-in-malacca-strait-3712849

Bito, Y. (2025). *Navigating Indo-Pacific Waters: Maritime Governance and Developing a National Unified Maritime Strategy for Japan*. East-West Center.

Brzezinski, I., & Arick, R. (2025). *Issue brief: A NATO strategy for countering Russia*. Atlantic Council.

Buchanan, E. (2023). Future Arctic Histories. In *Red Arctic: Russian Strategy Under Putin* (pp. 147–158). Brookings Institution Press. DOI: 10.5040/9780815750482.ch-007

Calcagno, E., Daga, G., & Ahmed, F. (2025). *Deterrence and Diplomacy in the Red Sea: Recommendations for Italy*. Istituto Affari Internazionali (IAI).

Canyon, D., Allen, E., Long, M., & Brown, C. (2021). *Policy recommendations for combatting overfishing and fisheries crime*. Daniel K. Inouye Asia-Pacific Center for Security Studies.

Collin, K. S. L. (2025). The Malacca Strait Patrols: Finding Common Ground. Bakrie Center Foundation. https://bcf.or.id/the-malacca-strait-patrols-finding-common-ground/

Council of the European Union. (2023). *Revised EU Maritime Security Strategy (EUMSS) and its Action Plan.* https://www.consilium.europa.eu/media/67499/st14280-en23.pdf

Council of the European Union. (2025). *Red Sea: Council prolongs the mandate of Operation ASPIDES.* https://www.consilium.europa.eu/en/press/press-releases/2025/02/14/red-sea-council-prolongs-the-mandate-of-operation-aspides/#:~:text=EUNAVFOR%20ASPIDES%20was%20established%20in,Red%20Sea%20and%20the%20Gulf

Courdingly, D. (2006). *Under the Black Flag: The Romance and the Reality of Life among the Pirates*. Random House.

Crowell, B., & Turvold, W. (2020). Illegal, Unreported, And Unregulated Fishing and The Impacts on Maritime Security. In Vuving, A. L. (Ed.), *Hindsight, Insight, Foresight: Thinking About Security in the Indo-Pacific* (pp. 209–216). Daniel K. Inouye Asia-Pacific Center for Security Studies.

Cunningham, A. (2024). Shifting Ice: How the Russian Invasion of Ukraine has changed Arctic Circle Governance and the Arctic Council's Path Forward. https://www.thearcticinstitute.org/shifting-ice-russian-invasion-ukraine-arctic-circle-governance-arctic-councils-path-forward/

Dalton, T., & Levite, A. (2023). AUKUS as a Nonproliferation Standard? *Arms Control Today*, *53*(6), 6–11.

Dams, T., van Schaik, L., & Stoetman, A. (2020). The power of presence: engaging China in the Arctic. In *Presence before power: China's Arctic strategy in Iceland and Greenland* (pp. 38–42). Clingendael Institute.

Deb, S., & Dutta, A. (2023). India's Role in Capability Development Measures for Maritime Security in the Indian Ocean Under SAGAR. *The Journal of Territorial and Maritime Studies*, *10*(2), 47–59. DOI: 10.2307/48750351

Dirhamsyah, D., Umam, S., & Arifin, Z. (2022). Maritime law enforcement: Indonesia's experience against illegal fishing. *Ocean and Coastal Management*, *229*, 229. DOI: 10.1016/j.ocecoaman.2022.106304

Dixon, J. (2024). What Drives Maritime Piracy in Southeast Asia? A Regional- and Country-Specific Analysis. *The Journal of East Asian Affairs*, *37*(2), 45–74.

Doumbouya, A., Camara, O. T., Mamie, J., Intchama, J. F., Jarra, A., Ceesay, S., Guèye, A., Ndiaye, D., Beibou, E., Padilla, A., & Belhabib, D. (2017). Assessing the effectiveness of monitoring control and surveillance of illegal fishing: The case of West Africa. *Frontiers in Marine Science*, *4*, 50. DOI: 10.3389/fmars.2017.00050

DSCC. (2019). DSCC Position Statement on Deep Seabed Mining. https://deep-sea-conservation.org/wp-content/uploads/2024/01/DSCC-Position-Statement-on-Deep-Seabed-Mining_July2019-1.pdf

DSCC. (2025). Voices calling for a moratorium. https://deep-sea-conservation.org/solutions/no-deep-sea-mining/momentum-for-a-moratorium/governments-and-parliamentarians/

Duffey, A., Mallett, R., Irvine, P. J., Tsamados, M., & Stroeve, J. (2023). Arctic amplification's contribution to breaches of the Paris Agreement. *Earth System Dynamics: ESD*, *14*(6), 1165–1169. DOI: 10.5194/esd-14-1165-2023

Edwards, S. (2022). Fragmentation, Complexity and Cooperation: Understanding Southeast Asia's Maritime Security Governance. *Contemporary Southeast Asia*, *44*(1), 87–121. DOI: 10.1355/cs44-1d

EIA. (2023). *Emerging Hydrocarbon Producers in Africa*. https://www.eia.gov/international/analysis/special-topics/Emerging_Hydrocarbon_Producers_in_Africa

Elmi, A. A., Affi, L., Knight, W. A., & Mohamed, S. (2015). Piracy in the Horn of Africa Waters: Definitions, History, and Modern Causes. *African Security*, *8*(3), 147–165. DOI: 10.1080/19392206.2015.1069118

EU. (2025a). *Copernicus Document Library*. https://www.copernicus.eu/en/copernicus-documents-library?keywords=&c=-1

EU. (2025b). EU Naval Force Operation ATALANTA. https://eunavfor.eu

European Parliament. (2025). EU Strategy For Offshore Renewable Energy Sources. https://www.europarl.europa.eu/legislative-train/carriage/offshore-wind/report?sid=9701

FAO. (2016). *Agreement on Port State Measures to Prevent, Deter and Eliminate Illegal, Unreported and Unregulated Fishing. Revised Edition*. https://www.fao.org/port-state-measures/resources/detail/en/c/1111616/

FAO. (2022). *The state of world fisheries and aquaculture 2022*. FAO.

FAO. (2025). *Parties to the PSMA.* https://www.fao.org/port-state-measures/background/parties-psma/en/

Faruk, O. (2024). Pirates demand $10 million in ransom after hijacking Chinese-owned fishing vessel with 18 captives. *The Associated Press.* https://apnews.com/article/somalia-piracy-chinese-fishing-vessel-91d071c26f4a45abf2942eafdb2626d6

Gaibulloev, K., & Sandler, T. (2016). Decentralization, institutions, and maritime piracy. *Public Choice, 169*(3/4), 357–374. DOI: 10.1007/s11127-016-0346-3

GAO. (2022). National Security Space: Overview of Contracts for Commercial Satellite Imagery. *U.S. Government Accountability Office.* https://www.gao.gov/assets/gao-23-106042.pdf

Global Fishing Watch. (2025). Revolutionizing Ocean Monitoring and Analysis. https://globalfishingwatch.org

Goward, D. (2017). GPS Spoofing Incident Points to Fragility of Navigation Satellites. *National Defense, 102*(766), 18–19.

Graham, B., Reilly, W. K., Beinecke, F., Boesch, D. F., Garcia, T. D., Murray, C. A., & Ulmer, F. (2011). *Deep Water: Deep Water: The Gulf Oil Disaster and The Future of Offshore Drilling - Report to the President (BP Oil Spill Commission Report).* https://www.govinfo.gov/content/pkg/GPO-OILCOMMISSION/pdf/GPO-OILCOMMISSION.pdf

Greenberg, A. (2018). The Untold Story of NotPetya, the Most Devastating Cyberattack in History. *Wired.* https://www.wired.com/story/notpetya-cyberattack-ukraine-russia-code-crashed-the-world/

Greenberg, A. (2019). *Sandworm: A New Era of Cyberwar and the Hunt for the Kremlin's Most Dangerous Hackers.* Doubleday.

Greenpeace. (2025). Deep Deception - How The Deep Sea Mining Industry is Manipulating Geopolitics to Profit from Ocean Destruction. https://www.greenpeace.org/static/planet4-usa-stateless/2025/07/9b462ecb-deep-deception-how-the-deep-sea-mining-industry-is-manipulating-geopolitics-to-profit-from-ocean-destruction-greenpeace-usa-report.pdf

Greenwood, F. (2020). Data colonialism, surveillance capitalism and drones. In Specht, D. (Ed.), *Mapping Crisis: Participation, Datafication and Humanitarianism in the Age of Digital Mapping* (pp. 89–118). University of London Press.

Grobarcik, D., Loomis, W., Poznansky, M., & Smith, F. (2022). *Wargaming to Find a Safe Port in a Cyber Storm.* Atlantic Council.

Haenlein, C. (2017). IUU Fishing and Security. In *Below the Surface: How Illegal, Unreported and Unregulated Fishing Threatens our Security* (pp. 31–36). Royal United Services Institute (RUSI).

Harithas, B. (2024). The Export Control Evasion Pipeline. In Mapping the Chip Smuggling Pipeline and Improving Export Control Compliance (pp. 4–18). Center for Strategic and International Studies (CSIS).

Hicks, K. H., & Metrick, A. (2018). Maritime Domain Awareness: Today and Tomorrow. In *Contested Seas: Maritime Domain Awareness in Northern Europe* (pp. 12–21). Center for Strategic and International Studies (CSIS).

Hillman, J. E. (2020). Game of Loans: Sri Lanka. In *The Emperor's New Road: China and the Project of the Century* (pp. 151–170). Yale University Press.

Hirayama, N. (2025). Decolonizing or Recolonizing? Chinese Historical Epistemologies in the New Cold War Climate. *China Review*, *25*(3), 1–27.

Horden, P., & Purcell, N. (2000). *The Corrupting Sea: A Study of Mediterranean History*. Wiley-Blackwell.

Hurley, G., Wilkinson, E., Weaver, J., & Kasanawaqa, V. (2025). *Turning the tide: Enhancing ocean equity for Small Island Developing States*. ODI.

IMB. (2012). *Piracy attacks in East and West Africa dominate world report*. https://icc-ccs.org/piracy-attacks-in-east-and-west-africa-dominate-world-report/

IMB. (2024). Maritime piracy dropped in 2024, but crew safety remains at risk. *International Chamber of Commerce*. https://iccwbo.org/news-publications/news/maritime-piracy-dropped-in-2024-but-crew-safety-remains-at-risk/International Chamber of Shipping. (2021). *COVID-19: Lessons for shipping and global trade*. London.

IMB. (2025). Piracy & Armed Robbery Map. *International Chamber of Commerce*. https://icc-ccs.org/map/

Imide, I. O., & Osadume, R. C. (2025). Port Infrastructural Investments and Maritime Industry Performance: Evidence From Nigeria. *Ibadan Journal of Sociology*, *16*(1), 1–13.

IMO. (2013). Yaoundé Code of Conduct. *Code of Conduct Concerning the Repression of Piracy, Armed Robbery Against Ships, and Illicit Maritime Activity in West and Central Africa*. https://wwwcdn.imo.org/localresources/en/OurWork/Security/Documents/code_of_conduct%20signed%20from%20ECOWAS%20site%20(2).pdf

IMO. (2017). Djibouti Code of Conduct. *Revised Code of Conduct Concerning The Repression of Piracy, Armed Robbery Against Ships, and Illicit Maritime Activity in The Western Indian Ocean and The Gulf of Aden Area*. https://wwwcdn.imo.org/localresources/en/OurWork/Security/Documents/A2%20Revised%20Code%20Of%20Conduct%20Concerning%20The%20Repression%20Of%20Piracy%20Armed%20Robbery%20Against%20Ships%20Secretariat.pdf

IMO. (2024). IMO Annual Report 2024. https://indd.adobe.com/view/publication/2c90e551-f240-4719-a018-377211df66f2/9u15/publication-web-resources/pdf/IMO_AR_wave_15_digital.pdf

IMO. (2025). Guidelines on Maritime Cyber Risk Management. https://wwwcdn.imo.org/localresources/en/OurWork/Security/Documents/MSC-FAL.1-Circ.3-Rev.3.pdf

Indo-Pacific Studies Center. (2025). From SAGAR to MAHASAGAR: India's Strategic Evolution in the Indian Ocean Region. https://www.indo-pacificstudiescenter.org/commentaries/india-mahasagar-initiative

Institute for Economics & Peace. (2024). *Ecological Threat Report 2024*. https://www.visionofhumanity.org/wp-content/uploads/2024/10/ETR-2024-web.pdf

International Energy Agency. (2019). *Offshore wind outlook 2019*. IEA.

International Seabed Authority. (2025a). The Mining Code. https://isa.org.jm/the-mining-code/

International Seabed Authority. (2025b). International Seabed Authority holds a landmark 30th session: the Assembly adopts historic decisions as negotiations on the Mining Code progress. https://isa.org.jm/news/international-seabed-authority-holds-a-landmark-30th-session-the-assembly-adopts-historic-decisions-as-negotiations-on-the-mining-code-progress/

IPCC. (2019). *Special Report on the Ocean and Cryosphere in a Changing Climate*. https://www.ipcc.ch/srocc/

Jacobsen, K. L., & Høy-Carrasco, J. (2018). Section I: Current Trends in the Maritime Crime Complex off the Horn of Africa. In *Navigating Changing Currents: A forward-looking evaluation of efforts to tackle maritime crime off the Horn of Africa* (p. page 13-page 43). Centre for Military Studies.

Jaehyon, L. (2025). *Strategies of the Philippines, China, and the US in the South China Sea in 2025 and Future U.S. Policies on Alliances and the Indo-Pacific Region*. Asan Institute for Policy Studies.

Jayawardhan, S. (2017). Vulnerability and Climate Change Induced Human Displacement. *Consilience*, *17*, 103–142.

Kaplan, R. D. (2021). *Asia's cauldron: The South China Sea and the end of a stable Pacific*. Random House.

Kasanawaqa, V. (2024). Deep-sea mining on the global stage. In *Keep the Ocean Blue: The opportunity cost of deep-sea mining and compensation for ecosystem services provided by Pacific Small Island Developing States* (pp. 6–6). ODI.

Kraska, J. (2011). The Law of the Sea Convention: A National Security Success - Global Strategic Mobility Through the Rule of Law. *The George Washington International Law Review*, *39*, 543–572.

Larsen, J., & Nissen, C. (2017). Protecting the Shipping Industry. In *Learning from Danish Counter-Piracy off The Coast of Somalia* (pp. 37–45). Danish Institute for International Studies.

Licker, R., Ekwurzel, B., Sideris, J., & Frumhoff, P. (2019). *Tracing Fossil Fuel Companies' Contribution to Climate Change and Ocean Acidification*. Union of Concerned Scientists.

Lloyd's List. (2025). Reported Houthi attacks in the Red Sea and Gulf of Aden. https://www.lloydslist.com/hot-topics/red-sea-risk/map-and-list-of-attacks

Lodge, M. (2017). The International Seabed Authority and Deep Seabed Mining. UN Chronicle. https://www.un.org/en/chronicle/article/international-seabed-authority-and-deep-seabed-mining

Lynch, A. H., Norchi, C. H., & Li, X. (2022). The interaction of ice and law in Arctic marine accessibility. *Proceedings of the National Academy of Sciences of the United States of America*, *119*(26), 1–3. DOI: 10.1073/pnas.2202720119 PMID: 35727968

McCarthy, R. (2020). Deep Sea Rush: With valuable metals on the ocean floor, speculators are circling. *The Baffler*, *54*, 114–124.

McVey, R. A. (2022). *Russian Strategic Interest in Arctic Heats Up as Ice Melts*. LSE IDEAS.

Melvin, N. (2019). *The New External Security Politics of The Horn of Africa Region*. Stockholm International Peace Research Institute.

Mendez-Parra, M., Agarwal, P., Luke, D., Vutula, N., Magooa, A., El-Aidi, I., Parshotam, A., Iacovou, E., & Chapman, F. (2025). Geopolitical dynamics. In *Impact of geopolitical dynamics on global trade: The case of the African automotives sector* (pp. 16–46). ODI.

Ministry of Foreign Affairs of Japan. (2020). *Free and Open Indo-Pacific vision*. MOFA.

Missbach, A., & Hoffstaedter, G. (2025). Survival and Deferred Place-Making at Sea: Onboard Socialities of Vietnamese and Rohingya Boatpeople. In Winters, N., Drotbohm, H., & González, Y. G. (Eds.), *Un)Settling Place: Diverse and Divergent Place-Making of People on the Move* (1st ed., Vol. 14, pp. 154–174). Berghahn Books.

Mitra, S. (2025). Somali pirates have used the Red Sea Crisis and war in Gaza to stage a return. *CNN*. https://edition.cnn.com/2025/02/22/world/somali-pirates-red-sea-gaza-dg#:~:text=At%20their%20peak%20in%202011,grenades%20in%20those%2012%20months

Mohammed, A. (2023). Maritime Security Governance in the Gulf of Guinea: Opportunities and Challenges of States' Responses. *The Journal of Territorial and Maritime Studies*, *10*(2), 60–77. DOI: 10.2307/48750352

Monaghan, S., Palmer, A., & Park, C. H. (2025). Policy Recommendations for Improving U.S. Multinational Strategic Planning. In *Fighting with Allies: Transforming Multinational Strategic Planning in the U.S. Department of Defense* (pp. 22–31). Center for Strategic and International Studies (CSIS).

MSCIO. (2025). What is the MSCIO? The Maritime Security Centre Indian Ocean (MSCIO), European Union Naval Force (EUNAVFOR) ATALANTA. https://www.mscio.eu/about-us/what-is-the-mscio/#:~:text=With%20up%20to%2047%25%20(EUROSTAT,welfare%20and%20prosperity%20of%20nations.

Myers, E., & Yozell, S. (2018). The Security Dimensions of IUU Fishing. In Civil-Military Cooperation to Combat Illegal, Unreported, and Unregulated (IUU) Fishing: A Summary of the September 2017 National Maritime Interagency Advisory Group Meeting (pp. 13–16). Stimson Center.

Myers, L. (2023). China's Economic Security Challenge: Difficulties Overcoming the Malacca Dilemma. *Georgetown Journal of International Affairs*. https://gjia.georgetown.edu/2023/03/22/chinas-economic-security-challenge-difficulties-overcoming-the-malacca-dilemma/

Myers, R. A., Hutchings, J. A., & Barrowman, N. J. (1997). Why do Fish Stocks Collapse? The Example of Cod in Atlantic Canada. *Ecological Applications*, *7*(1), 91–106. DOI: 10.1890/1051-0761(1997)007[0091:WDFSCT]2.0.CO;2

NASA. (2025). *CSDA Vendor - Spire*. https://www.earthdata.nasa.gov/about/csda/vendor-spire

Norbert, F. S. I. (2017). Insecurity in the Gulf of Guinea (GoG): Reinventing the Past to Explain the Origin and Development of Maritime Insecurity. *The Journal of Territorial and Maritime Studies*, *4*(2), 32–51. DOI: 10.2307/26664152

Oceana. (2024). Global Fishing Watch. https://usa.oceana.org/globalfishingwatch/

Oceans Beyond Piracy. (2012). *The economic cost of Somali piracy*. One Earth Future Foundation.

OECD. (2020). *Ocean economy in 2030*. OECD.

Ogunniyi, O. J., & Akpu, J. O. (2019). The Challenge of Drug Trafficking to Democratic Governance and Human Security in West Africa: A Historical Reflection. *Africa Development. Afrique et Developpement*, *44*(4), 29–50.

Orgill, M. J., & Moon, T. D. (2024). Use Autonomous Technology to Enable the Coast Guard. *U.S. Naval Institute*. https://www.usni.org/magazines/proceedings/2024/july/use-autonomous-technology-enable-coast-guard

Pauli, G. (2010). *The Blue Economy. 10 Years, 100 Innovations, 100 Million Jobs*. Paradigm Publication.

Pedersen, E. J., Thompson, P. L., Ball, R. A., Fortin, M. J., Gouhier, T. C., Link, H., Moritz, C., Nenzen, H., Stanley, R. R. E., Taranu, Z. E., Gonzalez, A., Guichard, F., & Pepin, P. (2017). Signatures of the collapse and incipient recovery of an overexploited marine ecosystem. *Royal Society Open Science*, *4*(7), 170215. DOI: 10.1098/rsos.170215 PMID: 28791149

Permanent Court of Arbitration. (2016). *The South China Sea Arbitration (The Republic of Philippines v. The People's Republic of China)*. https://pca-cpa.org/cn/cases/7/

Peters, B. C., & Paoli, L. (2025). Are Fishers the Forgotten Victims of Maritime Piracy? An Exploratory Global Analysis. *Fish and Fisheries*, *26*(5), 866–877. DOI: 10.1111/faf.70004

Poling, G. (2020). The Conventional Wisdom on China's Island Bases Is Dangerously Wrong. *War on The Rocks*. https://warontherocks.com/2020/01/the-conventional-wisdom-on-chinas-island-bases-is-dangerously-wrong/

Pomeroy, R., Arango, C., Lomboy, C. G., & Box, S. (2020). Financial inclusion to build economic resilience in small-scale fisheries. *Marine Policy*, *118*, 118. DOI: 10.1016/j.marpol.2020.103982

Poudineh, R., & Apostolopoulou, D. (2025). Coordination levels: horizontal versus vertical. In *The complex challenge of coordination in liberalised and carbon-constrained energy systems* (pp. 3–8). Oxford Institute for Energy Studies.

Rengelink, H. (2012). Tackling Somali piracy. *Trends in Organized Crime*, *15*(2-3), 180–197. DOI: 10.1007/s12117-012-9171-3

Shangri-La Dialogue. (2009). Gates says watershed in Indo-US relations. *International Institute for Strategic Studies (IISS).* https://www.domain-b.com/defence/defence-general/shangri-la-dialogue-gates-says-watershed-in-indo-us-relations

Shaver, A., & Yozell, S. (2018). Security Threats of IUU Fishing. In *Casting a Wider Net: The Security Implications of Illegal, Unreported, and Unregulated Fishing* (pp. 7–18). Stimson Center.

Silver, J. J., Gray, N. J., Campbell, L. M., Fairbanks, L. W., & Gruby, R. L. (2015). Blue Economy and Competing Discourses in International Oceans Governance. *Journal of Environment & Development*, *24*(2), 135–160. DOI: 10.1177/1070496515580797

Singapore, M. I. N. D. E. F. (2016). Malacca Straits Patrol: Member States Commemorate 10 Years of Cooperation. Ministry of Defence, Government of Singapore. https://www.mindef.gov.sg/news-and-events/latest-releases/2016apr21-news-releases-01782

Sodero, S. (2022). Transport Resilience. In *Under the Weather: Reimagining Mobility in the Climate Crisis* (Vol. 12, pp. 145–158). McGill-Queen's University Press.

Statista. (2024). *Number of pirate attacks against ships worldwide from 2010 to 2024.* https://www.statista.com/statistics/266292/number-of-pirate-attacks-worldwide-since-2006/?srsltid=AfmBOoofvdKmAPyp8xFa0l-ROctOCr7KOGPeVnhV4k-MJEZHgwagsGBp-

Steinberg, P., & Kristoffersen, B. (2018). Building a Blue Economy in the Arctic Ocean: Sustaining the Sea or Sustaining the State? In Gad, U. P., & Strandsbjerg, J. (Eds.), *Politics of sustainability in the Arctic: reconfiguring identity, space, and time* (pp. 136–148). Routledge. DOI: 10.4324/9781351031981-9

Stewart, E. C. D., Wiklund, H., Neal, L., Bribiesca-Contreras, G., Drennan, R., Boolukos, C. M. B., King, L. D., Rabone, M., Valls Domedel, G., Serpell-Stevens, A., Arias, M. B., Dahlgren, T. G., Horton, T., & Glover, A. G. (2025). Impacts of an industrial deep-sea mining trial on macrofaunal biodiversity. *Nature Ecology & Evolution*. Advance online publication. DOI: 10.1038/s41559-025-02911-4 PMID: 41350401

Strain, M. R. (2025). *The Economic Benefits of a Strong US Military, Forward Presence, & International Alliances*. IZA - Institute of Labor Economics. DOI: 10.2139/ssrn.5443444

Strating, R. (2020). *Defending the Maritime Rules-Based Order: Regional Responses to the South China Sea Disputes*. East-West Center.

The White House. (2025a). Temporary Withdrawal of All Areas on the Outer Continental Shelf from Offshore Wind Leasing and Review of the Federal Government's Leasing and Permitting Practices for Wind Projects. https://www.whitehouse.gov/presidential-actions/2025/01/temporary-withdrawal-of-all-areas-on-the-outer-continental-shelf-from-offshore-wind-leasing-and-review-of-the-federal-governments-leasing-and-permitting-practices-for-wind-projects/

The White House. (2025b). Unleashing America's Offshore Critical Minerals and Resources. https://www.whitehouse.gov/presidential-actions/2025/04/unleashing-americas-offshore-critical-minerals-and-resources/

The White House. (2025c). Restoring America's Maritime Dominance. https://www.whitehouse.gov/presidential-actions/2025/04/restoring-americas-maritime-dominance/

Thompson, D., & Chong, B. (2020). *Built for Trust, Not for Conflict: ASEAN Faces the Future*. US Institute of Peace.

TMC. (2025). TMC and Nauru Announce Updated Sponsorship Agreement for Nauru Ocean Resources Inc. (NORI). https://investors.metals.co/news-releases/news-release-details/tmc-and-nauru-announce-updated-sponsorship-agreement-nauru-ocean

Toukan, A. (1991). The Gulf War and The Environment: The Need For a Treaty Prohibiting Ecological Destruction As a Weapon Of War. *The Fletcher Forum of World Affairs*, *15*(2), 95–100.

Townsend, J., & Kendall-Taylor, A. (2021). Russian and Chinese Priorities in the Arctic and Prospects for Their Cooperation. In *Partners, Competitors, or a Little of Both?: Russia and China in the Arctic* (pp. 6–11). Center for a New American Security.

Transport Security International. (2023). Seeing the Unseeable: Today's Cargo Screening. https://tsi-mag.com/seeing-the-unseeable-todays-cargo-screening/

UNCTAD. (2023). *Review of maritime transport 2023*. Geneva: *United Nations Conference on Trade and Development*.

UNESCO. (2025). Mangrove ecosystem in the Caribbean shutterstock.com Mangrove restoration as a nature-based solution in biosphere reserves in Latin America and the Caribbean. *MangRes Project.* https://www.unesco.org/en/mab/mangres

UNHCR. (2011). Convention and Protocol Relating to The Status of Refugees. www.unhcr.org/media/1951-refugee-convention-and-1967-protocol-relating-status-refugees

United Nations. (1982). *United Nations Convention on the Law of the Sea (UNCLOS).* https://www.un.org/depts/los/convention_agreements/texts/unclos/unclos_e.pdf

United Nations. (2015). Paris Agreement. https://treaties.un.org/pages/ViewDetails.aspx?src=TREATY&mtdsg_no=XXVII-7-d&chapter=27&clang=_en

United Nations. (2022). *The ocean and the SDGs*. United Nations.

United Nations. (2023). *Agreement under the United Nations Convention on the Law of the Sea on the Conservation and Sustainable Use of Marine Biological Diversity of Areas beyond National Jurisdiction.* https://treaties.un.org/doc/Treaties/2023/06/20230620%2004-28%20PM/Ch_XXI_10.pdf

United Nations. (2024). *United Nations Fish Stocks Agreement: A Guide for raising awareness, furthering understanding and strengthening implementation of its provisions.* https://www.un.org/oceancapacity/sites/www.un.org.oceancapacity/files/advance_and_unedited_unfsa_guide.pdf

UNSC. (2021). *The situation with respect to piracy and armed robbery at sea off the coast of Somalia. S/2021/920.* https://docs.un.org/en/S/2021/920

Vakulenko, S. (2025). *Global Energy Competition in the Arctic: A Reassessment.* Carnegie Endowment for International Peace.

Van Loon, K., Bekkering, B., Verschoor, Y., & Zandee, D. (2025). Geopolitical developments in the Arctic. In *Cold calculations: The future of Arctic security and the role of the Netherlands* (pp. 9–23). Clingendael Institute.

Verbeek, M. (2020). Seas at Risk. https://www.un.org/sites/un2.un.org/files/2020/03/seas_at_risk_input_to_zero_draft_declaration_un_oceans_conference_2020.pdf

Wilson, D. (2021). The Slave Trading Lobby and Piracy in West Africa. In *Suppressing Piracy in the Early Eighteenth Century: Pirates, Merchants and British Imperial Authority in the Atlantic and Indian Oceans* (pp. 145–174). Boydell & Brewer.

Wintour, P. (2025). Europe is scrambling to form a united front and regain relevance in the Iran crisis. *The Guardian.* https://www.theguardian.com/world/2025/jul/04/europe-is-scrambling-to-form-a-united-front-and-regain-relevance-in-the-iran-crisis

World Bank. (2017). *The blue economy: Opportunities and challenges*. World Bank.

Xianqiang, M., Tao, H., Feng, H., Youkai, X., & Yubing, G. (2023). Greenhouse Gas Emission Reduction in Beijing: Goals, Actions, and Recommendations. In Spiegel-Feld, D., Wyman, K. M., & Coughlin, J. J. (Eds.), *Global Sustainable Cities: City Governments and Our Environmental Future* (pp. 220–233). NYU Press.

Zandee, D., Kruijver, K., & Stoetman, A. (2020). Arctic governance. In *The future of Arctic security: The geopolitical pressure cooker and the consequences for the Netherlands* (pp. 41–48). Clingendael Institute.

ADDITIONAL READING

Attri, V. N., & Bohler-Muller, N. (Eds.). (2018). *The Blue Economy Handbook of the Indian Ocean Region*. Africa Institute of South Africa. DOI: 10.2307/j.ctvgc60f0

Beaumont, J., Hampton, S., & Willima, D. (2025). *Africa's blue economy and climate change: Status and emerging pathways*. Institute for Security Studies.

Langlet, D., & Rayfuse, R. (Eds.). (2019). *The Ecosystem Approach in Ocean Planning and Governance: Perspectives from Europe and Beyond* (Vol. 87). Brill. DOI: 10.1163/9789004389984

McGinnis, M. V. (2016). *Science and Sensibility: Negotiating an Ecology of Place* (1st ed.). University of California Press. DOI: 10.1525/9780520960756

Nordquist, M. H., & Long, R. (Eds.). (2021). *Marine Biodiversity of Areas beyond National Jurisdiction*. Brill. DOI: 10.1163/9789004422438

O'Donnell, F. (2025). *Addressing Maritime Security in Smaller South Asian States: Opportunities for US, Indian, and International Partnerships* (Limaye, S., Ed.). East-West Center.

Rodger, N. A. M., & Buchet, C. (Eds.). (2017). *The Sea in History - The Modern World* (NED-New edition). Boydell & Brewer.

KEY TERMS AND DEFINITIONS

Blue Economy: Blue economy refers to economic activities that are based on the sea, the ocean, as well as coastal resources. It is formulated as a way to reconcile growth with sustainability. It goes beyond fishing and shipping; it goes up to renewable energy, marine biotechnology, as well as coastal tourism. It is aiming to develop economic opportunities while preserving coastal ecosystems so that future generations may also be beneficiaries of the resources of the sea.

Climate Change and Maritime Security: Climate change is fast being viewed as a "threat multiplier" in the ocean. Rising sea levels put coastal naval bases at risk, as do rising storms that even threaten shipping as much as coastal societies. Climate-linked scarcities, like shrinking supplies of fish, cause additional tension as countries contest the resources of the ocean. In that sense, climate change is as big an environmental threat as it is a security threat.

Global Ocean Governance: It refers to international laws, agreements, and institutions that regulate the use of the oceans. The United Nations Convention on the Law of the Sea (UNCLOS) provides the primary legal handbook, while organizations such as the International Maritime Organization (IMO) verify shipping regulations. However, enforcement is sometimes weak, and political disputes undermine the effectiveness of the global systems.

Hybrid Threats: Hybrid threats in the sea regions employ both conventional and unconventional tactics. It may involve cyber attacks on port infrastructure, propagation of misinformation about shipping lanes, or the use of civilian ships to advance military objectives. Hybrid tactics blur the line between peace and war, and they present legal as well as operational issues to international organizations as well as states.

Illegal, Unreported, and Unregulated (IUU) Fishing: IUU fishing is a global problem that robs legitimate businesses of billions of dollars while also destroying the stocks. It means fishing where it has been prohibited, failing to report the quantity, or fishing without authorization. Its impacts extend beyond finances: it diminishes food security, undermines the integrity of the regulatory bodies, as well as being usually associated with organized crime syndicates on the high seas.

Marine Protected Areas (MPAs): Marine protected areas are places where people cannot do certain activities to protect ecosystems and wildlife. Although they are mainly for the environment, MPAs also help with safety because they control how resources are used and lessen fights between different groups. How well they work often relies on how well rules are enforced, which connects to larger issues of managing the sea.

Maritime Chokepoints: Maritime chokepoints are narrow routes where many ships pass, making them very important for trade and energy. Some examples are the Strait of Hormuz, where almost one-fifth of the world's oil goes through, and the Strait of Malacca, which connects the Indian and Pacific Oceans. Because of their importance, these areas can become places for political rivalry and possible conflict.

Maritime Domain Awareness (MDA): Maritime domain awareness is the capability to sense, track, and perceive activity within the maritime domain. It balances the integration of technology, intelligence sharing, and regional coordination to establish a sense-based, real-time image of activity off the coast. MDA is as important to stopping criminal activities like smuggling or piracy as it is to responding to natural disasters like an oil spill. **Piracy:** Piracy is one of the oldest and most long-lasting naval security threats. While sometimes sensationalized in film, the reality of today's piracy is that the armed gangs are hijacking vessels, taking the crew hostage, and demanding ransom. Hotspots like the Gulf of Guinea and the Horn of Africa have shown the impact piracy has on destabilizing economies, increasing insurance premiums, as well as endangering significant shipping corridors.

Maritime Security: Security off the coast is no longer just about the protection of coastlines from naval raids; it also encompasses many other aspects, including piracy, terrorism, smuggling, illegal fishing, as well as environmental crimes. It also is about energy security since most of the world's oil and gas is transported across the sea. Maritime security, then, is ultimately about making the shipping lanes secure for commerce, protecting the resources of the sea, and upholding the international rule of the high seas.

Ocean Sustainability: Ocean sustainability means keeping the health of marine ecosystems good for a long time. This is important for economic development and human survival. It combines conservation, responsible resource use, and adapting to climate change in maritime policies. Sustainable practices make sure that oceans can still provide food, jobs, and support for biodiversity despite increasing pressures.

Regional Maritime Cooperation: No one country can control the seas by itself, especially in areas with problems caused by international crime. Working together—like through the Djibouti Code of Conduct in the Red Sea or ASEAN-led efforts in the Indo-Pacific—helps build trust, share information, and create similar policies. Teamwork can also lower the chance of military mistakes in disputed waters.

ENDNOTES

1 For greenhouse gas emission reduction in China please see: (Xianqiang et al., 2023).

2 Environmentally Displaced Persons (EDPs). They are not protected by the 1951 Refugee Convention and 1967 Protocol under the roof of United Nations High Commissioner for Refugees (UNHCR, 2011).

3 For The Mining Code, please see (International Seabed Authority, 2025a)

4 For DSCC Position Statement on Deep Seabed Mining, please see (DSCC, 2019)

5 In the broad sense: unmanned aerial systems (UAS).

6 Launched in 2008 to counter Somali piracy, it has evolved into a broader maritime security operation, including the protection of humanitarian aid deliveries. It illustrates how a regional bloc can project naval power for collective security (EU, 2025b).

7 Though underfunded, it represents progress in building African-led security structures.

8 Southeast Asian states have adopted pragmatic cooperation despite sovereignty sensitivities. The Malacca Strait Patrols show that joint patrols and intelligence-sharing can drastically reduce piracy. However, ASEAN's consensus-based model often limits bolder initiatives (Thompson & Chong, 2020).

9 With melting ice opening new routes, the Arctic Council provides a forum for dialogue among Arctic states, indigenous communities, and observers. While not a security organization, it exemplifies cooperative management of a fragile environment (Zandee et al., 2020).

10 The treaty will be effective as of the beginning of 2026. Official name of treaty is as follows: Agreement under the United Nations Convention on the Law of the Sea on the Conservation and Sustainable Use of Marine Biological Diversity of Areas beyond National Jurisdiction.

11 Security and Growth for All in the Region. For a new term as MAHASAGAR, great ocean, please see: (Indo-Pacific Studies Center. (2025).

12 India is trying to be member of Malacca Strait Patrols though some reluctancies within the member states. For more information please see: (Bhaumik, 2025).

13 Australia, India, Japan, U.S.

14 Australia, UK, U.S.

15 Canada, Denmark, Finland, Iceland, Norway, Russia, Sweden, and the U.S.

16 The number of worldwide pirate attacks decreased from 445 in 2010 to 116 in 2024 (Statista, 2024). .

17 And it raised again in Somalia from 0 since 2019 to 8 incidents after the Red Sea tension (Mitra, 2025)

Chapter 11
Blue Economy, Maritime Law, and SDGs:
Converging AI and Climate-Smart Policies for a Sustainable Future

Jipson Joseph
https://orcid.org/0009-0005-1795-5051
Christ University, India

ABSTRACT

Oceans cover 72% of the Earth's surface, constituting 99% of the living space by volume. By absorbing almost 25% of the carbon dioxide, it tries to prevent global warming. Oceans and maritime resources are becoming key players in international economic interactions, which has developed the idea of the blue economy. However, the increased economic interactions cause various insecurities, like interstate disputes, maritime zone confusions, piracy, terrorism, pollution, and illegal and unregulated exploitation of marine resources. This necessitated the advancement of maritime law with more possibilities and enforcement mechanisms. The interplay between the blue economy and maritime law is essential to realize Sustainable Development Goal No. 14. Their mutual dependence and empowerment are vital for conserving and promoting marine resources. From this perspective, this chapter, using qualitative methodology's critical analytical method, explores the possibilities of converging AI and climate-smart policies with the blue economy and maritime law framework to realize a sustainable future.

DOI: 10.4018/979-8-3373-5238-1.ch011

1 INTRODUCTION

In the second decade of the 21st century, the international community has made a historic step by initiating the Sustainable Development Goals (SDGs) to transform the world into a sustainable society. Compared to the Millennium Development Goals (MDGs), the SDGs adopted an all-inclusive approach with 17 goals and 169 targets that cover all vital areas of human interactions. Among them, SDG 14 specifically stands for conserving and sustainably using the oceans, seas, and maritime resources for sustainable development. Life below water has thus become a central concern of the global policy agenda. This goal is not against using oceans and maritime resources, but they should be used responsibly, as emphasized by SDG 12. If the oceans, seas, and maritime resources are not used responsibly, it may lead to the deterioration of oceanic resources, climate change, and related issues. Thus, when SDG 13 recommends taking urgent action against climate change and its impacts, it also involves actions that protect maritime resources.

From the dawn of civilization, oceans, seas, and maritime resources have been policymakers' crucial concerns to ensure economic growth. The systematic, responsible, and sustainable use of marine resources will promote sustained, inclusive, and sustainable economic development (SDG 8). From this perspective, the blue economy is a crucial field in sustainable, human-centered development. While emphasizing the importance of sustainably utilizing maritime resources for economic growth, improving livelihoods, and creating employment opportunities, the blue economy also requires protecting the health of the marine ecosystem. Since maritime law encompasses financial concerns and interactions on navigable waters, marine commerce, and other activities, the blue economy is closely tied to maritime law. The interplay between maritime law and the blue economy becomes more effective and efficient within the framework of SDGs.

As envisioned by SDG 9, the globe builds resilient infrastructure and fosters innovation, which extends its influence on the blue economy and maritime law. The incorporation and active use of artificial intelligence (AI) and climate-smart policies have contributed to optimizing and enhancing the efficiency and resilience of the blue economy. The chapter examines the significance of maritime resources in global efforts toward a sustainable future. Generally, policymakers and the public focus on reducing poverty, promoting education, expanding employment opportunities, providing healthcare facilities, and protecting the ecosystem to ensure a people-centered development. However, they neglect the relevance and protection of maritime resources (UN, n.d.).

Under the SDGs' framework, the ocean conservation efforts focused on protecting 30% by 2030. However, it could classify only 8.4% of the ocean as marine protected areas (UN, 2025). The continuous flow of pollutants into the ocean poses a significant threat to its conservation and the aspirations of sustainable development among people (Donnelly, 2025). The economic impact of the ocean is immeasurable. Besides providing many resources essential for human survival, it also strengthens global trade and economic interactions (UN, 2025). Both the blue economy and maritime law are relevant in this context. However, the extent to which they can become effective in enhancing economic growth and ocean protection is disputable based on the present statistics.

1.1 Significance of the Study and Methodology

The ocean is the most significant aspect of the Earth, as it covers three-quarters of its surface, contains 97% of its water sources, and represents 99% of the living space by volume (Arora et al., 2023). The oceans and the maritime resources are intrinsic to human lives, providing food, medicine, biofuels, and other vital resources. The oceans are immensely beneficial to humans and other organisms by balancing the Earth's temperature (absorbing 90% of the world's heat), supporting natural climate change, and absorbing carbon dioxide from the Earth's surface (UN, n.d.). It facilitates domestic and international trade, strengthening economic growth. Although the advantages are innumerable, humans focus very little on conserving oceans and maritime resources. The oceans face mounting threats from pollution, excessive fishing, biodiversity loss, and climate change (UN, 2025). It is estimated that about 75 to 199 million tonnes of plastic are clogging the ocean. Approximately 92% of microplastics, which can cause cancer, infertility, and nervous system damage, are found in 60% of the fish that humans consume yearly. Although about 3 billion people depend on the ocean for their livelihood, the increased pollution has caused adverse impacts (Donnelly, 2025). The recovery and conservation of the ocean is indispensable for the survival of humans and the ecosystem. It is also vital for sustainable development and economic growth. This requires responsible commitment from the state and non-state actors. Strong fisheries management, ecosystem-based approaches, blue finance, and efficient maritime law are necessary to achieve this ambitious and strategic conservation (UN, 2025). As the world increasingly benefits from adopting AI and smart policies, this chapter explores the possibility of incorporating AI and climate-smart policies in realizing SDG 14.

The chapter employs the qualitative methodology's analytical, exploratory, descriptive, and critical tools to advance an AI-supported interplay between maritime law and the blue economy policy framework, ensuring a sustainable future through the achievement of SDGs. The first part of the chapter uses descriptive method, the second part adopts analytical method, the third part utilizes exploratory method, and the fourth part adopts critical method.

2 THEORETICAL BACKGROUND

This chapter revolves around five significant terminologies: blue economy, maritime law, SDGs, AI, and climate-smart policies. They are interconnected and mutually enriching in the global march towards a sustainable future. The importance of the SDGs in the discussions on marine resources and conservation lies in their recognition that life below water is a crucial policy goal in the international domain (Holm, 2022). As the globe faces an increasingly unpredictable security threat due to ocean warming, acidification, overfishing, and illegal practices that threaten maritime ecosystems and challenge the livelihoods of small-scale fishermen, incorporating the life below water into the framework of the SDGs is laudable (UN, 2025). However, even after nearly ten years of its initiation, SDG 14 has had only a limited impact. The existing blue economy and maritime law were believed to benefit from the SDGs policy framework. Even when some benefits could be credited, they lack proper orientation and implementation strategies. In such a scenario, a need arises to strengthen them with the support of the latest technological advancements, like AI and climate-smart policies. Al and climate-smart policies are actively used in realizing many other SDGs, such as poverty eradication (Goralski & Tan, 2022), zero hunger (Hiywotu, 2025), healthcare (Bajwa et al., 2021), clean water and sanitation (Turyasingura et al., 2024), affordable and clean energy (Awogbemi et al., 2024), sustainable cities and communities (Filho et al., 2024), climate action (Akter et al., 2024), and life on land (Li et al., 2023). Similarly, many authors have explored the interplay between AI and life below water (Doorn, 2021; Isabelle & Westerlund, 2022), AI and maritime law (Chen, 2023; Durlik et al., 2024), AI and blue economy (Mukherjee et al., 2025), climate-smart policies and life below water (Zhang et al., 2023), climate-smart policies and maritime law (Olaniyi et al., 2024), and climate-smart policies and blue economy (Garcia & Carter, 2022). These pieces of literature affirm that AI and other intelligent mechanisms are being increasingly utilized to achieve the desired objective of the SDGs. Based on this existing knowledge and framework, this chapter aims to expand the discussions on the ocean and marine resources in relation to AI and climate-smart policies, exploring the interplay between maritime law and the blue economy, and underscoring the inspirational agenda of SDG 14.

2.1 Blue Economy

The blue economy is an umbrella term encompassing many activities, including fisheries, aquaculture, marine tourism, and marine trade (Ghazali, 2025). The blue economy refers to the economic interactions that utlilize marine resources for economic growth or the ocean as a medium. It emphasizes the sustainable and progressive use of marine and coastal areas and resources for economic development, improved livelihoods, and the enhancement of the ecosystem's health (Schutter et al., 2021). The definitional ambiguity of the blue economy is an explicit example of its broader scope. The blue economy thus affirms its commitment towards economic, social, and environmental sustainability. According to estimates, the world population is approximately 8.2 billion as of 2025. More than one-quarter of the world's population, i.e., almost 3 billion people, depend on the ocean for survival and a dignified standard of living (Donnelly, 2025). Furthermore, nearly 3% of the global GDP is derived from the blue economy (UNCTAD, 2022). It is also termed the seventh-largest economy in the world. This underlines the significance of the blue economy in the present globe.

A well-defined and globally accepted definition of the blue economy has yet to be finalized (Silver et al., 2015). International agencies, such as the UN and World Bank, have attempted to define it, which appears to be the prioritized definition in academic circles. Although Gunter Pauli had introduced the concept of the blue economy in 1994, it received wider attention with its acceptance at the Rio+20 Conference in 2012 (Rajeevan, 2023). Until then, the blue economy was part of the discussions of the green economy. Recognizing the importance of ocean resource management in achieving the sustainable development aspirations and economic growth, the world, under the leadership of UN agencies, has separated the blue economic concerns from the green economy. It helped the world address various challenges and complex questions concerning the marine and coastal economy (Das, 2023). Blue economy is the "sustainable use of ocean resources for economic growth, improved livelihoods, and jobs while preserving the health of ocean ecosystems" (Lee et al., 2020; World Bank, 2017). The UNEP's (2011, p. 16) definition of green economy also applies to blue economy. It defined green economy as an economy that "results in improved human well-being and social equity, while significantly reducing environmental risks and ecological scarcities." It helps developing and underdeveloped countries in coastal areas advance long-term development programs and policies, positively incorporating the benefits of marine resources. Several countries, including Australia, Brazil, Norway, Russia, the US, and the UK, have established progressive ocean policies with significant budgetary allocations to promote the interactions of the blue economy, giving a model to other countries (Rajeevan, 2023).

2.2 Maritime Law

The ocean has a decisive role in the lives of people and the ecosystem. Humans depend on the ocean for many necessities, like food, renewable energy, transportation, leisure, and tourism. The continuous involvement and utilization of marine and coastal areas and resources for socioeconomic growth necessitate a systematic legal framework to ensure their smooth functioning. The evolution of maritime law can be traced in such a background. It originated as *lex maritima* within the context of the 12th century *lex mercatoria*. The shipping industry's customs and practices have significantly impacted the evolution of *lex maritima*. It is a *jus commune* as it refers to a whole jurisdiction or more than one jurisdiction, consisting of broad and general principles. Mostly, they were unwritten laws and practices, which became codified later (Tetley, 1994). However, the codification of the United Nations Convention on the Law of the Sea (UNCLOS, 1982) is a historic step in the progress of maritime law.

Maritime law refers to the "maritime customs, codes, conventions, and practices from earliest times to the present, which have had no international boundaries and which exist in any particular jurisdiction unless limited or excluded by a particular statute" (Tetley, 1994, p. 108). It is a compilation of all the customary practices, codes, conventions, and agreements concerning maritime activities across the countries. It is a part of international law, as almost 72% of the Earth's surface is water (Rajeevan, 2023). However, it also has domestic applications. Additionally, if a particular law does not limit or exclude international maritime law within the boundary of a state, it will have its application. Maritime law and admiralty law are used interchangeably in modern times. Admiralty law originated in England to address trade and commerce on the high seas, and maritime law was later developed to protect the workers and seamen injured on the high seas (Rivkind Law, 2022). So basically, while admiralty law focused on economic activities, maritime law considered the personal injuries, concerns, and rights of the workers on the high seas. Such a differentiation does not exist today.

2.3 SDGs

From the second half of the 20th century, the globe prioritized sustainability in its policy discussions and economic development programs. However, most of these considerations were limited to the environmental protection strategies. They often contradicted the economic developmental aspirations, forcing the globe to prioritize sustainability only in theory (Kingsbury et al., 2004). With the 1987 report of the UN's World Commission on Environment and Development (WCED), widely known as the Brundtland Commission, the concept of sustainable development was officially introduced, bridging both developmental and environmental aspirations (Jørgensen, 2001). Based on this inspirational agenda, the UN introduced the SDGs in 2015 to transform the world towards a sustainable future by 2030.

The UN has accepted and proposed 17 well-defined goals with 169 targets to achieve sustainability. SDGs revolve around the five Ps: people (SDGs 1-5), planet (SDGs 6, 7, 13, 14, 15), prosperity (SDGs 8-12), peace (SDG 16), and partnership

(SDG 17). However, all five Ps ultimately concentrate on people's well-being, as the goals of planet, prosperity, peace, and partnerships are to ensure a meaningful life for the people. They also affirm the social, economic, and environmental dimensions of sustainability. The balance between society, economy, and environment is a central element of the SDGs (UN, 2015). These three dimensions are balanced to improve the quality of life (Elgazzar & El-Gazzar, 2017). Although it can be criticized for trying to enhance humans' quality of life, its genuine concern is to improve the quality of life for all living beings on the planet. From this perspective, the SDGs have a broader objective and policy framework.

2.4 Artificial Intelligence

The Turing test, developed in the 1950s to measure the intelligence of a computer system, was the first step towards artificial intelligence (AI). During the Dartmouth conference of 1956, John McCarthy first introduced the term AI. Although there are various definitions, none is universally accepted (Glasser et al., 2020). However, all these definitions point to the machine's pre-designed capacity to imitate the human capacity for intelligent activities, like learning, reasoning, and decision-making. An outsider might feel a sense of 'human-likeness' while experiencing the intelligent activities of an AI-guided machine. The provided data empowers machines to replicate human intelligence and make choices and decisions. This process is facilitated by inputs and training or the trial-and-error method (Donahue, 2018). It transforms the activities of a system into human-like activities. AI is "a comprehensive discipline that studies and develops theories, methods, technologies, and application systems for simulating, extending, and expanding humans and other animals and developing various machine intelligence and intelligent machines" (Ma & Hou, 2021, p.44). It undertakes various functions conventionally understood as distinctive features of humans and other animals, like visual perception, speech recognition, language translation, and decision-making (Solanki, 2021). AI is further expanded by the evolution of Machine Learning (ML). It provides machines with more capabilities as they can learn and improve with an AI application without direct instruction from human agents. The role of human instructors is being reduced with the advancements in AI technology, as it can learn from data, adapt to changing circumstances with experience, and improve its functioning formats over time. The incorporation of ML has made tremendous changes in AI, significantly transforming machines into an autonomous system (Ma & Hou, 2021). Nevertheless, AI and ML are not beyond criticism. They have some specific drawbacks, as they are not neutral. However, many scholars argue that their disadvantages are comparatively lesser than their benefits (Mittelstadt et al., 2016; Floridi & Cowls, 2019).

The US' National Artificial Intelligence Initiative Act (2020) explains the significant functions of AI systems: "a) perceive real and virtual environments; b) abstract such perceptions into models through analysis in an automated manner; and c) use model inference to formulate options for information or action" (Sec. 5002(3)). The core functions of AI are perception, abstraction, and formulation of action. They are interconnected intelligent activities. From the Industrial Revolution, machines helped people automate routine tasks easily and perfectly. The evolution of AI has enabled people to automate non-routine and knowledge-based tasks, enlarging people's choices and opportunities (Armour & Sako, 2020). AI can systematically and effortlessly combine social, technical, economic, and environmental features to achieve sustainability (Ma & Hou, 2021). This has led to massive investments in AI from public and private actors, expanding the possibilities of its growth and application in various fields, such as the blue economy. The advantages of AI for economic growth are undoubtedly proven. However, its alignment with ethical and legal concerns remains under investigation. Various authors have systematically studied the significance of AI in enhancing the possibilities of maritime law (Chen, 2023; Durlik et al., 2024) and blue economy (Mukherjee et al., 2025) for economic growth and a sustainable future. Just as almost all fields of human interaction are leveraging the incorporation of AI, the blue economy and maritime law are also exploring the possibilities of increased AI use to ensure alignment with the objectives of SDG 14.

2.5 Climate-Smart Policies

Climate-smart policies are closely linked to the issue of climate change (Lipper & Zilberman, 2017). Gupta (2010) notes five phases in the evolution of climate change policies from 1977 to 2010. The first phase, the pre-1990 period, is well-known for establishing the Intergovernmental Panel on Climate Change (IPCC, 1979) and the Brundtland Commission Report (1987). This period necessitated global action to stabilize greenhouse gas (GHG) emissions and affirmed the interplay between climate change and sustainable development. The second period (1991-1996) marked the first attempts to establish a global policy framework through the adoption of Agenda 21. This period is notable for the establishment of the UN Framework Convention on Climate Change in 1994. The responsibility of developed countries to support the eco-friendly activities of developing countries, as part of the mitigation process, has been established. Civil society also started to play a key role in the discussion on climate change. However, while the Northern organizations emphasized environmental protection, the Southern organizations advocated for development policies

(Lipper & Zilberman, 2017). The third period (1997-2001) underlines the significant growth towards the first international agreement on climate change, the Kyoto Protocol. It affirmed the sharing of responsibilities to meet the targets of climate and environmental security. The final period began with the 2008 financial crisis, which underscored that developing countries also have mitigating responsibilities alongside developed countries (Gupta, 2010).

Climate-smart policies evolved within the context of the SDGs and the integration of AI to effectively achieve a sustainable future. The concept of climate-smart policies was first launched in 2009 (Lipper & Zilberman, 2017) within the context of agriculture. The UN's Food and Agriculture Organization (FAO) in the Hague Conference on Agriculture, Food Security, and Climate Change (2010) proposed the urgent need for climate-smart agriculture policies. It affirmed that the climate-smart agriculture policies can enhance the sustainable production of agricultural products, reduce the burden on the workers, and ensure food security. Thereafter, many stakeholders contributed significantly to its development, recognizing the importance of such proactive concepts for achieving better results. Climate-smart policies aim to provide globally applicable principles that can manage and guarantee food security and economic growth under the complexities of climate change (Lipper & Zilberman, 2017). These were developed in response to a significant need to bridge the gap between climate change and sustainable development concerning both green and blue economies.

3 INTERPLAY BETWEEN BLUE ECONOMY AND MARITIME LAW

As explained above, the blue economy and maritime law explicitly consider activities related to marine resources. They focus on the people who live in coastal areas and socioeconomically benefit from marine resources, the sustainability of marine species and ecosystems, the eradication of pollutants like plastic thrown into the ocean, and the removal of threats that can cause climate change. Both terms underline the significance of the ocean and marine resources for the well-being of people and the planet. This section examines the interplay between the blue economy and maritime law, as well as the role of the SDGs in enhancing this interconnection. The concept of the blue economy was particularly developed by the Small Island Developing States (SIDS), which have extensive ocean jurisdictions and vital marine industries for their economic growth (Silver et al., 2015). The financial activities of the coastal states are massively dependent on the ocean and marine resources.

If they do not have sovereign power over their marine resources up to a specified ocean area, their economic growth will face severe repercussions.

The UNCLOS (1982) has solved this issue by systematically demarcating the ocean area, where the states have full sovereignty. The involvement of maritime law through international agreements and the codification of a specific zone of freedom and rights affirms the interplay between the blue economy and maritime law. The emergence of the blue economy cannot be examined without situating it within the context of interstate politics. The growth of the blue economy depends on the interaction of several actors without being restricted to a specific territorial boundary. The interstate interaction helps to transform the oceanic space into developmental space (Schutter et al., 2021). When the involvement and interaction of interstate actors are essential for the growth of the blue economy, an adequate legal framework, such as maritime law, is also necessary for the effective coordination and implementation of the agreements and understandings.

3.1 Co-Evolution

Although maritime law originated centuries ago, its modern resurgence began with the evolution of the blue economy. Due to globalization, neoliberalism, and open market dynamics, international economic interactions have increased substantially globally. Most global economic activities use marine transportation for supply chain management. Additionally, the use of marine resources for food, medicine, cosmetics, and other essential fields has also contributed to the growth of the blue economy. The increased marine economic activities necessitated an efficient legal framework to maintain and guarantee sustainability and order. It affirms that, alongside the growth of the blue economy, maritime law has also evolved to address the issues and complexities that arise in the ocean concerning marine economic activities. In such a scenario, it can be said that maritime law is a subsect of the blue economy (Voyer et al., 2018). At the same time, it can also be argued that the maritime law's focus in the present times is not merely to maintain, coordinate, and enforce order in economic interactions. It has many other responsibilities, like environmental protection, coordination of transportation, and interaction with the naval forces. Accordingly, the scope of the maritime law is broader than economic matters. However, authors also argue that the blue economy significantly addresses all these areas. Although there is a tendency to limit its scope to economic activities alone, it covers a wide range of areas, including fisheries, transportation, tourism, aquaculture, renewable energy, seabed extracting activities, coastal protection, waste disposal, climate change, and food security (Rajeevan, 2023). It demonstrates that the blue economy and maritime law are inextricably linked in SDG 14.

3.2 Mutual Dependence

Maritime law facilitates the blue economy by providing an efficient inter-state mechanism and a legal framework for navigation and economic activities. The UNCLOS (1982) explains this mutual support system. It underlines 12 nautical miles and 200 nautical miles (from the baseline), which are extremely important for the blue economic activities. The 12 nautical miles refer to a country's internationally agreed territorial sea, where the state has full sovereignty and jurisdiction over the seabed, subsoil, and airspace. The ships of other countries have the right to "innocent passage," respecting the laws and regulations of that particular state's territorial integrity. However, the 200 nautical miles refer to the sovereign rights of the coastal states within this specified boundary to explore, conserve, use, and manage the living and non-living marine resources. This specified ocean area is called the Exclusive Economic Zone (EEZ). Although the coastal states have resource rights within the 200 nautical miles, other states have the right of navigation and overflight. Such an international agreement helped the coastal states to secure their marine resources from the encroachment of more powerful actors. It helps increase the economic activities of coastal states (the blue economy). At the same time, it supports other states' economic activities as they can freely navigate through the EEZ, overflight vessels and aircraft, and lay submarine pipelines and cables for international trade, transportation, and communication (Voyer et al., 2018).

A 1984 report of the FAO confirms that almost 90% of the marine fish and shellfish are caught within the 200 nautical miles (Schurman, 1998). It was also found out that nearly 87% of the global submarine oil deposits are within this 200 nautical miles (Churchill & Lowe, 1999). When the EEZ has more resources, it benefits coastal countries, as they can strengthen their economic growth by systematically utilizing their resources.

3.3 SDGs and the Intersection

Compared to the MDGs, the SDGs provided a broader global action framework, addressing all crucial fields of sustainability. Although many concerns about environmental protection were part of the MDG7, it failed to focus on the ocean and marine resources. However, SDG 14, with its ten targets and indicators, concentrates explicitly on conserving, promoting, and sustainably using ocean and marine resources for sustainable development. Target 14.1 underlines the importance of preventing and reducing marine pollution from land-based activities by 2025. The indicators of this target are a) the index of coastal eutrophication (ICE), and b) plastic debris density. ICE is measured on the loads and ratios of nitrogen, phosphorus, and silica delivered by rivers to coastal waters, the excess of which can cause the growth of potentially harmful algae (Garner et al., 2010). While certain types of algae are beneficial for sustainable development, others are harmful to the growth of the blue economy (Ismail & El Zokm, 2025). Plastic pollution is seen as one of the significant challenges to the ocean (Donnelly, 2025). The concentration of harmful algae and microplastics in the ocean harms marine resources and affects the growth of the blue economy. This necessitates a practical maritime law framework that can prevent such a scenario. The same interplay can also be attributed to the following targets.

Target 14.2 aims to restore and achieve healthy and productive oceans by sustainably managing and protecting marine and coastal ecosystems, avoiding adverse impacts, and strengthening their resilience. The UN (2012) affirms in its report the importance of healthy marine ecosystems in ensuring a standard of living for millions of people and other beings. Target 14.3 focuses on enhancing national and international scientific cooperation to minimize the impacts of ocean acidification. However, after critically examining 90 legislative documents from 17 European countries, Galdies et al. (2020) conclude that attempts to minimize the adverse impacts of ocean acidification are uncoordinated. This requires a greater commitment to collaboration among countries to strengthen the blue economy, supported by maritime law.

Target 14.4 emphasizes the importance of regulating harvesting, ending overfishing, and preventing illegal, unreported, and unregulated (IUU) fishing to ensure sustainability. According to the FAO (2001), IUU refers to all types of fishing that violate national, regional, and international laws concerning the conservation and utilization of fisheries. IUU is one of the significant challenges to SDG 14. Since it is primarily an economic activity, it comes under the parameters of the blue economy. Since it can adversely impact the inspirational agenda of sustainable development, it should be prevented (Ma, 2020). Target 14.5 explicitly links the conservation of at least 10% of coastal marine areas to maritime law, arguing that conservation efforts should adhere to national and international standards. This is an essential requirement for the protection of marine biodiversity. Recognizing the importance of conservation, many countries have committed to protecting at least 30% of their coastal areas by 2030 (Rechberger et al., 2025).

Target 14.6 aims at reducing and eliminating subsidies for overfishing and IUU fishing. It also provides an option for enhancing subsidies for developing and underdeveloped countries to support sustainable fishing practices. Subsidies have played a significant role in improving fishing activities and promoting the growth of the blue economy in the past. However, they harm the blue economic market, international marine cooperation, and small-scale fisheries. At the same time, adequate subsidies should be provided to SIDS and the least developed countries for sustainable fishing to increase their economic growth (de Gama, 2025). Target 14.7, in this line, aims to improve the financial benefits of SIDSs and the least developed countries from the sustainable use of marine resources by 2030. However, the desired objective is yet far from being realized. Examining the Seychelles experience, Benzaken (2024) suggests practical measures that the international community should undertake to realize this objective.

Target 14.a focuses on increasing scientific knowledge, developing research capacity, and transferring marine technology from developed to SIDSs and the least developed countries. The role of AI and climate-smart policies can be situated here. Most technological advancements primarily occur in developed countries, strengthening their economies. There should be effective coordination between developed and developing countries to transfer these technological advancements and strengthen the latter's efforts in sustainable development (Doorn, 2021). Target 14.b provides access to small-scale artisanal fishers to marine resources and global markets. Preferential access should be given to small-scale fisheries to enhance their standard of living. This can be ensured by demarcating a preferential access area, which helps protect small-scale fisheries and coastal regions (Basurto et al., 2024). Target 14.c again explicitly refers to maritime law, affirming its interconnection with the blue economy in realizing SDG 14. It prioritizes enhancing the conservation and sustainable use of oceans and their resources by implementing international law as reflected in UNCLOS and affirmed in paragraph 158 of *The Future We Want* (UN, 2012).

4 IMPACT OF AI ON THE BLUE ECONOMY AND MARITIME LAW

Due to many factors, the oceans, blue economy, and maritime law are crucial factors for discussion in the current scenario. Firstly, as mentioned above, oceans cover almost 72% of the Earth's surface, marine resources are essential to the growth of coastal communities and the global economy, and deep-sea mining and other innovative areas are rapidly growing. Secondly, most ocean spaces are beyond the territorial jurisdiction of states, demanding the active involvement of international law and the mutual cooperation of states. Thirdly, the interest of both public and private actors in marine resources is rapidly increasing (Silver et al., 2015). Scientific and technological advancements, such as AI, have significantly influenced all fields of human interactions. Socioeconomic life has become more positive and progressive with the incorporation of AI. This is also visible in relation to the blue economy. Pre-pandemic estimates suggest that the blue economy contributed nearly US$2.5 trillion to the global economy, and is projected to reach US$3 trillion by 2030. The demand for marine resources is increasing globally, and numerous public and private actors are making significant investments in the blue economy (Benzaken et al., 2024). The increased utilization of AI's advantages can facilitate this objective.

The continuous monitoring and coordination of maritime resources, as well as protecting the marine ecosystem, may not be easy for humans, as it generates a large amount of data and requires a constant support system. Big data and cloud computing can help with data storage. Embedded data-driven and AI-guided sensors can facilitate the monitoring and coordination process. Digital twinning helps augment the operational perfection of ships and sensors (Sakhuja, 2023). However, it is also worth noting that efficiency equivalence and choice equivalence are worth considering when adopting the latest technologies. Efficiency equivalence refers to the technical substitution of devices, materials, methods, and equipment. Experiments can refine and make them suitable for adapting to the system. On the contrary, choice equivalence points to an intelligent technology's choices. It examines whether these choices are equivalent to those made by a rational person in such a context (Chen, 2023). These choices are significant because if they fail to resemble human intelligence, they may bring chaos to the system.

4.1 Sustainable Energy Generation

Energy generation is one of the critical necessities of modern times. Renewable energy sources are increasingly desirable in the current scenario, given the rapid rise in population and a comparatively lower availability of fossil fuels (Rajeevan, 2023). Although people began to depend on the ocean for energy from the 18th century onwards, its large-scale production, particularly wave and tidal energy, started only in the last quarter of the 20th century. Currently, oil and gas extraction, deep-sea mining operations, water desalination, and maritime renewable energy sources such as tidal, wind, and wave energy are receiving wider attention across the globe. This forced countries to extend their coastal areas and maritime boundaries for energy conservation, economic growth, and control of resources. The maritime law, as seen above, has delimited the states' naval space extension initiatives to protect the rights of all countries. However, within the EEZ, countries have full power over their resources. At the same time, concerning energy sources in the high seas, maritime law has provided a means for consensus among concerned parties, protecting the rights of one another and supporting the development of the blue economy (Voyer et al., 2018).

The potential of the renewable blue energy sources is unimaginable. A progressive commitment by coastal states to generate renewable energy from wind, wave, tidal, thermal, and biomass sources will alter the value of the blue economy in global economic interactions. However, utilizing these resources requires incorporating the latest scientific and technological tools. The deep-sea mining mission may not be easy for humans, but AI tools can serve as a substitute for it (Rajeevan, 2023). Utilizing AI techniques to generate more sustainable and renewable energy from the ocean enhances global aspirations for economic growth and environmental sustainability.

4.2 Effective Maritime Transport

The blue economy faces the most complex threats related to shipping, transportation, ports, and marine tourism. These threats create tension on the stability and sustainability of the blue economy. Maritime transport is the backbone of the global economy, and it requires cross-border functioning, necessitating an international legal framework for its effectiveness (Voyer et al., 2018). Maritime tourism, transport, and shipping are also vulnerable to transnational crimes, like terrorism, piracy, and smuggling. AI can empower them with tracking possibilities, identification tools, and other preventive mechanisms. For instance, face recognition and biometric tools have provided a broader framework for easily identifying suspects or illegal materials in shipments, ensuring security and order. AI-enabled autonomous vessels are under development, and they can enhance efficiency, reduce downtime, and harmonize operations that are beyond the reach of humans. Since they require less or no human personnel on the board, the conventional system of allotting living space for humans can be substituted with cargo, augmenting the economic benefits. Many believe smart vessels can reduce ship accidents, which are conventionally considered to happen due to human error (Sakhuja, 2023). Although many authors disagree with excluding human actors from ship operations, there are significant opportunities for enhancing marine tourism, transportation, and shipments. However, the wide incorporation of fully autonomous ships requires updating maritime law, as it had never imagined such a scenario (Chen, 2023). This implies that maritime law should adapt to changing circumstances.

4.3 Efficient Management of Fisheries and Coastal Areas

Fishing is a vital economic activity related to the ocean. Fishing at biologically sustainable levels will help balance the marine resources without damage. However, overfishing can become a massive threat to the sustainability objective. It was estimated that, in 2021, while 64.5% of marine fish stocks were fished in a manner that respects the sustainability parameter, 35.5% were overfished. This indicates that the overfishing rate is increasing, with an annual increase of almost 1% (UN, 2025). Overfishing is a complex problem that requires urgent and effective fisheries management. This can be effectively done with the support of AI and other smart intelligence mechanisms. Satellite monitoring and computerized reporting help maintain accountability and transparency in the management of fisheries and coastal areas (Holm, 2022).

To secure sustainable fish stocks for future generations is a responsibility of the present generation. It requires effective fisheries management. Regarding tuna fishing, vital management efforts and cooperation appear to be in place, affirming that they can also be applied to other areas. However, illegal, unreported, and unregulated (IUU) fishing exists in almost all regions, posing complex challenges to the sustainability objective of fisheries, disrupting the maritime ecosystem, endangering food security, and threatening economic growth. IUU fishing is estimated to account for nearly 15% of the total catches, harming coastal communities, reducing fish availability, and contributing to increased poverty. Although various international agreements and treaties, such as the Agreement on Port State Measures (PSMA), have been established, IUU fishing persists, necessitating more committed intervention from national and international agencies (UN, 2025).

4.4 Protection of Vulnerable Communities

Oceans are crucial for the survival and economic growth of coastal communities. The reduction of poverty and sustainable development of coastal communities depend on the specific attention given to protecting and promoting marine resources. The work of small-scale fishers should be valued and respected with adequate remuneration and access to international markets. It also requires sufficient strategies to prevent IUU fishing, which can endanger the coastal communities' standard of living. Prioritization should be given to the rights of coastal communities over marine resources rather than promoting the interests of multinational and transnational corporations (Silver et al., 2015).

4.5 Data-Driven Economic Growth

Oceans and marine resources continuously support the population's economic, social, and environmental aspirations. They are indispensable for a meaningful, sustainable development and economic growth. The utilization of marine resources requires smart ports, along with smart vessels, to drive economic growth. Using data-driven decision-making tools, they ensure operational efficiency. They can reduce the operational time at the port and promote the fast movement of people and cargo. Supply chain management has become easier and perfect by incorporating AI tools. Customs, port authorities, and shipping companies can easily exchange data with each other, making the process significantly transparent and efficient. Many scholars thus opine that only smart ports will survive (Sakhuja, 2023).

4.6 Protection of Ocean Territories

International agencies, such as the UNCLOS, have specified domestic and international ocean territories to promote the sustainable development of all countries. The EEZ and domestic coastal areas give high powers and rights to the coastal states. Nevertheless, the high seas are considered international territory, where all countries have equal rights preserved through consensus. Some initiatives, such as the 'One Belt One Road' initiative, promote integrating territorial rights of transport and shipping to advance economic growth. It shows that economic growth has become the primary interest of almost all countries, even beyond the national interest of security. However, it highlights the key principles of the blue economy and maritime law, like ocean-based prosperity, naval security, inclusive growth, and collaborative governance (Voyer et al., 2018). However, these innovative ideas operate within the framework of maritime law, which safeguards the interests of states in the international scenario.

Maritime law is also responsible for protecting the ocean's environmental health. Although it was not part of its conventional assignment, the present times highly demand this responsibility. This is achieved explicitly through the implementation of environmental protection laws against the dumping of plastic and other pollutants into the ocean. The navy and the coastal guard assist in implementing these environmental protection measures outlined in maritime law. They also play key roles in protecting ocean territories from pirates, accidents, and oil leakages. Besides implementing environmental laws and regulations, the blue economy and maritime law also focus on enhancing research and scientific knowledge on the marine ecosystem and biodiversity. They also concentrate on predicting, preventing, and responding to environmental disasters that can cause adverse repercussions on the blue economy (Voyer et al., 2018).

4.7 Enhanced Legal Framework

Towards the end of the 20th century, with an increased interest in the economic opportunities of marine resources, a pressing need arose to define and safeguard maritime space. Since both public and private actors have shown a growing interest in marine resources, the protection of maritime space has become beyond the everyday concerns of national security. This applied to domestic and international maritime space to ensure an effective surveillance system and enforcement mechanism, as these threats often crossed multiple maritime jurisdictions. The question of jurisdiction is essential concerning mineral resources and other vital elements of the blue economy (Voyer et al., 2018). The increased adoption and application of AI and other latest technological advancements necessitate expanding the framework of maritime law to include all the possible complexities and confusion (Isabelle & Westerlund, 2022). The autonomous functioning of AI may raise liability issues, questioning the viability of vicarious liability (Sakhuja, 2023). Several ethical concerns will also become irrelevant when machines replicate human intelligence (Chen, 2023). All these questions require attention from a legal perspective.

5 CLIMATE-SMART POLICIES FOR A SUSTAINABLE BLUE ECONOMY

The ocean is the largest ecosystem on Earth. It is our planet's life support system, regulating the global climate. It helps the Earth address unpredictable climate change by controlling its temperature and absorbing carbon dioxide (UN, n.d.). If any adverse events occur in the oceans, they can have severe repercussions for the world population, economic growth, and the climate system. The irresponsible exploitation of marine resources without recognizing their critical importance to human life and the Earth has led to alarming insecurity and degradation, negatively affecting the global climate system and blue economy initiatives. The world must invest in the research and development of non-polluting substitutes to protect the ocean and marine resources. The commitment to the circular economy, reducing dependence on plastic, and developing new industries with adequate waste management solutions can enhance the resilience of the ocean (UNCTAD, 2022). Climate-smart policies have evolved in the context of protecting and promoting the ocean, the Earth's surface, and its ecosystems. They provide globally accepted principles and guidelines for managing marine resources to ensure food, economic, and environmental security with a people-centered perspective (Lipper & Zilberman, 2017).

Although the blue economy primarily focuses on economic activities, protecting the marine ecosystem, biodiversity, and marine resources, as well as promoting various climate-smart policies that prevent climate change, also fall under its responsibilities. It also prioritizes research and development, believing many marine resources, like oil, gas, and deep-sea minerals, are still unexplored and untapped. Such research should be conducted systematically and scientifically without harming biodiversity and the marine ecosystem (Schutter et al., 2021).

5.1 Marine Spatial Planning

Marine Spatial Planning (MSP) can be termed as a subsect of the environmental security program. MSP is a systematic "process of analyzing and allocating the spatial and temporal distribution of human activities in marine areas to achieve ecological, economic, and social objectives that have been specified through a political process"

(Rajeevan, 2023, p. 4). It is a science-based tool that coastal countries use to manage issues and challenges concerning the ocean and marine resources. It proposes various methods for inclusive growth, sustainable development, employment generation, and the conservation of marine resources to ensure a sustainable future (Ehler & Douvere, 2009). It is essential to address human-human and human-environment conflicts and advance ecosystem services by maintaining and promoting adequate management strategies (Rajeevan, 2023). Since all environmental protection measures focus on people-centred development, they underline the social dimension of MSP. It also has several environmental benefits, including identifying biologically and ecologically important marine areas, protecting biodiversity, and allocating protected areas for conserving marine resources, which are aimed at the future. Its economic benefits are also manifold: greater access to new private actors in the blue economy, greater investment opportunities, identification of areas of development, adoption of emerging technologies, and the promotion of the efficient use of marine resources (Ehler & Douvere, 2009).

5.2 Marine Protected Areas

Marine conservation is essential for human well-being and ecological integrity (Silver et al., 2015). Maritime law promotes the designation of Marine Protected Areas (MPAs) and maintains fisheries regulation that can ensure sustainability in the blue economy and prevent unpredictable climate changes (Voyer et al., 2018). According to the latest statistics, only 46% marine key biodiversity areas are under formal protection (UN, n.d.). It affirms that the globe must increase its endeavors with more positivity and commitment.

5.3 Climate-Resilient Infrastructure and Activities

The latest statistics show that the ocean's average pH is 8.1. It appears to be nearly 30% more acidic than during the pre-industrial period. The increase in ocean acidification harms the survival of marine life, challenges the functioning of the food web, and endangers the services provided by the ocean to living beings on Earth and below water (UN, n.d.). Since injudicious human activities are its major causes (Arora et al., 2023), a systematic focus on policies that prevent ocean acidification is required. Climate-smart policies and AI mechanisms can provide early warnings and detection of harmful materials, leading to ocean acidification and pollution. They can also offer optimal solutions to complex environmental problems that contribute to unpredictable climate change issues (Doorn, 2021). A healthy, resilient, and productive marine ecosystem is often considered essential for all beings that live below the water and people who live in coastal areas. It is also indispensable for all species living on land, including humans, as it is the balancer of the global climate (Arora et al., 2023).

5.4 Sustainable Smart Governance

Almost 8 to 10 million tonnes of plastic are estimated to enter the ocean annually. It shows that nearly 0.5% of plastic is reaching the ocean (Donnelly, 2025). UNEP (2017) states that by 2050, oceans will have more plastic than fish. The increased concentration of plastic and other pollutants in the sea causes adverse impacts on its resources. As a result, the people and economic activities that depend on them are also massively affected. It increases the oceanic temperature; the last decade was estimated to be the warmest since 1800 (EPA, 2025). This has resulted in a nearly 40% global rise in sea level, causing a threat to all countries (IOC-UNESCO, 2024). Such a scenario demands more collaborative efforts from the state and non-state actors. The third United Nations Ocean Conference (2025) offers a renewed understanding and a challenge to the entire world concerning the sustainable governance of the ocean. Recognizing the ocean's growing threats to humanity due to pollution and rising sea levels, the member countries of the Conference adopted strong decisions, declarations, and more than 800 voluntary commitments to safeguard the value and depth of the ocean. It focused on ocean protection, pollution control, marine smart governance, and collaborative support for vulnerable nations to address challenges and advance ocean protection measures. Although the Conference could ensure 19 more signatories to the most-awaited Agreement on Marine Biological Diversity of Areas beyond National Jurisdiction (BBNJ Agreement), it still lacks the ratification of 10 more states to reach the required number of 60, needed for the Agreement to enter into force (UN, 2025). Compared to many other areas, marine protection has not yet been prioritized in the task list of many countries.

International and national actors should prioritize controlling and eradicating the illicit trade in marine resources. They must focus on the global trends that exclude vulnerable and marginalized people, as well as small-scale fishermen, from the policymaking and decision-making processes related to marine resources (UNCTAD, 2022). Individuals and communities living in coastal areas with marine resources have unique methods and techniques to protect those resources. Their traditional knowledge should be systematically used with AI and other smart technologies to guarantee SDG 14. It requires effective coordination and management, which can only be achieved by a truly guided welfare government.

6 CONCLUSION

The world requires a systematic and committed practice of blue justice that respects the value of the ocean and marine resources, emphasizes fair distribution of the benefits and responsibilities associated with them, and advocates the inclusion of marginalized sections of society, especially the small-scale fishermen, in the decision-making bodies in relation to maritime law, blue economy, and the implementation of SDG14 (Ghazali, 2025). Based on the above critical analysis, it can be argued that blue justice should encompass environmental, social, and economic sustainability, incorporating the advantages of AI and smart policies, in support of the SDG-guided march towards a sustainable future. Global economic growth, food security, climate change stability, human well-being, and societal welfare require a substantial interplay between maritime law and the blue economy. The latest scientific advancements, such as AI, have presented numerous opportunities to achieve the objectives of SDG 14. More notably, society requires people with ocean literacy who know the enormous contributions oceans provide to humans (Garcia & Carter, 2022). Even when there are a myriad of scientific and technological advancements and opportunities to assess the importance of the blue economy and adequate legal systems (maritime law) to prevent pollution, nothing can substitute a well-formed individual who is highly aware of the importance of the oceans. If people are well-formed to protect the ocean, sustainably use its resources, conserve the integrity of the coastal areas, and promote collaborative efforts (SDG 17) to maintain the value of the high seas, the latest advancements like AI and climate-smart policies can easily facilitate their efforts to augment the blue economy, implement the provisions of maritime law, and strengthen the march towards a sustainable future. Although AI-guided methods and techniques are comparatively less prevalent in the blue economy and discussions on marine resources, they are attempting to catch up with the pace of the rest of the global economic interactions for increased sustainability.

REFERENCES

Akter, S., Babu, M. M., Hani, U., Sultana, S., Bandara, R., & Grant, D. (2024). Unleashing the Power of Artificial Intelligence for Climate Action in Industrial Markets. *Industrial Marketing Management, 117*, 92–113. DOI: 10.1016/j.indmarman.2023.12.011

Armour, J., & Sako, M. (2020). AI-Enabled Business Models in Legal Services: From Traditional Law Firms to Next-Generation Law Companies? *Journal of Professions and Organization*, 7(1), 27–46. DOI: 10.1093/jpo/joaa001

Arora, N. K., Mishra, I., & Arora, P. (2023). SDG 14: Life Below Water – Viable Oceans Necessary for a Sustainable Planet. *Environmental Sustainability*, *6*(4), 433–439. https://link.springer.com/article/10.1007/s42398-023-00299-0. DOI: 10.1007/s42398-023-00299-0

Awogbemi, O., von Kallon, D. V., & Kumar, K. S. (2024). Contributions of Artificial Intelligence and Digitization in Achieving Clean and Affordable Energy. *Intelligent Systems with Applications*, *22*, 200389. DOI: 10.1016/j.iswa.2024.200389

Bajwa, J., Munir, U., Nori, A., & Williams, B. (2021). Artificial Intelligence in Healthcare: Transforming the Practice of Medicine. *Future Healthcare Journal*, *8*(2), e188–e194. https://pmc.ncbi.nlm.nih.gov/articles/PMC8285156/. DOI: 10.7861/fhj.2021-0095 PMID: 34286183

Basurto, X., Virdin, J., Franz, N., DeLand, S., Smith, B., Cleary, J., Vegh, T., & Halpin, P. (2024). A Global Assessment of Preferential Access Areas for Small-Scale Fisheries. *npj Ocean Sustainability, 3*, 56. https://www.nature.com/articles/s44183-024-00096-0

Benzaken, D., Adam, J. P., Virdin, J., & Voyer, M. (2024). From Concept to Practice: Financing Sustainable Blue Economy in Small Island Developing States: Lessons Learnt from the Seychelles Experience. *Marine Policy*, *163*, 106772. DOI: 10.1016/j.marpol.2024.106072

Chen, L. (2023). Maritime Rights, Obligations, and Liabilities of Intelligent Ships from the Perspective of Risk Distribution. *Journal of International Maritime Safety, Environmental Affairs, and Shipping*, *7*(4), 2264566. DOI: 10.1080/25725084.2023.2264566

Churchill, R., & Lowe, A. (1999). *The Law of the Sea*. Manchester University Press.

Das, J. (2023). Blue Economy, Blue Growth, Social Equity, and Small-Scale Fisheries: A Global and National Level Review. *Studies in Social Science Research*, *4*(1), 38–82. https://www.scholink.org/ojs/index.php/sssr/article/view/8669. DOI: 10.22158/sssr.v4n1p38

De Gama, M. (2025). The Impact of Fisheries Subsidies on Sustainable Development. In Rolland, S. E. (Ed.), *Research Handbook on Trade Law and Development* (pp. 148–161). Elgar., https://www.elgaronline.com/edcollchap/book/9781035325962/chapter10.xml DOI: 10.4337/9781035325962.00022

Donahue, J. (2018). *A Primer on Using Artificial Intelligence in the Legal Profession.* JOLT Digest. https://jolt.law.harvard.edu/digest/a-primer-on-using-artificial-intelligence-in-the-legal-profession

Donnelly, C. (2025). *Ocean Pollution Facts: Statistics & Rankings* (2025). GreenMatch. https://www.greenmatch.co.uk/ocean-pollution-facts

Doorn, N. (2021). Artificial Intelligence in the Water Domain: Opportunities for Responsible Use. *The Science of the Total Environment*, *755*, 142561. DOI: 10.1016/j.scitotenv.2020.142561 PMID: 33039891

Durlik, I., Miller, T., Kostecka, E., & Tuński, T. (2024). Artificial Intelligence in Maritime Transportation: A Comprehensive Review of Safety and Risk Management Applications. *Applied Sciences (Basel, Switzerland)*, *14*(18), 8420. https://www.mdpi.com/2076-3417/14/18/8420. DOI: 10.3390/app14188420

Ehler, C., & Douvere, F. (Eds.). (2009). *Marine Spatial Planning: A Step-by-Step Approach toward Ecosystem-Based Management.* UNESCO., https://repository.oceanbestpractices.org/bitstream/handle/11329/459/186559e.pdf?sequence=1&isAllowed=y

Elgazzar, R. F., & El-Gazzar, R. (2017). Smart Cities, Sustainable Cities, or Both? – A Critical Review and Synthesis of Success and Failure Factors. *Proceedings of the 6th International Conference on Smart Cities and Green ICT Systems*, 250-257. https://www.scitepress.org/Link.aspx?doi=10.5220/0006307302500257

EPA (US Environmental Protection Agency). (2025). *Climate Change Indicators: Sea Surface Temperature.* https://www.epa.gov/climate-indicators/climate-change-indicators-sea-surface-temperature#

FAO. (2001). *International Plan of Action to Prevent, Deter, and Eliminate Illegal, Unreported, and Unregulated Fishing (IPOA-IUU).* https://openknowledge.fao.org/server/api/core/bitstreams/a80c3bfb-1d5b-4ee6-9c85-54b7e83986a2/content

Filho, W. L., Mbah, M. F., Dinis, M. A. P., Trevisan, L. V., de Lange, D., Mishra, A., Rebelatto, B., Hassen, T. B., & Aina, Y. A. (2024). The Role of Artificial Intelligence in the Implementation of the UN Sustainable Development Goal 11: Fostering Sustainable Cities and Communities. *Cities (London, England)*, *150*, 105021. DOI: 10.1016/j.cities.2024.105021

Floridi, L., & Cowls, J. (2019). A Unified Framework of Five Principles for AI in Society. *Harvard Data Science Review*, *1*(1). https://hdsr.mitpress.mit.edu/pub/l0jsh9d1/release/8

Galdies, C., Bellerby, R., Canu, D., Chen, W., Garcia-Luque, E., Gašparović, B., Godrijan, J., Lawlor, P. J., Maes, F., Malej, A., Panagiotaras, D., Romera, B. M., Reymond, C. E., Rochette, J., Solidoro, C., Stojanov, R., Tiller, R., Torres de Noronha, I., Uścinowicz, G., & Guerra, R. (2020). European Policies and Legislation Targeting Ocean Acidification in European Waters – Current State. *Marine Policy*, *118*, 103947. DOI: 10.1016/j.marpol.2020.103947

Garcia, O., & Carter, C. (2022). Life Below Water: Challenges for Tourism Partnerships in Achieving Ocean Literacy. *Journal of Sustainable Tourism*, *30*(10), 2428–2447. https://www.tandfonline.com/doi/epdf/10.1080/09669582.2020.1850747?needAccess=true. DOI: 10.1080/09669582.2020.1850747

Garner, S. R., Neff, B. D., & Bernards, M. A. (2010). Dietary carotenoid levels affect carotenoid and retinoid allocation in female Chinook salmon Oncorhynchus tshawytscha. *Journal of Fish Biology*, *76*(6), 1474–1490. DOI: 10.1111/j.1095-8649.2010.02579.x PMID: 20537026

Ghazali, F. (2025). The Wealth Beneath the Ocean Floor: Potentials and Challenges. In de Pablos, P. O., Lytras, M. D., & Malik, S. (Eds.), *Securing Sustainable Futures Through Blue and Green Economies* (pp. 1–15). IGI Global., https://www.igi-global.com/chapter/the-wealth-beneath-the-ocean-floor/369767# DOI: 10.4018/979-8-3693-7893-9.ch001

Glasser, N. M., Green, F. M., Zumwalt, G., & Wold, I. (2020). *AI and Ethics in the Legal Profession*. Epstein Becker Green. https://www.ebglaw.com/assets/htmldocuments/eltw/eltw385/AI-and-Ethics-in-the-Legal-Profession-Epstein-Becker-Green.pdf

Goralski, M. A., & Tan, T. K. (2022). Artificial Intelligence and Poverty Alleviation: Emerging Innovations and Their Implications for Management Education and Sustainable Development. *International Journal of Management Education*, *20*(3), 100662. DOI: 10.1016/j.ijme.2022.100662

Gupta, J. (2010). A History of International Climate Change Policy. *Wiley Interdisciplinary Reviews: Climate Change, 1*(5), 621–763. https://wires.onlinelibrary.wiley.com/doi/full/10.1002/wcc.67. DOI: 10.1002/wcc.67

Hiywotu, A. M. (2025). Advancing Sustainable Agriculture for Goal 2: Zero Hunger – A Comprehensive Overview of Practices, Policies, and Technologies. *Agroecology and Sustainable Food Systems, 49*(7), 1027–1055. DOI: 10.1080/21683565.2025.2451344

Holm, P. (2022). SDG14 – Exploiting and Managing the Alien and Unseen World Below Water. In Gutmann, M., & Gorman, D. (Eds.), *Before the UN Sustainable Development Goals: A Historical Companion* (pp. 425–447). Academic., DOI: 10.1093/oso/9780192848758.003.0015

IOC-UNESCO. (2024). *State of the Ocean Report.* https://unesdoc.unesco.org/ark:/48223/pf0000390054.locale=en

Isabelle, D. A., & Westerlund, M. (2022). A Review and Categorization of Artificial Intelligence-Based Opportunities in Wildlife, Ocean, and Land Conservation. *Sustainability (Basel), 14*(4), 1979. https://www.mdpi.com/2071-1050/14/4/1979. DOI: 10.3390/su14041979

Ismail, M. M., & El Zokm, G. M. (2025). Algae as Keystone for Blue Economy: Sustainability and Challenges. *Discover Sustainability, 6*(1), 366. https://link.springer.com/article/10.1007/s43621-024-00746-w. DOI: 10.1007/s43621-024-00746-w

Jørgensen, U. (2001). Greening of Technology and Ecotechnology. *International Encyclopedia of the Social and Behavioral Sciences*, 6393-6396.

Kingsbury, D., Remenyi, J., McKay, J., & Hunt, J. (2004). *Key Issues in Development.* Palgrave Macmillan.

Lee, K.-H., Noh, J., & Khim, J. S. (2020). The Blue Economy and the United Nations' Sustainable Development Goals: Challenges and Opportunities. *Environment International, 137*, 105528. DOI: 10.1016/j.envint.2020.105528 PMID: 32014791

Li, X., Han, Q., Wang, J., & Xu, W. (2023). Harnessing Big Data to Track Progress Towards SDG 15: Life on Land. *International Journal of Digital Earth, 16*(2), 4597–4600. DOI: 10.1080/17538947.2023.2273661

Lipper, L., & Zilberman, D. (2017). A Short History of the Evolution of the Climate Smart Agriculture Approach and Its Links to Climate Change and Sustainable Agriculture Debates. In Lipper, L., McCarthy, N., Zilberman, D., Asfaw, S., & Branca, G. (Eds.), *Climate Smart Agriculture: Building Resilience to Climate Change* (pp. 13–30). Springer., https://link.springer.com/chapter/10.1007/978-3-319-61194-5_2#Bib1

Ma, B., & Hou, Y. (2021). Artificial Intelligence Empowers the Integrated Development of Legal Education: Challenges and Responses. *Future Human Image, 16*, 43-54. https://www.fhijournal.org/16-4/

Ma, X. (2020). An Economic and Legal Analysis of Trade Measures Against Illegal, Unreported, and Unregulated Fishing. *Marine Policy*, *117*, 103980. DOI: 10.1016/j.marpol.2020.103980

Mittelstadt, B. D., Allo, P., Taddeo, M., Wachter, S., & Floridi, L. (2016). The Ethics of Algorithms: Mapping the Debate. *Big Data & Society*, *3*(2), 2053951716679679. https://journals.sagepub.com/doi/10.1177/2053951716679679. DOI: 10.1177/2053951716679679

Mukherjee, S., Wamane, G., & Pundkar, V. (2025). Role of Artificial Intelligence in the Study of the Blue Economy and Marine Ecosystem: Use of AI in Preservation of Marine Ecosystem. In Singh, B., Kaunert, C., Lal, S., Arora, M. K., & Raghav, A. (Eds.), *Developing Smart Sustainability with Artificial Intelligence* (pp. 215–238). IGI Global., https://www.igi-global.com/chapter/role-of-artificial-intelligence-in-the-study-of-the-blue-economy-and-marine-ecosystem/367616

Olaniyi, E. O., Solarte-Vasquez, M. C., & Inkinen, T. (2024). Smart Regulations in Maritime Governance: Efficacy, Gaps, and Stakeholder Perspectives. *Marine Pollution Bulletin*, *202*, 116341. DOI: 10.1016/j.marpolbul.2024.116341 PMID: 38608427

Rajeevan, M. (2023). Blue Economy: India's Pathway to Sustainable, Secure, and Resilient Economy. In Ganeshan, S. (Ed.), *Blue Economy: India's Pathway to Sustainable, Secure, and Resilient Economy* (pp. 2–10). TERI., https://www.teriin.org/sites/default/files/2022-07/Blue_Economy_Publication.pdf

Rechberger, K. D., Mayorga, J., Booth, M., & Sala, E. (2025). A Pathway to Protect 30% of Coastal Waters by 2030. *Marine Policy*, *180*, 106773. DOI: 10.1016/j.marpol.2025.106773

Rivkind Law. (2022). *Are Admiralty Law and Maritime Law the Same Thing?* https://www.rivkindlaw.com/are-admiralty-law-and-maritime-law-the-same-thing/

Sakhuja, V. (2023). Blue Economy, Industry 4.0 Technologies, and Market Trends. In Ganeshan, S. (Ed.), *Blue Economy: India's Pathway to Sustainable, Secure, and Resilient Economy* (pp. 59–66). TERI., https://www.teriin.org/sites/default/files/2022-07/Blue_Economy_Publication.pdf

Schutter, M. S., Hicks, C. C., Phelps, J., & Waterton, C. (2021). The Blue Economy as a Boundary Object for Hegemony Across Scales. *Marine Policy*, *132*, 104673. DOI: 10.1016/j.marpol.2021.104673

Silver, J. J., Gray, N. J., Campbell, L. M., Fairbanks, L. W., & Gruby, R. L. (2015). Blue Economy and Competing Discourses in International Oceans Governance. *Journal of Environment & Development*, *24*(2), 135–160. DOI: 10.1177/1070496515580797

Solanki, S. C. (2021). Artificial Intelligence and Legal Profession. *International Journal of Law*, *7*(5), 173–175. https://www.lawjournals.org/assets/archives/2021/vol7issue5/7-5-45-715.pdf

Tetley, W. (1994). The General Maritime Law: The Lex Maritima. *Syracus, 20*, 105-146. https://heinonline.org/HOL/AuthorProfile?action=edit&search_name=%20Tetley,%20William&collection=journals

Turyasingura, B., Bekana, D., Niwagaba, C. B., Dejene, S. W., & Ayiga, N. (2024). Climate-Smart Water Management Practices for Sustainable Agriculture in Uganda. *Journal of Water and Climate Change*, *15*(7), 2940–2960. DOI: 10.2166/wcc.2024.471

UN. (2012). *The Future We Want: Outcome Document of the United Nations Conference on Sustainable Development*. https://sustainabledevelopment.un.org/content/documents/733FutureWeWant.pdf

UN. (2015). *Transforming Our World: The 2030 Agenda for Sustainable Development*. https://sdgs.un.org/2030agenda

UN. (2025). *The Sustainable Development Goals Report 2025*. https://unstats.un.org/sdgs/report/2025/The-Sustainable-Development-Goals-Report-2025.pdf

UN. (n.d.). *Goal 14: Conserve and Sustainably Use the Oceans, Seas, and Marine Resources*. UN Sustainable Development Goals. https://www.un.org/sustainabledevelopment/oceans/#

UNCTAD. (2022). *5 Global Actions Needed to Build a Sustainable Ocean Economy*. https://unctad.org/news/5-global-actions-needed-build-sustainable-ocean-economy

UNEP. (2017). *UN Declares War on Ocean Plastic*. https://www.unep.org/news-and-stories/press-release/un-declares-war-ocean-plastic-0#

Voyer, M., Schofield, C., Azmi, K., Warner, R., McIlgorm, A., & Quirk, G. (2018). Maritime Security and the Blue Economy: Intersections and Interdependence in the Indian Ocean. *Journal of the Indian Ocean Region*, *14*(1), 28–48. DOI: 10.1080/19480881.2018.1418155

World Bank. (2017). *The Potential of the Blue Economy: Increasing Long-Term Benefits of the Sustainable Use of Marine Resources for Small Island Developing States and Coastal Least Developed Countries*. World Bank.

Zhang, Y., Li, Y., & Liu, J. (2023). Global Decadal Assessment of Life Below Water and on Land. *iScience*, *26*(4), 106420. DOI: 10.1016/j.isci.2023.106420 PMID: 37035006

KEY TERMS AND DEFINITIONS

Artificial Intelligence: Artificial Intelligence refers to machines' capacity to replicate human intelligence's functioning.

Blue Economy: The blue economy refers to the socioeconomic and environmental interactions that use marine resources for human well-being and societal welfare.

Climate Action: Climate action refers to activities undertaken to mitigate global warming and other complexities leading to unpredictable climate change.

Maritime Law: Maritime law is the legal framework that has evolved over many centuries to regulate, protect, and promote order in marine interactions, including the demarcation of maritime zones, transportation, shipping, fishing, mineral extraction, and environmental sustainability.

People-Centered Policies: People-centered policies refer to efforts that prioritize human interests and concerns within the national and international policy frameworks.

SDGs: The SDGs refer to the UN-designed policy objectives aimed at ensuring a sustainable global future by 2030.

Sustainable Development: Sustainable Development means that the developmental and environmental aspirations of the present generations should not undermine those of future generations.

Chapter 12
Ethics and Compliance:
Navigating Data Privacy in AI Applications

Anant Prakash Awasthi
https://orcid.org/0009-0004-0980-7019
Amity University, Noida, India

Niraj Kumar Singh
https://orcid.org/0000-0002-6037-4771
Amity University, Noida, India

Biswajit Mondal
Amity University, Noida, India

ABSTRACT

Artificial intelligence (AI) systems increasingly shape decisions in health care, finance, employment, and public services—often by learning from vast, sensitive datasets. This chapter examines the ethical and compliance landscape for data privacy in AI. We ground the discussion in major ethical theories (utilitarianism, deontology, virtue ethics, and care ethics) and connect them to practical challenges such as algorithmic bias, surveillance, and opacity. We then synthesize global legal regimes (GDPR, CCPA/CPRA, LGPD, PIPL, PDPA) and extract the core protection principles that matter most for AI builders. Governance tools including Data Protection Impact Assessments (DPIAs), privacy-by-design/default, and post-deployment monitoring are translated into concrete practices. Four case studies illustrate recurring pitfalls and lessons learned. Finally, we survey privacy-enhancing technologies (federated learning, differential privacy, homomorphic encryption, and secure multi-party computation) and outline emerging trends.

DOI: 10.4018/979-8-3373-5238-1.ch012

1. INTRODUCTION

1.1 Overview of Ai and Data Privacy

Artificial intelligence (AI) is the imitation or simulation of human cognition (learning, perception, language understanding, and decision-making) by deriving patterns from data using statistical learning and optimisation. Modern AI is mostly data-driven: instead of rules developed manually, models such as gradient-boosted decision trees, transformers, and diffusion models are trained on both labelled and unlabelled corpora, as well as on telemetry and interaction logs. The growth of capabilities is accompanied by an increase in reliance on large, and even sensitive, datasets, with corresponding risks to individual privacy and social values (California Legislature, 2018, 2020).

In reality, AI development is a data lifecycle which includes data sourcing and collection, model labelling and curation, model training, model evaluation, model deployment, model monitoring and eventual retirement or retraining. In these phases, personal data can appear in the form of direct identifiers, indirect or contextual identifiers, representations derived from data, such as embeddings, and high-level model inferences. Recent regulatory and technical directions indicate that privacy risk is growing due to inferential analytics and model behaviour, rather than explicit identifiers. Empirical research and rules of governance prove that people can be re-identified or substantially profiled based on model results, supporting data, or system-level links, despite the conventional methods of anonymisation. This means that modern AI governance strategies are not based on the principle of relying on de-identification only but focused on the lifecycle-oriented risk assessment, accountability, and ongoing monitoring of data usage and model behaviour (El Emam et al., 2021; Carlini et al., 2023; NIST, 2023; European Union, 2024).

Figure 1. Privacy-by-design and privacy-by-default controls across the AI lifecycle

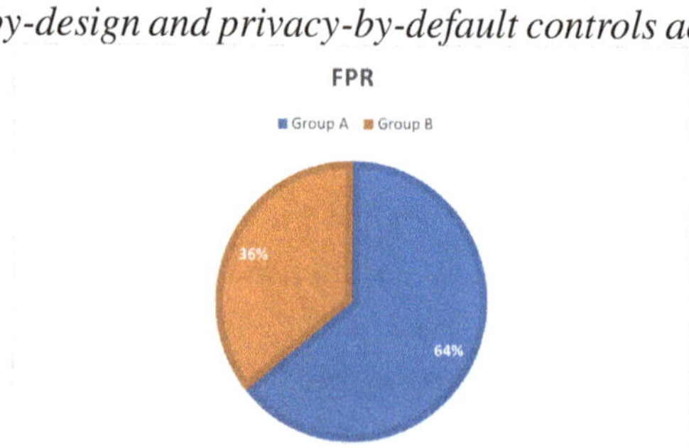

Source: Authors' illustration, informed by Cavoukian (2009), European Union (2016), ISO/IEC 42001:2023 and NIST (2023).

The figure shows privacy risks appear at all AI lifecycle stages: data collection, labelling, deployment, monitoring, and retirement. Technical, organisational, and legal controls should be maintained throughout the lifecycle, not just added at the end as a compliance measure. Privacy risks may affect each step of the AI lifecycle, including data collection and labeling, training, evaluation, deployment, monitoring, and even retirement, and a comprehensive set of technical, organizational, and legal controls must be implemented at all levels of the AI lifecycle instead of being implemented at the end of the AI lifecycle to accommodate compliance. These are in line with the privacy-by-design/default and lifecycle governance models (Cavoukian, 2009; European Union, 2016; ISO/IEC 42001, 2023; National Institute of Standards and Technology, 2023).

Table 1. AI lifecycle privacy controls, risks, and required evidence (audit-ready)

AI Lifecycle Stage	Key Privacy Risks	Mandatory Controls	Evidence / Artefacts
Collect	Excess data, unlawful basis	Lawful basis, minimization, consent/notice	Data inventory, legal basis note, DPIA trigger record
Label	Vendor leakage, access misuse	NDA/DPA, RBAC, pseudonymization	Vendor agreement, access logs, labeling SOP
Train	Memorization, leakage	Secure env, encryption, restricted datasets	Training pipeline logs, security controls, dataset versioning
Evaluate	Hidden bias, privacy attacks	Privacy attack testing, subgroup testing	Test reports, model card draft, risk register updates
Deploy	Over-sharing, misuse	Release gate, monitoring plan, transparency notice	Approval log, user notice, API policy
Monitor/Retire	Drift, retention violations	Retention limits, incident response, deletion/unlearning	Monitoring reports, incident log, deletion proof

Source: Authors' compilation aligned with GDPR principles (European Union, 2016), NIST AI RMF (National Institute of Standards and Technology, 2023), and (ISO/IEC, 2023) AI management guidance (ISO/IEC 42001:2023).

1.2 Importance of Data Privacy in Ai Applications

In numerous fields of healthcare, finance, employment, and even public services, AI processes personal and special-category data regularly (e.g., health, biometrics). Weak protections put at risk personal harms, i.e., identity theft, economic damages, dignitary harms, and unfair denial of services, as well as social harms, i.e., ubiquitous monitoring, chilling effects, and discrimination (Barocas and Selbst, 2016; Buolamwini and Gebru, 2018; Solove, 2006). Privacy has the ethical advantage of personal autonomy and equal concern on the part of privacy guards, and, at an organisational level, it reduces organisational risk and reinforces stakeholder confidence (Held, 2006; NIST, 2023).

The AIs also exhibit exceptional privacy failure modes. One, the risk of re-identification is high in high-dimensional data (Ohm, 2010; Narayanan and Shmatikov, 2008). Second, representational harms arise when models are historically biased, leading to unequal errors or burdens (Barocas and Selbst, 2016; Buolamwini and Gebru, 2018; Obermeyer et al., 2019). Third, any secondary use and general-purpose creep may undermine initial expectations, particularly when data are used to personalise, promote advertisements or surveil (Solove, 2006; European Union, 2016). Fourth, it is opaque, making it difficult to understand how and when personal data influence outcomes, thereby complicating consent, contestability, and redress (Wachter et al., 2017; Raji et al., 2020). In light of such dynamics, privacy, as a design constraint and performance requirement rather than just a legal checkbox, has to be enforced within the organisation.

1.3 Scope and Purpose of The Chapter

The chapter fills the gap between ethical theory, data protection law, and engineering practice, enabling practitioners, researchers, and policymakers to create and use AI systems that are effective and respectful of individuals' privacy. We (a) base the choice of privacy on normative theory – utilitarian, deontological, virtue and care ethics – to elucidate the trade-offs between collective benefit, individual rights, institutional character, and relational responsibilities (Held, 2006); (b) synthesise cross-jurisdictional obligation and the GDPR and other legislation and explain how they are reflected in engineering controls (European Union, 2016; California Legislature, 2018, 2020; OECD, 2024);

In practice, the chapter also provides checklists and patterns of incorporating privacy into product specifications, data contracts, and ML operations (MLOps) and mentions examples of failures where governance failures generated both legal and ethical damages (Barocas and Selbst, 2016; Angwin et al., 2016; Obermeyer et al., 2019). It will provide readers with decision-useful information: how to discover privacy threats at an early stage, select mitigation that is fit-for-purpose, document it, and hold oneself accountable over time (NIST, 2023).

1.4 Structure of The Chapter

The current state of affairs in the world suggests a major transformation toward risk-based, enforceable regulation of artificial intelligence and data privacy. The instruments such as the European Union Artificial Intelligence Act, the OECD AI Principles (amended), the Recommendation on the Ethics of Artificial Intelligence by UNESCO, and the NIST AI Risk Management Framework (UNESCO, 2021; OECD, 2024; NIST, 2023; European Union, 2024) put emphasis on the lifecycle accountability and transparency of AI systems, as well as on human-centred data governance of AI systems. Section 2 provides the ethical premise and outlines the position of these in privacy decisions in the design, implementation and management (Held, 2006). Section 3 unifies international law regimes and identifies the principles most relevant to AI creators (European Union, 2016; California Legislature, 2018, 2020; OECD, 2024). The fourth part is the governance-in-practice translation, including DPIAs, privacy-by-design/default, records of processing, records of vendor management, documentation, and paths to explanation and contestation (Wachter et al., 2017; Raji et al., 2020). Part 5 consists of the risk management playbook (threat modelling, PETs selection, access control, retention, and monitoring). The new trends of the convergence of ethics and privacy standards in AI, the evolution of PETs, and the growing necessity to demystify and set control over AI have been identified in Section 7 (NIST, 2023; OECD, 2024). At the end of section 8, there are certain recommendations and priorities for the implementation. The foreshadowings of the new out-of-abstract-ethical direction and actual, lifecycle-based regulation of AI data practices are the new regulatory measures, the European Union Artificial Intelligence Act (2024), new amendments to the OECD AI Principles (2024), the global AI ethics framework by UNESCO, and the NIST AI Risk Management Framework (European Union, 2024; OECD, 2024; UNESCO, 2021; NIST, 2023).

2. THEORETICAL FOUNDATIONS OF ETHICS IN AI AND DATA PRIVACY

2.1 Ethical Theories Relevant to AI

In this chapter, the authors focus on useful AI governance paradigms that will turn the ethical values into enforceable regulatory, organizational, and technical rules of real-world AI. To tackle the threats posed by massive data utilization, the use of automatism in inferential and decision-making processes, large-scale data consumption is more frequently discussed in the framework of modern AI ethics as a set of operational principles human agency, fairness, transparency, accountability, and sustainability (UNESCO, 2021; OECD, 2024).

Such foundations as bioethical principles (autonomy, beneficence, non-maleficence, justice) applied to high-stakes areas (Beauchamp and Childress, 2019), the capability approach that connects privacy to human opportunity and well-being (Sen, 1999; Nussbaum, 2011), application frameworks like AI4People and Value Sensitive Design which introduce privacy and stakeholder values into system requirements (Floridi et al., 2018; Friedman and Hendry, 2019) These points of view are interrelated and relate privacy to dignity, equality, and contextual integrity, which are complementary in their approach to responsible AI (Held, 2006; Nissenbaum, 2010).

2.2 Defining Data Privacy in AI

The privacy of AI data deals with the regulation of individual information under the model lifecycle, i.e., collection, labelling, training, evaluation, deployment, monitoring, and retirement; therefore, such applications are legal, fair, transparent, necessary, secure, and contextually appropriate (European Union, 2016; Nissenbaum, 2010).

The most significant definitional tensions arise from the fact that modern ML has been trained on high-dimensional representations (embeddings) and produces downstream predictions (e.g., health risk, creditworthiness). The existence of identifiers does not prevent opponents from transferring records by linking them to other related datasets (Ohm, 2010; Narayanan and Shmatikov, 2008). Other attacks, such as model inversion and membership inference, show that even trained models can reveal information about individuals observed during training (Fredrikson et al., 2015; Shokri et al., 2017). Consequently, privacy controls should be applied not only to raw data but also to model artefacts (weights, gradients, logs), evaluation results, and monitoring telemetry.

In practice, an effective definition of AI privacy is:

- Scopes Direct and indirect identifiers, sensitive attributes, inferences, and signals generated by the model (European Union, 2016; Fredrikson et al., 2015).
- Actors and roles: controllers, processors, sub-processors, and vendors of foundation models; explicit contracts and audit rights.
- Purposes & boundaries: Clearly defined, explicit purposes with boundaries against functional creeps (Solove, 2006).
- Technical (e.g., differential privacy, federated learning, access control), organisational (e.g., DPIAs, training, sanctions) and procedural (e.g., rights handling, incident response) (Dwork, 2006; McMahan et al., 2017).

2.2.1 Global AI governance efforts (20212024)

The privacy and accountability requirements of AI systems are increased by the recent AI governance initiatives in the world. The EU Artificial Intelligence Act puts forward a more regulated risk-based model with the prerequisites of compliance with high-risk systems AI, privacy-by-design, and auditability are formed as the key compliance requirements (European Union, 2024). The OECD AI Principles (the last update was made in 2024) as a supplement to binding regulation revolve around reliable AI, responsibility, transparency, and robustness, which prefer policy cross-border interoperability (OECD, 2024). At the global level of ethics, within the Recommendation on the Ethics of AI, UNESCO goes further offering the human-rights-grounded protection, inclusiveness, and governance considerations to the conditions of the public interest and sustainability (UNESCO, 2021; UNESCO, 2024). Together, these efforts complement the contemporary foundation of AI data privacy control and simplify the chapter in accordance with the existing global norms and demands.

2.3 AI-Related Big Data Ethical Challenges

Trade-offs between fairness and algorithmic bias. Historical patterns serve as training data and may lead to disparate errors or burdens on protected groups (Barocas and Selbst, 2016; Buolamwini and Gebru, 2018). Formal labour demonstrates the irreconcilability of fairness standards: you cannot typically be perfectly calibrated and have even error rates when there are differences in base rates (Kleinberg et al., 2017; Chouldechova, 2017). Consequently, fairness is to be put in context and negotiated, and explicit metric options (e.g., equalised odds) and domain-specific reasons should be presented (Hardt et al., 2016). The example of healthcare triage and recidivism risk shows how the benefit of group-level trade-offs can be an individual detriment (Obermeyer et al., 2019).

Figure 2. Algorithms Fairness tensions in algorithmic decision-making: (A) Concept diagram comparing the trade-offs between typical fairness measures when the base rates of subgroups vary; (B) illustrative subgroup error rates (i.e. false positives/ false negatives) where the trade-off between one fairness measure and a different one may be inversely correlated.

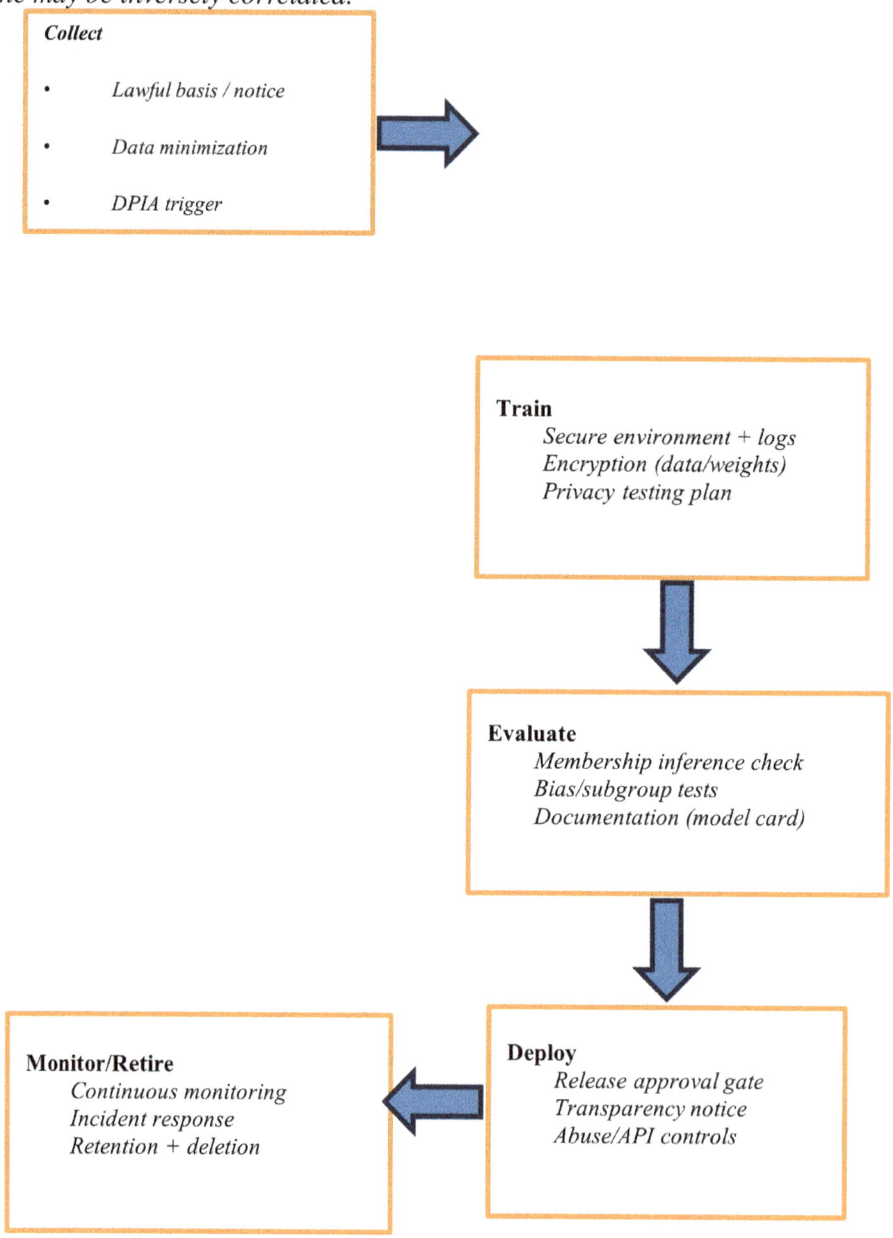

Reason: The authors used their illustration based on fairness trade-off literature (Hardt et al., 2016; Chouldechova, 2017; Kleinberg et al., 2017).

Opacity and explainability: Deep models make accountants, auditors, and regulators inscrutable, which sabotages accountability and contestability (Ananny and Crawford, 2018). The more believable strategy, in high-stakes cases, is to apply interpretable models where feasible and give explanations and documentation artefacts that make sense to the domain (Rudin, 2019; Raji et al., 2020). Surveillance and social chilling effects: Ubiquitous sensing, AdTech identity graphs, and face/voice recognition can enable mass surveillance, mould behaviour, and diffuse harms that are difficult to detect or repair (Zuboff, 2019). Even the seemingly harmless analytics can be chilling when individuals censor themselves under the appearance of being monitored (Solove, 2006).

Asymmetries of power and consent fatigue: It is possible to suggest acquiescence to platform and big data controllers through the use of dark patterns (e.g., confusing toggles, default opt-ins), which compromise meaningful choice and autonomy (Mathur et al., 2019). In practice, the principles of privacy protection imply mitigating these asymmetries through the principle of protection by default, stronger controls, and avenues for contesting automated decisions (European Union, 2016; Wachter et al., 2017).

Re-identification and leakage: anonymisation may not be as robust as it seems, even with good intentions, especially when working with sparse or rich datasets (Ohm, 2010; Narayanan and Shmatikov, 2008). Recent attack models adopt trained models as porous boxes, which incentivise the ideas of differential privacy, robust access control, and sensitive dataset and model release policies (Dwork, 2006; Shokri et al., 2017).

Function creep and secondary use: Data gathered for quality improvement or safety can later be reused for personalisation, pricing, or surveillance, which violates expectations of contextual integrity and purpose restriction (Nissenbaum, 2010; Solove, 2006). To be strong in governance, purpose-binding and change-management reviews, and, in some cases, renewed consent or other legal foundations, are needed (European Union, 2016). Collectively, these issues highlight that privacy-proximate AI is a socio-technical project: advancement will require technical regulation, institutional responsibility, humanistic design, and transparent agreements to make (Raji et al., 2020; Beauchamp and Childress, 2019).

Summary of Key Points

Not only raw data, but also derived artefacts (features, logs, outputs) cause privacy risk throughout the lifecycle, created by AI systems.

private-by-design demands minimisation, purpose restriction, access control and retention discipline starting at the intake stage.

The re-identification, inference, and leakage risks are considered as attack surfaces, therefore privacy should be considered on both the data and model levels. Good privacy governance must be documented and be accountable (who gathered what, why,/where stored, who accessed it and how long held). There is a lifecycle perspective (collect label train evaluate deploy monitor/retire), which assists in mapping controls to the corresponding stage and eliminates gaps.

3. INTERNATIONAL DATA PRIVACY LAW AND COMPLIANCE IN ARTIFICIAL INTELLIGENCE

3.1 Background of Legalisation of World Data Protection

The use of AI systems is slowly moving towards accountability and risk-management principles, rather than a normative approach to data protection. New regulatory frameworks are intended to ensure compliance retention through documentation, technical protection, and ongoing control of the AI life cycle, especially for high-risk and automated decision-making systems (European Union, 2024; NIST, 2023; ISO/IEC 42001, 2023). Omnibus vs. sectoral: Most jurisdictions regulate personal information through omnibus legislation with more stringent requirements for sensitive information and automated processing, but the United States still uses sectoral (e.g., HIPAA for health, GLBA for the banking industry) and state privacy regulations (California Legislature, 2018, 2020). The Omnibus regimes are characterised by a tendency to interlink AI applications across industries and to extraterritorial use, in which organisations target or surveil inhabitants. European Union (GDPR): GDPR is the universal, extraterritorial, full-fledged, complete data-subject rights, full fines and penalties, and also clear privacy-by-design/default, DPIA, records of processing, processor contracts, and breach notification (European Union, 2016). Automated decision-making and profiling have greater protection and transparency, particularly when the outcomes of the decision-making have legal implications and other similar consequences (European Union, 2016).

3.2 Key Principles of Data Protection in AI

Principles such as lawfulness, fairness, and transparency; purpose limitation; minimum data; accuracy; storage limitation; integrity and confidentiality; and accountability are common across all jurisdictions (European Union, 2016). These are interpreted as physical control to AI constructors:

- Lawfulness and legal grounds: Choose and declare a valid legal ground (e.g., consent, contract, legitimate interests, vital/public interest). Where there are special-category data, there should be high exceedances (European Union, 2016; PRC, 2021).
- Fairness and non-discrimination: Compare and reduce the disparate effects across categories; equalise the model's purposes with situational assumptions and rights expectations (Barocas and Selbst, 2016).
- Transparency: Publish issue-layered notices, model-level documentation (e.g., model cards, data sheets), and interpretable descriptions of automated outputs with a significant influence on individuals (European Union, 2016).
- Purpose limitation and minimisation Tie features and inputs to clear, valid purposes, deny function creep, and make identifiability and exposure small with PETs (Dwork, 2006; McMahan et al., 2017).
- Precision and quality of data: possess data governance, control of lineage, validation, labelling quality and monitoring of drift.
- Design Storage constraint/deletion Design retention by purpose and legal retention; execute deletion/withdrawal actions to raw data and derived artefacts (features, embeddings, logs).

3.3 Compliance Obligations for AI Applications

High-impact decisions by AI systems need to be transparent, reviewable by humans, and have an appeal (EU, 2016; PRC, 2021). Use cases of high risks must be documented in the DPIA containing legal basis, necessity, key risk, and mitigations (EU, 2016). Manage the vendors/foundation models through effective DPAs, due diligence/governance documentation (EU, 2016, 2024). In case of a cross-border transfer, approved mechanisms (approved adequacy/SCCs/BCRs or local equivalents) are to be used and should be accompanied with transfer risk assessment (EC, 2021; CJEU, 2020; PRC, 2021). Enforce security + breach response, provide additional protection to the minors, and retain audit evidence based on standards (ISO/NIST) (EU, 2016; ISO, 2019/2022/2023; NIST, 2023; UK, 2018; California, 2020).

4. ETHICAL DECISION-MAKING AND GOVERNANCE IN AI DATA PRIVACY

4.1 Role of Ethics Committees and AI Governance

AI governance has been transformed into an organised organisational activity that includes ethical considerations, legal standards, and technical risk management.

New standards also address how formal governing institutions, the approval process, and accountability systems may be applied to the development, implementation, and monitoring of AI across organisational boundaries (ISO/IEC 42001, 2023; NIST, 2023; OECD, 2024). Socio-environmental responsibilities, such as stakeholder participation, community data rights, and environmental justice, are also featured in recent AI ethics discourse. These views are especially applicable to AI systems used in the public sector and sustainability-oriented contexts, where decisions in data governance can have group and long-term consequences (UNESCO, 2021; OECD, 2024). Proper AI governance is based on a multidisciplinary ethics committee with the authority to make decisions, define escalation paths, and document responsibilities throughout the model lifecycle (OECD, 2024; NIST, 2023). Companies usually have cross-functional ethics committees with formal charters that define scope, decision-making authority, and escalation procedures (ISO, 2023a).

Figure 3. Governance operating model across the AI lifecycle

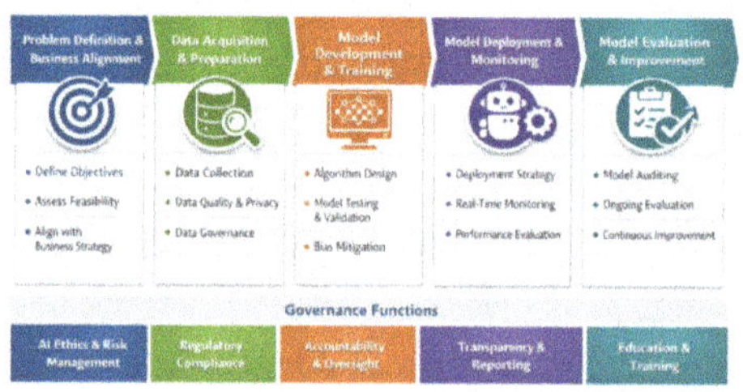

As illustrated in Figure 3, AI governance structures integrate ethical oversight, legal compliance, and technical risk management across the system lifecycle.

Table 3. Governance responsibilities (RACI) across AI privacy activities

Activity	Product/ML (1st line)	Privacy/Legal (2nd line)	Audit/Assurance (3rd line)	Leadership
Use-case intake & data mapping	R	C	I	A
DPIA execution & sign-off	C	R/A	I	A
Vendor due diligence	R	A	C	I
Risk testing (privacy + bias)	R	C	C	A
Release approval	R	A	C	A
Monitoring & incident response	R	C	C	A

Every regime is characterized by the common core of operations of AI systems transparency, limitation to purposes, security, and reported risk assessment. The significant differences remain in the lawful grounds and transfer judgments, and so the organisations should adopt a standard set of controls and declare specific transfer and consent terms under the necessary conditions depending on the jurisdictions (European Union, 2016; OECD, 2024).

4.2 Data Privacy Impact Assessments (DPIAs)

A DPIA is a systematic evaluation of risks to human rights and liberties and the ways to mitigate them (European Union, 2016). One of the 10-step DPIA templates, which reconciles privacy and AI risk management, is:

1. Screening: large-scale sensitive data, systematic monitoring, biometric identification, profiling and significant effects (large-scale).
2. Explain processing (purposes, types of data/data sources, roles – controller or processor, data flows, data retention, transfers between countries).
3. Legal justification & necessity (legal basis according to type of data; to legitimate interests, carry out LIA balancing test).
4. Proportionality & alternatives (why the approach is needed; less intrusive options have been thought of).
5. Stakeholders and consultation (DPO, impacted users; no residual risk, pay attention to the supervisory authority).
6. Risk analysis (probability x result of: harm to privacy, discrimination, chilling effects, loss of opportunity, dignity).
7. Minimisation, de-identification/pseudonymisation, differential privacy, federated learning, encryption, RBAC/MFA, and human-in-the-loop (Dwork, 2006; McMahan et al., 2017).
8. Model, data documentation (data sheets, model cards, subgroup evaluation; failure modes that are known) (Gebru et al., 2021; Mitchell et al., 2019; Raji et al., 2020).
9. Unplanned risk & decision (accept, reduce further, or discontinue; allocate owners and due dates).
10. Review frequency (re-examine drift, scope variation, new information source, or incident).

4.2.1 Privacy-conserving AI is AI controls as DPIA mitigation

DPIAs must directly assess privacy-saving AI controls where business objectives need analytics and the reduction of disclosure. Differential privacy has the ability to decrease disclosure risk in statistical results and learning, as well as the contribution of individual records are bounded, whereas federated learning can train the model through cooperation between organizations without centralizing raw data (Kairouz et al., 2021; Pan et al., 2024; Zhao and Chen, 2022). The DPIAs are expected to report the selected privacy parameters, threat model assumptions, residual risks, and governance controls including access restrictions, logging audit and independent verification.

4.3 Corporate Social Responsibility (CSR) and Ethical AI Development

AI governance has become a well-organised organisational role that incorporates the ethical considerations, legal requirements and technical risk management. New standards focus on the formal governance organisations, approval procedures, and responsibility systems to manage AI creation, implementation, and surveillance (ISO/IEC 42001, 2023; NIST, 2023; OECD, 2024).

4.4 The AI System Accountability and Transparency

Evidence of accountability is needed: policies, training documentation, TIAs, DPIAs, RoPA, processor due diligence, testing procedures, incident documentation, and logs documenting the decision-making process and the identity of the decision-maker (European Union, 2016; ISO, 2022). For transparency, combine:

- P.- Explainability: interpretable models should be used in high-stakes areas; in cases where black-box models are required, provide area-relevant explanations and disclaimers (Rudin, 2019; Miller, 2019).
- Contestability and redress: noticeable usage of automated processing; available means to seek manual consideration; counterfactual description to make people perceive any actionable alteration (Wachter et al., 2018; Veale and Zuiderveen Borgesius, 2021).
- Ongoing assessment: subgroup monitoring of performance, detecting drift, misuse/abuse test, and publicising known limitations, in particular where it had already been found harmful in prior studies (Angwin et al., 2016; Raji et al., 2020).

Its aim is realistic transparency: the amount of information that allows people and regulators to learn about the effects of personal data on the outcomes and to request redress without sacrificing security and trade secrets (Veale & Zuiderveen Borgesius, 2021; Miller, 2019).

5. MANAGING RISK AND ENSURING COMPLIANCE IN AI SYSTEMS

5.1 Determining the risks of privacy in AI

Privacy threats are experienced throughout the data and model lifecycle and entail both direct and inferential disclosures:

- Training-data leakage and memorisation: It is possible for models to memorise and regurgitate rare strings (e.g., names, medical notes) or passages under copyright, enabling the extraction of training data (Carlini et al., 2021). When trained on large foundation models, they can memorise and reproduce rare or sensitive training examples, enabling the extraction of training data and the exposure of privacy leaks. Recent empirical research shows that both generative and multimodal models are prone to memorisation, especially when trained on large, heterogeneous datasets and not privacy-constrained (Carlini et al., 2021; Carlini et al., 2023; NIST, 2023).
- Unauthorised access and insufficient isolation: Raw data, features, embeddings, gradients and logs are exposed to inadequate access controls, broad service accounts, shared credentials and insufficient tenant isolation (ISO, 2022).
- Re-identification of the anonymised datasets: According to more recent evidence, even when direct identifiers have been removed, linkage attacks and high-dimensional data structures can still enable re-identification. Synthetic data generation has also shown that privacy leakage may occur when models overfit or lack formal privacy guarantees, especially with health and mobility data (European Union, 2024; NIST, 2023; El Emam et al., 2021).
- Expectations and creep of profiling: Driven profiling can have a secondary use that goes beyond the user's original expectations, e.g., behavioural prediction, dynamic pricing, or surveillance. Recent regulatory guidance has proposed contextual integrity, purpose restriction, and transparency as important protections against function creep in automated systems (European Union, 2024; OECD, 2024; NIST, 2023).
- Unequal effects of biased data or purpose: Audits of AI systems still find unequal error rates and discriminant effects caused by biased training data, proxy variables and optimisation decisions. The new governance systems demand written bias testing, subgroup testing, and mitigation approaches to the AI high-risk applications (OECD, 2024; NIST, 2023; European Union, 2024).

- Poisoning of data and loops: Distorted or biased inputs may change a model's behaviour over time, exposing more privacy and amplifying discriminatory effects. The recent safety and governance reviews have identified the need for effective data validation, monitoring, and incident response systems to reduce such risks in deployed AI systems (Weidinger et al., 2022; NIST, 2023).
- The risks of cross-border transfers and vendor chains: Multifaceted AI supply chains, such as cloud providers, foundation-model vendors, and data annotators, impose jurisdictional and disproportionate privacy interventions. The latest regulatory changes focus on risk-based assessment of transfers, vendor responsibility, and records of cross-border data flows in AI systems (European Union, 2024; OECD, 2024).

Risks to human-computer interaction: Human-computer interface risks. Meaningful user control and redress may be compromised by manipulative/, or insufficiently explained interface designs, as well as by insufficient explanation of interface behaviour. Recent AI governance policies emphasise the role of transparent notices, contestability, and human-centred design to reduce such risks (OECD, 2024; European Union, 2024).

Table 2. AI privacy risk register: scenarios, impact, detection, and mitigation

Risk Scenario	Example (AI context)	Impact	Detection / Signals	Mitigation Controls
Training data leakage	Model repeats sensitive text	High	Red-team prompts, extraction tests	DP/regular testing, access restrictions, release gating
Membership inference	Identify if user was in training set	High	MI attack evaluation	DP-SGD, limit model access, logging
Re-identification	Link "anonymous" data with other sets	High	Linkage tests, k-anonymity checks	Minimization, tokenization, strict sharing rules
Function creep	Data used beyond original purpose	Medium–High	Change requests, scope drift	Purpose binding, governance approvals, re-consent where needed
Vendor chain exposure	Annotator/cloud leakage	High	Vendor audit findings	DPA, audit rights, least privilege access
Biometric misuse	Faceprints for surveillance	Very High	Complaint signals, audit triggers	High-risk gate, strict consent + retention, legal review

Source: Authors' compilation informed by NIST AI RMF risk framing (National Institute of Standards and Technology, 2023) and contemporary privacy risk research on re-identification and model leakage (El Emam et al., 2021; Carlini et al., 2023).

5.1.1 The Privacy, AI Security, and Safety intersection are the aspects that are one and the same

The privacy risks in AI systems are getting increasingly similar to security and safety risks. Indicatively, it is possible to inject and misuse AI interfaces and unwillingly reveal sensitive information by outputs, logging, tool-calls or retrieval-augmented generation pipelines. On the same note, the risks of model misuse and data leaving may occur when the enemies take advantage of the weak access controls, insecure APIs, or inadequate monitoring. In reality, privacy safety, thus, needs compounded controls threat modelling, secure deployment plan, red-teaming, access control and ongoing monitoring. In line with this, privacy controls must be aligned to risk management practices and privacy enforced as a business activity, rather than as a legal mandate (National Institute of Standards and Technology, 2023).

5.2 Risk Mitigation Strategies for Data Privacy

A risk-appropriate programme is a set of technical, organisational and procedural controls, which are coordinated:

Technical controls

1. Good key management of encrypted at-rest/in-transit data, features, embeddings and logs, secret isolation and model; least privilege (ISO, 2022).
2. Pseudonymisation/anonymisation with re-identification tests; tokenisation or keyed hashing of sensitive data elements (pseudo-identifiers); store the re-identification keys in a different location (Ohm, 2010).
3. PETs are the technical solutions that reduce the risk of data exposure and inference within the AI systems in the event of their use in a system of governance. The new guidelines emphasise the proportionate use of techniques such as differential privacy and federated learning, evaluated on the basis of auditability, scalability, and risk correspondence rather than on mathematical guarantees alone (NIST, 2023; ISO/IEC 42001, 2023).
4. JIT/JEA-level role-based access control (RBAC) and multi-factor authentication (MFA), JIT/JEA-level, and privileged session logging (ISO, 2022).

5.3 Continuous Surveillance and Auditing of AI Systems

Privacy is a dynamic one; surveillance should be frequent and hazardous: • Telemetry & access monitoring: Collect access logs in a central, immutable location;

model queries; export events and administrative operations; and reject anomalous queries (ISO, 2022).

- Behaviour modelling and drift measure the performance of relevant population cohorts and monitor changes in distribution, which may reflect discriminatory performance or a threat to privacy, consistent with risk-based AI monitoring expectations (NIST, 2023; OECD, 2024).
- Independent assurance: Schedule internal audits and critical system independent testing in accordance with modern AI governance and security policies, such as ISO/IEC 27001, ISO/IEC 42001, and the NIST AI Risk Management Framework (ISO/IEC 42001, 2023; NIST, 2023).

5.4 Collaboration Between Stakeholders in Risk Management

Effective privacy risk management is cross-functional and participatory:

- **RACI across the three lines:** product/ML teams own controls; privacy/legal provides independent challenge; internal audit provides assurance; executives/boards receive risk dashboards (NIST, 2023).
- **DPO and regulators:** In high-risk cases, involve the DPO early; where residual risk remains high, consult the supervisory authority (European Union, 2016).
- **Vendors & data suppliers:** Advanced PETs (e.g., differential privacy and federated learning) should be selected instead of considering them universal default in accordance with the threat models, feasibility, and auditability (National Institute of Standards and Technology, 2023; ISO/IEC 42001, 2023). The technicality of aspects can be cited in the sources of expertise but what is discussed in this chapter is the incorporation of governance.
- **Users and affected communities:** In sensitive domains such as healthcare, employment, and public services, recent AI governance guidance emphasises participatory design, accessible explanations, and grievance mechanisms to improve legitimacy and surface hidden harms (OECD, 2024; UNESCO, 2021).

5.5 Obstacles to Implementation and Costs

Although the implementation of privacy-by-design is highly encouraged, it is subject to real-life challenges. First of all, privacy governance introduces direct costs, including staffing (DPO, privacy engineers, auditors), tooling (logging, access management, data catalogues), due diligence of the vendor and ongoing audit/

assurance. Second, better privacy enhancing technologies may impose overhead on computation and also engineering which may slow down the number of iteration cycles or even raise the cost of deployment in a resource constrained setting. Third, the constraints of organisational maturity, including the inability to comprehend data ownership, the failure of adopting a solid documentation culture, the absence of the data lineage, and the non-uniform responsibility, are likely to slow down the efficient implementation of DPIA and repeated control measures.

To conquer these challenges, organisations need not make a sudden step. The initial maturity milestone should consider the simplest controls (data minimisation, purpose limitation, access control, retention governance, and DPIA workflows) initially and address the advanced PETs. Privacy compliance, particularly of high risk AI systems, can be made more feasible and sustainable by clearly assigning roles (product owner, ML lead, DPO, security, audit), and through evidence based governance (risk registers, approval gates, monitoring metrics) (National Institute of Standards and Technology, 2023).

- Minor or no initial steps: inventory of data, retention policy, access control, DPIA template.
- Medium level controls: controls on vendors, audit documentation, standard documentation.
- Advanced steps, which are risk-justified: DP/FL/HE.

6. CASE STUDIES ON DATA PRIVACY AND COMPLIANCE IN AI

6.1 GDPR Enforcement and AI-Linked Data Processing (Google/CNIL)

The mass profiling and cross-context identifiers form the basis of the online advertising structures. The GDPR states that there must be some legal basis to such processing, the fulfilment of the conditions of transparency and purpose limitation, and effective user control (European Union, 2016). The CNIL of France fined Google (a) because the inaccessibility, layering, and understandability of the information on the uses of data were considered insufficient; (b) there was a lack of explicit and unambiguous consent to the personalisation of ads on the services and devices; and (c) the forms of consent were not presented as required by GDPR in a manner that was granular and free choice (CNIL, 2019). This decision dealt with 5(1)(a)-(b) (lawfulness, fairness, transparency; limitation of purpose), 6 and 7 (lawful basis and consent) and 12-14 (transparent information). Consent controls for AI-based personalisation must be purpose-specific and auditable. Although AI processing may be technically lawful in certain circumstances, the lack of transparency in an AI system could result in significant regulatory penalties.

6.2. Health Data, Clinical AI, and Legal Processing (Royal Free/DeepMind Streams)

Health data is a special-category data, hence the use of AI has to be based on a clear legal basis, minimization, transparency, and proportionality (European Union, 2016). The data-sharing between Royal Free and DeepMind regarding Streams was inappropriate in the eyes of UK ICO as the transparency of patients and the reasons why such access should be performed on a large scale were insufficient (ICO, 2017). Therefore, DPIAs and clear disclosure should be used to support clinical AI.

6.3 COMPAS Algorithms: Criminal Justice Bias

The risk tools may produce fairness and due-process concerns when the rates of errors vary among groups (Barocas and Selbst, 2016). A bias in the results of the COMPAS was reported by ProPublica (Angwin et al., 2016), and subsequent research revealed that metrics of fairness may contradict each other when base rates are dissimilar (Chouldechova, 2017; Kleinberg et al., 2017). Use of high-stakes must record trade-offs, and must be manned by human beings.

6.4 Facial Recognition and Biometric Privacy (Clearview AI)

Biometric data (e.g., faceprints) are inherently identifying and commonly considered sensitive. An example of such a statute is the Biometric Information Privacy Act (BIPA) of Illinois, which requires notice, consent, and retention and provides a private right of action. In 2022, under BIPA, the ACLU sued Clearview AI over its use of a facial-recognition database, and the case was settled with restrictions on how the company could provide its service to most private entities in the country, as well as other injunctive limitations (ACLU, 2022). The case highlighted that scraping publicly accessible images did not absolve some biometric-privacy duties and that massive faceprint databases pose unique dangers of surveillance and abuse. Biometric privacy requirements are not denied by publicly available data.

Facial recognition technologies must be approached as high-risk AIs that require special scrutiny under governance.

Cross-Case Takeaways

- The factors that are decisive in AI privacy enforcement are transparency and the limit of purpose.
- Developed AI systems that involve high risks should have fairness and contestability documented.
- The focus of regulatory defensibility is on DPIAs and governance documentation.
- Programmatic accountability minimises the harm caused by ethics and legal liability.

6.5 Marine biodiversity monitoring and ethical data sharing

There is an increase in the use of AI in marine biodiversity surveillance to integrate satellite-based data, acoustic images, and in-the-field measurements on the risks to an ecosystem and facilitate conservation activities. Though such systems may enable climate-sensitive policies, they have privacy and equity concerns when datasets contain human-related patterns of places (e.g. small-scale fishers, coastal communities) or of operational telemetry. The governance plan would unite information decrease, DPIAs, permission control, transparency logs, and audit of risk on a regular basis based on the provisions of AI risk administration (National Institute of Standards and Technology, 2023; ISO/IEC, 2023a). Federated learning and privacy-preserving aggregation can be employed to reduce the disclosure in addition to share insights in cases that require a multi-institutional collaboration (Kairouz et al., 2021; Pan et al., 2024).

7. EMERGING TRENDS AND FUTURE DIRECTIONS IN AI DATA PRIVACY

7.1 Innovations in Data-Privacy Protection for AI

European Union and NIST are requiring AI-related responsibility which requires improved risk reporting, lifecycle management and high-impact system assurance after deployment, increasingly (European Union, 2024; NIST, 2023). There is also the possibility that the procurement also passes to standardized and auditable documentation as a minimum threshold (OECD, 2024). In the meantime, the PETs including federated learning and differential privacy are leaving the realm of research but enter practice, although they need to be incorporated into lifecycle governance rather than being a solution in itself (NIST, 2023). The initial action in this direction is the EU AI Act because, in the case of high-risk AI, data regulation, documentation, as well as post-market surveillance would be a requirement (European Union, 2024).

Figure 4. Privacy–AI governance stack for responsible deployment (from principles to assurance).

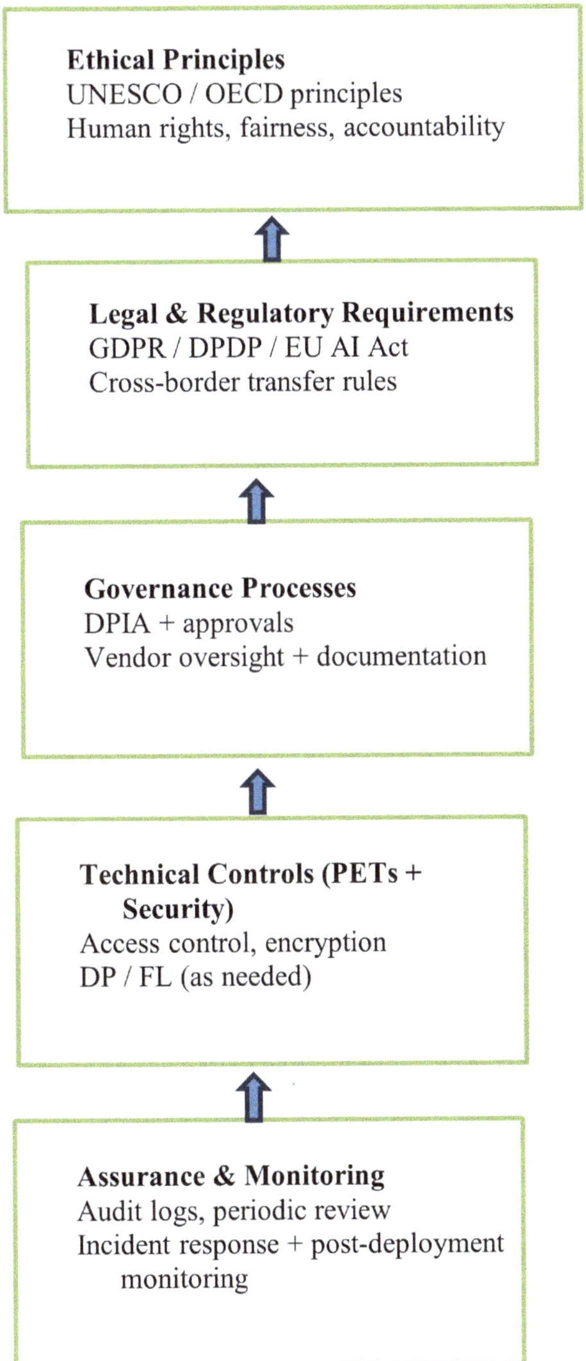

Source: Authors' illustration (aligned with NIST, 2023 and ISO/IEC 42001, 2023).

7.2 Global Convergence

The cross-jurisdictional alignment is converging at higher levels, with legality, fairness, transparency, accountability, security, and human checks emerging as new international policy instruments and risk-driven systems of governance. Nevertheless, material differences exist in legitimate foundations and personal liberty, as well as in the terms of cross-border transfers, and AI privacy packages should be cautiously localised (OECD, 2024; European Union, 2024; PRC, 2021). The modern standards and guidelines, such as the ISO/IEC 27001 of security baselines, the ISO/IEC 23894 and the ISO/IEC 42001 of the AI risk and management systems, the NIST AI Risk Management Framework, etc., are even more supportive of convergence, as they all share similar governance scaffolding even where statutory requirements vary (ISO/IEC 42001, 2023; ISO/IEC 23894, 2023; NIST), even more supportive of convergence, as they all share similar governance scaffolding even where statutory requirements vary (ISO/IEC 42001, 2023; ISO/IEC 23894, 2023; National Institute of Standards and Technology, 2023).

7.3 Cross-border data governance within blue-economy AI applications

The implementation of AI-powered marine ecosystem surveillance, including vessel location, illegal fishing detection, port management, and biodiversity monitoring, is based on volumes of sensor, satellite, and geospatial data that often cross national and jurisdictional boundaries. Such data flows pose unique privacy and governance issues, especially when data related to individuals or communities (e.g., the location of small-scale fishers, their livelihoods, or their movement patterns, etc.) is cross-border processed with no apparent accountability mechanisms.

The cross-border systems of AI should then be provided with transfer protection and participation governance models that ensure transparency, benefit sharing, and adherence to local norms (UNESCO, 2021; OECD, 2024; European Union, 2024). Instead of decreasing exposure by limiting data sharing to a specific purpose, sharing data via federation or edge, or using risk-based governance systems, reduced exposure can be achieved without sacrificing the societal value of marine AI systems.

The necessity of integrating privacy-protective AI into ocean governance systems underscores the importance of transitioning toward compliance, with the ultimate goal of promoting an inclusive and sustainable data ecosystem in which environmental monitoring and the protection of individual rights complement one another rather than undermining each other.

Table 4. Blue-economy AI data sharing: governance, sovereignty, and safeguards

Data Type	Blue-economy Use	Privacy/Ethical Concern	Cross-border Challenge	Recommended Safeguards
AIS vessel/ location data	Illegal fishing detection	Community surveillance risk	Transfer legality, jurisdiction conflicts	Aggregation, purpose limitation, access tiers, DPIA
Coastal community catch data	Fisheries planning	Livelihood profiling	Secondary reuse by commercial entities	Community consent, benefit-sharing, retention limits
Port logistics/ worker telemetry	Port optimization	Workplace monitoring	Processor/sub-processor chain	RBAC, minimization, transparency notice, audit rights
Marine biodiversity sensors	Ecosystem monitoring	Indirect human inference	Multi-institution sharing	Federated learning option, privacy-preserving aggregation
Satellite imagery + analytics	Vessel detection	Re-identification via linkage	Data provenance issues	Provenance logs, restricted access, TIA/DPIA evidence

Source: Authors' compilation aligned with cross-border transfer safeguards and risk-based governance expectations (European Union, 2016; European Union, 2024; OECD, 2024; UNESCO, 2021).

7.3.1 International exchange of marine and geospatial information regulation

The blue-economy AI may often be based on the transnational river of AIS vessel signals, satellite imagery, port logistic information and coastal-related community information. Regarding the transfer of personal information, an international regulation of the legal transfer, responsibility and security should exist in case of such an incident. Organisations are expected to record transfer bases, hang contractual controls with processors /sub-processors and implement transfer risk assessments and other actions where necessary. Role mapping (controller/processor), retention boundaries, and access controls are very essential because marine datasets usually present location indicators which have livelihood patterns that can re-distinguish a person or the community.

7.3.2 Data ownership/data sovereignty Ocean/ community data ownership

In addition to the composition of formal compliance, the blue-economy data governance will be obliged to take into account the ocean data sovereignty and the community data ownership. With patterns of movement, economical behaviour or dependability upon assets being uncovered through the tracking systems, the fishing communities along coastal and smaller regions may be affected in a disproportionate manner. These governance should have participatory decision making, transparency in the use of the data, sharing of benefits and grievance. The threat of surveillance, isolation, or unfair treatment must be well assessed with the assistance of the moral examination and engagement with stakeholders must be presented as the main defence mechanism somewhat of a chance.

7.3.3 Data sharing stored by AI in blue-economy

The techniques of privacy saving AI may be used to aid in collective surveillance and decrease the risk of leakage. Federated learning can facilitate joint model training across ports, agencies, or research institutes, and will facilitate safe sharing of aggregated insights, and, since it is not necessary to aggregate raw datasets, one can use differential privacy to facilitate safe sharing of aggregated insights. Practically, these means should be put in place together with the evidence of governance (design selections, budgets/parameters, access rules, audit logs) and re-identification and inference risks evaluated to be put in place.

7.3.4 Mini-case examples

Case 1: Deterring the detection of illegal fishing operations: The model of vessels activity that are united with the AIS, satellite photographs, and enforcement data poses a threat of privacy to the small-scale fishers. These include such restrictions as purpose limitation (enforcement vs research), limited access to the location hotspots trails and anti-privacy publication of aggregated hotspots.

Case 2: sustainable fisheries management- Community-related catch data is able to provide the trend of income and dependency on resources. Such forms of ethical protection encompass community consenting procedures, benefit-sharing, re-using by commercial profiling and transparency reports on the way the data is being used to make sustainability choices.

8. CONCLUSION

8.1 Summary of Key Points

This chapter discussed the intersection of ethics, data protection law, and technical governance in the design and deployment of artificial intelligence systems. Based on existing ethical principles and privacy regulations, as well as new AI governance standards, it showed how privacy risks arise throughout the lifecycle, including data collection and model training, deployment, monitoring, and retirement. The analysis revealed that lawful, fair, and responsible AI requires more than regulatory compliance; it requires proactive regulatory mechanisms, such as privacy-by-design, impact assessment, documentation, and ongoing monitoring (European Union, 2016; OECD, 2024; NIST, 2023). Combined with other trends, they represent a broader movement toward risk-based, auditable, and human-centred AI data governance, underpinned by new tools such as the EU AI Act, NIST AI RMF, ISO/IEC 42001, and UNESCO ethics recommendations.

8.2 Final Reflections

Strong data privacy governance cannot exist without ethical AI, as the principle of informational self-determination underpins person autonomy, dignity, and trust in automated systems (Solove, 2006; Held, 2006). With the ongoing rise in the use of AI systems to influence access to services, opportunities, and other resources available to the masses, privacy failures may cause not only personal harm but also societal risks. In line with this, organisations need to consider privacy as a fundamental design constraint and responsibility, and as an ethical concern in technical designs, organisational decision-making, and lifecycle policies, rather than following a post hoc compliance reaction (European Union, 2016; NIST, 2023).

8.2.1 Blue-economy AI governance lessons learnt

This chapter suggests that the proper regulation of AI privacy is beyond the compliance checklists, it needs to be operational with DPIAs, technical protection, monitoring, and ethics corresponding to the stakeholders. Governance in blue-economy settings should further take into account cross border data sharing, community data rights, and sustainability goals without causing transparency and contestability. An integrated solution based on GDPR-compatible privacy settings, risk-based AI regulation, and ongoing audit is a viable way to achieve the credible and sustainable deployment of AI (European Union, 2016, 2024; National Institute of Standards and Technology, 2023; UNESCO, 2021).

8.3 Blue-Economy AI Privacy Practical Implementation Checklist

The implementation success will require operationalization of privacy controls in a risk-based approach. Organisers utilising AI applications in blue-economy applications ought to begin with a comprehensive inventory of data (sources, legal foundation, retention) and map controls in the AI lifecycle. The process of DPIA screening is expected to be used on high-impact use cases, and the risks, mitigation, and residual risk acceptance should be documented. The implementation of advanced PETs should be preceded by technical protection that should focus on controlling accessibility, encryption, records, and enforcing retention. The vendor and sub-processor control should be implemented by contractual, evidence checks, and least-privilege controls as a lot of failures arise in the data supply chains. Monitoring of drift, new threats and new flows of data after deployment should be followed with the help of incident response playbooks and periodic re-reviews. Lastly, the human-centred privacy must be enhanced by providing clear notices, meaningful consent where asked and mechanisms of complaint management and contestability in high-stakes outcomes.

REFERENCES

ACLU. (2022). *ACLU settlement places limits on Clearview AI's face recognition database*. American Civil Liberties Union. https://www.aclu.org/press-releases

Ananny, M., & Crawford, K. (2018). Seeing without knowing: Limitations of the transparency ideal and its application to algorithmic accountability. *New Media & Society*, *20*(3), 973–989. DOI: 10.1177/1461444816676645

Angwin, J., Larson, J., Mattu, S., & Kirchner, L. (2016, May 23). *Machine bias: There's software used across the country to predict future criminals. And it's biased*. ProPublica. https://www.propublica.org/article/machine-bias-risk-assessments-in-criminal-sentencing

Barocas, S., & Selbst, A. D. (2016). Big data's disparate impact. *California Law Review*, *104*(3), 671–732.

Beauchamp, T. L., & Childress, J. F. (2019). *Principles of biomedical ethics* (8th ed.). Oxford University Press.

Buolamwini, J., & Gebru, T. (2018). Gender shades: Intersectional accuracy disparities in commercial gender classification. *Proceedings of Machine Learning Research*, *81*, 1–15.

California Legislature. (2018). *California Consumer Privacy Act of 2018 (AB 375)*. State of California. https://leginfo.legislature.ca.gov

California Legislature. (2020). *California Privacy Rights Act of 2020 (CPRA)*. State of California. https://leginfo.legislature.ca.gov

Carlini, N., Tramer, F., Wallace, E., Jagielski, M., Herbert-Voss, A., Lee, K., Roberts, A., Brown, T., Song, D., Erlingsson, Ú., Oprea, A., & Raffel, C. (2021). Extracting training data from large language models. *arXiv*. https://arxiv.org/abs/2012.07805

Carlini, N., Tramer, F., Wallace, E., Jagielski, M., Herbert-Voss, A., Lee, K., Roberts, A., Brown, T., Song, D., Erlingsson, Ú., Oprea, A., & Raffel, C. (2023). Extracting training data from large language models. *Communications of the ACM*, *66*(10), 66–74. DOI: 10.1145/3595112

Cavoukian, A. (2009). *Privacy by design: The 7 foundational principles*. Information and Privacy Commissioner of Ontario.

Chouldechova, A. (2017). Fair prediction with disparate impact: A study of bias in recidivism prediction instruments. *Big Data*, *5*(2), 153–163. DOI: 10.1089/big.2016.0047 PMID: 28632438

Commission nationale de l'informatique et des libertés (CNIL). (2019). *Délibération SAN-2019-001 du 21 janvier 2019 prononçant une sanction pécuniaire à l'encontre de la société GOOGLE LLC*. CNIL.

Court of Justice of the European Union. (2020). *Data Protection Commissioner v Facebook Ireland Ltd and Maximillian Schrems (Case C-311/18)*. https://curia.europa.eu

Dwork, C. (2006). Differential privacy. In Bugliesi, M., Preneel, B., Sassone, V., & Wegener, I. (Eds.), *Automata, languages and programming* (pp. 1–12). Springer., DOI: 10.1007/11787006_1

El Emam, K., Mosquera, L., & Bass, J. (2021). Evaluating the risk of re-identification of anonymised data. *The Journal of Privacy and Confidentiality*, *11*(2), 1–29. DOI: 10.29012/jpc.778

European Commission. (2021). *Commission implementing decision (EU) 2021/914 of 4 June 2021 on standard contractual clauses for the transfer of personal data to third countries pursuant to Regulation (EU) 2016/679*. Official Journal of the European Union.

European Union. (2016). *Regulation (EU) 2016/679 (General Data Protection Regulation)*. Official Journal of the European Union.

European Union. (2024). *Regulation (EU) 2024/1689 (Artificial Intelligence Act) (OJ L, 2024/1689, 12.7.2024)*. Official Journal of the European Union.

Floridi, L., Cowls, J., Beltrametti, M., Chatila, R., Chazerand, P., Dignum, V., Luetge, C., Madelin, R., Pagallo, U., Rossi, F., Schafer, B., Valcke, P., & Vayena, E. (2018). AI4People—An ethical framework for a good AI society: Opportunities, risks, principles, and recommendations. *Minds and Machines*, *28*(4), 689–707. DOI: 10.1007/s11023-018-9482-5 PMID: 30930541

Fredrikson, M., Jha, S., & Ristenpart, T. (2015). Model inversion attacks that exploit confidence information and basic countermeasures. In *Proceedings of the 22nd ACM SIGSAC Conference on Computer and Communications Security (CCS '15)* (pp. 1322–1333). Association for Computing Machinery. https://doi.org/DOI: 10.1145/2810103.2813677

Friedman, B., & Hendry, D. G. (2019). *Value sensitive design: Shaping technology with moral imagination*. MIT Press. DOI: 10.7551/mitpress/7585.001.0001

Gebru, T., Morgenstern, J., Vecchione, B., Vaughan, J. W., Wallach, H., Daumé, H.III, & Crawford, K. (2021). Datasheets for datasets. *Communications of the ACM*, *64*(12), 86–92. DOI: 10.1145/3458723

Hardt, M., Price, E., & Srebro, N. (2016). Equality of opportunity in supervised learning. *arXiv*. https://arxiv.org/abs/1610.02413

Held, V. (2006). *The ethics of care: Personal, political, and global.* Oxford University Press.

Information Commissioner's Office. (2017). *Royal Free—Google DeepMind trial failed to comply with data protection law.* ICO. https://ico.org.uk

ISO. (2019). *ISO/IEC 27701:2019—Security techniques—Extension to ISO/IEC 27001 and ISO/IEC 27002 for privacy information management.* International Organization for Standardization.

ISO. (2022). *ISO/IEC 27018:2019—Code of practice for protection of personally identifiable information (PII) in public clouds acting as PII processors* (or latest edition, if used). International Organization for Standardization.

ISO/IEC. (2022). *ISO/IEC 27001:2022—Information security, cybersecurity and privacy protection—Information security management systems.* International Organization for Standardization.

ISO/IEC. (2023). *ISO/IEC 23894:2023—Information technology—Artificial intelligence—Risk management.* International Organization for Standardization.

ISO/IEC. (2023). *ISO/IEC 42001:2023—Information technology—Artificial intelligence—Management system requirements.* International Organization for Standardization. https://www.iso.org/standard/81230.html

Kairouz, P., McMahan, H. B., Avent, B., Bellet, A., Bennis, M., Bhagoji, A. N., Bonawitz, K., Charles, Z., Cormode, G., Cummings, R., D'Oliveira, R. G. L., Eichner, H., El Rouayheb, S., Evans, D., Gardner, J., Garrett, Z., Gascón, A., Ghazi, B., Gibbons, P. B., & Yu, F. X. (2021). Advances and open problems in federated learning. *Foundations and Trends in Machine Learning*, *14*(1–2), 1–210. DOI: 10.1561/2200000083

Kleinberg, J., Mullainathan, S., & Raghavan, M. (2017). Inherent trade-offs in the fair determination of risk scores. *arXiv*. https://arxiv.org/abs/1609.05807

Mathur, A., Acar, G., Friedman, M. J., Lucherini, E., Mayer, J., Narayanan, A., & Chetty, M. (2019). Dark patterns at scale: Findings from a crawl of 11K shopping websites. *Proceedings of the ACM on Human-Computer Interaction, 3*(CSCW), Article 81, 1–32. https://doi.org/DOI: 10.1145/3359183

McMahan, H. B., Moore, E., Ramage, D., Hampson, S., & Arcas, B. A. (2017). Communication-efficient learning of deep networks from decentralized data. In *Proceedings of the 20th International Conference on Artificial Intelligence and Statistics (AISTATS)* (pp. 1273–1282). PMLR.

Miller, T. (2019). Explanation in artificial intelligence: Insights from the social sciences. *Artificial Intelligence*, *267*, 1–38. DOI: 10.1016/j.artint.2018.07.007

Mitchell, M., Wu, S., Zaldivar, A., Barnes, P., Vasserman, L., Hutchinson, B., Spitzer, E., Raji, I. D., & Gebru, T. (2019). Model cards for model reporting. In *Proceedings of the Conference on Fairness, Accountability, and Transparency (FAT '19)* (pp. 220–229). Association for Computing Machinery. https://doi.org/DOI: 10.1145/3287560.3287596

Narayanan, A., & Shmatikov, V. (2008). Robust de-anonymization of large sparse datasets. In *2008 IEEE Symposium on Security and Privacy* (pp. 111–125). IEEE. https://doi.org/DOI: 10.1109/SP.2008.33

National Institute of Standards and Technology. (2023). *Artificial intelligence risk management framework (AI RMF 1.0) (NIST AI 100-1)*. U.S. Department of Commerce. https://www.nist.gov/publications/artificial-intelligence-risk-management-framework-ai-rmf-10

Nissenbaum, H. (2010). *Privacy in context: Technology, policy, and the integrity of social life*. Stanford University Press.

Nussbaum, M. C. (2011). *Creating capabilities: The human development approach*. Harvard University Press. DOI: 10.4159/harvard.9780674061200

Obermeyer, Z., Powers, B., Vogeli, C., & Mullainathan, S. (2019). Dissecting racial bias in an algorithm used to manage the health of populations. *Science*, *366*(6464), 447–453. DOI: 10.1126/science.aax2342 PMID: 31649194

OECD. (2024). *AI principles (updated in 2024)*. Organisation for Economic Co-operation and Development. https://www.oecd.org/en/topics/ai-principles.html

Ohm, P. (2010). Broken promises of privacy: Responding to the surprising failure of anonymization. *UCLA Law Review. University of California, Los Angeles. School of Law*, *57*(6), 1701–1777.

Pan, K., Ong, Y.-S., Gong, M., Li, H., Qin, A. K., & Gao, Y. (2024). Differential privacy in deep learning: A literature survey. *Neurocomputing*, *589*, 127663. DOI: 10.1016/j.neucom.2024.127663

Raji, I. D., Smart, A., White, R. N., Mitchell, M., Gebru, T., Hutchinson, B., Smith-Loud, J., Theron, D., & Barnes, P. (2020). Closing the AI accountability gap. In *Proceedings of the 2020 Conference on Fairness, Accountability, and Transparency (FAccT '20)* (pp. 33–44). Association for Computing Machinery. https://doi.org/ DOI: 10.1145/3351095.3372873

Rudin, C. (2019). Stop explaining black box machine learning models for high stakes decisions and use interpretable models instead. *Nature Machine Intelligence*, *1*(5), 206–215. DOI: 10.1038/s42256-019-0048-x PMID: 35603010

Sen, A. (1999). *Development as freedom*. Alfred A. Knopf.

Shokri, R., Stronati, M., Song, C., & Shmatikov, V. (2017). Membership inference attacks against machine learning models. In *2017 IEEE Symposium on Security and Privacy* (pp. 3–18). IEEE. https://doi.org/DOI: 10.1109/SP.2017.41

Solove, D. J. (2006). A taxonomy of privacy. *University of Pennsylvania Law Review*, *154*(3), 477–564. DOI: 10.2307/40041279

Standing, P. R. C. Committee of the National People's Congress. (2021). *Personal Information Protection Law of the People's Republic of China*.

Treasury Board of Canada Secretariat. (2019). *Directive on automated decision-making*. Government of Canada. https://www.canada.ca/en/government/system/digital-government/modern-emerging-technologies/directive-automated-decision-making.html

UNESCO. (2021). *Recommendation on the ethics of artificial intelligence*. UNESCO. https://unesdoc.unesco.org/ark:/48223/pf0000380455

UNESCO. (2024, September 26). *Recommendation on the ethics of artificial intelligence* (Last update: 26 September 2024). UNESCO. https://www.unesco.org/en/articles/recommendation-ethics-artificial-intelligence

Veale, M., & Zuiderveen Borgesius, F. (2021). Demystifying the GDPR's "right to explanation". *International Data Privacy Law*, *11*(4), 1–22.

Wachter, S., Mittelstadt, B., & Russell, C. (2018). Counterfactual explanations without opening the black box. *Harvard Journal of Law & Technology*, *31*(2), 841–887.

Weidinger, L., Mellor, J., Rauh, M., Griffin, C., Uesato, J., Huang, P.-S., Cheng, M., Glaese, A., Balle, B., Kasirzadeh, A., Mirza, M., & Gabriel, I. (2022). Ethical and social risks of harm from language models. In *Proceedings of the ACM Conference on Fairness, Accountability, and Transparency (FAccT '22)* (pp. 213–229). Association for Computing Machinery. https://doi.org/DOI: 10.1145/3531146.3533158

Zhao, Y., & Chen, J. (2022). A survey on differential privacy for unstructured data content. *ACM Computing Surveys, 54*(10s). *ACM Computing Surveys, 207*(10s), 1–28. DOI: 10.1145/3490237

Zuboff, S. (2019). *The age of surveillance capitalism: The fight for a human future at the new frontier of power*. PublicAffairs.

KEY TERMS AND DEFINITIONS

AI Governance: Organisational, legal, and technical policies and frameworks that interfere with the accountable production, execution, and administration of AI systems.

Algorithmic Bias: Consistent and unreasonable inequalities in the output of AI, which are brought about by biased data, design choices or deployment conditions.

Artificial Intelligence (AI): This is a form of computing designed to execute functions typically associated with human intelligence, e.g., learning, reasoning, and decision-making.

Data Protection Impact Assessment (DPIA): This is a formal process that involves identifying privacy threats associated with data processing activities, assessing them, and managing them.

Explainability: The extent to which the activities and the output of an AI system can be understood and described by individuals.

High-Risk AI Systems: AI applications that can potentially cause significant changes to fundamental rights, safety, or even access to essential services should be more strictly regulated.

Privacy-Enhancing Technologies (PETs): The technical solutions that reduce the revelation of personal information and enable analysis of the information and production of artificial intelligence models.

Chapter 13

Impact of Renewable Energy Technologies on Sustainable Development

Arzu Alvan
https://orcid.org/0000-0003-2183-0209
Cyprus Aydın University, Turkey

Sina Kısacık
https://orcid.org/0000-0002-3603-6510
Cyprus Aydın University, Turkey

Nalan Gelirli
https://orcid.org/0000-0003-2391-4967
İstanbul Bilgi University, Turkey

ABSTRACT

Renewables have moved from the sidelines to the center of policy and investment. Evidence from IEA, IRENA, REN21, and recent studies shows steep cost drops in solar and wind, record capacity growth, and more jobs. Environmental gains (lower GHGs, cleaner air and water, smarter land use) come with economic and social upsides (jobs, energy security, rural access via decentralized systems). But real constraints persist: grid congestion, curtailment, flexibility gaps; high financing costs and regulatory uncertainty in emerging markets; supply chain risks; and community concerns over siting and land rights. The challenge has shifted from tech costs to system integration and governance—durable policies, markets that value flexibility, quicker permitting, and just transition tools. Priorities ahead: operating

DOI: 10.4018/979-8-3373-5238-1.ch013

high VRE under extremes, long-duration storage and sector coupling (hydrogen, heat, industry), critical minerals and circularity, resilience to climate and cyber risks, and effective de-risking in the Global South.

BACKGROUND

Over the past two decades, the global energy sector has experienced a profound transformation, largely driven by the urgent need to address climate change, ensure energy security, and stabilize economies vulnerable to fossil fuel price volatility. The increasing frequency and severity of climate-related disasters, such as wildfires, droughts, and floods, have underscored the necessity for a rapid transition away from carbon-intensive energy sources (Stern, 2021; IPCC, 2022). In parallel, the depletion of easily accessible fossil fuel reserves and the geopolitical risks associated with energy imports have prompted both developed and developing nations to diversify their energy portfolios (Goldthau & Sovacool, 2016).

Renewable energy technologies—including solar, wind, hydro, and bioenergy—have emerged as central pillars of this transition. According to the International Energy Agency (IEA, 2023), renewables accounted for nearly 30% of global electricity generation in 2022, with solar and wind leading the expansion. The economic rationale for renewables is increasingly compelling: the levelized cost of electricity (LCOE) for solar photovoltaics and onshore wind has fallen by more than 80% since 2010, making them cost-competitive or even cheaper than new fossil fuel plants in many regions (IRENA, 2023; Lazard, 2023).

This shift is not merely technological but also economic and social. The renewable energy sector has become a major source of job creation, with over 13 million people employed worldwide as of 2022 (IRENA, 2023). Moreover, renewables offer opportunities for rural development, energy access, and industrial innovation. However, the transition is not without challenges: integrating variable renewables into existing grids, managing the decline of fossil fuel industries, and ensuring a just transition for affected workers remain critical issues (Sovacool et al., 2021).

The global energy transformation is measurable through concrete indicators showing both scale and pace of change across regions, as presented in the data that follows.

Table 1. Renewable Energy Capacity Growth by Major Region (2010-2023)

Region/Country	2010 Capacity (GW)	2023 Capacity (GW)	CAGR (%)	2023 Share (%)
China	252	1,453	13.5	39.2
United States	93	425	12.0	11.5
European Union	185	638	9.9	17.2
India	32	186	14.1	5.0
Brazil	71	192	7.9	5.2
Japan	28	132	12.3	3.6
Rest of World	565	680	1.4	18.3
Global Total	1,226	3,706	8.9	100.0

Source: IRENA Renewable Capacity Statistics (2024), IEA World Energy Outlook (2024)

Table 2. Global Renewable Energy Employment by Technology and Region (2022)

Technology	Asia	Europe	Americas	Africa/ME	Total (millions)
Solar PV	3.47	0.31	0.48	0.09	4.35
Wind Energy	0.98	0.34	0.25	0.02	1.59
Hydropower	1.52	0.11	0.32	0.08	2.03
Bioenergy	2.78	0.58	0.62	0.14	4.12
Solar Heating	0.68	0.09	0.03	0.01	0.81
Geothermal	0.11	0.03	0.04	0.03	0.21
Total	10.54	1.46	1.74	0.37	14.11

Source: IRENA (2023), Renewable Energy and Jobs: Annual Review 2023

Table 3. Levelized Cost of Electricity Trends by Technology (2010-2023, USD/MWh)

Technology	2010	2015	2020	2023	Reduction (%)
Solar PV (Utility)	378	122	57	48	87.3
Onshore Wind	89	66	39	33	62.9
Offshore Wind	162	154	84	75	53.7
Conc. Solar Power	346	243	126	108	68.8
Geothermal	76	72	69	68	10.5
Hydropower	52	51	49	47	9.6
Coal (new build)	75	85	102	112	-49.3
Gas (CCGT)	65	68	74	81	-24.6

Source: Lazard LCOE Analysis (2023), IRENA Renewable Power Generation Costs (2024)

Objectives and Scope

This chapter sets out to provide a comprehensive and critical analysis of the economic dimensions of renewable energy technologies within the broader context of sustainable development. The primary objectives are: (1) to trace the evolution and current status of renewable energy technologies, (2) to evaluate their economic, environmental, and social impacts, and (3) to identify the principal challenges and opportunities associated with their integration into global and regional energymarkets.

The scope of this chapter is intentionally broad, encompassing both developed and developing economies, and drawing on empirical evidence from recent peer-reviewed studies, international energy reports, and policy analyses. Special emphasis is placed on the economic viability of renewables, the mechanisms of market integration, and the role of policy frameworks in accelerating the energy transition. The analysis also considers the distributional effects of the transition, including the implications for employment, investment, and regional development. By synthesizing insights from economics, policy, and technology, the chapter aims to inform both academic debate and practical policymaking.

LITERATURE REVIEW

Evolution of Renewable and Conventional Energy Technologies

Daniel Yergin represents one of the most noteworthy intellectuals who sought out to speak on the disputed notion of energy security thru the 1980s, when he offered his explanation as "to Guarantee acceptable, consistent supplies of energy at inexpensive charges and in conducts that do not put at risk foremost nationwide standards and purposes". Therefore, the most substantial coercions and risks to energy security as specified by Yergin, remain shocks, stoppages along with management of supplies that may bring about unforeseen and forceful increase in the background of charges, that will implement supplementary economic and political weight on the state. Through specifying the energy security as the corporal presence of plentiful supplies at inexpensive charges, Yergin has highlighted that there happen plentiful characteristics of energy security. The physical security represents the initial one that comprises preserving the possessions, set-up, chains of supply and trade directions and enabling the fast substitutions and exchange when required. Second one remains the serious meaning of accession into the energy which incorporates the competence to develop and acquire energy supplies in corporal, predetermined and marketable conducts. The third one stands that energy security means an arrangement composed of countrywide strategies and worldwide establishments that endure established for responding in a synchronized technique to disruptions, displacements added to emergency circumstances counting the aid to permit the continual stream of supplies. Finally, and peremptorily, provided that lengthier period unsurprisingly remains the investment. Energy security requires approaches and a practicable setting which appeal investment and progression along with invention to pledge that adequate supplies and system will stay obtainable, in the right time in the approaching times. When measured from oil and gas importers, they do deliberate in terms of supply security. Instead, the energy exporters have been lining up "demand security" in terms of their oil and gas exports since they remain dependent on it for backup the economic progress along with a very enormous share of government profits and for conserving the solidity of public. They command to concede that the markets will occur there accordingly they remain up to plan their budgets and vindicate upcoming phases of the investments. The world bank experts' group have compatibly articulated a portrayal of energy security in 2005 as "guaranteeing states are able to sustainably generate as well as benefit from energy at sensible charge to ease economic progress, along with refining the eminence of people life by expanding accession into contemporary energy service". However, notwithstanding this inclusive account, the precise portrayal of the concept as supposed by the Bank's experts separates from state to another based on three groups of criteria: the phase of economic progress, the size of reachable energy resources added to the plausible impact of worldwide energy order. In this background, states as mentioned by the world bank stand categorized into five groups diverge according to the three previous variables that stand:

One of the most all-encompassing leading ingenuities for the clarification of hypothesizing energy has been set forth by the Asia-Pacific energy research Center by 2007 as "The skill of an economy to assure the obtainability of energy resource supply in a maintainable and appropriate means with the energy charge standing at a stage that will not harmfully disturb the economic routine of the economy". For this account, there occur five structures marking energy supply security which remain as: (1) availability of reasonable fuel deposits from both homegrown and external deposits. (2) the ability of an economy to demand supply to reply the predicted demand for energy. (3) the phase of an economy's energy resource difference and energy supplier differentiation. (4) obtainability to fuel resources, by extra convenience of energy substructure and energy conveyance system. (5) the geopolitical panics associated with resources acquisition practice (Kısacık & Güçyetmez, 2022: 98-102).

Within the times when the world has reached intermediate in terms of the energy transition, particularly fiscal and societal prospects of worldwide oil and gas corporations stand being interrogated. It can stand securely presumed that international oil companies (IOCs) upkeep more about their monetary returns than the upcoming of the world. Yet, it would stand very mistaken to ignore these companies, which enjoy huge quantity of international capital, from the transition course. They certain embrace the superiority to accomplish the mandatory savings decisive for the fruitful achievement of the conversion. Aware of this state of affairs, IOCs possess divergent boldness vis-à-vis the process. Intrinsically, within its report entitled "International Oil Companies and the Energy Transition" International Renewable Energy Agency (IRENA) labels the IOCs in two groupings as "corporations that commenced transitioning" and "corporations that select to remain fossil-attentive". The most important drive for the IOCs tangled in transitioning process, naturally, remains conceivable foremost monetary paybacks to stand added from original commerce and action zones of the evolving renewable energy segment (Erkan & Yılmaz, 2022: 128-129).

Worldwide energy request would be augmented by 2.2% in 2024 – quicker than the usual rate over the earlier ten years. Claim for entire fuels and technologies has extended in 2024. The rise would occur under the leadership of the power sector as electricity claim poured by 4.3%, well directly above the 3.2% progress in worldwide GDP, motivated by greatest temperatures, electrification and digitalization. Renewables have been responsible for most of the progress in international energy supply (38%), trailed by gas (28%), coal (15%), oil (11%) accompanied by nuclear (8%). Worldwide electricity usage has risen by approximately 1 100 terawatt-hours (TWh) in 2024, over twice the yearly average upsurge over the past decade. The rise – in excess of Japan's yearly electricity usage – has been the principal ever, separate of terms when the worldwide economy bounced back from recession. China has comprised over half of the international upsurge in electricity claim, nonetheless the growth has been broad-based, with progress of 4% in other emergent and rising economies. Electricity claim has accessed into an original topmost stage in innovative economies. Growing universal electricity usage has been impacted by aspects for instance snowballing freezing claim ensuing from risky temperatures, mounting usage by industry, the electrification of transportation, and the extension of the data centre sector. Electricity consumption within buildings has been responsible for virtually 60% of total progress by 2024. The connected volume of data centers has internationally augmented by a projected 20%, or about 15 gigawatts (GW), frequently within the United States and China. Temporarily, the constant progress in the interest of electric vehicles has given rise within electricity consumption within transportation. International sales of electric cars have increased by more than 25%, exceeding 17 million units and constituting one-fifth of entire car sales, corresponding to the IEA's predictions for 2024. By 2024, 80% of the progress in international electricity output would be realized by renewable sources and nuclear power. Together, they would contribute 40% of overall production for the first time, with renewables alone responsible for 32%. Original renewables fixings have accessed into the topmost stages for the 22nd consecutive year, with about 700 GW of entire renewable capacity additional in 2024, almost 80% of which remained solar PV. Production originating from solar PV and wind has augmented by a top 670 TWh, while production originating from gas has risen by 170 TWh and coal by 90 TWh. Within the European Union, the portion of production originating from solar PV and wind has exceeded the collective portion of coal and gas for the first time within the history. Within the United States, solar PV and wind's portion has increased into 16%, passing that of coal. In terms of China, solar PV and wind have constituted approximately 20% of entire production. By 2024, in excess of 7 GW of nuclear power capacity has been realized operative, 33% over by 2023. The original capacity included has stayed the fifth-highest level in the preceding thirty years. Electricity production originating from nuclear as of 2024 has risen by 100 TWh, equating the principal growth this century separate of the post-Covid reverberation. Construction beginnings for nuclear power plants have augmented by 50% in 2024, completely benefiting from Chinese and Russian plans (International Energy Agency, 2025, pp.5-9).

Diverse basics of the world's energy scheme has seen very dissimilar rates of progress by 2024, shimmering both the influence of short-term features and deeper mechanical tendencies. Worldwide energy claim has grown by 2.2% within 2024, a remarkably quicker rate than the yearly average of 1.3% realized between 2013 and 2023. This uptick has remained moderately by reason of the consequence of life-threatening climate, which we guess additional 0.3 percentage points to the 2.2% progress. Notwithstanding this, energy claim has grown more gradually when compared with the international economy, which extended by 3.2% within 2024, nearby to its long-standing average. Electricity claim has grown more speedily when compared with both total energy claim and GDP, snowballing by 4.3% within 2024. The complete upsurge in terms of claim has stayed the principal ever documented (exclusive of the hedges within terms when the international economic system has improved from downturn). This echoes operational inclinations for instance mounting entree to electricity concentrated appliances like air conditioning and an alteration on the way to electricity-concentrated manufacturing, along with snowballing power claim from digitalization, data centers and AI, and the snowballing electrification of final consumptions. Overall, the power sector has constituted three-fifths of the overall upsurge within international energy claim. Renewables have been responsible for the principal portion of the progress in whole energy supply (38%), trailed by natural gas (28%), coal (15%), oil (11%) and also nuclear (8%). The energy concentration of the universal economic system has enhanced by a simple 1%, enduring the stoppage realized within the past terms. The growth in energy-associated CO2 emissions has decelerated to 0.8%, in comparison with 1.2% by 2023 (International Energy Agency, 2025, pp.5-9).

Sustainable Development Goals and Energy

The UN's 2030 Agenda for Sustainable Development, established in 2015, comprises 17 Sustainable Development Goals (SDGs), with SDG7 emphasizing access to affordable, reliable, sustainable, and modern energy for all (Chhunji Sherpa et al., 2022). Research underscores the critical role energy plays in addressing global challenges such as poverty, health, and education. Approximately 1.2 billion people lack access to electricity while nearly 3 billion, mostly in Sub-Saharan Africa and South Asia, rely on traditional biomass for cooking, factors that exacerbate poverty, indoor air pollution, and educational barriers (Hostettler, 2015). Studies highlight SDG 7's interconnections with other goals. For example, energy access supports poverty reduction (SDG 1) by enabling economic activities; it improves health (SDG 3) by reducing indoor air pollution; and it enhances education (SDG 4) by delivering reliable power for schools (Ahuja & Tatsutani, 2009; Thiam, 2011). These findings shine a light on the cross-cutting impact energy provision has on sustainable development.

Despite progress thanks to a 10% reduction in global electricity access gaps since 2010, significant barriers persist (Hügel & Pardo, 2017). Scholars identify economic constraints like the high cost of rural energy infrastructure, and political challenges such as weak regulatory frameworks as key obstacles (Kettner et al., 2019). Social resistance to renewable energy projects such as wind farms, further complicates implementation (Ahuja & Tatsutani, 2009). The estimated $30 trillion funding gap for clean energy, primarily in developing countries, remains a critical challenge (Kettner et al., 2019). However, research points out how successful interventions like Ghana's energy-efficient lighting program and Malawi's solar kiosks, attest to the potential value of innovative financing and decentralized renewable systems (Acheampong et al., 2019; Frame et al., 2019). The literature suggests a need for further studies on scalable policy frameworks and financing mechanisms to accelerate SDG 7 progress, particularly in underrepresented regions.

Having established the centrality of energy access within the SDG framework, we now examine the current research landscape addressing how renewable technologies translate development commitments into measurable outcomes across diverse contexts.

Current State of Research

Recent research on Sustainable Development Goal 7, which aims to ensure access to affordable, reliable, sustainable modern energy, underscores the critical role it can play in advancing global sustainability objectives (Chhunji Sherpa et al., 2022). As noted above, studies highlight persistent energy access gaps, where some 1.2

billion people lack electricity and 2.8 billion have to rely on traditional biomass for cooking (Hostettler, 2015). In literature, energy access is seen as a synergistic goal with other goals such as poverty reduction (SDG 1) through economic activities, health improvement (SDG 3) through reduction of indoor air pollution, and education enhancement (SDG 4) through reliable power supply (Ahuja & Tatsutani, 2009; Thiam, 2011). Recent analyses also explore the impact of global disruptions, such as COVID-19, which have slowed progress by straining energy infrastructure and funding in developing countries (Chhunji Sherpa et al., 2022). Scholars advocate for renewable energy adoption, noting that investments in solar and wind by countries like China and India have outpaced those in developed nations (Hügel & Pardo, 2017).

Despite such progress, the research identifies significant barriers that remain if SDG 7 is to be achieved. These include a $30 trillion funding gap for clean energy, primarily in developing regions (Kettner et al., 2019). Economic constraints, weak policy frameworks, and social resistance to projects like wind farms hinder progress (Ahuja & Tatsutani, 2009). Case studies, such as the previously cited Ghanian energy-efficient lighting initiative and Malawi's solar kiosks, show the potential decentralized renewable systems and innovative financing can offer. However, their scalability remains underexplored (Acheampong et al., 2019; Frame et al., 2019). The literature also reveals there is insufficient focus on energy efficiency and clean cooking solutions, with slow progress in reducing reliance on solid fuels (Hügel & Pardo, 2017). Research gaps include the need for longitudinal studies on the long-term impacts of renewable energy interventions and the effectiveness of policy frameworks in diverse regional contexts, particularly in least developed countries.

The conversion stands still within the beginning phases – Universal energy-correlated emissions stand still mounting and stand only expected to topmost in 2024. That has been successfully the point upon which the changeover commences, albeit transversely countless countries and societies, energy-associated discharges have by now been in progress to decrease. Between 2017 and 2022, renewables have encountered 51% of original energy claim and fossil sources 49%. In complete standings, fossil-fuel consumption stands still upward. The 'grasp for gas' in the aftermath of Russia's attack of Ukraine, and the interruption of the oil market, has commanded to high charges and a flow in original oil and gas ventures. High gas charges have similarly realized numerous nations strengthen coal-fired power production over the preceding 18 months, pouring discharges hitherto higher. Natural gas stands losing its standing as a 'connecting fuel' for the conversion. Renewables surpass fossils from the mid-2020s. The conversion comprises both the totalling of renewables and the elimination of fossil sources. It will be occurring within the forthcoming nearly thirty years to change the energy mixture from the contemporary 80% fossil 20% non-fossil split to a 48%:52% ratio by mid-21st century. Starting from 2025 ahead, virtually entire clear original volume supplementary stays non-fossil. Wind and solar will be growing ten-times and 17-times, correspondingly, amid 2022 and 2050. Within the forthcoming ten year-term, original fossil generation within low and medium-income states will principally stand invalidated by decreases inside the high-income states. Coal consumption has reached its topmost level within 2014 nonetheless has reached nearby to that stage within past few terms. Yet, its portion of chief energy reduces from 26% nowadays to 10% as of 2050. Fossil key energy claim drops from 490 EJ to 314 EJ as of 2050. Cumulatively, the fossil energy not benefited in comparison with contemporary consumption reaches into 1,673 EJ or 275,000 million barrels of oil equal as of 2050 (DNV, 2023).

About 97 mb/d of oil and 4 150 bcm of natural gas would be used universally in 2022. This has brought about just over 18 Gt CO2 emissions, nearly half of whole energy-associated CO2 emissions. Fresh impetus in installing clean energy technologies comes into meaning that oil and gas claim crowning beforehand 2030 in the Stated Policies Scenario (STEPS), but the decays after these points are not sharp adequate to attain the world's weather objectives. Clear zero conversions necessitate an enormous quickening in clean energy technology positioning and quicker decreases in oil and gas consumption. In the Announced Pledges Scenario (APS), oil and gas claim decay by about 2% every term on average towards 2050 (to 55 mb/d and 2 400 bcm) and in the Net Zero Emissions as of 2050 (NZE) Scenario they drop by in excess of 5% yearly on average towards 2050 (to 24 mb/d and 920 bcm). Consideration on the oil and gas industry habitually emphasizes on the big worldwide oil and gas corporations (the "majors"), nonetheless they possess not more than 13% of international oil and gas generation and deposits. To compare, nation-wide oil corporations hold in excess of half of generation and approximately to 60% of deposits. Universal liquefied natural gas (LNG) commerce observes robust near-term progress, nevertheless trade points in the APS beforehand 2035 and the operation of exportation stations drips; in the NZE Scenario, claim for LNG can be encountered in cumulative by still-operative facilities. Countless manufacturers have set off why they consider their deposits ought to stand favoured for progress in net zero conversions. Some mention that they do enjoy the lowermost generation charges or emission concentrations; others do allege that they stand a healthier choice for energy security; and some designate that innovative oil and gas progresses stand required to advance well-being. In the demand environment of the NZE Scenario, any original oil and gas resource advances would require to stand harmonised by generation discounts somewhere else to circumvent oversupply and fossil fuel lock-in. Both over and underinvestment in fossil fuels do encompass dangers for protected and reasonably priced conversions. Sequencing the decay in oil and gas investment and the upsurge in clean energy investment stands vibrant to side-step destructive charge spikes or supply surpluses. Right now, dangers seem to stand calculated further in the direction of overinvestment in comparison with the reverse (McGlade et. al, 2023).

KEY RENEWABLE ENERGY TECHNOLOGIES

The sustained quickening scenario stands grounded on a bottom-up valuation of the step of positioning for wind, solar and electric vehicles (EVs) which possibly will stand accomplished if these know-hows endure to scale speedily up the S-curve. This stays completed at the nation-wide and local level, when accumulated to a worldwide level. Wind and solar are going to constitute nearly 37% of international electricity production as of 2030 and in excess of 80% of production as of 2050, if their placement lingers to quicken up these arcs. For the meantime, EVs are expected to reach on the way to 90% of international car sales as of 2030, if they endure to step up the S-curve corresponding to historical tendencies and are anticipated to approach 100% of international sales as of 2035. The energy change-over stands proceeding, and its impetus stands turning out to be overwhelming. In terms of fixed capacity, wind and solar grasp 7.5 TW by 2030 under this scenario. This stays generally adjacent to the heights of wind and solar detected in the IEA's publicised pledges scenario. This displays that wind and solar positioning stands already mounting sufficiently quick to answer states' existing countrywide aims and initiates. Overall renewable volume by 2030 grasps about 9 TW in our scenario, about 2.5 times contemporary stages. While this stands not a boosting of renewable volume, it stands not far away from accomplishing this. As the stride of the energy conversion has enhanced, predominantly in the last two years, generalising onwards the stride of transformation in just the last two years would result in even advanced volume numbers than our line. A triplicating of renewable volume as of 2030, motivated by turbocharged wind and solar placement, stands still fit inside grasp. Transversely 2023 and 2024, the constant quickening scenario realizes about 1 TW of wind and solar fixed cumulatively. This remains approximately associated with the IEA's newest predictions of what possibly will reasonably stand accomplished in 2023 and 2024, of about 950 GW (Fyson et. al, 2023: 14-20).

By 2024, the quick progress in zero-carbon technologies within this scenario has been envisaged as sufficiently to completely answer the progress in total energy claim, and commences to erode at claim for fossil fuels. This signifies whole fossil fuel claim may well grasp its highest within 2023, and then pass in a passé of quick deterioration as of 2025. Coal claim would reach its highest levels within 2023 under this scenario and decrease afterwards. In the meantime, fossil gas claim, for which the IEA remains envisioning to reach its highest point as of 2029, highest points by 2024 and instigates to decay speedily subsequently. Oil claim would reach its highest point by 2025 under this scenario. By 2030, EVs within overall are going to relocate over 5 million barrels of oil per day of claim. Much of this shift stands by now comprise in the STEPS scenario (~3.5 million barrels per day), consequently the extra effect of an augmented EV conversion stays minor. Yet, the sustained quickening scenario is envisaging to observe distinguishing highest points within entire three fossil fuels as of 2025. Yet, governments stand cooperatively still scheduling to hugely increase fossil fuel generation. When all strategies and forecasts stay combined, they are going to result in an upsurge in coal generation by 2030, and within oil and gas generation till 2050. However, act to increase fossil fuel supply may perhaps lessen inclination to quicken climate action, by embedding bestowed benefits in the energy system and dropping fossil fuel charges. Contemporary fossil fuel increase strategies characterize both a vast climate and economic danger that does not require to stay taken. As fossil fuel claim highest points and also decays, opening with coal, thus besides ensure fossil CO2 emissions. Within the continued acceleration scenario, fossil CO2 emissions has reached its highest point by 2023 and pass in a passé of speedy weakening, therefore as of 2030 they remain about 2 GtCO2/yr lesser than within the STEPS scenario. Whole CO2 emissions still raise in 2023 by about 0.5%, which stands within the vagueness array of existing estimations. The massive common of the emissions savings originate from the power sector, where wind and solar principally relocate coal, but also fossil gas and oil-fired production within some geographies (e.g., the Middle East and Africa). Out to 2035, the effect of a fast-tracked EV conversion seems to stand comparatively unimportant. The contemporary Internal Combustion Engine (ICE) fleet stands accountable for about 4 GtCO2 of international emissions – considerable, nonetheless only a third of international power sector emissions. Seeing the comprehensive conversion to electric mobility possibly will upsurge emissions savings more. Secondly, there stands a variance between sales and stocks. The quick-tempered development in EVs stands at present in the portion of whole car sales, more willingly than stocks. As worldwide car sales characterize unevenly 6% of the international car stock, the increasing supremacy of EVs in complete car sales does not interpret proximately into large-scale emissions savings, as it necessitates some times for high EV sales to stream via into variations within the total car stock. (Fyson et. al, 2023: 14-20).

SUSTAINABLE DEVELOPMENT IMPACTS

Environmental Benefits

Renewable energy technologies like solar, wind, hydro, biomass, and geothermal are gamechangers for sustainable development. They offer low-carbon solutions that cut down on greenhouse gas emissions and help us tackle climate change head on (Shukla et al., 2022). This aligns perfectly with global goals like that of the Paris Agreement to keep climate warming below 1.5–2°C and supports SDG 13 (Climate Action) (United Nations, 2023). It's not just about numbers -- it's about ensuring that the planet is liveable for future generations.

One of the biggest wins is cleaner air. Burning coal, oil, or gas pumps out nasty pollutants like sulfur dioxide, nitrogen oxides, and particulate matter, which wreak havoc on our health, causing everything from asthma to heart disease, even acid rain (Jacobson et al., 2015). Renewables, especially solar panels and wind turbines, produce power with barely any of these toxins, resulting in healthier communities and clearer skies. Just imagine - millions fewer hospital visits all because of switching to cleaner energy. That's the kind of impact we're talking about.

Table 4. Comparative Environmental Impact: Renewable vs. Fossil Fuel Technologies

Impact Category	Coal	Natural Gas	Solar PV	Wind	Hydro
CO_2 Emissions (g/kWh)	820	490	45	11	24
SO_2 Emissions (mg/kWh)	2,500	12	0.5	0.2	0.8
NO_X Emissions (mg/kWh)	1,800	320	0.8	0.3	0.6
Water Use (L/kWh)	2.1	0.7	0.02	0.001	22
Land Use (m^2/GWh/yr)	12,000	3,600	4,500	72,000	1,500
Waste (kg/GWh)	150,000	850	45	12	28

Source: IPCC (2014), Jacobson et al. (2015), NREL Life Cycle Assessment (2021)

Then, there's water. Fossil fuel plants guzzle massive amounts of it for cooling and processing, putting stress on already scarce resources in dry regions (Mekonnen & Hoekstra, 2016). Solar and wind? They barely touch water once they're up and running, which is a huge step toward SDG 6 (Clean Water and Sanitation) (United Nations, 2023). For example, solar panels might need a quick rinse now and then, but wind turbines don't use a drop of water. This matters significantly in places where every liter counts.

Biodiversity gets boosted too. Mining and drilling for fossil fuels often tear forests and habitats apart (Leung & Yang, 2012). But renewables, when planned thoughtfully like offshore wind farms or solar arrays on already cleared land, can coexist with nature and even help restore ecosystems. Picture a solar farm on degraded land sprouting native plants underneath, giving wildlife a new home.

In short, renewables aren't just about cutting carbon. They clean our air, save our water, and protect our ecosystems, making them a cornerstone of a sustainable future.

Economic Implications

The economic implications of renewable energy deployment are complex and far-reaching, touching on macroeconomic stability, job creation, industrial competitiveness, and fiscal policy. The global renewable energy sector has become a significant engine of economic growth, employing over 13 million people worldwide as of 2022, with projections suggesting this figure could surpass 40 million by 2050 if current policy ambitions are realized (IRENA, 2023; IEA, 2023). Solar photovoltaics, wind, and bioenergy are the leading sources of new employment, with Asia—particularly China and India—dominating manufacturing and installation activities (IRENA, 2023).

Investment in renewables has also proven resilient, even in the face of global economic shocks. In 2022, global investment in renewable power and fuels reached a record $495 billion, outpacing investments in fossil fuel-based power generation for the first time (BloombergNEF, 2023). This shift is not only a response to environmental imperatives but also to the declining costs of renewable technologies. The levelized cost of electricity (LCOE) for utility-scale solar and onshore wind has dropped by more than 80% since 2010, making renewables the cheapest source of new power in many markets (Lazard, 2023; IRENA, 2023).

However, the economic benefits of renewables are not distributed evenly. Regions with abundant renewable resources and supportive policy environments tend to capture greater value, while others may face challenges related to grid integration, financing, and workforce transition (Sovacool et al., 2021). The transition also poses risks for workers and communities dependent on fossil fuel industries, highlighting the need for just transition policies, retraining programs, and targeted investment in affected regions (ILO, 2022; Newell & Simms, 2020).

Moreover, renewables can enhance energy security by reducing dependence on imported fuels and insulating economies from price volatility in global energy markets (IEA, 2023). The multiplier effects of renewable energy investments—through supply chains, local manufacturing, and service industries—further amplify their economic impact. Yet, realizing these benefits requires coordinated policy action, robust regulatory frameworks, and sustained investment in innovation and infrastructure.

Table 5. Global Renewable Energy Investment Flows by Region (2015-2023, USD Billions)

Region	2015	2018	2020	2022	2023	CAGR (%)
China	102	91	134	165	180	7.4
United States	44	46	73	141	150	16.6
Europe	58	63	82	101	108	8.0
India	10	13	15	18	22	10.4
Brazil	7	4	7	12	14	9.0
Rest of World	45	53	65	58	74	6.4
Global Total	266	270	376	495	548	9.6

Source: BloombergNEF Energy Transition Investment (2024), IEA World Energy Investment (2024)

Social and Community Effects

Renewable energy isn't just good for the planet it's a lifeline for people and communities, especially in places where power is scarce. It's about bringing light, opportunity, and fairness to those who need it most, tying directly into SDG 7 (Affordable and Clean Energy) (United Nations, 2023).

Take energy access. Over a billion people still live without electricity, especially in rural areas of developing countries (Sovacool, 2012). Off-grid solar systems and small-scale hydropower projects are changing that, powering homes, schools, and clinics with affordable, clean energy. In African villages, for instance, solar lanterns mean kids can study after dark, and families can ditch dangerous kerosene lamps, saving money and breathing easier. It's not just power -- it's a chance for a better life.

Jobs are another big winner. The renewable energy sector -- wind farms, solar panel factories, and the associated maintenance -- creates tons of work, from engineers to local installers (Wei et al., 2010). In places like Spain and Portugal, wind energy has generated thousands of jobs, giving local economies a real boost. These "green jobs" aren't just temporary; they offer steady careers and build skills for the future, supporting SDG 8 (Decent Work and Economic Growth).

But it's not enough to simply build the projects communities need in order to be part of the process. When locals have a say, projects are more likely to succeed and less at risk of facing pushback (Wüstenhagen et al., 2007). Look at Brazil's wind and water cooperatives where the authorities have created jobs and even strengthened local governance by involving communities in decision-making. That said, challenges over land disputes or cultural concerns can arise, but open, honest planning can play a key role in ensuring that everyone remains on board.

Renewables also empower women and marginalized groups, tying into SDG 5 (Gender Equality). In many places, women spend hours gathering firewood, time they could use for education or work. Clean energy solutions like solar cookers or small-scale biogas systems free up that time, opening doors to new opportunities (Sovacool, 2012). Moreover, community-led energy projects give people control over their own resources, reducing dependence on large, distantly located fossil fuel companies.

In the end, renewable energy builds stronger, fairer communities by providing power, jobs, and a voice. It's about creating a world where everyone has a shot at a better, more sustainable future.

CHALLENGES AND OPPORTUNITIES

Policy and Regulatory Frameworks

The success of renewable energy sources like solar, wind, or geothermal hinges on the policies and regulations behind it. A solid, predictable policy framework is like a green light for investors, sparking confidence and paving the way for a sustainable energy future. But realistically speaking getting these policies exactly right is no easy feat, and plenty of countries experience hitting roadblocks (International Energy Agency [IEA], 2021).

Regional Policy Performance: A Comparative Analysis

Policy frameworks exhibit substantial regional variation, reflecting different institutional architectures, resource endowments, and political economies. The European Union has pursued binding renewable targets embedded in climate legislation, achieving 42% renewable electricity by 2023. The policy mix combines feed-in tariffs (phasing out), competitive auctions, and carbon pricing through the EU Emissions Trading System (ETS), where allowance prices reached €85/tCO_2 in 2024. Germany's Energiewende demonstrated both achievements and frictions: renewable capacity tripled between 2010 and 2023, yet grid congestion costs exceeded €1.4 billion in 2022, and curtailment reached 8.1 TWh. The revised Renewable Energy Directive (RED III) mandates 45% renewable energy by 2030, introducing streamlined permitting via designated 'go-to areas' and enhanced cross-border cooperation mechanisms (European Commission, 2023).

The United States relies on a federalist structure where national tax incentives—the Investment Tax Credit (ITC) and Production Tax Credit (PTC)—operate alongside state-level Renewable Portfolio Standards (RPS) and clean energy mandates.

The Inflation Reduction Act (2022) extended these credits for a decade and added technology-neutral pathways, unlocking an estimated $370 billion in investment. California achieved 31% renewable generation in 2023, supported by aggressive RPS targets and storage mandates; Texas reached 26% despite lacking formal mandates, driven by favorable wind resources and competitive wholesale markets. However, fragmented permitting across federal (FERC, BLM, DOE), state, and local authorities imposes delays averaging 5-8 years for transmission projects, constraining optimization across NERC regions (LBNL, 2023; EIA, 2024).

China combines centralized capacity mandates in Five-Year Plans with provincial renewable quotas and competitive procurement. China installed 55% of global renewable capacity in 2023 (310 GW), achieving 34% non-fossil electricity. Priority dispatch rules and massive grid infrastructure investment reduced national curtailment from 17% (2016) to 3.2% (2023), though integration challenges persist in northwestern provinces. India's auction-based system attracted record-low bids—solar at $17/MWh, wind at $22/MWh—but payment delays by state distribution companies (averaging 3-6 months) and land acquisition bottlenecks have stalled 15 GW of awarded capacity. The introduction of sovereign payment guarantees and bundled hydro-solar tenders aim to address these constraints (CEEW, 2023; IEA, 2024).

Latin America and Sub-Saharan Africa exhibit institutional fragility alongside pockets of success. Brazil's hybrid auction system mobilized $35 billion in renewable investment since 2009, achieving 85% clean electricity, though hydro dependence (60%) creates vulnerability to drought. South Africa's REIPPPP catalyzed 7.7 GW across six bidding rounds, demonstrating that transparent contracts and development commitments can attract private capital despite utility distress; yet fiscal constraints and grid instability delayed Round 5. Kenya achieved 90% renewable electricity through geothermal development and IPP frameworks backed by concessional finance (World Bank, AfDB); Morocco's integrated strategy combined auctions, regional interconnection, and green hydrogen planning to reach 37% renewable share (IRENA, 2023; World Bank, 2024).

Table 6. Regional Renewable Energy Policy Performance Comparison (2024)

Region	Policy Instrument	2023 RE Share	Carbon Price	Avg. Auction (USD/MWh)	Key Barrier
EU-27	Auctions + ETS + Mandates	42%	Yes (~€85/ tCO_2)	35-55	Grid flexibility
United States	Tax Credits + State RPS	22%	Limited (RGGI only)	25-45	Permitting delays
China	Quotas + Auctions + Priority Dispatch	34%	Yes (national ETS)	15-28	Regional curtailment

continued on following page

Table 6. Continued

Region	Policy Instrument	2023 RE Share	Carbon Price	Avg. Auction (USD/MWh)	Key Barrier
India	Competitive Auctions + RPO	19%	No	17-30	Payment security
Brazil	Hybrid Auctions	85%*	No	20-38	Hydro dependence
South Africa	REIPPPP (IPP auctions)	10%	Carbon tax	38-62	Utility finances
Kenya	Feed-in + IPP frameworks	90%	No	45-75	Infrastructure limits
Morocco	Integrated RE Plan + Auctions	37%	No	40-70	Finance access

Source: IEA World Energy Outlook (2024), IRENA Global Landscape (2024), World Bank ESMAP (2024).
*Brazil figure includes large hydro.

The biggest headaches come from inconsistent rules, short-sighted planning, and incentives that change faster than the weather. These issues scare private companies and slow down innovation, especially in developing countries where red tape and limited government capacity can stall projects for years (REN21, 2022). It can resemble trying to build a house on shifting sand without a stable foundation -- things don't necessarily move forward.

On the flip side, smart policies can work wonders. Tools like feed-in tariffs, renewable portfolio standards, or auction systems have turbocharged solar and wind growth in places that get it right (Shukla et al., 2022). Countries with clear permit processes and steady incentives such as streamlined land approval systems or reliable subsidies see renewable projects boom and bloom like wildflowers (International Renewable Energy Agency (IRENA), 2020). When the rules are clear, investors and innovators show up ready to play.

Tying these policies to bigger goals like the UN's Sustainable Development Goals or the Paris Agreement opens doors to global partnerships and climate funding, according to the United Nations Environment Programme (UNEP) in 2021. It goes beyond energy. It's about weaving renewables into plans for transportation, industry, even city design, so as to create a system that works for everyone.

In other words, while negotiating policy hurdles can feel like climbing a mountain, they also offer a chance to rethink how we do things. Governments need to roll up their sleeves and build flexible, inclusive regulations that keep pace with renewable tech and make the energy transition fair for all.

Market Integration and Economic Viability

The successful integration of renewable energy technologies into existing energy markets is a cornerstone of their long-term economic viability. As renewables—particularly solar and wind—are inherently variable and decentralized, their large-scale adoption requires significant changes in market design, grid management, and regulatory frameworks (IEA, 2023; Jenkins et al., 2018). Flexible electricity markets, advanced storage solutions, and digital grid technologies are essential to accommodate the intermittency of renewables and maintain system reliability (REN21, 2023).

In many regions, renewables are now the lowest-cost source of new electricity generation, but their integration is often hindered by legacy infrastructure, regulatory inertia, and insufficient market incentives (IRENA, 2023). Real-time pricing, demand response programs, and cross-border electricity trade are increasingly being adopted to enhance market flexibility and support the integration of variable renewables (IEA, 2023). The internalization of externalities—such as carbon pricing and emissions trading schemes—further improves the economic case for renewables by reflecting their true societal value (Stiglitz et al., 2017). Despite these advances, challenges remain. Grid congestion, curtailment, and the need for backup capacity can erode the economic benefits of renewables if not properly managed (Jenkins et al., 2018). Moreover, the transition requires substantial investment in transmission infrastructure, digitalization, and workforce development. Policymakers must ensure that market rules are technology-neutral, transparent, and supportive of innovation, while also safeguarding affordability and equity for consumers (IEA, 2023; REN21, 2023).

FUTURE RESEARCH DIRECTIONS

Trends and Projections

The future of renewable energy is bright, but it's not without its puzzles. As technologies like solar, wind, and geothermal reshape the global energy landscape, researchers need to zero in on trends that will drive sustainable development forward. The big question is how to make renewables not just widespread but also equitable and efficient enough to meet soaring energy demands while attaining sustainability goals (International Energy Agency [IEA], 2024).

One major trend is the rapid drop in costs for solar and wind power, which is making them competitive with fossil fuels in many regions (IRENA, 2023). But there's a catch: more research is needed if we are push these costs even lower, especially for emerging tech like green hydrogen or advanced energy storage. Batteries, for instance, are critical for balancing intermittent renewable sources, but their materials can be pricey and environmentally tricky (Sovacool et al., 2020). Figuring out sustainable, scalable storage solutions is a research goldmine waiting to be tapped.

Another pressing need is integrating renewables into smart grids. These high-tech systems use real time data to balance energy supply and demand, but they're complex to design and deploy, especially in developing countries where infrastructure lags (REN21, 2023). Research into affordable, adaptable smart grid tech could unlock huge gains, letting communities tap renewables without massive upfront costs. Think of rural areas powering up with solar microgrids game-changing for energy access and hitting SDG 7 (Affordable and Clean Energy).

On top of this, there's the social angle. Renewables can't just be about tech --they need to work for people. Studies show that community pushback over land use and/or cultural concerns can stall projects (Wüstenhagen et al., 2007). Future research should focus on how to design inclusive policies that bring locals on board, ensuring that such associated benefits as jobs and clean energy reach everyone, marginalized groups in particular. This relates specifically to SDG 10 (Reduced Inequalities).

Looking ahead, aligning research with global goals like the Paris Agreement is vital. We need data-driven projections on how renewables can decarbonize sectors like transportation and industry, not just electricity (IEA, 2024). The challenge is big, but so is the opportunity. Research can light the way to a future where clean energy powers a fairer, greener world.

6.2 Research Gaps and Future Directions

Despite rapid advances in renewable energy technologies and their growing integration into power systems, several critical research gaps remain at the intersection of technology, policy, finance, and social outcomes. Addressing these gaps is pivotal for translating current momentum into durable, equitable progress toward the Paris Agreement and the 2030 Agenda.

System integration at high variable renewable penetration

Gap: The literature increasingly documents technical feasibility of power systems with high shares of variable renewables, yet operational playbooks for ≥70–90% VRE share under diverse geographies, climates, and market rules remain underdeveloped, especially beyond a handful of OECD and Chinese provincial cases (IEA, 2023; Jenkins et al., 2018).

Future research: Comparative, open-data system studies that stress-test flexibility portfolios—short- and long-duration storage, demand response, interconnection, grid-enhancing technologies, and clean firm resources—under extreme weather, correlated renewable droughts, and cyber-physical risks are needed (Blonsky et al., 2023; Sepulveda et al., 2018). Priority should be given to operational metrics (loss-of-load probability, ramping scarcity, inertia constraints) and market designs that remunerate fast, accurate flexibility (Hirth & Ziegenhagen, 2015).

Long-duration storage and seasonal balancing

Gap: While battery costs have fallen significantly, techno-economic pathways for seasonal storage and multi-day balancing (e.g., hydrogen, ammonia, compressed-air, pumped storage in constrained geographies) remain uncertain in terms of roundtrip efficiency, siting, and system value at scale (IEA, 2024; Bogdanov et al., 2021).

Future research: System co-optimization models linking power, hydrogen, and heat networks with realistic infrastructure constraints and permitting timelines; field-scale demonstration data on multi-year degradation, safety, and performance; and robust valuation frameworks for system adequacy contributions (Allison et al., 2023; Victoria et al., 2019).

Critical minerals, supply chains, and circularity

Gap: Decarbonization scenarios often assume unconstrained availability of critical minerals (lithium, nickel, cobalt, rare earths) and manufacturing capacity. Yet supply risks, geopolitical concentration, environmental impacts of mining, and end-of-life management are under-characterized in integrated assessments (IEA, 2021; Hund et al., 2020).

Future research: End-to-end material flow analyses that quantify demand under differing technology mixes; policy designs for recycling, repairability, and substitution; social and environmental safeguards in mining regions; and scenario stress tests of supply disruptions (Harper et al., 2019; Nuñez-López & Moskal, 2019).

Equity, just transition, and distributional outcomes

Gap: Evidence on employment gains is robust, but longitudinal evaluations of worker transitions, wage trajectories, and community-level fiscal impacts in fossil-dependent regions are scarce, especially in the Global South (ILO, 2022; Newell & Simms, 2020).

Future research: Mixed-methods studies tracking displaced workers over time; design and evaluation of place-based industrial strategies; and fiscal federalism approaches to stabilize local tax bases during coal, oil, and gas phase-downs (Caldecott et al., 2021; Sovacool et al., 2021). Research should prioritize gendered and intersectional impacts of energy access and clean cooking transitions (Gannon et al., 2020).

Finance mobilization and de-risking in emerging and developing economies

Gap: The cost of capital remains the single largest determinant of LCOE differentials across regions, yet empirical evidence on which de-risking instruments most effectively lower weighted average cost of capital (WACC) at scale is limited (Bhattacharya et al., 2022; IRENA, 2023).

Future research: Quasi-experimental and transaction-level analyses of blended finance, sovereign guarantees, currency hedges, and policy-based lending; standardized project data rooms to reduce due diligence frictions; and macroprudential frameworks that align green taxonomies with bank capital rules (OECD, 2023; World Bank, 2023).

Permitting, social license, and land-use governance

Gap: Time-to-permit has become a binding constraint in many jurisdictions, but rigorous, comparative evidence on the effects of permitting reforms, community benefit agreements, and participatory siting on both speed and legitimacy is nascent (Wüstenhagen et al., 2007; Rüdinger et al., 2023).

Future research: Natural experiments and process-tracing on reform packages (e.g., one-stop shops, spatial planning with biodiversity safeguards), cumulative impact assessments, and frameworks to reconcile conservation with scale-up via agrivoltaics, repowering, and brownfield siting (Hernandez et al., 2015; IPBES, 2019).

Coupling power with end-use sectors

Gap: Deep decarbonization requires electrification and fuel switching in hard-to-abate sectors (industry, heavy transport, aviation). Empirical system studies integrating renewables with green hydrogen, direct electrification, and carbon management are limited outside model exercises (Bataille, 2020; IEA, 2024).

Future research: Sector-coupled optimization with hourly resolution linking renewables to electrolyzers, district heat, and industrial clusters; infrastructure corridor planning for CO2, hydrogen, and electricity; and techno-economics of synthetic fuels with lifecycle accounting (Glenk & Reichelstein, 2019; Mac Dowell et al., 2017).

Resilience under compounding climate and cyber risks

Gap: Most planning studies inadequately incorporate compounding hazards—heatwaves coincident with droughts, wildfire smoke reducing solar output, or simultaneous line outages—nor do they fully address cyber-physical attack surfaces in digitalized grids (Chattopadhyay et al., 2021; Mills, 2021).

Future research: Probabilistic risk frameworks coupling climate downscaling with asset-level fragility functions; resilient design standards for inverters, substations, and DER aggregators; and market products that value resilience attributes (Palmintier et al., 2022).

Measurement, data, and transparency

Gap: Heterogeneous data standards impede cross-country learning on costs, performance, curtailment, and socio-environmental outcomes. Many impact claims remain difficult to verify (REN21, 2023; IEA, 2023).

Future research: Open, machine-readable datasets on hourly system operations, standardized socio-environmental indicators for project impacts, and shared evaluation protocols for policy instruments and auctions (Aklin & Urpelainen, 2018).

Clean cooking and thermal energy

Gap: Electricity access has advanced faster than access to clean cooking and thermal energy services, with persistent health and gender inequities (IEA et al., 2024; Rosenthal et al., 2018).

Future research: Scalable business models for modern fuels (LPG transition pathways to electric/induction, biogas, e-cooking), behavioral and cultural determinants, and integration with distributed solar to reduce energy poverty and indoor air pollution (Pillarisetti et al., 2021).

Governance of negative emissions and nature–energy interfaces

Gap: Net-zero strategies increasingly rely on carbon dioxide removal (CDR) and bioenergy with carbon capture and storage (BECCS), yet land, water, and biodiversity trade-offs remain poorly bounded empirically (Fuss et al., 2018; Smith et al., 2016).

Future research: Regional land-system models that co-optimize food, energy, carbon, and biodiversity; governance frameworks for MRV (monitoring, reporting, verification) of CDR; and guardrails that prioritize restoration and high-integrity offsets (IPCC, 2022).

Taken together, these gaps suggest a pivot from technology-centric questions to system, institutional, and justice-centered inquiries. The most urgent research is translational: converting targets into bankable projects, aligning flexibility services with market value, ensuring that communities benefit, and building institutional capacity to plan, permit, and operate resilient, deeply decarbonized energy systems. A research agenda that is interdisciplinary, comparative across regions, and anchored in high-quality open data will be central to ensuring that renewable energy technologies deliver not only rapid decarbonization but also broad-based development gains.

CONCLUSION

This chapter has examined the multifaceted relationship between renewable energy technologies and sustainable development, tracing a decade of accelerating change in energy systems, markets, and policy frameworks. Three findings stand out.

First, renewables have moved from the periphery to the center of global energy strategies. Rapid cost declines in solar photovoltaics and onshore wind, together with maturing supply chains and record annual capacity additions, have fundamentally altered investment calculus. In many markets, renewables now represent the cheapest source of new electricity, and their diffusion is increasingly driven by economics as much as by climate imperatives. The evidence reviewed—rising shares of zero-carbon generation, record annual additions of solar and wind, and expanding global employment—underscores that the energy transition is not aspirational but underway. The macro-level trend is clear: in scenarios consistent with current pledges and near-term market dynamics, fossil fuel demand plateaus and begins to decline, while renewables, complemented by nuclear in some systems, take on a growing share of generation. Although the speed varies by region, the direction of travel is broadly consistent across advanced, emerging, and developing economies.

Second, the transition is reshaping the contours of economic opportunity and social welfare. Renewable deployment is creating new jobs across manufacturing, construction, operations, and services, while strengthening energy security by reducing exposure to imported fuels and price shocks. These benefits, however, are unevenly distributed. Countries with supportive and predictable policy environments, robust grids, and access to affordable finance capture disproportionate gains. Communities tied to fossil fuel value chains confront dislocation risks that must be addressed through just transition policies, reskilling programs, and place-based industrial strategies. Social acceptance remains a decisive variable: projects that meaningfully engage local stakeholders, respect cultural and environmental sensitivities, and share tangible benefits are more likely to succeed. In low- and middle-income settings, distributed renewables are proving transformative for energy access, public health, and livelihoods—advancing SDG 7 while delivering co-benefits across SDGs 1, 3, 4, 5, 8, 10, and 13.

Third, systems and governance now form the frontier of the transition. As variable renewables scale, the challenges are increasingly about integration rather than technology cost: congestion, curtailment, flexibility needs, and market design.

Addressing these requires investments in transmission and distribution, advanced metering and controls, storage, demand-side flexibility, and cross-border interconnection. Market reforms that reward flexibility and internalize externalities—through carbon pricing, well-designed auctions, and transparent interconnection rules—are central to unlocking further investment. Policy durability and regulatory clarity are equally important: abrupt policy shifts and permitting bottlenecks impose hidden costs that undermine competitiveness and investor confidence, particularly in developing economies where institutional capacity constraints are pronounced.

Environmental outcomes reinforce the case for acceleration. Replacing coal and unabated gas generation with solar, wind, hydro, geothermal, and sustainable bioenergy reduces greenhouse gas emissions while delivering immediate local benefits: cleaner air, lower water withdrawals, and reduced ecological disruption when projects are sited and operated responsibly. These gains, however, hinge on careful planning to minimize land-use conflicts, protect biodiversity, and manage the material footprints of new technologies, including critical minerals and battery value chains. Lifecycle approaches, circular economy practices, and environmental safeguards should therefore move from the margins to the mainstream of sector planning.

Looking forward, three priorities emerge for research and practice. The first is deep integration: aligning technology deployment with market design, system operations, and digitalization to ensure reliability at high shares of variable renewables. The second is equity and inclusion: designing policies that expand access, distribute benefits fairly, and support workers and regions in transition. The third is institutional capability: strengthening planning, permitting, and regulatory functions to convert targets into bankable projects at scale. Progress on these fronts will determine whether the current momentum translates into durable, system-wide decarbonization consistent with the Paris Agreement and the 2030 Agenda.

In sum, renewable energy technologies have matured into engines of economic development, social progress, and environmental stewardship. They are not a singular solution, but they are indispensable to any credible pathway toward a resilient, low-carbon future. Realizing their full potential will depend less on technological breakthroughs than on policy coherence, infrastructural readiness, and the political will to manage change fairly. If these conditions are met, the transition can do more than decarbonize power—it can anchor a broader development model that is more secure, more inclusive, and more sustainable.

REFERENCES

Acheampong, M., Yu, Q., Cansu Ertem, F., Deba Enomah Ebude, L., Tanim, S., Eduful, M., Vaziri, M., & Ananga, E. (2019). Is Ghana ready to attain sustainable development goal (SDG) number 7?—A comprehensive assessment of its renewable energy potential and pitfalls. *Energies*, *12*(3), 408. DOI: 10.3390/en12030408

Ahuja, D., & Tatsutani, M. (2009). Sustainable energy for developing countries. SAPI EN. S. *Surveys and Perspectives Integrating Environment and Society*, *2*, 1.

Aklin, M., & Urpelainen, J. (2018). *Renewables: The politics of a global energy transition*. MIT Press. DOI: 10.7551/mitpress/11112.001.0001

Allison, T. L., Cole, W. J., & Frazier, A. W. (2023). The value of long-duration storage in decarbonized power systems. *Joule*, *7*(5), 1033–1052. DOI: 10.1016/j.joule.2023.03.012

Bataille, C. (2020). Physical and policy pathways to net-zero industry. *One Earth*, *3*(3), 258–265. DOI: 10.1016/j.oneear.2020.08.015

Bhattacharya, A., Callen, T., & Ralston, D. (2022). Financing a big investment push in emerging markets and developing economies. Brookings Institution. https://www.brookings.edu/

Blonsky, M., Hodge, B.-M., & Krishnamurthy, D. (2023). Ensuring reliability in grids with high variable renewable energy: A review of flexibility metrics and market mechanisms. *Renewable & Sustainable Energy Reviews*, *182*, 113394. DOI: 10.1016/j.rser.2023.113394

Bloomberg, N. E. F. (2023). Energy Transition Investment Trends 2023. https://about.bnef.com/energy-transition-investment/

Bogdanov, D., Ram, M., Aghahosseini, A., Gulagi, A., Oyewo, A. S., Child, M., Caldera, U., Sadovskaia, K., Farfan, J., De Souza Noel Simas Barbosa, L., Fasihi, M., Khalili, S., Traber, T., & Breyer, C. (2021). Low-cost renewable electricity as the key driver of the global energy transition towards sustainability. *Energy*, *227*, 120467. DOI: 10.1016/j.energy.2021.120467

Caldecott, B., Clark, A., & Zenghelis, D. (2021). Investing in a just transition. Grantham Research Institute. https://www.lse.ac.uk/granthaminstitute/

Chattopadhyay, D., Anandarajah, G., & McCalley, J. D. (2021). Power system planning with increasing variable renewable penetration: A review of emerging issues. *IEEE Systems Journal*, *15*(3), 4200–4211. DOI: 10.1109/JSYST.2020.3045647

Chhunji Sherpa, K. C., Satpati, G. G., Mal, N., Khalko, A. S., & Rajak, R. C. (2022). Effect of the COVID-19 on access to affordable and clean energy. In *COVID-19 and the Sustainable Development Goals* (pp. 79–104). Elsevier. DOI: 10.1016/B978-0-323-91307-2.00011-0

DNV. (2023). Energy Transition Outlook 2023 Executive Summary: A global and regional forecast to 2050. https://www.connaissancedesenergies.org/sites/connaissancedesenergies.org/files/pdf-actualites/DNV_ETO_Main_Report_2023_highres_comp.pdf

Erkan, A. Ç., & Yılmaz, H. (2022). Energy transition era and China. In Kısacık, S., & Güçyetmez, F. (Eds.), *Global Energy and Geopolitical Transformation* (pp. 119–141). İdeal Kültür Yayıncılık.

Frame, D., Dauenhauer, P., Eales, A., & Galloway, S. (2019, October). Sustainability of solar PV energy kiosks for off-grid energy access: Malawi case study. In *2019 IEEE Global Humanitarian Technology Conference (GHTC)* (pp. 1–8). IEEE. DOI: 10.1109/GHTC46095.2019.9033449

Fuss, S., Canadell, J. G., & Peters, G. P.. (2018). Negative emissions—Part 2: Costs, potentials and side effects. *Environmental Research Letters*, *13*(6), 063002. DOI: 10.1088/1748-9326/aabf9f

Fyson, C., Grant, N., Das, N., Maxwell, V., Reynolds, C., Rogelj, J. C.-F., Schleußner, J.-C., & Waterton, O. (2023). When will global greenhouse gas emissions peak? Climate Analytics. https://ca1-clm.edcdn.com/assets/When-will-global-greenhouse-gas-emissions-peak.pdf?v=1700638534

Gannon, K. E., Dubash, N. K., & Rajan, S. C. (2020). Making climate policy work for the poor. *Annual Review of Environment and Resources*, *45*, 29–59. DOI: 10.1146/annurev-environ-012320-083807

Glenk, G., & Reichelstein, S. (2019). Economics of converting renewable power to hydrogen. *Nature Energy*, *4*(3), 216–222. DOI: 10.1038/s41560-019-0326-1

Goldthau, A., & Sovacool, B. K. (2016). Energy technology, politics, and interpretative frames: Shale gas, energy security and the politics of expert knowledge. *Energy Policy*, *82*, 179–189. DOI: 10.1016/j.enpol.2015.03.004

Harper, G., Sommerville, R., Kendrick, E., Driscoll, L., Slater, P., Stolkin, R., Walton, A., Christensen, P., Heidrich, O., Lambert, S., Abbott, A., Ryder, K., Gaines, L., & Anderson, P. (2019). Recycling lithium-ion batteries from electric vehicles. *Nature*, *575*(7781), 75–86. DOI: 10.1038/s41586-019-1682-5 PMID: 31695206

Hernandez, R. R., Easter, S. B., & Murphy-Mariscal, M. L.. (2015). Environmental impacts of utility-scale solar energy. *Nature Climate Change*, *5*, 989–995. DOI: 10.1038/nclimate2637

Hirth, L., & Ziegenhagen, I. (2015). Balancing power and variable renewables. *Renewable & Sustainable Energy Reviews*, *50*, 1035–1051. DOI: 10.1016/j.rser.2015.04.180

Hostettler, S. (2015). Energy challenges in the Global South. In *Sustainable Access to Energy in the Global South: Essential Technologies and Implementation Approaches* (pp. 3–9). Springer International Publishing. DOI: 10.1007/978-3-319-20209-9_1

Hügel, J., & Pardo, R. (2017). Access to clean and efficient energy in developing countries: The need for EU action to implement SDG7. EPC Report.

Hund, K., La Porta, D., Fabregas, T. P., Laing, T., & Drexhage, J. (2020). Minerals for climate action: The mineral intensity of the clean energy transition. World Bank. https://documents.worldbank.org/

IEA. (2021). The role of critical minerals in clean energy transitions. International Energy Agency. https://www.iea.org/

IEA. (2023). Electricity market report 2023. International Energy Agency. https://www.iea.org/

IEA. (2023). World Energy Investment 2023. International Energy Agency. https://www.iea.org/reports/world-energy-investment-2023

IEA. (2023). World Energy Outlook 2023. International Energy Agency. https://www.iea.org/reports/world-energy-outlook-2023

IEA. (2024). World energy outlook 2024. Paris: IEA. https://www.iea.org/reports/world-energy-outlook-2024

IEA. IRENA, WHO, World Bank, & UN. (2024). Tracking SDG7: The energy progress report 2024. World Bank. https://trackingsdg7.esmap.org/

ILO. (2022). Skills for a Greener Future: A Global View. https://www.ilo.org/global/publications/books/WCMS_732214/lang--en/index.htm

International Energy Agency. (2021). Renewables 2021: Analysis and forecast to 2026. Paris: IEA. https://www.iea.org/reports/renewables-2021

International Renewable Energy Agency. (2020). Renewable energy policies in a time of transition: Heating and cooling. Abu Dhabi: IRENA. https://www.irena.org/publications

International Renewable Energy Agency. (2023). Renewable Energy and Jobs: Annual Review 2023. https://www.irena.org/publications/2023/Sep/Renewable-Energy-and-Jobs-Annual-Review-2023

International Renewable Energy Agency. (2023). World Energy Transitions Outlook 2023. https://www.irena.org/publications/2023/Mar/World-Energy-Transitions-Outlook-2023

International Renewable Energy Agency. (2023). World energy transitions outlook 2023: 1.5°C pathway. Abu Dhabi: IRENA. https://www.irena.org/publications

IPBES. (2019). *Global assessment report on biodiversity and ecosystem services*. Intergovernmental Science-Policy Platform on Biodiversity and Ecosystem Services.

IPCC. (2022). [Mitigation of climate change] [WG III] [. Cambridge University Press. https://www.ipcc.ch/]. *Climatic Change*, *2022*, •••.

Jacobson, M. Z., Delucchi, M. A., Bazouin, G., Bauer, Z. A., Heavey, C. C., Fisher, E., Morris, S. B., Piekutowski, D. J. Y., Vencill, T. A., & Yeskoo, T. W. (2015). 100% clean and renewable wind, water, and sunlight (WWS) all-sector energy roadmaps for the 50 United States. *Energy & Environmental Science*, *8*(7), 2093–2117. DOI: 10.1039/C5EE01283J

Jenkins, J. D., Luke, M., & Thernstrom, S. (2018). Getting to zero carbon emissions in the electric power sector. *Joule*, *2*(12), 2498–2510. DOI: 10.1016/j.joule.2018.11.013

Jenkins, J. D., Zhou, Z., Ponciroli, R., Vilim, R. B., Ganda, F., De Sisternes, F. J., & Sepulveda, N. A. (2018). The benefits of nuclear flexibility in power system operations with renewable energy. *Applied Energy*, *222*, 872–884. DOI: 10.1016/j.apenergy.2018.03.002

Kettner, C., Kletzan-Slamanig, D., Köppl, A., Littig, B., & Zielinska, I. (2019). Monitoring sustainable energy development: A cross-country comparison of selected EU members (No. 575). WIFO Working Paper.

Kısacık, S., & Güçyetmez, F. (2022). European energy security: Can the balanced energy mix within the European Union be accomplished in the 21st century? In Kısacık, S., & Güçyetmez, F. (Eds.), *Global Energy and Geopolitical Transformation* (pp. 93–118). İdeal Kültür Yayıncılık.

Lazard. (2023). Levelized Cost of Energy+, Version 17.0. https://www.lazard.com/research-insights/2023-levelized-cost-of-energyplus/

Leung, D. Y., & Yang, Y. (2012). Wind energy development and its environmental impact: A review. *Renewable & Sustainable Energy Reviews*, *16*(1), 1031–1039. DOI: 10.1016/j.rser.2011.09.024

Mac Dowell, N., Fajardy, M., Sádaba, I., & Fantuzzi, A. (2017). The role of CO2 capture and utilization. *Annual Review of Chemical and Biomolecular Engineering*, *8*, 277–294. DOI: 10.1146/annurev-chembioeng-060816-101334

McGlade, C., Gould, T., Bennett, S., Bredariol, T. D. O., Grimal, P., Hilaire, J., & Zeniewski, P. (2023). The Oil and Gas Industry in Net Zero Transitions. IEA WEO Special Report. https://iea.blob.core.windows.net/assets/f065ae5e-94ed-4fcb-8f17-8ceffde8bdd2/TheOilandGasIndustryinNetZeroTransitions.pdf

Mekonnen, M. M., & Hoekstra, A. Y. (2016). Four billion people facing severe water scarcity. *Science Advances*, *2*(2), e1500323. DOI: 10.1126/sciadv.1500323 PMID: 26933676

Mills, A. (2021). Impacts of smoke and haze on solar photovoltaic generation. *Solar Energy*, *221*, 88–96. DOI: 10.1016/j.solener.2021.04.005

Newell, P., & Simms, A. (2020). Towards a fossil fuel non-proliferation treaty. *Climate Policy*, *20*(8), 1043–1054. DOI: 10.1080/14693062.2019.1636759

Nuñez-López, V., & Moskal, E. (2019). Potential of CO2-EOR for near-term decarbonization. *Greenhouse Gases: Science and Technology*, *9*(6), 1096–1113. DOI: 10.1002/ghg.1935

OECD. (2023). Scaling up climate finance in developing countries. Organisation for Economic Co-operation and Development. https://www.oecd.org/

Palmintier, B., O'Neill, R., & Hodge, B.-M. (2022). Valuing resilience in electricity systems. *The Electricity Journal*, *35*(2), 107023. DOI: 10.1016/j.tej.2022.107023

Pillarisetti, A., Jamison, D. T., & Smith, K. R. (2021). Household energy, health, and climate. *Annual Review of Public Health*, *42*, 145–164. DOI: 10.1146/annurev-publhealth-090419-102255

REN21. (2022). Renewables 2022 global status report. Paris: REN21 Secretariat. https://www.ren21.net/reports/global-status-report/

REN21. (2023). Renewables 2023 Global Status Report. https://www.ren21.net/reports/global-status-report/

Rosenthal, J., Balakrishnan, K., & Bruce, N.. (2018). Clean cooking and the SDGs. *Energy for Sustainable Development*, *42*, 152–159. DOI: 10.1016/j.esd.2017.11.003 PMID: 29861575

Rüdinger, A., Berghmans, N., & Colombier, M. (2023). Accelerating renewable deployment through permitting reform in the EU. *Energy Policy*, *176*, 113488. DOI: 10.1016/j.enpol.2023.113488

Sepulveda, N. A., Jenkins, J. D., de Sisternes, F. J., & Lester, R. K. (2018). The role of firm low-carbon electricity resources. *Joule*, *2*(11), 2403–2420. DOI: 10.1016/j.joule.2018.08.006

Sherpa, K. C., Satpati, G. G., Mal, N., Khalko, A. S., & Rajak, R. C. (2022). Effect of the COVID-19 on access to affordable and clean energy. In *COVID-19 and the Sustainable Development Goals* (pp. 79–104). Elsevier. DOI: 10.1016/B978-0-323-91307-2.00011-0

Shukla, P. R., Skea, J., Slade, R., Al Khourdajie, A., van Diemen, R., McCollum, D., & Malley, J. (2022). [Mitigation of climate change. Contribution of Working Group III to the Sixth Assessment Report of the Intergovernmental Panel on Climate Change.]. *Climatic Change*, •••, 2022.

Smith, P., Davis, S. J., Creutzig, F., Fuss, S., Minx, J., Gabrielle, B., Kato, E., Jackson, R. B., Cowie, A., Kriegler, E., van Vuuren, D. P., Rogelj, J., Ciais, P., Milne, J., Canadell, J. G., McCollum, D., Peters, G., Andrew, R., Krey, V., & Yongsung, C. (2016). Biophysical and economic limits to negative CO2 emissions. *Nature Climate Change*, *6*(1), 42–50. DOI: 10.1038/nclimate2870

Sorrell, S. (2015). Reducing energy demand: A review of issues, challenges and approaches. *Renewable & Sustainable Energy Reviews*, *47*, 74–82. DOI: 10.1016/j.rser.2015.03.002

Sovacool, B. K. (2012). The political economy of energy poverty: A review of key challenges. *Energy for Sustainable Development*, *16*(3), 272–282. DOI: 10.1016/j.esd.2012.05.006

Sovacool, B. K., Hook, A., Martiskainen, M., & Baker, L. (2021). The decarbonisation divide: Contextualizing landscapes of low-carbon exploitation and inequality. *Global Environmental Change*, *70*, 102357. DOI: 10.1016/j.gloenvcha.2021.102357

Sovacool, B. K., Ryan, S. E., Stern, P. C., Janda, K., Rochlin, G., Spreng, D., & Pasqualetti, M. J. (2015). The clean energy revolution: Case studies in leadership from the coal, nuclear, and renewable energy transitions. *Energy Policy*, *77*, 1–6. DOI: 10.1016/j.enpol.2014.12.011

Stern, N. (2021). G7 leadership for sustainable, resilient and inclusive economic recovery and growth. Grantham Research Institute, LSE. https://www.lse.ac.uk/granthaminstitute/publication/g7-leadership-for-sustainable-resilient-and-inclusive-economic-recovery-and-growth/

Stiglitz, J. E., Stern, N., Duan, M., Edenhofer, O., Giraud, G., Heal, G., . . . Winkler, H. (2017). Report of the High-Level Commission on Carbon Prices. World Bank. https://www.carbonpricingleadership.org/report-of-the-highlevel-commission-on-carbon-prices

Thiam, D. R. (2011). Renewable energy, poverty alleviation and developing nations: Evidence from Senegal. *Journal of Energy in Southern Africa*, *22*(3), 23–34. DOI: 10.17159/2413-3051/2011/v22i3a3219

United Nations. (2023). *The Sustainable Development Goals Report 2023*. United Nations Publications.

United Nations Environment Programme. (2021). Global trends in renewable energy investment 2021. Nairobi: UNEP. https://www.unep.org/resources

Victoria, M., Zhu, K., & Brown, T.. (2019). Solar photovoltaics are ready to power a sustainable future. *Joule*, *3*(10), 2498–2513. DOI: 10.1016/j.joule.2019.09.018

Wei, M., Patadia, S., & Kammen, D. M. (2010). Putting renewables and energy efficiency to work: How many jobs can the clean energy industry generate in the US? *Energy Policy*, *38*(2), 919–931. DOI: 10.1016/j.enpol.2009.10.044

World Bank. (2023). Scaling solar and storage: Delivering bankable projects in emerging markets. World Bank Group. https://www.worldbank.org/

Wüstenhagen, R., Wolsink, M., & Bürer, M. J. (2007). Social acceptance of renewable energy innovation: An introduction to the concept. *Energy Policy*, *35*(5), 2683–2691. DOI: 10.1016/j.enpol.2006.12.001

ADDITIONAL READING

Derman, G. S. (Ed.). (2022). *International Energy Policies*. Cambridge Scholars Publishing.

Pamir, N. (2017). *Enerjinin İktidarı: Enerji Kaynaklarını Elinde Tutan Dünyayı Elinde Tutar! Güncellenmiş ve Genişletilmiş 3. Baskı*. Hayykitap.

Sevim, C. (2019). *Küresel Enerji Stratejileri ve Jeopolitik. Güncellenmiş 4. Baskı.* Seçkin Yayıncılık.

Yergin, D. (2020). *The New Map: Energy, Climate, and the Clash of Nations.* Penguin Press.

REFERENCES

Acheampong, M., Yu, Q., Cansu Ertem, F., Deba Enomah Ebude, L., Tanim, S., Eduful, M., Vaziri, M., & Ananga, E. (2019). Is Ghana ready to attain sustainable development goal (SDG) number 7?—A comprehensive assessment of its renewable energy potential and pitfalls. *Energies, 12*(3), 408. DOI: 10.3390/en12030408

Ahuja, D., & Tatsutani, M. (2009). Sustainable energy for developing countries. SAPI EN. S. *Surveys and Perspectives Integrating Environment and Society, 2*, 1.

Aklin, M., & Urpelainen, J. (2018). *Renewables: The politics of a global energy transition.* MIT Press. DOI: 10.7551/mitpress/11112.001.0001

Allison, T. L., Cole, W. J., & Frazier, A. W. (2023). The value of long-duration storage in decarbonized power systems. *Joule, 7*(5), 1033–1052. DOI: 10.1016/j.joule.2023.03.012

Bataille, C. (2020). Physical and policy pathways to net-zero industry. *One Earth, 3*(3), 258–265. DOI: 10.1016/j.oneear.2020.08.015

Bhattacharya, A., Callen, T., & Ralston, D. (2022). Financing a big investment push in emerging markets and developing economies. Brookings Institution. https://www.brookings.edu/

Blonsky, M., Hodge, B.-M., & Krishnamurthy, D. (2023). Ensuring reliability in grids with high variable renewable energy: A review of flexibility metrics and market mechanisms. *Renewable & Sustainable Energy Reviews, 182*, 113394. DOI: 10.1016/j.rser.2023.113394

Bloomberg, N. E. F. (2023). Energy Transition Investment Trends 2023. https://about.bnef.com/energy-transition-investment/

Bogdanov, D., Ram, M., Aghahosseini, A., Gulagi, A., Oyewo, A. S., Child, M., Caldera, U., Sadovskaia, K., Farfan, J., De Souza Noel Simas Barbosa, L., Fasihi, M., Khalili, S., Traber, T., & Breyer, C. (2021). Low-cost renewable electricity as the key driver of the global energy transition towards sustainability. *Energy, 227*, 120467. DOI: 10.1016/j.energy.2021.120467

Caldecott, B., Clark, A., & Zenghelis, D. (2021). Investing in a just transition. Grantham Research Institute. https://www.lse.ac.uk/granthaminstitute/

Carbon Pricing Chattopadhyay, D., Anandarajah, G., & McCalley, J. D. (2021). Power system planning with increasing variable renewable penetration: A review of emerging issues. *IEEE Systems Journal*, *15*(3), 4200–4211. DOI: 10.1109/JSYST.2020.3045647

Chhunji Sherpa, K. C., Satpati, G. G., Mal, N., Khalko, A. S., & Rajak, R. C. (2022). Effect of the COVID-19 on access to affordable and clean energy. In *COVID-19 and the Sustainable Development Goals* (pp. 79–104). Elsevier. DOI: 10.1016/B978-0-323-91307-2.00011-0

Critical Minerals, D. N. V. (2023). Energy Transition Outlook 2023 Executive Summary: A global and regional forecast to 2050. https://www.connaissancedesenergies.org/sites/connaissancedesenergies.org/files/pdf-actualites/DNV_ETO_Main_Report_2023_highres_comp.pdf

Derman, G. S. (Ed.). (2022). *International Energy Policies*. Cambridge Scholars Publishing.

Energy Access Energy Systems Integration Erkan, A. Ç., & Yılmaz, H. (2022). Energy transition era and China. In Kısacık, S., & Güçyetmez, F. (Eds.), *Global Energy and Geopolitical Transformation* (pp. 119–141). İdeal Kültür Yayıncılık. Flexibility.

Frame, D., Dauenhauer, P., Eales, A., & Galloway, S. (2019, October). Sustainability of solar PV energy kiosks for off-grid energy access: Malawi case study. In *2019 IEEE Global Humanitarian Technology Conference (GHTC)* (pp. 1–8). IEEE. DOI: 10.1109/GHTC46095.2019.9033449

Fuss, S., Canadell, J. G., & Peters, G. P.. (2018). Negative emissions—Part 2: Costs, potentials and side effects. *Environmental Research Letters*, *13*(6), 063002. DOI: 10.1088/1748-9326/aabf9f

Fyson, C., Grant, N., Das, N., Maxwell, V., Reynolds, C., Rogelj, J. C.-F., Schleußner, J.-C., & Waterton, O. (2023). When will global greenhouse gas emissions peak? Climate Analytics. https://ca1-clm.edcdn.com/assets/When-will-global-greenhouse-gas-emissions-peak.pdf?v=1700638534

Gannon, K. E., Dubash, N. K., & Rajan, S. C. (2020). Making climate policy work for the poor. *Annual Review of Environment and Resources*, *45*, 29–59. DOI: 10.1146/annurev-environ-012320-083807

Glenk, G., & Reichelstein, S. (2019). Economics of converting renewable power to hydrogen. *Nature Energy*, *4*(3), 216–222. DOI: 10.1038/s41560-019-0326-1

Goldthau, A., & Sovacool, B. K. (2016). Energy technology, politics, and interpretative frames: Shale gas, energy security and the politics of expert knowledge. *Energy Policy*, *82*, 179–189. DOI: 10.1016/j.enpol.2015.03.004

Harper, G., Sommerville, R., Kendrick, E., Driscoll, L., Slater, P., Stolkin, R., Walton, A., Christensen, P., Heidrich, O., Lambert, S., Abbott, A., Ryder, K., Gaines, L., & Anderson, P.Grid Congestion and Curtailment Harper. (2019). Recycling lithium-ion batteries from electric vehicles. *Nature*, *575*(7781), 75–86. DOI: 10.1038/s41586-019-1682-5 PMID: 31695206

Hernandez, R. R., Easter, S. B., & Murphy-Mariscal, M. L.. (2015). Environmental impacts of utility-scale solar energy. *Nature Climate Change*, *5*, 989–995. DOI: 10.1038/nclimate2637

Hirth, L., & Ziegenhagen, I. (2015). Balancing power and variable renewables. *Renewable & Sustainable Energy Reviews*, *50*, 1035–1051. DOI: 10.1016/j.rser.2015.04.180

Hostettler, S. (2015). Energy challenges in the Global South. In *Sustainable Access to Energy in the Global South: Essential Technologies and Implementation Approaches* (pp. 3–9). Springer International Publishing. DOI: 10.1007/978-3-319-20209-9_1

Hügel, J., & Pardo, R. (2017). Access to clean and efficient energy in developing countries: The need for EU action to implement SDG7. EPC Report.

Hund, K., La Porta, D., Fabregas, T. P., Laing, T., & Drexhage, J. (2020). Minerals for climate action: The mineral intensity of the clean energy transition. World Bank. https://documents.worldbank.org/

IEA. (2021). The role of critical minerals in clean energy transitions. International Energy Agency. https://www.iea.org/

IEA. (2023). Electricity market report 2023. International Energy Agency. https://www.iea.org/

IEA. (2023). World Energy Investment 2023. International Energy Agency. https://www.iea.org/reports/world-energy-investment-2023

IEA. (2023). World Energy Outlook 2023. International Energy Agency. https://www.iea.org/reports/world-energy-outlook-2023

IEA. IRENA, WHO, World Bank, & UN. (2024). Tracking SDG7: The energy progress report 2024. World Bank. https://trackingsdg7.esmap.org/

IEA. (2024). World energy outlook 2024. Paris: IEA. https://www.iea.org/reports/world-energy-outlook-2024

IEA. (2025). Global Energy Review 2025. https://iea.blob.core.windows.net/assets/5b169aa1-bc88-4c96-b828-aaa50406ba80/GlobalEnergyReview2025.pdf

ILO. (2022). Skills for a Greener Future: A Global View. https://www.ilo.org/global/publications/books/WCMS_732214/lang--en/index.htm

International Energy Agency. (2021). Renewables 2021: Analysis and forecast to 2026. Paris: IEA. https://www.iea.org/reports/renewables-2021

International Renewable Energy Agency. (2020). Renewable energy policies in a time of transition: Heating and cooling. Abu Dhabi: IRENA. https://www.irena.org/publications

International Renewable Energy Agency. (2023). Renewable Energy and Jobs: Annual Review 2023. https://www.irena.org/publications/2023/Sep/Renewable-Energy-and-Jobs-Annual-Review-2023

International Renewable Energy Agency. (2023). World Energy Transitions Outlook 2023. https://www.irena.org/publications/2023/Mar/World-Energy-Transitions-Outlook-2023

International Renewable Energy Agency. (2023). World energy transitions outlook 2023: 1.5°C pathway. Abu Dhabi: IRENA. https://www.irena.org/publications

IPBES. (2019). *Global assessment report on biodiversity and ecosystem services*. Intergovernmental Science-Policy Platform on Biodiversity and Ecosystem Services.

IPCC. (2022). [Mitigation of climate change] [WG III] [. Cambridge University Press. https://www.ipcc.ch/]. *Climatic Change*, *2022*, •••.

Jacobson, M. Z., Delucchi, M. A., Bazouin, G., Bauer, Z. A., Heavey, C. C., Fisher, E., Morris, S. B., Piekutowski, D. J. Y., Vencill, T. A., & Yeskoo, T. W. (2015). 100% clean and renewable wind, water, and sunlight (WWS) all-sector energy roadmaps for the 50 United States. *Energy & Environmental Science*, *8*(7), 2093–2117. DOI: 10.1039/C5EE01283J

Jenkins, J. D., Luke, M., & Thernstrom, S. (2018). Getting to zero carbon emissions in the electric power sector. *Joule*, *2*(12), 2498–2510. DOI: 10.1016/j.joule.2018.11.013

Jenkins, J. D., Zhou, Z., Ponciroli, R., Vilim, R. B., Ganda, F., De Sisternes, F. J., & Sepulveda, N. A. (2018). The benefits of nuclear flexibility in power system operations with renewable energy. *Applied Energy*, *222*, 872–884. DOI: 10.1016/j.apenergy.2018.03.002

Kettner, C., Kletzan-Slamanig, D., Köppl, A., Littig, B., & Zielinska, I. (2019). Monitoring sustainable energy development: A cross-country comparison of selected EU members (No. 575). WIFO Working Paper.

Kısacık, S., & Güçyetmez, F. (2022). European energy security: Can the balanced energy mix within the European Union be accomplished in the 21st century? In Kısacık, S., & Güçyetmez, F. (Eds.), *Global Energy and Geopolitical Transformation* (pp. 93–118). İdeal Kültür Yayıncılık.

Lazard. (2023). Levelized Cost of Energy+, Version 17.0. https://www.lazard.com/research-insights/2023-levelized-cost-of-energyplus/

Leung, D. Y., & Yang, Y. (2012). Wind energy development and its environmental impact: A review. *Renewable & Sustainable Energy Reviews*, *16*(1), 1031–1039. DOI: 10.1016/j.rser.2011.09.024

Mac Dowell, N., Fajardy, M., Sádaba, I., & Fantuzzi, A. (2017). The role of CO2 capture and utilization. *Annual Review of Chemical and Biomolecular Engineering*, *8*, 277–294. DOI: 10.1146/annurev-chembioeng-060816-101334

McGlade, C., Gould, T., Bennett, S., Bredariol, T. D. O., Grimal, P., Hilaire, J., & Zeniewski, P. (2023). The Oil and Gas Industry in Net Zero Transitions. IEA WEO Special Report. https://iea.blob.core.windows.net/assets/f065ae5e-94ed-4fcb-8f17-8ceffde8bdd2/TheOilandGasIndustryinNetZeroTransitions.pdf

Mekonnen, M. M., & Hoekstra, A. Y. (2016). Four billion people facing severe water scarcity. *Science Advances*, *2*(2), e1500323. DOI: 10.1126/sciadv.1500323 PMID: 26933676

Mills, A. (2021). Impacts of smoke and haze on solar photovoltaic generation. *Solar Energy*, *221*, 88–96. DOI: 10.1016/j.solener.2021.04.005

Newell, P., & Simms, A. (2020). Towards a fossil fuel non-proliferation treaty. *Climate Policy*, *20*(8), 1043–1054. DOI: 10.1080/14693062.2019.1636759

Nuñez-López, V., & Moskal, E. (2019). Potential of CO2-EOR for near-term decarbonization. *Greenhouse Gases: Science and Technology*, *9*(6), 1096–1113. DOI: 10.1002/ghg.1935

OECD. (2023). Scaling up climate finance in developing countries. Organisation for Economic Co-operation and Development. https://www.oecd.org/

Palmintier, B., O'Neill, R., & Hodge, B.-M. (2022). Valuing resilience in electricity systems. *The Electricity Journal*, *35*(2), 107023. DOI: 10.1016/j.tej.2022.107023

Pamir, N. (2017). *Enerjinin İktidarı: Enerji Kaynaklarını Elinde Tutan Dünyayı Elinde Tutar! Güncellenmiş ve Genişletilmiş 3. Baskı*. Hayykitap.

Pillarisetti, A., Jamison, D. T., & Smith, K. R. (2021). Household energy, health, and climate. *Annual Review of Public Health*, *42*, 145–164. DOI: 10.1146/annurev-publhealth-090419-102255

REN21. (2022). Renewables 2022 global status report. Paris: REN21 Secretariat. https://www.ren21.net/reports/global-status-report/

REN21. (2023). Renewables 2023 Global Status Report. https://www.ren21.net/reports/global-status-report/

Renewable Energy Technologies Resilience Rosenthal, J., Balakrishnan, K., & Bruce, N.. (2018). Clean cooking and the SDGs. *Energy for Sustainable Development*, *42*, 152–159. DOI: 10.1016/j.esd.2017.11.003 PMID: 29861575

Rüdinger, A., Berghmans, N., & Colombier, M. (2023). Accelerating renewable deployment through permitting reform in the EU. *Energy Policy*, *176*, 113488. DOI: 10.1016/j.enpol.2023.113488

Sector Coupling Sepulveda, N. A., Jenkins, J. D., de Sisternes, F. J., & Lester, R. K. (2018). The role of firm low-carbon electricity resources. *Joule*, *2*(11), 2403–2420. DOI: 10.1016/j.joule.2018.08.006

Sevim, C. (2019). *Küresel Enerji Stratejileri ve Jeopolitik. Güncellenmiş 4. Baskı*. Seçkin Yayıncılık.

Shukla, P. R., Skea, J., Slade, R., Al Khourdajie, A., van Diemen, R., McCollum, D., & Malley, J. (2022). [Mitigation of climate change. Contribution of Working Group III to the Sixth Assessment Report of the Intergovernmental Panel on Climate Change. Small, often customer-sited assets—rooftop solar, behind-the-meter batteries, smart loads, EVs—that can supply, store, or modulate energy and provide grid services.]. *Climatic Change*, •••, 2022.

Smith, P., Davis, S. J., Creutzig, F., Fuss, S., Minx, J., Gabrielle, B., Kato, E., Jackson, R. B., Cowie, A., Kriegler, E., van Vuuren, D. P., Rogelj, J., Ciais, P., Milne, J., Canadell, J. G., McCollum, D., Peters, G., Andrew, R., Krey, V., & Yongsung, C. (2016). Biophysical and economic limits to negative CO2 emissions. *Nature Climate Change*, *6*(1), 42–50. DOI: 10.1038/nclimate2870

Sorrell, S. (2015). Reducing energy demand: A review of issues, challenges and approaches. *Renewable & Sustainable Energy Reviews*, *47*, 74–82. DOI: 10.1016/j.rser.2015.03.002

Sovacool, B. K. (2012). The political economy of energy poverty: A review of key challenges. *Energy for Sustainable Development*, *16*(3), 272–282. DOI: 10.1016/j.esd.2012.05.006

Sovacool, B. K., Hook, A., Martiskainen, M., & Baker, L. (2021). The decarbonisation divide: Contextualizing landscapes of low-carbon exploitation and inequality. *Global Environmental Change*, *70*, 102357. DOI: 10.1016/j.gloenvcha.2021.102357

Sovacool, B. K., Ryan, S. E., Stern, P. C., Janda, K., Rochlin, G., Spreng, D., & Pasqualetti, M. J. (2015). The clean energy revolution: Case studies in leadership from the coal, nuclear, and renewable energy transitions. *Energy Policy*, *77*, 1–6. DOI: 10.1016/j.enpol.2014.12.011

Stern, N. (2021). G7 leadership for sustainable, resilient and inclusive economic recovery and growth. Grantham Research Institute, LSE. https://www.lse.ac.uk/granthaminstitute/publication/g7-leadership-for-sustainable-resilient-and-inclusive-economic-recovery-and-growth/

Stiglitz, J. E., Stern, N., Duan, M., Edenhofer, O., Giraud, G., Heal, G., . . . Winkler, H. (2017). Report of the High-Level Commission on Carbon Prices. World Bank. https://www.carbonpricingleadership.org/report-of-the-highlevel-commission-on-carbon-prices

Thiam, D. R. (2011). Renewable energy, poverty alleviation and developing nations: Evidence from Senegal. *Journal of Energy in Southern Africa*, *22*(3), 23–34. DOI: 10.17159/2413-3051/2011/v22i3a3219

United Nations. (2023). *The Sustainable Development Goals Report 2023*. United Nations Publications.

United Nations Environment Programme. (2021). Global trends in renewable energy investment 2021. Nairobi: UNEP. https://www.unep.org/resources

Victoria, M., Zhu, K., & Brown, T.. (2019). Solar photovoltaics are ready to power a sustainable future. *Joule*, *3*(10), 2498–2513. DOI: 10.1016/j.joule.2019.09.018

Wei, M., Patadia, S., & Kammen, D. M. (2010). Putting renewables and energy efficiency to work: How many jobs can the clean energy industry generate in the US? *Energy Policy*, *38*(2), 919–931. DOI: 10.1016/j.enpol.2009.10.044

World Bank. (2023). Scaling solar and storage: Delivering bankable projects in emerging markets. World Bank Group. https://www.worldbank.org/

Wüstenhagen, R., Wolsink, M., & Bürer, M. J. (2007). Social acceptance of renewable energy innovation: An introduction to the concept. *Energy Policy, 35*(5), 2683–2691. DOI: 10.1016/j.enpol.2006.12.001

Yergin, D. (2020). *The New Map: Energy, Climate, and the Clash of Nations.* Penguin Press.

KEY TERMS AND DEFINITIONS

Carbon Pricing: Taxes or trading systems that assign a cost to greenhouse gas emissions, aligning private incentives with climate goals and interacting with renewable policies.

Circularity (Circular Economy in Energy): Lifecycle strategies—design for longevity, repair, reuse, high-yield recycling—to reduce material footprints for batteries, PV modules, and wind components.

Clean Cooking: Solutions that cut indoor air pollution and climate impacts versus traditional biomass, including LPG transition, electric/induction, biogas, and advanced stoves.

Critical Minerals: Minerals vital to clean tech (lithium, nickel, cobalt, rare earths) with supply, environmental, and social risks; priorities include diversification, recycling, and substitution.

Distributed Energy Resources (DERs): Small, often customer-sited assets—rooftop solar, behind-the-meter batteries, smart loads, EVs—that can supply, store, or modulate energy and provide grid services.

Energy Access: Reliable, affordable, modern energy for electricity and clean cooking; measured by quality, affordability, safety, and support for productive uses.

Energy Systems Integration: Coordinated planning and operation of power, heat, transport, and industry with grids, storage, and digital platforms to efficiently integrate variable renewables.

Flexibility: The system's capability to respond to rapid supply–demand changes via storage, demand response, interconnection, grid-enhancing tech, and clean firm generation.

Grid Congestion and Curtailment: Insufficient transmission capacity and the resulting reduction of renewable output, signaling needs for upgrades, storage, flexible demand, or market reform.

Just Transition: Policies ensuring the shift from fossil fuels is fair and inclusive, covering worker protection, reskilling, community revitalization, and equitable cost–benefit sharing.

Levelized Cost of Electricity (LCOE): Average lifetime cost per unit of electricity from a technology; useful for comparisons but excludes system integration costs and values.

Long-Duration Energy Storage (LDES): Storage delivering energy for many hours to seasons (pumped hydro, flow batteries, hydrogen/ammonia, compressed air) to balance prolonged variability.

Renewable Energy Technologies: Technologies that convert replenishing resources (sun, wind, water, geothermal, sustainable biomass) into energy, including utility-scale and distributed assets often paired with digital controls and storage.

Resilience: The energy system's ability to withstand and recover from shocks—extreme weather, cyber-physical attacks—through hardening, diversification, and valuing resilience.

SDG 7 (Affordable and Clean Energy): Goal to ensure universal access to affordable, reliable, sustainable, modern energy, covering electricity and clean cooking access, efficiency, and renewable share.

Sector Coupling: Linking electricity with buildings, transport, and industry (heat pumps, EVs, green hydrogen) to raise flexibility, efficiency, and emissions reductions.

Social License to Operate: Ongoing community acceptance of projects earned through transparent engagement, fair benefits, and respect for environmental, land, and cultural rights.

Sustainable Development: Development that meets present needs without compromising future generations, integrating environmental integrity, economic prosperity, and social equity in line with the SDGs and Paris Agreement.

Variable Renewable Energy (VRE): Weather-dependent resources, chiefly wind and solar PV, requiring forecasting, flexibility, storage, fast-ramping supply, and transmission to manage variability.

Chapter 14
Global Pathways to Net Zero:
Comparative Analysis of Corporate Carbon Performance Across Sectors and Countries

İhsan Erdem Kayral
https://orcid.org/0000-0002-8335-8619
Başkent University, Turkey

Melike Aktaş Bozkurt
https://orcid.org/0000-0001-5207-0615
Ostim Technical University, Turkey

Tiago Silveira Gontijo
https://orcid.org/0000-0003-2636-899X
Federal University of São João del-Rei, Brazil

Gustavo de Souza Groppo
https://orcid.org/0000-0003-1302-4987
Sanitation Company of Minas Gerais (Copasa), Brazil

Leonardo Goliatt
https://orcid.org/0000-0002-2844-9470
Universidade Federal de Juiz de Fora, Brazil

ABSTRACT

This study evaluated corporate alignment with the 2050 Net Zero Emissions targets, a cornerstone of the global transition to a low-carbon economy as mandated by the

DOI: 10.4018/979-8-3373-5238-1.ch014

Paris Agreement. Utilizing the Transition Pathway Initiative (TPI) framework, as endorsed by OECD reports, we analyzed the carbon performance of 2,028 firms across 24 sectors and 28 countries. Companies are scored on a 0–5 scale, with 5 indicating the highest alignment with net-zero goals. Through descriptive statistics, statistical analyses, and hierarchical clustering, we assessed performance variations by firm size, sector, and countries, comparing their alignment with Paris-Compliant Pathways. Results reveal that larger firms, sectors such as telecommunications and oil & gas, and European countries like Denmark and France exhibit superior carbon management, while smaller firms, carbon-intensive sectors like coal mining, and regions like China and the Middle East lag. These findings reveal sectoral and regional disparities, offering actionable insights to guide decarbonization efforts by policymakers and investors.

INTRODUCTION

The concept of corporate carbon performance has evolved significantly since the early 1990s with the establishment of the United Nations Framework Convention on Climate Change (UNFCCC) in 1992, which laid the groundwork for international climate action. The Kyoto Protocol in 1997 introduced binding emission reduction targets for developed countries, prompting corporations to begin tracking and reporting greenhouse gas (GHG) emissions. The Paris Agreement in 2015 marked a pivotal shift by emphasizing voluntary nationally determined contributions ((s) and encouraging corporate alignment with global goals. Initiatives like the Science Based Targets initiative (SBTi) in 2015 and the Transition Pathway Initiative (TPI) in 2017 emerged to provide standardized methodologies for assessing and guiding corporate decarbonization efforts, reflecting a progression from awareness to strategic implementation in response to escalating climate urgency.

Achieving the Paris Agreement's climate goals of limiting global warming to well below 2°C above pre-industrial levels, with efforts to cap the increase at 1.5°C (United Nations, 2015), demands unprecedented reductions in GHG emissions across all sectors. In the short term, these efforts align with Sustainable Development Goal 13 (SDG 13: Climate Action), which emphasizes urgent actions to combat climate change and its impacts through enhanced adaptive capacity, strengthened resilience, and reduced emissions (United Nations General Assembly, 2015). Achieving the stabilization of global temperatures in line with the Paris Agreement goal requires net GHG emissions to decline at unprecedented rates (IPCC, 2018). The urgency of this transition is underscored by the need for global net-zero emissions by 2050, a target that requires corporations to align their strategies with science-based pathways to avoid ambiguous or insufficient commitments (Rogelj et al., 2021). SDG

13 underscores the importance of immediate corporate actions, such as integrating climate risk management and reporting into business practices, to lay the foundation for long-term decarbonization goals. Recent assessments highlight that global progress toward net-zero remains uneven, with corporate contributions critical to closing the emissions gap. As Blanchard et al. (2025) argue, many voluntary net-zero frameworks lack the scientific rigor needed to drive authentic decarbonization. They caution against overreliance on carbon offsets and advocate for stricter corporate climate standards that prioritize direct emissions reductions, as well as improved transparency and accountability. Companies have a key role to play in achieving this goal. However, many companies are unable to significantly reduce their greenhouse gas emissions due to the high costs of transitioning to a low-carbon economy (Kuramochi et al., 2020). Estimates suggest that global spending on net-zero carbon initiatives could reach US$275 trillion by 2050, equivalent to approximately 7.5% of global GDP (McKinsey & Company, 2022). These financial burdens are compounded by operational complexities, particularly in carbon-intensive sectors like coal mining and cement, where technological and infrastructural barriers limit rapid decarbonization (Li et al., 2022). These financial challenges underscore the need for robust frameworks to guide corporate action and align business strategies with global climate targets (Kayral & Aktaş Bozkurt, 2025). Recent work by Bendig et al. (2025) emphasize the importance of utilizing standardized, multidimensional metrics, including absolute emissions, intensity, and verification procedures, to accurately assess corporate carbon performance. The authors emphasize that reliance on unverified self-reporting can lead to significant misjudgments, underscoring the need for transparent and credible carbon reporting frameworks to guide effective climate action.

To effectively address climate change, there needs to be increased awareness of climate risks among corporate executives, regulators, investors and other stakeholders, and that companies are managed in a way that will remain profitable during the transition (Millar et al., 2018). Despite growing commitments, with over 4,000 companies globally pledging net-zero targets by 2022, inconsistencies in ambition and implementation persist, particularly across countries and sectors (Hale et al., 2022). Ensuring the credibility of these commitments remains a challenge, as many firms in high-emission sectors overstate their progress without sufficient action. A consistent, transparent and defensible framework is also needed to translate global climate targets into actionable, company-specific emissions reduction pathways. However, as shown by Jiang et al. (2025), nearly one-third of companies silently abandoned their publicly announced emissions targets, with minimal market or reputational consequences. This accountability gap highlights a pressing need for rigorous monitoring systems, such as TPI, to ensure that corporate pledges are not only ambitious but also effectively enforced.

Several methodologies have been developed to translate global climate targets into specific emissions trajectories for each company. The Sectoral Decarbonization Approach (SDA) was the first methodology adopted. It was co-developed by the Science Based Targets initiative in 2015 (an initiative of the Carbon Disclosure Project, Worldwide Fund for Nature, UN Global Compact and World Resources Institute). The SDA adopts a sectoral approach, using the International Energy Agency (IEA) decarbonization pathways (OECD/IEA, 2017), given that different sectors of the economy face different challenges regarding the transition to a low-carbon economy. The SDA compares companies in the same sector with each other and with sectoral benchmarks, establishing the performance of an average company aligned with the international emissions targets under the Paris Agreement. For instance, Abdo et al. (2025) demonstrate that even within high-emitting sectors such as oil and gas, there are sharp disparities in preparedness, with smaller firms facing acute risks of asset impairments and write-downs as they transition to net-zero, a reality often overlooked in broad-brush corporate climate disclosures.

While defining Paris-Compliant Pathways (PCPs) and their operational requirements is critical, it is insufficient without metrics to evaluate corporate performance. There is, therefore, a need to create metrics for this assessment. There are two recent initiatives that aim to do this and assess the company's transition performance using the SDA method: i)TPI; (Dietz et al., 2018; Jahn et al., 2023) and ii) Low Carbon Transitions Assessment (Accelerate Climate Transitions (ACT), 2025).

TPI was created in January 2017 and is a global initiative led by asset owners and supported by asset managers. In 2022, the TPI Centre (2022) was established at the London School of Economics and Political Science (LSE), within the Grantham Research Institute on Climate Change and the Environment. The TPI evaluates companies on the following aspects: i) the quality of governance and management of greenhouse gas emissions by companies and the risks and opportunities related to the low carbon transition and ii) carbon emissions by companies against different climate scenarios consistent with the UN Paris Agreement, comparing companies in high-emission sectors with each other and with sector-specific benchmarks (Jahn et al., 2023). Recent refinements to TPI's methodology have enhanced its ability to assess alignment with 1.5°C pathways, providing investors with more granular insights into corporate transition plans. The assessment of management quality focuses on processes, while the assessment of performance focuses on commitment to decarbonization. Together, they aim to provide a measure of companies' progress in the low-carbon transition.

Considering the 2050 Net Zero Emissions objectives, this study compared the carbon performance of corporations in various industries and countries. To accomplish this goal, we use the TPI levels, a crucial tool for evaluating carbon performance in OECD reports (2024a, 2024b). The OECD's "Review on Aligning Finance with

Climate Targets" (2024a) emphasizes the significance of thorough, evidence-based assessments to guide climate-aligned policy and investment. By leveraging TPI's standardized metrics, this study addresses the gap in understanding how corporate performance varies across diverse sectoral and regional contexts, providing insights critical for policymakers and investors to drive equitable and effective decarbonization strategies (Li et al., 2022). Furthermore, cross-country evidence by Jiang et al. (2025) suggest that the national context, including legal systems, media freedom, and disclosure mandates, strongly influences whether firms achieve, delay, or abandon their emissions targets, making country-level benchmarking a vital component in evaluating global carbon alignment.

While existing literature provides valuable insights into corporate climate commitments, accountability gaps, and sector-specific challenges, there remains a significant research gap in conducting comprehensive, cross-sectoral and cross-country comparative analyses using standardized metrics like TPI scores. Prior studies, such as those by Jiang et al. (2025) and Abdo et al. (2025), often focus on specific sectors or regions, limiting the understanding of global patterns and disparities in carbon performance. This study fills this gap by leveraging TPI data to systematically compare performance across 24 sectors and 28 countries, highlighting variations that can inform targeted decarbonization strategies.

This study is grounded in theoretical frameworks such as the Resource-Based View (RBV), which posits that firms' competitive advantages stem from unique resources like advanced carbon management capabilities, and Stakeholder Theory, which emphasizes the role of external pressures from investors, regulators, and society in driving corporate climate actions. Additionally, Institutional Theory explains variations in performance across countries due to differing regulatory and normative pressures. These frameworks underpin the analysis of TPI scores, illustrating how internal resources and external contexts influence decarbonization strategies.

The foundations of corporate carbon performance rest on science-based targets derived from IPCC reports and IEA scenarios, which provide benchmarks for emissions reductions. Tools like the SDA and TPI build on these by translating global goals into sector-specific pathways, ensuring comparability and accountability. This study leverages these foundations to evaluate firm alignment with 1.5°C pathways.

Assessing corporate carbon performance is crucial for mitigating climate risks, ensuring long-term profitability, and contributing to global sustainability goals. It enables investors to identify low-risk opportunities, policymakers to design effective regulations, and companies to enhance resilience against transition risks, ultimately supporting the achievement of net-zero emissions by 2050.

Specifically, the study's objectives are to:

i) Classify the companies assessed by the TPI based on size, country of origin, and economic sector, using data available as of January 2025.
ii) Provide descriptive statistics on the carbon performance of these classified companies, offering an initial overview of the observed trends and patterns.
iii) Apply statistical analyses to compare carbon performance across different sectors and countries based on the TPI classifications.
iv) Evaluate the usefulness of the TPI framework as a tool for comparative analysis in the context of low-carbon transition, particularly for policy development and investment decision-making.

This chapter is structured as follows: Section 1 introduces the study and contextualizes the topic and its objectives. Section 2 presents literature review, Section 3 describes the databases used, focusing on TPI, and details the methodological design adopts to conduct the analyses. Section 4 presents the results and discussions; Section 5 includes conclusion.

LITERATURE REVIEW

Recent studies highlight a growing gap between corporate climate commitments and credible action. For instance, Blanchard et al. (2025) argued that many voluntary net-zero pledges and frameworks lack scientific rigor, often allowing companies to rely excessively on carbon offsets rather than achieving real emissions reductions. These authors call for more stringent standards that prioritize direct emissions reductions and improved transparency. The concern is that without robust standards, net-zero promises may amount to "greenwashing", giving firms public relations benefits without guaranteeing meaningful decarbonization.

Empirical evidence from recent studies reinforces these concerns about accountability. Jiang et al. (2025) analyzed the outcomes of 1,041 corporate emissions-reduction targets that were due by 2020, providing one of the first large-scale assessments of how firms perform on their climate goals. Their analysis revealed that while about 60% of companies met their 2020 targets, roughly 9% outright failed, and another 31% quietly ceased reporting progress, effectively abandoning their pledges. Crucially, Jiang and colleagues found no significant market penalties or reputational consequences for firms that missed or dropped their climate goals: only a handful of the failures received any media coverage, and there were no notable declines in stock prices, media sentiment, or ESG ratings following missed targets (Jiang et al., 2025). In contrast, simply announcing ambitious targets tended to boost a company's media image and sustainability ratings. This "accountability gap" suggests that many firms have enjoyed the positive press of climate pledges without the risk of punishment for non-delivery. The highest target-meeting success rates have been observed in companies headquartered in countries with stronger press freedom and stakeholder scrutiny, underscoring how external accountability can drive better climate performance. Overall, Jiang et al. conclude that current oversight is insufficient and call for stronger institutional mechanisms, akin to auditing and reporting standards in financial markets, to ensure firms are held accountable for climate promises.

Given these gaps, researchers emphasize the need for rigorous metrics and verification to track corporate decarbonization progress. Bendig et al. (2025) introduce a comprehensive framework for assessing corporate carbon performance (CCP) that standardizes the measurement of emissions outcomes across firms. Their study defines CCP along multiple dimensions, including total emissions, emissions intensity (emissions per economic output), and the robustness of a company's carbon data and disclosures, to make performance truly comparable. Using this framework, Bendig et al. evaluate companies' alignment with climate targets and find that data quality is a critical issue: verified emissions data are far more reliable than self-reported numbers, which often contain errors or optimistic estimates. They demonstrate that firms using unverified, self-reported emissions can appear to be performing better than they are, leading to misjudgments about progress. The study's findings have important implications for practitioners and policymakers. Bendig et al. urge companies and investors to use multiple metrics and independent verification to track carbon performance, rather than relying on a single self-reported indicator. They argue that only through transparent reporting and third-party validation can we obtain an accurate picture of which firms are genuinely on track to meet climate goals. Regulators, accordingly, are encouraged to mandate more rigorous disclosure and verification practices to improve data quality and uphold the integrity of net-zero commitments. By highlighting measurement inconsistencies, this research underscores the importance of robust assessment tools (such as the Transition Pathway Initiative or Science-Based Targets) for distinguishing "apples from oranges" in corporate climate performance (Bendig et al., 2025).

Beyond measuring commitments, scholars are also examining the actions underlying corporate climate pledges to understand why many firms remain off-track. Cenci et al. (2023) developed a novel text-analysis approach to systematically track the climate actions of major companies and assess their alignment with the Paris Agreement. Analyzing disclosures and reports from thousands of publicly traded firms, they found a notable disconnect between companies' stated climate goals, the actions they implemented, and their actual emissions outcomes. This study suggests that many firms focus disproportionately on risk mitigation and incremental efficiency improvements, measures that protect the business or optimize processes, rather than on transformative initiatives such as clean technology innovation or cross-sector collaboration. This imbalance helps explain why even with an influx of corporate climate pledges in recent years, the emissions trajectories of roughly 80% of large companies remain misaligned with the Paris temperature targets. Cenci et al. reported that as of 2020, only about 19% of ~13,600 large companies had emissions pathways consistent with limiting warming to well below 2°C. The majority are falling behind, likely because they prioritize easier or short-term measures over the innovative, systemic changes needed to cut emissions substantially. These findings highlight that simply setting a target is not enough; the nature of corporate actions matters. Firms investing in deeper decarbonization strategies (such as renewable energy projects, low-carbon product innovation, or supply chain collaboration) are more likely to achieve real emissions reductions. In contrast, those who treat climate goals as a risk-management checkbox may see little change in outcomes. The research by Cenci et al. provides a framework for monitoring corporate climate behavior and urges investors and policymakers to redirect capital and incentives toward firms taking genuinely impactful actions. In summary, this work illuminates the root causes of the implementation gap. It suggests that a reorientation toward innovation and bolder action is required for corporate pledges to translate into measurable reductions in emissions.

Studies have also explored how these dynamics play out in specific high-emitting sectors and regions. For example, Abdo et al. (2025) examine the oil and gas industry's financial readiness for net-zero transitions, using a combination of content analysis of company reports and in-depth interviews with industry insiders. Focusing on UK oil and gas firms, they found that many companies, especially smaller operators with older fossil fuel assets, are underestimating the economic risks posed by the clean energy transition. These firms often do not fully account for impending asset impairments and write-downs that could occur as climate policies tighten and demand for hydrocarbons declines. The study highlights several looming challenges for the sector, including reduced access to capital, higher borrowing costs, and the prospect of stranded assets as carbon-intensive operations become less viable. Notably, Abdo et al. point out a transparency deficit: current financial reporting and sustainability disclosures by oil and gas companies rarely convey the accurate scale of climate transition risks. Standard climate disclosure frameworks (such as the TCFD guidelines or the ISSB standards) provide general guidance. However, they often fail to capture the unique asset-level and revenue risks that fossil-fuel companies face in a zero-carbon future. Interviewees in the study called for much more precise, granular disclosures, for instance, detailing remaining reserves, expected write-downs, and how business models are shifting, so that investors and stakeholders can accurately gauge a firm's long-term viability in a decarbonizing world. In their findings, Abdo et al. emphasize that smaller oil and gas firms with limited diversification are most vulnerable to these transition pressures. Without frank acknowledgement of these risks, such companies may be overvalued and ill-prepared, ultimately posing threats not only to their shareholders but also to the broader energy sector and even public revenues. This sector-specific insight complements the broader literature by illustrating how climate performance and risk can vary dramatically even within the same industry. It underlines the importance of tailored, sector-specific metrics and transition plans; a one-size-fits-all approach to corporate climate responsibility may obscure critical vulnerabilities in certain industries. Moreover, the work by Abdo et al. reinforces calls for stronger disclosure rules: by making firms openly account for transition-related financial impacts, markets can better distinguish leaders from laggards and channel investment accordingly, thereby facilitating a more orderly transition in high-risk sectors (Abdo et al., 2025).

In summary, the recent literature paints a sobering yet instructive picture of corporate climate efforts. There is a consensus that while corporate climate commitments have proliferated, implementation and accountability remain inconsistent. Top-tier studies stress the need for greater rigor in climate target-setting (such as clearer standards and the avoidance of dubious offsets), stronger accountability mechanisms to ensure firms follow through on their pledges, and improved metrics and transparency to measure progress accurately. Researchers are increasingly leveraging large datasets and innovative methods, from statistical analyses of emissions targets to textual analysis of sustainability reports, to evaluate not just what companies promise, but what they deliver. The literature also makes it clear that context matters: differences in regulatory environments, stakeholder pressures, firm size, and sectoral challenges all influence how well companies align with global climate goals. These insights set the stage for evaluating new empirical analyses that compare corporate climate performance across firms, sectors, and countries. By building on the gaps and recommendations identified in recent research, we can better understand where corporations stand on the path to net-zero and how to strengthen policy and industry practices to bridge the current ambition–action gap.

DATA AND METHODOLOGY

Data

This study aims to conduct a comparative analysis of sectors and countries by focusing on significant companies within the framework of the 2050 Net Zero Emissions targets. To achieve this purpose, we utilize TPI levels, a key resource evaluated for assessing carbon performance in the OECD reports (2024a, 2024b), which underscore the importance of robust, evidence-based assessments for informing climate-aligned investments and policymaking, as emphasized in the OECD's "Review on Aligning Finance with Climate Goals" (2024a).

The TPI framework, first developed by the London School of Economics (LSE) in 2017, provides a standardized measure of corporate carbon performance, and this study employs the most recent values calculated as of January 2025. The OECD (2024b) recognizes TPI as a reliable tool, citing its "State of Transition in the Banking Sector Report" for its contribution to understand climate alignment in finance, while the broader purpose of the TPI is to support investors and companies in transition to a low-carbon economy aligned with the 1.5°C target of the Paris Agreement.

The TPI provides a comprehensive framework for assessing companies' preparedness for the transition to a low-carbon economy. TPI levels are computed for companies with a global impact, assigning each a corporate score ranging from 0 to 5. A score of 0 indicates the lowest carbon management quality, while 5 represents the highest alignment with the 2050 Net Zero Emissions targets. These scores are determined based on the following criteria:

Level 0 – Unaware (or does not recognize): climate change as a business issue,
Level 1 – Recognizes climate change as a business issue: These changes present risks and/or opportunities for the business and companies have a responsibility to manage their emissions,
Level 2 – Building capacity: The company develops its capabilities, systems and management processes and begins to report on practices and performance,
Level 3 – Integration into operational decision-making: The company improves its operational practices, holds senior management/board accountable for changes and provides comprehensive disclosures on its practices and performance,
Level 4 – Strategic assessment: The company develops a strategic understanding of the risks and opportunities related to the low-carbon transition, integrating this into the corporate business strategy,
Level 5 – Transition planning and implementation: The company uses its strategic understanding to create a detailed transition plan that aligns business practices and capital expenditure decisions with its decarbonization goals (Jahn et al., 2023) (see Figure 1).

Figure 1. TPI Scoring Framework for Corporate Climate Transition

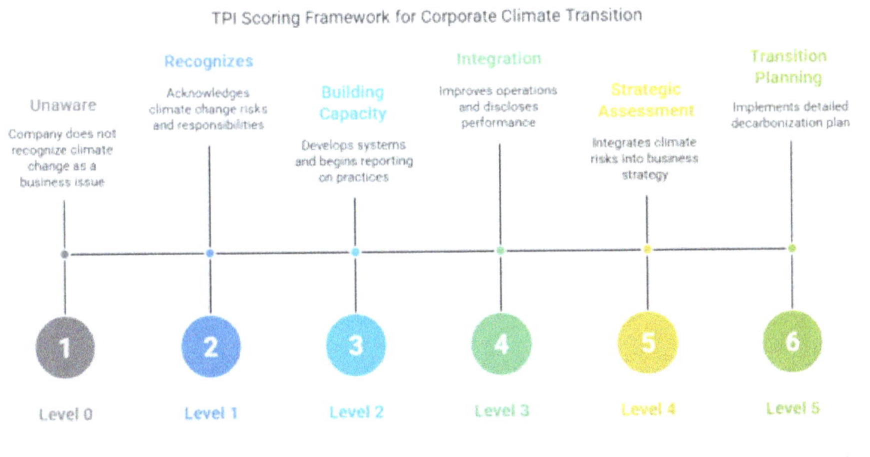

In this study, we analyzed 2,028 firms with available TPI scores, selected from the 2,066 firms reviewed by the LSE based on data availability as of January 2025. Using these scores, we classified the companies by size, country, and sector, as presented in the dataset. Descriptive statistics are derived from this classification to provide an initial overview of performance trends. Additionally, we apply statistical analyses using the R programming language to compare carbon performance across sectors and countries, enhancing the depth of our findings.

Methodology

Statistical analyses are performed using the R programming language to ensure rigor and reproducibility (R Core Team, 2024). The methodology encompasses both descriptive and inferential techniques to assess differences in TPI scores across sectors and countries. The Shapiro-Wilk test evaluates normality, followed by the Kruskal-Wallis test for group differences. Then, hierarchical clustering groups sectors and countries based on TPI score similarity, using Euclidean distance and complete linkage criteria.

Descriptive Statistics

Initial insights are drawn from mean TPI scores, standard deviations, and frequency distributions of firms across TPI levels (0–5), segmented by firm size, sector, and country. These statistics highlight performance trends and variability, providing a foundation for subsequent analyses (Moore et al., 2017).

Normality Testing

The Shapiro-Wilk Normality Test is applied to assess whether TPI scores for sectors and countries follow a normal distribution (Shapiro & Wilk, 1965). In the Equation (1), the test statistic SW is calculated as:

$$SW = \frac{\left(\sum_{i=1}^{n} a_i x_{(i)}\right)^2}{\sum_{i=1}^{n} (x_i - \bar{x})^2} \tag{1}$$

, where $x_{(i)}$ are the ordered observations, a_i are constants derived from the expected values of ordered statistics, x_i are the sample observations, and $\bar{x}$ is the sample mean. The null hypothesis assumes normality, and a p-value below 0.05 indicates a non-normal distribution, guiding the selection of non-parametric tests.

Kruskal-Wallis Chi-Squared Test

Given the non-normal distribution of TPI scores, the Kruskal-Wallis Chi-Squared Test is employed to test for significant differences in TPI scores across multiple groups (sectors or countries) (Kruskal & Wallis, 1952). In the Equation (2), the test statistic H is computed as:

$$H = \frac{12}{N(N+1)} \sum_{i=1}^{k} \frac{R_i^2}{n_i} - 3(N+1) \tag{2}$$

, where k is the number of groups, n_i is the number of observations in the i-th group, N is the total number of observations, and R_i is the sum of ranks for the i-th group. Ranks are assigned by combining all observations and ordering them from smallest to largest. The null hypothesis assumes identical distributions across groups. The statistic H follows an approximate chi-squared distribution with $k-1$ degrees of freedom, and a p-value below 0.05 indicates significant differences between at least one sub-group and others.

Hierarchical Clustering

To group sectors and countries into performance tiers based on TPI scores, hierarchical clustering is applied (Everitt et al., 2011). The method uses a distance metric, typically Euclidean distance, defined in Equation (3) for two observations x_i and x_j as:

$$d(x_i, x_j) = \sqrt{\sum_{m=1}^{p} (x_{im} - x_{jm})^2} \tag{3}$$

, where p is the number of variables (e.g., mean TPI scores). In Equation (4), the clustering algorithm employs a linkage criterion, such as complete linkage, which defines the distance between clusters C_a and C_b as:

$$D(C_a, C_b) = \max_{x_i \in C_a, x_j \in C_b} d(x_i, x_j) \tag{4}$$

The algorithm iteratively merges clusters to minimize the increase in within-cluster variance, producing a dendrogram. The dendrogram is cut at an appropriate height to form distinct clusters, grouping sectors or countries with similar TPI score distributions.

RESULTS & DISCUSSIONS

Descriptive Statistics

The distribution of companies by size reveals that large companies exhibit the highest average TPI score (3.11), followed by medium-sized (2.92) and small (2.65) companies, with unlisted companies having the lowest average (2.52) (see Figure 2). This trend suggests that larger companies tend to demonstrate superior carbon management quality, possibly due to their extensive resources and complex organizational structures, greater access to resources, and adherence to stricter regulatory requirements.

Figure 2. TPI Levels by Firm Size

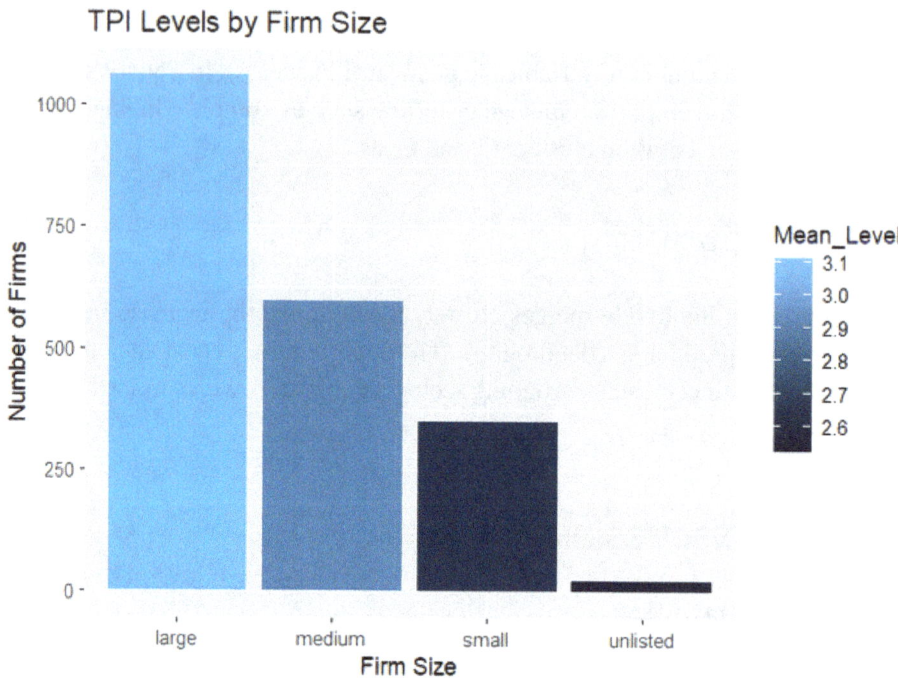

Table 1 presents detailed insights into this distribution, showing the number of firms at each TPI level within each size category. For large companies, which number 1,060 in the sample, the distribution is heavily skewed toward higher TPI levels. Notably, 555 firms are at Level 3, 180 at Level 4, and 127 at Level 5, with only 1 firm placing at Level 0 and 117 at Level 1. This concentration at Levels 3 through 5 explains the high mean score of 3.11. It suggests that large firms are not only more likely to engage in carbon management but also to achieve advanced stages of integration and strategic alignment with climate goals, possibly driven by greater regulations.

Table 1. TPI Levels by Firm Size

Classification	Number of Firms	Level 0	Level 1	Level 2	Level 3	Level 4	Level 5	Mean
Large	1060	1	117	80	555	180	127	3.11
Medium	595	6	74	48	335	97	35	2.92
Small	350	2	49	59	208	25	7	2.65
Unlisted	23	0	6	4	10	1	2	2.52

Medium-sized companies, totaling 595 firms, show a similar trend, with a lower mean TPI score of 2.92. Here, 335 firms are at Level 3, 97 at Level 4, and 35 at Level 5, while the lower end includes 6 firms at Level 0 and 74 at Level 1. This distribution indicates that medium-sized firms still perform well, though they fall short of the consistency seen in large firms. Their slightly lower mean score could reflect constraints in resources or less pressure from external stakeholders compared to their larger counterparts, yet they remain competitive in adopting carbon management practices.

Small companies, with 350 firms and a mean TPI score of 2.65, exhibit greater variability. Table 1 shows 208 firms at Level 3, but only 25 at Level 4 and 7 at Level 5, with 49 at Level 1 and 59 at Level 2. This spread suggests that while many small firms are making efforts, fewer achieve the advanced practices of Levels 4 and 5. Smaller firms with limited organizational capacity and financial resources struggle to match the carbon management quality of larger entities.

Finally, unlisted companies, though a small group of 23, have the lowest mean TPI score of 2.52. Table 1 reveals 10 firms at Level 3, but only 1 at Level 4 and 2 at Level 5, with 6 at Level 1 and 4 at Level 2.

According to Table 2, the total sample of 2,028 firms yields an overall mean TPI score of 2.97, with 1,108 firms at Level 3, 303 at Level 4, and 171 at Level 5. This broad distribution, likely summarized in Figure 2, suggests that while many firms achieve moderate carbon management (Level 3), fewer reach the advanced stages (Levels 4 and 5), a pattern influenced by sector-specific dynamics.

Table 2. TPI Levels by Sector

Sector	Total Firms	Level 0	Level 1	Level 2	Level 3	Level 4	Level 5	Mean
Airlines	39	0	6	3	12	11	7	3.26
Aluminium	31	0	6	1	13	8	3	3.03
Autos	38	0	4	6	17	6	5	3.05
Basic Materials	42	0	13	1	21	3	4	2.62
Cement	71	4	9	7	38	7	6	2.75
Chemicals	136	1	26	9	85	4	11	2.72
Coal Mining	53	0	13	8	24	6	2	2.55
Consumer Goods	138	0	25	12	71	20	10	2.84
Consumer Services	163	1	19	18	94	27	4	2.85
Diversified Mining	31	0	1	4	16	4	6	3.32
Electricity Utilities	150	0	17	14	68	27	24	3.18
Financials	161	0	9	18	86	39	9	3.13

continued on following page

Table 2. Continued

Sector	Total Firms	Level 0	Level 1	Level 2	Level 3	Level 4	Level 5	Mean
Food Producers	81	0	7	7	58	7	2	2.88
Health Care	105	2	8	15	59	12	9	2.93
Industrials	325	0	40	33	192	47	13	2.88
Oil & Gas	96	0	4	7	54	15	16	3.33
Oil & Gas (Other)	32	0	5	4	18	3	2	2.78
Oil & Gas Distri.	27	0	1	5	14	2	5	3.19
Paper	36	0	5	1	21	5	4	3.06
Shipping	33	0	6	3	17	3	4	2.88
Steel	77	0	12	9	43	9	4	2.79
Technology	110	1	9	2	60	24	14	3.26
Telecommunications	39	0	1	2	18	13	5	3.49
Utilities	14	0	0	2	9	1	2	3.21
Total	**2028**	**9**	**246**	**191**	**1108**	**303**	**171**	**2.97**

It has been determined that, of the 24 sectors listed in Table 2, 12 have mean TPI scores above 3, while 12 have mean TPI scores below 3. The sector-wise distribution indicates that telecommunications (3.49), oil and gas (3.33), and diversified mining (3.32) sectors achieve the highest average TPI scores, reflecting superior carbon performance, possibly due to stringent regulations or significant investments in innovation, while the coal mining (2.55) and basic materials (2.62) record the lowest averages, pointing to specific operational or environmental challenges within these industries.

Table 2 details the performance of each sector, including the total number of firms and their distribution across TPI levels. For telecommunications, with 39 firms and a mean of 3.49, the data shows a strong tilt toward higher levels: 18 firms at Level 3, 13 at Level 4, and 5 at Level 5, with only 1 at Level 1 and 2 at Level 2.

This concentration at the upper end of the TPI scale suggests that telecommunications firms, possibly improved by innovation and less emissions-intensive operations, integrating carbon management into their strategies. Similarly, the oil and gas sector (96 firms, mean 3.33) has 54 firms at Level 3, 15 at Level 4, and 16 at Level 5, reflecting robust performance despite its carbon-intensive nature, perhaps driven by significant investments in cleaner technologies or stringent regulatory oversight.

Diversified mining (31 firms, mean 3.32) follows telecommunications and oil & gas sectors, with 16 firms at Level 3, 4 at Level 4, and 6 at Level 5, indicating that diversification may mitigate some environmental challenges compared to more specialized extractive industries.

In contrast, coal mining (53 firms, mean 2.55) and basic materials (42 firms, mean 2.62) position the lower end. Coal mining has 24 firms at Level 3 but only 6 at Level 4 and 2 firms at Level 5, with 13 at Level 1 and 8 at Level 2, underscoring the sector's struggle to move beyond basic compliance, likely due to its heavy reliance on fossil fuels and limited decarbonization options. Basic materials show a similar pattern, with 21 firms at Level 3 but just 3 at Level 4 and 4 at Level 5, and a notable 13 at Level 1, pointing to operational constraints or weaker incentives for advancing carbon management.

Figure 3. TPI Levels by Sector

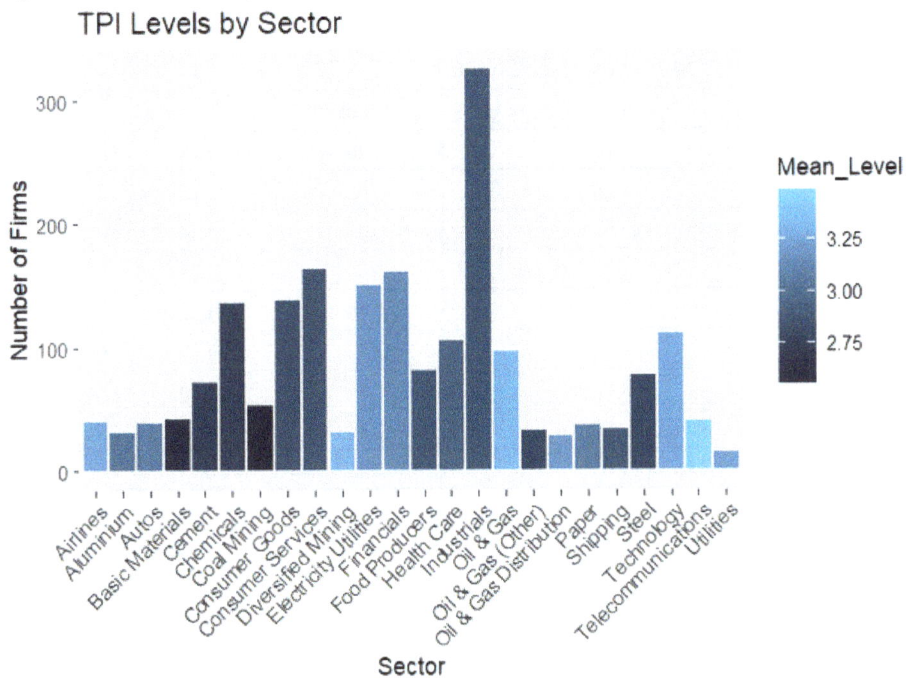

Figure 3 captures this disparity, likely plotting the mean TPI scores across sectors. This graphical representation in Figure 3 provides the sectoral leaders and laggards. Carbon management quality varies significantly by industry, possibly influenced by regulatory pressures, technological capabilities, or market expectations.

Table 3 shows depth by including standard deviations, revealing variability within sectors. Telecommunications (mean 3.49, s. dev. 0.89) and financials (mean 3.13, S. Dev. 0.89) exhibit low dispersion, indicating consistent performance across firms, possibly due to uniform regulatory frameworks or industry-wide adoption of best practices. Food producers (mean 2.88, s. dev. 0.78) also show tight clustering, suggesting a standardized approach to carbon management within this sector.

Conversely, airlines (mean 3.26, s. dev. 1.29) and basic materials (mean 2.62, s. dev. 1.27) display high variability. Airlines sector (39 firms) shows 12 firms at Level 3, 11 at Level 4, and 7 at Level 5, but also 6 at Level 1 and 3 at Level 2, reflecting the performance where some firms excel while others lag, possibly due to differing efficiencies or investment capacities. Basic materials' high standard deviation aligns with its scattered TPI distribution in Table 3, highlighting inconsistent adoption of carbon strategies across firms, perhaps tied to varying sub-sector challenges or resource availability.

Table 3. Descriptive Statistics of TPI Sectors

Sector	Mean	S. Dev.	N
Airlines	3.26	1.29	39
Aluminium	3.03	1.22	31
Autos	3.05	1.14	38
Basic Materials	2.62	1.27	42
Cement	2.75	1.23	71
Chemicals	2.72	1.09	136
Coal Mining	2.55	1.10	53
Consumer Goods	2.84	1.11	138
Consumer Services	2.85	0.94	163
Diversified Mining	3.32	1.05	31
Electricity Utilities	3.18	1.16	150
Financials	3.13	0.89	161
Food Producers	2.88	0.78	81
Health Care	2.93	1.04	105
Industrials	2.88	0.94	325
Oil & Gas	3.33	0.98	96
Oil & Gas (Other)	2.78	1.04	32
Oil & Gas Distribution	3.19	1.08	27
Paper	3.06	1.09	36
Shipping	2.88	1.19	33
Steel	2.79	1.02	77
Technology	3.26	1.05	110
Telecommunications	3.49	0.89	39
Utilities	3.21	0.89	14

Table 4 provides the raw data to substantiate observations, offering a detailed distribution of firms per country across TPI levels (0 to 5), which allows us to assess the trends. Analysis by country demonstrates that European nations such as Denmark (4.14), France (3.79), and Germany (3.61), achieve the highest average TPI scores, likely due to stringent regulatory environments and effective governance practices, while developed countries, including Japan (3.38), Canada (3.27), and Australia (3.30) show solid but intermediate performance. Emerging economies like Brazil (3.21) and India (2.76) exhibit greater variability, indicative of diverse structural challenges. In contrast, China (1.86) and Middle East countries such as Saudi Arabia (1.59) and Qatar (1.20), record the lowest averages, possibly reflecting weaker regulatory standards. The United States (2.99) displays average performance with significant dispersion among companies. Notably, countries like Colombia (4.00) and Greece (4.00) show high scores tempered by small number of firms analyzed.

Table 4. TPI Levels by Countries

Country	Total Firms	Level 0	Level 1	Level 2	Level 3	Level 4	Level 5	Mean
Australia	56	0	1	5	33	10	7	3.30
Austria	5	0	0	0	1	3	1	4.00
Belgium	3	0	0	0	2	1	0	3.33
Brazil	29	0	1	3	17	5	3	3.21
Canada	62	0	2	3	38	14	5	3.27
Chile	8	0	1	1	4	1	1	3.00
China	330	1	151	73	104	0	1	1.86
Colombia	4	0	0	0	2	0	2	4.00
Czechia	1	0	0	0	0	1	0	4.00
Denmark	7	0	0	0	2	2	3	4.14
European Union*	1	0	0	0	1	0	0	3.00
Finland	10	0	0	0	3	4	3	4.00
France	47	0	0	0	23	11	13	3.79
Germany	46	0	1	0	25	10	10	3.61
Greece	3	0	0	0	1	1	1	4.00
Hong Kong	26	0	1	2	16	6	1	3.15
Hungary	1	0	0	0	0	1	0	4.00
India	72	0	13	5	42	10	2	2.76
Indonesia	17	0	3	5	9	0	0	2.35
Ireland	3	0	0	0	2	0	1	3.67

continued on following page

Table 4. Continued

Country	Total Firms	Level 0	Level 1	Level 2	Level 3	Level 4	Level 5	Mean
Israel	5	0	0	1	3	1	0	3.00
Italy	21	0	0	0	10	8	3	3.67
Japan	221	0	1	2	156	37	25	3.38
Kuwait	1	0	0	0	1	0	0	3.00
Malaysia	21	0	5	4	11	1	0	2.38
Mexico	16	1	2	0	11	0	2	2.81
Netherlands	15	0	0	0	9	4	2	3.53
New Zealand	5	0	0	0	2	3	0	3.60
Nigeria	1	0	0	1	0	0	0	2.00
Norway	8	0	0	0	4	2	2	3.75
Philippines	10	0	2	3	5	0	0	2.30
Poland	7	0	0	2	5	0	0	2.71
Portugal	5	0	0	0	2	1	2	4.00
Qatar	5	0	4	1	0	0	0	1.20
Romania	1	0	0	0	1	0	0	3.00
Saudi Arabia	22	4	9	3	5	0	1	1.59
Singapore	11	0	0	1	8	1	1	3.18
South Africa	26	0	1	0	12	6	7	3.69
South Korea	47	0	0	1	29	7	10	3.55
Spain	18	0	0	0	8	3	7	3.94
Sweden	20	0	1	0	8	8	3	3.60
Switzerland	22	0	0	0	16	3	3	3.41
Taiwan	58	0	2	5	39	5	7	3.17
Thailand	20	0	0	0	17	1	2	3.25
Turkey	10	0	0	1	3	4	2	3.70
United Arab Emirates	8	0	4	1	3	0	0	1.88
UK	64	0	0	0	30	17	17	3.80
USA	629	3	41	68	385	111	21	2.99

For Denmark, with 7 firms and a mean TPI score of 4.14, Table 4 shows 2 firms at Level 3, 2 at Level 4, and 3 at Level 5, with no firms below Level 3. This tight clustering at the upper end suggests that Denmark's strong environmental policies and small, manageable corporate landscape foster exceptional carbon management consistency. France (47 firms, mean 3.79) follows with 23 firms at Level 3, 11 at Level 4, and 13 at Level 5, and Germany (46 firms, mean 3.61) has 25 at Level 3,

10 at Level 4, and 10 at Level 5, both with minimal presence at lower levels. These distributions highlight how robust EU regulations and national commitments to climate goals likely drive firms toward advanced carbon strategies, as seen in the significant share of firms at Levels 4 and 5.

Other developed countries like Japan, Canada, and Australia show a different profile. Japan (221 firms, mean 3.38) has 156 firms at Level 3, 37 at Level 4, and 25 at Level 5, with only 3 firms below Level 3. Canada (62 firms, mean 3.27) has 38 at Level 3, 14 at Level 4, and 5 at Level 5, and Australia (56 firms, mean 3.30) records 33 at Level 3, 10 at Level 4, and 7 at Level 5. These intermediate scores, with a heavy weighting toward Level 3, suggest that while these countries enforce solid standards, fewer firms reach the cutting-edge practices of Levels 4 and 5, possibly due to less aggressive mandates compared to top European performers.

Emerging economies present a mixed picture. Brazil (29 firms, mean 3.21) has 17 firms at Level 3, 5 at Level 4, and 3 at Level 5, but also 4 firms at Levels 1 and 2, indicating variability that may stem from uneven industrial development or regulatory enforcement. India (72 firms, mean 2.76) shows 42 firms at Level 3, 10 at Level 4, and 2 at Level 5, but 13 at Level 1 and 5 at Level 2, reflecting a broader spread that likely ties to structural challenges like resource constraints or diverse sectoral priorities. In contrast, China (330 firms, mean 1.86) stands out with 151 firms at Level 1, 73 at Level 2, and 104 at Level 3, with just 1 firm each at Levels 0 and 5. This low average and concentration at lower levels suggest that despite its economic might, China's regulatory focus may prioritize growth over stringent carbon management, limiting firm advancement.

Middle East countries like Saudi Arabia (22 firms, mean 1.59) and Qatar (5 firms, mean 1.20) show even worse performance than China. Saudi Arabia has 4 firms at Level 0, 9 at Level 1, and 3 at Level 2, with only 5 at Level 3 and only 1 at Level 5, while Qatar has 4 firms at Level 1 and 1 at Level 2, with no firms above Level 2. These scores are likely to reflect a reliance on fossil fuel economies and weaker climate policies, constraining firms' carbon management efforts.

With the small number of firms (samples), Colombia (4 firms, mean 4.00) and Greece (3 firms, mean 4.00) achieve high averages, with Colombia having 2 firms at Level 3 and 2 at Level 5, and Greece showing 1 firm each at Levels 3, 4, and 5. However, their small sample sizes limit the generalization, as these figures may not represent broader national trends but rather the performance of a few number of firms.

Sector and Country Analysis Results

Following the descriptive statistics, we conduct statistical analyses to compare carbon performance across sectors and countries using the R programming language (R Core Team, 2024). We have analyzed the average TPI scores for 24 sectors, encompassing all 2,028 firms, and included 28 countries, each represented by at least 10 companies, in the comparative analyses.

Firstly, we test the assumption of normality using the Shapiro-Wilk test. The null hypothesis of normal distribution is rejected for both sectors (SW = 0.95204, $p < 0.001$) and countries (SW = 0.97371, $p < 0.001$), indicating that the data do not follow a normal distribution. Consequently, we employ the non-parametric Kruskal-Wallis chi-squared test to assess differences in TPI scores. The results, presented in Table 5, confirm significant differences among sectors ($\chi^2 = 83.692$, df = 23, $p < 0.001$) and countries ($\chi^2 = 630.47$, df = 27, $p < 0.001$).

Table 5. Statistical Tests for Sectors and Countries

Category	Test	Statistic	Value	Degrees of Freedom (df)	p-value
Sectors	Shapiro-Wilk Normality Test	SW	0.95204	-	<0.001
	Kruskal-Wallis Chi-Squared Test	Chi-squared	83.692	23	<0.001
Countries	Shapiro-Wilk Normality Test	SW	0.97371	-	<0.001
	Kruskal-Wallis Chi-Squared Test	Chi-squared	630.47	27	<0.001

We then apply hierarchical clustering, grouping sectors and countries into distinct clusters based on their TPI scores. Hierarchical clustering analyses reveal distinct performance tiers among sectors and countries.

Table 6 organizes 24 sectors into five distinct groups based on their average TPI scores, ranging from high-performing Group I (3.26–3.49) to low-performing Group V (2.55–2.62). Group I includes airlines, diversified mining, oil & gas, technology, and telecommunications, with TPI scores between 3.26 and 3.49. This top tier, anchored by telecommunications (3.49) and oil & gas (3.33), suggests that these sectors benefit from advanced carbon management strategies, possibly driven by technological innovation or heightened regulatory scrutiny. Telecommunications' high score may reflect its lower emissions footprint and ability to adopt digital efficiencies, while oil & gas's performance could stem from strategic investments in decarbonization despite its carbon-intensive nature.

Group II, with scores from 3.03 to 3.21, comprises aluminium, autos, electricity utilities, financials, oil & gas distribution, paper, and utilities. This middle-high tier indicates solid but less exceptional performance, likely reflecting a mix of regulatory compliance and sector-specific capabilities. Electricity utilities (3.18) and financials (3.13), for instance, may leverage structured reporting requirements and risk management frameworks to achieve these scores.

Table 6. Sector Clusters

Group	Sectors	TPI Scores
I	Airlines, Diversified Mining, Oil & Gas, Technology, Telecommunications	3.26 - 3.49
II	Aluminium, Autos, Electricity Utilities, Financials, Oil & Gas Distribution, Paper, Utilities	3.03 - 3.21
III	Consumer Goods, Consumer Services, Food Producers, Health Care, Industrials, Shipping	2.84 - 2.93
IV	Cement, Chemicals, Oil & Gas (Other), Steel	2.72 - 2.79
V	Basic Materials, Coal Mining	2.55 - 2.62

Group III, ranging from 2.84 to 2.93, includes consumer goods, consumer services, food producers, health care, industrials, and shipping. These sectors hover around the global average (approximately 2.97 from prior data), suggesting moderate efforts in carbon management that may be constrained by diverse operational demands or less direct exposure to climate pressures.

Group IV, with TPI scores of 2.72 to 2.79, encompasses cement, chemicals, oil & gas (other), and steel, pointing to challenges tied to energy-intensive production processes that limit progress beyond basic compliance, despite some mitigation efforts. Finally, Group V, consisting of basic materials (2.62) and coal mining (2.55), occupies the lowest tier. These sectors show the lowest performance, highlighting challenges such as reliance on carbon-intensive processes. These tiers suggest that sector-specific factors, such as technological innovation or market demands, significantly influence carbon outcomes.

Table 7 classifies 28 countries into five groups based on their average TPI scores, ranging from Group I (3.41–4.00) to Group V (1.59–1.86). Group I, predominantly European nations such as Denmark (4.14, though not listed here, aligning with the number of firms), France (3.79), Germany (3.61), and the UK (3.80), alongside others like South Korea (3.55) and South Africa (3.69), achieves the highest scores. This cluster's superior performance, likely reflects stringent environmental policies, effective governance, and a cultural emphasis on sustainability. The inclusion of Finland (4.00) and Spain (3.94) reinforces the European dominance, while South Africa's presence hints at emerging leadership in specific contexts, possibly driven by resource sector reforms.

Table 7. Country Clusters

Group	Countries	TPI Scores
I	Germany, Italy, Netherlands, South Africa, Spain, Sweden, Switzerland, Turkey, UK, South Korea, Finland, France	3.41 - 4.00
II	Australia, Brazil, Canada, Japan, Thailand	3.21 - 3.38
III	Hong Kong, India, Mexico, Singapore, Taiwan, USA	2.76 - 3.18
IV	Indonesia, Malaysia, Philippines	2.30 - 2.38
V	China, Saudi Arabia	1.59 - 1.86

Group II, with scores from 3.21 to 3.38, includes Australia (3.30), Brazil (3.21), Canada (3.27), Japan (3.38), and Thailand (3.25). This group represents a solid intermediate tier, where developed economies like Japan and Canada balance regulatory frameworks with industrial priorities, while Brazil and Thailand show promise among emerging markets, possibly buoyed by targeted climate initiatives. Group III, ranging from 2.76 to 3.18, comprises Hong Kong (3.15), India (2.76), Mexico (2.81), Singapore (3.18), Taiwan (3.17), and the USA (2.99). This diverse cluster, hovering around the global mean, reflects varied performance. The USA's average score masks significant internal dispersion, while India's lower score suggests structural challenges, offset by high performers like Singapore leveraging economic advantages.

Group IV, with scores of 2.30 to 2.38, includes Indonesia (2.35), Malaysia (2.38), and the Philippines (2.30). These emerging economies exhibit weaker performance, likely due to developmental priorities or less mature regulatory systems, though their clustering suggests a shared baseline of effort. Group V, consisting of China (1.86) and Saudi Arabia (1.59), records the lowest scores. This tier, likely reflects economic models prioritizing carbon-intensive industries (fossil fuels in Saudi Arabia and manufacturing in China) coupled with less rigorous climate policies, starkly contrasting with Group I's achievements.

European nations in Group I set a global benchmark, driven by cohesive policy frameworks, while Group V's lower scores signal a need for stronger incentives or international pressure to align with sustainability goals. These groupings in Tables 6 and 7 collectively suggest that both sectoral characteristics and national contexts, ranging from technological capacity to regulatory rigor, create distinct performance landscapes, offering critical insights for tailoring decarbonization strategies at both industry and country levels.

CONCLUSION

This study provides a comprehensive analysis of carbon performance across sectors and countries using TPI scores to assess alignment with the 2050 Net Zero Emissions targets. The results indicate that larger companies, sectors such as telecommunications and oil & gas, and European countries like Denmark, France, and Germany consistently achieve higher TPI scores. These high performers demonstrate advanced carbon management, likely driven by stringent regulations, technological innovation, and strong governance frameworks. In the short term, these leaders exemplify SDG 13's call for urgent climate action by integrating robust emissions tracking and climate risk management into their operations, setting a model for immediate progress. This observation echoes Rogelj et al. (2021), who emphasized the importance of scientifically credible, near-term milestones and the reduction of offset dependency in corporate net-zero commitments. These patterns also align with OECD assessments (2024a, 2024b) underscoring that standardized, evidence-based climate disclosure frameworks are essential to align corporate finance and investment decisions with Paris Agreement objectives. Conversely, smaller firms, carbon-intensive sectors like coal mining and basic materials, and certain regions such as China and the Middle East exhibit lower scores, reflecting structural operational, or policy-related barriers to decarbonization.

The clustering analysis further elucidates these disparities, grouping sectors and countries into performance tiers. High-performing sectors in Group I (e.g., telecommunications, oil & gas) benefit from innovation and regulatory pressures, while low-performing sectors in Group V (e.g., coal mining) face inherent challenges due to carbon-intensive operations. Similarly, Group I countries, predominantly European, set a global benchmark through cohesive climate policies, whereas Group V countries like China and Saudi Arabia lag, likely due to economic priorities favoring fossil fuels or manufacturing. These findings align with the emphasis on using credible, transparent data to combat greenwashing and inform climate-aligned finance. The lagging performance of coal and cement sectors is consistent with Kuramochi et al. (2020), who identified the financial and operational constraints that slow decarbonization in heavy industries. Similarly, Li et al. (2022) highlighted a persistent misalignment between climate-related discourse and real investment flows among major fossil-fuel corporations, reinforcing the need for verifiable action plans and transparent reporting. By adopting short-term measures such as interim emissions targets and enhanced climate disclosures, as advocated by SDG 13, companies in lagging sectors and regions can build resilience and accountability, paving the way for long-term decarbonization. The TPI framework proves valuable for benchmarking corporate progress, supporting policy development, and informing investment decisions, as it provides granular, sector-specific insights into decarbonization efforts.

The results of this study largely corroborate the patterns identified in the literature, while adding nuanced cross-country and cross-sector evidence. Using the TPI metric, we evaluated the carbon management performance of over 2,000 companies worldwide. We found significant disparities across sectors and national contexts, a finding that aligns closely with prior research. For example, our analysis shows that only about half of the 24 sectors assessed achieve a mean TPI score above 3 (on a 0–5 scale), and relatively few firms reach the highest TPI levels (4 or 5). This indicates that most companies are still in the middle of the pack (Level 3: integrating climate considerations in operations) and have not advanced to the most robust transition planning stages. This pattern is consistent with Cenci et al.'s observation that only a minority of firms truly align with Paris Agreement pathways, while most lag in incremental or moderate performance. In our study, even in traditionally progressive industries, many firms remain at an intermediate level, underscoring the implementation gap highlighted by prior authors.

Notably, we found that high-emitting sectors exhibit the widest performance spread, which is a concern raised by sector-specific analyses such as those of Abdo et al. (2025). In our data, the telecommunications and oil & gas sectors are paradoxically among the top performers in average TPI scores (with mean scores ~3.3–3.5, indicating relatively strong carbon management practices). In contrast, coal mining and basic materials sit at the bottom with averages in the mid-2's. The strong performance of oil and gas majors in our sample likely reflects the fact that many large, well-resourced firms in this sector have begun investing in decarbonization technologies and have set formal transition strategies, which boost their TPI assessments. However, this does not contradict Abdo et al.'s findings. Instead, it highlights a selection effect: our broad sample includes many leading multinational oil companies, whereas Abdo and colleagues focused on smaller UK firms that often lack the same level of preparedness. Their warning that smaller fossil-fuel players face acute transition risks remains highly relevant. Our results, showing oil & gas as a high-scoring sector on average, may mask internal disparities between climate-progressive giants and vulnerable smaller producers. This suggests that even within a sector that appears to be doing well by certain metrics, one must dig deeper (through disclosures or firm-level analysis) to identify which companies might be overexposed to future asset write-downs or "stranded" assets, an insight strongly advocated by Abdo et al. (2025). Meanwhile, the persistently low TPI scores of coals and basic materials companies in our study underscore the structural challenges these industries face in decarbonizing, aligning with Jiang et al.'s observation that material- and energy-intensive sectors have historically had the lowest target achievement rates. Together, these findings reinforce the literature's point that heavy industry requires urgent, tailored interventions, as market forces alone have yet to elevate most of those firms beyond basic compliance with climate issues.

Our cross-country analysis provides empirical support for the notion that national policy and transparency contexts critically influence corporate climate performance. We found that companies in Western Europe, for example, Denmark, France, and Germany, cluster at the upper end of the TPI scale, with a large share of firms attaining Level 4 or 5 (strategic climate integration or full transition planning). In Denmark, the average TPI score of sampled firms exceeds 4.1, with no company below Level 3. This aligns perfectly with the literature's indication that strong regulatory environments and public accountability drive better outcomes. Indeed, Jiang et al. (2025) observed that firms in countries with greater media freedom and stricter climate oversight were far more likely to meet their emissions targets. Our findings echo the idea that the prevalence of advanced climate governance among European firms reflects robust EU climate policies (such as mandatory emissions disclosure and aggressive emissions-reduction mandates), which create pressure to perform. By contrast, our study shows that companies in countries with weaker climate governance tend to lag. Notably, China's corporate sector scored an average TPI of only ~1.9, with an overwhelming majority of Chinese firms stuck at low levels (Levels 0–2) and virtually none reaching the top performance levels. This dramatic gap suggests that, as the literature intimates, rapid economic growth and lenient regulatory enforcement can hinder climate action, allowing many firms to prioritize expansion over decarbonization. Similarly, firms based in oil-dependent Middle East economies (Saudi Arabia, Qatar) showed some of the lowest performance of all, with most companies failing to move beyond rudimentary climate awareness. This finding aligns with global assessments that identify policy deficits in those regions; voluntary corporate initiatives alone appear insufficient in the absence of national net-zero strategies or transparency norms. On the other hand, our analysis identified a few outliers: small countries like Colombia and Greece have very high average TPI scores among the few firms assessed, though this likely reflects idiosyncratic cases rather than broad trends. Overall, the cross-country patterns in our results strongly reinforce the argument made by researchers such as Jiang et al. (2025) and others that the credibility of corporate climate action is heavily context dependent. It underlines the importance of public policy, media oversight, and civil society engagement in holding companies to their climate commitments. Where external accountability is strong, companies are pushed into more ambitious climate management; where it is weak, corporate promises may remain largely symbolic.

Additionally, our finding that larger companies tend to score higher on climate performance than smaller or privately held companies resonates with the broader sustainability literature. Larger firms often have more resources, stakeholder scrutiny, and public visibility, which can drive them to adopt more rigorous climate governance, as our data showed (large-cap firms averaged higher TPI levels, whereas many unlisted companies lagged at lower levels). This trend is implicitly supported by works like Blanchard et al. (2025), who noted that many corporate climate initiatives and standards are geared toward major companies and require substantial capacity to implement. It also complements Bendig et al.'s emphasis on the need for standardized metrics. When common yardsticks (like TPI or science-based targets) are applied, the superior performance of well-resourced firms becomes clear, suggesting a need to bring smaller companies up to speed through capacity-building or tailored frameworks. Our results, therefore, support the call for more inclusive and adaptable climate accountability tools that can be adopted by firms of all sizes, ensuring that mid-tier and smaller companies are not left behind in the net-zero transition.

The observed disparities underscore the need for tailored strategies to enhance carbon performance. High-performing sectors and countries can serve as models, sharing best practices in governance, technology adoption, and transition planning. For underperforming sectors like coal mining, targeted interventions—such as subsidies for clean technologies or stricter emissions standards—could accelerate progress. Similarly, low-scoring countries like China and Saudi Arabia may benefit from international collaboration or incentives to align with global sustainability goals. Such interventions align with SDG 13's emphasis on fostering international cooperation and capacity-building to address climate change in the near term. The USA's average performance, coupled with high variability, suggests a need for stronger federal policies to reduce disparities among firms and create a better facility for climate action.

The analysis presented in this study is constrained by the limitations of the TPI dataset used. The dataset, comprising 2,028 firms selected from 2,066 reviewed by the LSE as of January 2025, may not fully represent the global corporate landscape due to its focus on firms with available TPI scores, potentially excluding smaller or less transparent companies. The reliance on TPI scores, while standardized, may not capture the full spectrum of decarbonization efforts, such as qualitative aspects of corporate strategies or recent initiatives not yet reflected in the data. Additionally, the analysis is limited to 24 sectors and 28 countries, which may not account for nuances in other sectors or countries with fewer firms or less data availability. The small sample sizes for certain countries, such as Colombia and Greece, limit the generalizability of their high TPI scores, as these may reflect the performance of a few firms rather than broader national trends. Finally, the cross-sectional nature of the data restricts insights into temporal changes, which could affect the understanding of how firms' carbon performance evolves over time.

Future research should address the limitations of this study and explore the drivers of performance differences, such as technological adoption rates, regulatory enforcement mechanisms, or cultural factors influencing corporate behavior. Incorporating SDG 13's focus on short-term climate education and awareness could further enhance corporate engagement with decarbonization strategies. Additionally, longitudinal tracking TPI scores over time studies could assess how TPI scores evolve as companies implement transition plans, providing insights into the effectiveness of current strategies. Integrating TPI data with other frameworks, such as the ACT initiative, could further enhance the robustness of corporate climate assessments by incorporating additional metrics or qualitative indicators. Furthermore, expanding the dataset to include a broader range of firms, particularly smaller or unlisted companies, and underrepresented sectors or countries, would improve the generalizability of findings. Investigating the role of financial mechanisms, such as green bonds or carbon pricing, in driving higher TPI scores could also inform investment strategies. Finally, examining the socio-economic impacts of decarbonization efforts, including job creation or economic displacement in carbon-intensive sectors, could provide a holistic view of the transition to a low-carbon economy, guiding equitable and sustainable policy development.

The findings of this study have several key implications for policymakers, investors, and corporate leaders. For policymakers, the disparities in carbon performance across sectors and countries highlight the need for targeted regulations and incentives to support lagging regions and industries, such as enhanced subsidies for clean technologies in carbon-intensive sectors like coal mining and basic materials. This can accelerate global decarbonization efforts and align with SDG 13 by promoting urgent climate action through capacity-building and international cooperation. Investors can use the TPI-based insights to prioritize funding for high-performing firms, reducing risks associated with stranded assets and fostering sustainable finance practices. For corporate leaders, particularly in underperforming areas, the study underscores the importance of adopting advanced carbon management strategies, such as integrating climate risks into business models and enhancing transparency, to improve competitiveness and meet net-zero targets. Overall, these implications emphasize the role of standardized tools like TPI in driving accountable and equitable transitions to a low-carbon economy.

In conclusion, this study underscores the TPI framework's utility as a tool for evaluating corporate alignment with Paris-Compliant Pathways. By identifying leaders and laggards across sectors and countries, it contributes critical insights to the global effort to achieve net-zero emissions by 2050. Stakeholders can leverage these findings to design targeted policies and foster a sustainable, low-carbon future. Finally, as global climate governance evolves, tools like the TPI must not only inform but also enforce — ensuring that the path to net zero is not just aspirational, but operational.

REFERENCES

Abdo, H., Angwin, D., Ben Othman, H., & Owusu, F. B. (2025). Transitioning to net zero: Assessing the impacts on asset impairment, write-downs and the going concern of oil and gas companies operating in the UK. *Global Environmental Change*, *92*, 103004. DOI: 10.1016/j.gloenvcha.2025.103004

Accelerate Climate Transitions (ACT). (2025). *ACT Assessment Methodologies*. Retrieved from https://actinitiative.org/en/act-methodologies/ (Accessed 20 May 2025).

Bendig, D., Schulz, C., Erbar, F., & Heubeck, T. (2025). Apples to apples: Accurately assessing corporate carbon performance. *Journal of Cleaner Production*, *486*, 144338. DOI: 10.1016/j.jclepro.2024.144338

Blanchard, L., Lezak, S., & Anderegg, W. R. (2025). Update corporate climate standards to increase rigor and ambition. *One Earth*, *8*(6), 101312. Advance online publication. DOI: 10.1016/j.oneear.2025.101312

Cenci, S., Burato, M., Rei, M., & Zollo, M. (2023). The alignment of companies' sustainability behavior and emissions with global climate targets. *Nature Communications*, *14*(1), 7831. DOI: 10.1038/s41467-023-43116-2 PMID: 38052791

Dietz, S., Fruitiere, C., Garcia-Manas, C., Irwin, W., Rauis, B., & Sullivan, R. (2018). An assessment of climate action by high-carbon global corporations. *Nature Climate Change*, *8*(12), 1072–1075. DOI: 10.1038/s41558-018-0343-2

Everitt, B. S., Landau, S., Leese, M., & Stahl, D. (2011). *Cluster Analysis* (5th ed.). Wiley., DOI: 10.1002/9780470977811

Hale, T., Smith, S. M., Black, R., Cullen, K., Fay, B., Lang, J., & Mahmood, S. (2022). Assessing the rapidly-emerging landscape of net zero targets. *Climate Policy*, *22*(1), 18–29. DOI: 10.1080/14693062.2021.2013155

IPCC. (2018). *Global Warming of 1.5°C: An IPCC Special Report on the Impacts of Global Warming of 1.5°C Above Pre-industrial Levels and related global greenhouse gas emission pathways, in the context of strengthening the global response to the threat of climate change, sustainable development, and efforts to eradicate poverty* [Masson-Delmotte, V., P. Zhai, H.-O. Pörtner, D. Roberts, J. Skea, P.R. Shukla, A. Pirani, W. Moufouma-Okia, C. Péan, R. Pidcock, S. Connors, J.B.R. Matthews, Y. Chen, X. Zhou, M.I. Gomis, E. Lonnoy, T. Maycock, M. Tignor, & T. Waterfield (eds.)]. Cambridge University Press. DOI: 10.1017/9781009157940

Jahn, V., Inghan, R., & Dietz, S. (2023). *TPI's Methodology Report: Management Quality and Carbon Performance*. Transition Pathway Initiative.

Jiang, X., Kim, S., & Lu, S. (2025). Limited accountability and awareness of corporate emissions target outcomes. *Nature Climate Change*, *15*(3), 279–286. DOI: 10.1038/s41558-024-02236-3

Kayral, İ. E., & Aktaş Bozkurt, M. (2025). *Are Renewable Energy Firms a Viable Investment Alternative in the Wake of Global Crises and the Paris Climate Agreement? Finance of Renewable Energy in a Sustainable World, 115–126.* Nova Science Publishers.

Kruskal, W. H., & Wallis, W. A. (1952). Use of ranks in one-criterion variance analysis. *Journal of the American Statistical Association*, *47*(260), 583–621. DOI: 10.1080/01621459.1952.10483441

Kuramochi, T., Roelfsema, M., Hsu, A., Lui, S., Weinfurter, A., Chan, S., Hale, T., Clapper, A., Chang, A., & Höhne, N. (2020). Beyond national climate action: The impact of region, city, and business commitments on global greenhouse gas emissions. *Climate Policy*, *20*(3), 275–291. DOI: 10.1080/14693062.2020.1740150

Li, M., Trencher, G., & Asuka, J. (2022). The clean energy claims of BP, Chevron, ExxonMobil and Shell: A mismatch between discourse, actions and investments. *PLoS One*, *17*(2), e0263596. DOI: 10.1371/journal.pone.0263596 PMID: 35171938

McKinsey & Company. (2022). *The Net-Zero Transition: What It Would Cost, What It Could Bring*. Retrieved from https://www.mckinsey.com/capabilities/sustainability/our-insights/the-net-zero-transition-what-it-would-cost-what-it-could-bring (Accessed 20 May 2025).

Millar, R., Hepburn, C., Beddington, J., & Allen, M. (2018). Principles to guide investment towards a stable climate. *Nature Climate Change*, *8*(1), 2–4. DOI: 10.1038/s41558-017-0042-4

Moore, D. S., McCabe, G. P., & Craig, B. A. (2017). *Introduction to the Practice of Statistics* (9th ed.). W.H. Freeman and Company.

OECD/IEA. (2017). *Energy Technology Perspectives 2017: Catalysing Energy Technology Transformations*. Retrieved from https://www.oecd.org/content/dam/oecd/en/publications/reports/2017/06/energy-technology-perspectives-2017_g1g7a2d9/energy_tech-2017-en.pdf (Accessed 19 May 2025).

Organisation for Economic Co-operation and Development (OECD). (2024a). *Review on Aligning Finance with Climate Targets*.

Organisation for Economic Co-operation and Development (OECD). (2024b). *State of Transition in the Banking Sector Report*.

R Core Team. (2024). R: A Language and Environment for Statistical Computing. R Foundation for Statistical Computing. https://www.R-project.org/

Rogelj, J., Geden, O., Cowie, A., & Reisinger, A. (2021). Net-zero emissions targets are vague: Three ways to fix. *Nature*, *591*(7850), 365–368. DOI: 10.1038/d41586-021-00662-3 PMID: 33727725

Shapiro, S. S., & Wilk, M. B. (1965). An analysis of variance test for normality (complete samples). *Biometrika*, *52*(3/4), 591–611. DOI: 10.2307/2333709

Transition Pathways Initiative (TPI). (2022). *Sectoral Decarbonisation Pathways*. Retrieved from https://www.transitionpathwayinitiative.org/ (Accessed 19 April 2025).

United Nations. (2015). *Paris Agreement*. United Nations Framework Convention on Climate Change.

United Nations General Assembly. (2015). *Transforming our world: The 2030 Agenda for Sustainable Development*. A/RES/70/1. DOI: 10.1891/9780826190123.ap02

ADDITIONAL READING

Battiston, S., Monasterolo, I., Riahi, K., & van Ruijven, B. J. (2021). Accounting for finance is key for climate mitigation pathways. *Science*, *372*(6545), 918–920. DOI: 10.1126/science.abf3877 PMID: 34016742

Berrang-Ford, L., Ford, J. D., & Paterson, J. (2011). Are we adapting to climate change? *Global Environmental Change*, *21*(1), 25–33. DOI: 10.1016/j.gloenvcha.2010.09.012

Fankhauser, S., Smith, S. M., Allen, M., Axelsson, K., Hale, T., Hepburn, C., Kendall, J. M., Khosla, R., Lezaun, J., Mitchell-Larson, E., Obersteiner, M., Rajamani, L., Rickaby, R., Seddon, N., & Wetzer, T. (2022). The meaning of net zero and how to get it right. *Nature Climate Change*, *12*(1), 15–21. DOI: 10.1038/s41558-021-01245-w

Le Quéré, C., Korsbakken, J. I., Wilson, C., Tosun, J., Andrew, R., Andres, R. J., & Peters, G. P. (2019). Drivers of declining CO2 emissions in 18 developed economies. *Nature Climate Change*, *9*(3), 213–217. DOI: 10.1038/s41558-019-0419-7

KEY TERMS AND DEFINITIONS

Carbon Performance: A measure of a company's greenhouse gas emissions and their alignment with low-carbon transition goals, often assessed through metrics like TPI scores, focusing on emissions intensity and reduction commitments.

Greenhouse Gas (GHG) Emissions: Gases such as carbon dioxide, methane, and nitrous oxide released into the atmosphere, contributing to global warming, which companies are tasked with reducing to meet climate targets.

Low-Carbon Economy: An economic system that minimizes greenhouse gas emissions through the adoption of clean technologies, renewable energy, and sustainable practices, aiming to achieve net-zero emissions.

Net Zero Emissions: A state where the amount of greenhouse gases emitted into the atmosphere is balanced by an equivalent amount removed, resulting in no net increase in atmospheric greenhouse gas concentrations, typically targeted by 2050 to align with the Paris Agreement goals.

Paris-Compliant Pathways (PCPs): Emissions reduction trajectories for companies or sectors designed to align with the Paris Agreement's goal of limiting global warming to well below 2°C, ideally 1.5°C, above pre-industrial levels, often based on sector-specific benchmarks.

Sectoral Decarbonization Approach (SDA): A methodology that sets sector-specific emissions reduction targets based on International Energy Agency (IEA) pathways, comparing companies within the same sector to benchmarks aligned with global climate goals.

Transition Pathway Initiative (TPI): A global initiative assessing companies' preparedness for a low-carbon economy, evaluating their management quality and carbon performance against climate scenarios consistent with the Paris Agreement, using a 0–5 scoring scale.

Chapter 15
Green Finance and Blue Economy:
Economic Vision of the Future

Rüya Ataklı Yavuz
https://orcid.org/0000-0003-3147-333X
Çanakkale Onsekiz Mart University, Turkey

Fulya Kıroğlu
https://orcid.org/0000-0002-1007-1561
Independent Researcher, Turkey

ABSTRACT

The effects of the global climate crisis have necessitated radical transformations in economic policies. This study analyses the impact of green finance and blue economy concepts, which are shaped in line with sustainable development goals, on economic structures. While green finance encompasses financial instruments that reduce environmental risks and support sustainable projects, blue economy aims at the sustainable utilisation of marine and ocean resources for economic development. This chapter analyses the impacts of green financing mechanisms and blue economy models on economic development. It also discusses how artificial intelligence technologies support the green and blue economy in areas such as data analytics, resource management and climate risk prediction. In conclusion, the necessity of a holistic economic vision that adapts to the climate crisis and supports economic growth is emphasised. Developing green and blue economy-based strategies for policymakers is a critical step towards achieving sustainable development goals.

DOI: 10.4018/979-8-3373-5238-1.ch015

INTRODUCTION: THE RISE OF THE SUSTAINABLE ECONOMY

Climate change impacts the global economy both directly and indirectly. Factors such as temperature increases, extreme weather events, water scarcity, and sea level rise reduce productivity in agriculture, energy, infrastructure, and healthcare sectors, threatening countries' economic growth (IPCC, 2023). For example, according to the Columbia Climate School (2022), the total economic cost of climate-related disasters in the United States alone over the past three years has exceeded US$415 billion. The United Nations Environment Programme (UNEP, 2021a) reports that approximately 26 million people are being pushed back into climate-related poverty each year. Furthermore, according to a study by Burke, Hsiang, and Miguel (2015), a 2°C increase in global temperatures could result in a decline in global GDP of between 2.5% and 7.5% by the end of the century.

Climate change and environmental degradation are among the fundamental problems threatening economic sustainability and development. One of the global efforts to address these problems was the United Nations Climate Change Conference (Copenhagen Summit) held in 2009. This summit, which aimed to reduce global warming and greenhouse gas emissions, took the first steps toward establishing financial markets to contribute to combating climate and environmental problems (Liu, 2022, p. 288). Subsequently, at the 2010 United Nations Climate Change Conference held in Cancun, Mexico, the establishment of the "Green Climate Fund" was agreed upon by a joint decision of 194 countries to support developing countries in reducing greenhouse gas emissions and adapting to climate change. With the establishment of the fund, the concept of "green finance" began to appear frequently in policy documents and reports of both international organizations and national governments (Zhang et al., 2019, p. 425). The United Nations 2030 Agenda for Sustainable Development, adopted in 2015, includes 17 sustainable development goals (SDGs) that integrate environmental protection, social equity, and economic growth (UN, 2015). The SDGs are shown in Figure 1.

Figure 1. United Nations Sustainable Development Goals

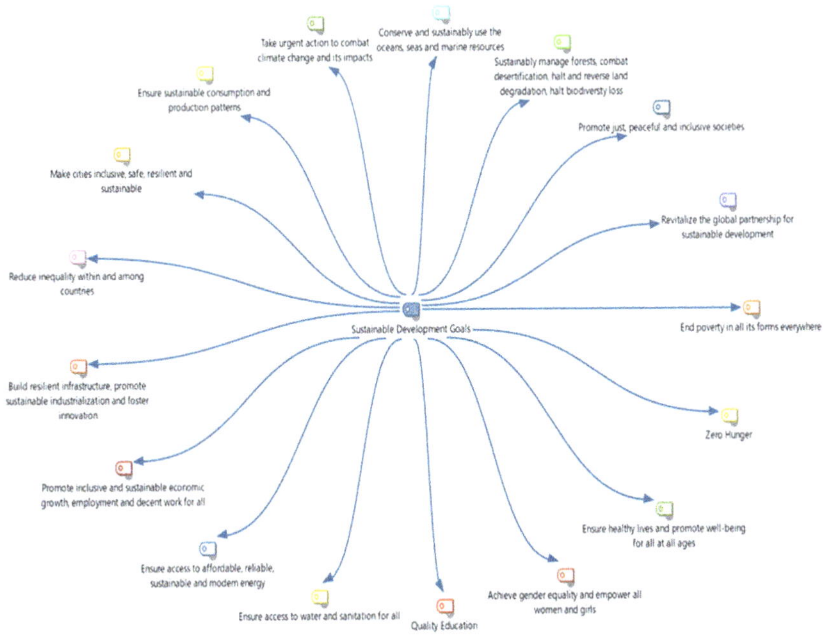

Source: *It was drawn by the authors using the United Nations (2025) data. https://www.un.org/sustainabledevelopment/sustainable-development-goals/*

The SDGs have paved the way for concepts such as the green and blue economy to gain prominence in policy and investment decisions. Of these goals, SDG 13: Climate Action, in particular, aims to strengthen countries' capacity to take action against the climate crisis, adapt, and develop resilient development strategies. Analyses published by the OECD (2021) and the IEA (2021) demonstrate that the green transformation not only slows climate change but also has the potential to generate employment and economic growth. It is projected that if policies based on clean energy and resource efficiency are implemented, global GDP could increase by 0.2% by 2040. Sustainable economic transformation involves not only reducing environmental risks but also ensuring social justice and increasing economic efficiency (IPCC, 2023). Green financing mechanisms, clean energy investments, and innovative technologies play a key role in this process (IEA, 2021).

This chapter is based on a conceptual and critical literature review. The research design involves the systematic analysis of academic articles, international organization reports, and policy documents that evaluate the relationship between green finance, blue economy, and artificial intelligence from a sustainable development perspective. Literature selection was based on the following criteria: direct relationship with climate change and sustainable finance, reference by international policymakers and financial institutions, and presentation of measurable environmental and economic impact (OECD, 2020a; UNEP FI, 2021). The main objective of the study is to present the economic impacts, governance structure, and application limitations of green and blue finance instruments within a holistic framework.

Consequently, the need for a holistic economic vision that adapts to the climate crisis and supports economic growth is emphasized. For policymakers, developing strategies based on green and blue economies is a critical step toward achieving sustainable development goals. The green and blue economy paradigm is a concrete reflection of this necessity in the economic sphere, and this study is evaluated within this framework.

THE ROLE AND IMPORTANCE OF GREEN FINANCE

Green finance is a broad concept encompassing investments in commercial and industrial activities that support the sustainable use of natural and societal resources such as land, water, biodiversity, air, and human health. This concept encompasses a variety of products, services, tools, and mechanisms, including the management of climate and environmental financial risks (Sachs et al., 2019). Green finance can be defined as a set of financial activities aimed at financing environmentally friendly projects, combating climate change, and supporting environmentally compatible development processes (OECD, 2021). This process refers to a set of strategies and methods encompassing the collection, acquisition, and allocation of various funds, including both private and public sector resources and philanthropic contributions (Sachs et al., 2019).

The primary goal of green finance is to provide financing for environmentally focused projects to protect the natural environment, prevent pollution, and support sustainable development (Chang et al., 2022, p. 1). In this context, financing models focused on areas such as renewable energy, energy efficiency, technologies that reduce carbon emissions, and sustainable transportation are gaining prominence (UNEP FI, 2022). Green finance, a newly emerging and increasingly important intersection in finance, aims to adapt traditional financing methods to a low-carbon economy by incorporating trust-based roles in sustainable development, combating climate change, and developing innovative technologies. In this respect, green finance is a multidisciplinary financing approach that aims to achieve a balance between economic growth and environmental protection.

Green finance plays a critical role in achieving sustainable development goals and aims to produce effective solutions to environmental problems. As emphasized by the United Nations Environment Programme (UNEP), infrastructures for sustainable development are both economically sound and necessary (UNEP 2021a). These infrastructures will help countries meet their climate action commitments under the Paris Agreement, make their nationally determined contributions, and achieve the seventeen United Nations Sustainable Development Goals during the 2021–2030 period (Volz, 2018; Nawaz et al., 2021). However, achieving these goals requires interdisciplinary technology development, policy development, business practices, designing legal and regulatory frameworks, ensuring multi-stakeholder and investor participation, as well as developing collaboration and new delivery models at the local, regional, national, and international levels.

As the concept of sustainable development gains strength, approaches aimed at minimizing environmental degradation are increasingly prioritized. The development process no longer requires solely economic growth; it also requires a holistic assessment of environmental, social, and economic dimensions. In this context, the focus on green and innovative financing instruments has increased significantly. Green financing practices consist of various financial instruments, mechanisms, and methods developed to support environmental sustainability. The main types of green financing include green bonds, sustainability-linked loans, carbon markets and emissions trading, green funds and investment funds, grants and incentives, microfinance, insurance, and risk management products.

Green Bonds: Green bonds, one of the approaches that aim to simultaneously achieve environmental protection and economic development, emerged in global markets in 2007. Green bonds are debt instruments allocated to the financing of specific environmental projects. They are used in areas such as renewable energy, energy efficiency, and clean water projects. Green bonds offer investors the opportunity to participate in sustainable projects, while carbon markets create market-based mechanisms for reducing greenhouse gas emissions. Green bonds are issued to finance environmental investments such as projects to combat climate change, the development of renewable energy sources, and resource efficiency. The funds raised through this instrument are intended to be directed to renewable energy projects. For example, the European Union Emissions Trading System (EU ETS) is one of the most widely used models for reducing carbon emissions (European Commission, 2021). Sustainable investment funds also incentivize companies to invest based on environmental, social, and governance (ESG) criteria.

The World Bank began its green bond program in 2008 in line with its sustainable development goals and, as of the end of fiscal year 2024 (FY24), had issued a total of US$19.4 billion in green bonds. Green bonds issued in FY24 alone totaled US$1.3 billion. As of June 2024, the amount of green bonds in circulation stood at US$10.3 billion. To date, 146 green bond-supported projects have been added to the list, and these projects have been implemented in 37 countries. The World Bank has made a total financial commitment of US$24.9 billion since 2008 (World Bank 2024).

Table 1. shows a comparative comparison of the actual and expected outcomes of the World Bank's green bond projects as of 2024. The data presented is that based on key indicators developed in accordance with the Harmonized Framework for Impact Reporting for Green Bonds developed by the International Capital Markets Association (ICMA). An annual emission reduction of 41.5 million tons of CO_2 equivalent has been achieved. A reduction of 26.7 million tons of CO_2 equivalent is expected. 126.3 million people have directly benefited from the projects. 73.6 million more people are expected to benefit. 37.5 million MWh of annual energy savings have been achieved. 466,200 MWh of additional savings are expected. 2,521 MW of renewable energy capacity has been installed. 5,338 MW of additional capacity is planned to be added. 24.2 million people have benefited from sustainable transportation systems. 170,600 more people are expected to benefit. 1.5 million people have benefited from measures taken against disasters such as floods and droughts. 17.1 million people will benefit. 45 million m^3 of annual water savings have been achieved. 17 million m^3 of additional savings are expected. 23.1 million m^3 of wastewater has been treated, reused, or prevented. 408.7 million m^3 is targeted. 3.3 million tons of waste has been prevented, reduced, or recycled. An additional 243,300 tons are expected. 20.3 million hectares have been brought under sustainable management. An additional 1.2 million hectares will be covered. Soil conservation and regenerative practices have been implemented on 39,500 hectares of agricultural and forest land, with a target of 21.6 million hectares more. Another 205,000 hectares will be managed under climate resilience and disaster risk reduction programs. Accordingly, while significant progress has been made in some categories (e.g., energy conservation and greenhouse gas reduction), large-scale improvements are expected in other areas (especially wastewater management, soil conservation, and climate adaptation).

Table 1. World Bank Green Bond projects (as of June 30, 2024):

Category	Achieved	Expected
GHG Emissions Reduced ($MtCO_2$ eq/year)	41.5	26.7
Beneficiaries (million people)	126.3	73.6
Energy Savings (MWh/year)	37,500,000.00	466,200.00
Renewable Energy Capacity (MW)	2,521.00	5,338.00
Sustainable Transport Beneficiaries (people)	24,200,000.00	170,600.00
Flood/Drought Mitigation Beneficiaries (people)	1,500,000.00	17,100,000.00
Annual Water Savings (m^3)	45,000,000.00	17,000,000.00
Wastewater Treated/Reused/Avoided (m^3/year)	23,100,000.00	408,700,000.00

continued on following page

Table 1. Continued

Category	Achieved	Expected
Waste Prevented/Recycled (tons/year)	3,300,000.00	243,300.00
Biodiversity - Managed Area (hectares)	20,300,000.00	1,200,000.00
Soil/Agroforestry Practices (hectares)	39,500.00	21,600,000.00
Climate Resilience Area (hectares)	0	205,000.00

Note: It was created by the authors using The World Bank, Impact Report 2024:16.

Sustainability-Linked Loans (SLLs): Sustainability-Linked Loans (SLLs) are an innovative financing tool developed to enhance the financial sector's contribution to environmental and social sustainability. This type of loan aims to provide improved credit terms (e.g., interest rates) if the borrower achieves predetermined sustainability performance targets (SPTs) (LMA, APLMA & LSTA, 2023). Although SLLs are loans with no specific purpose, they contribute to strengthening corporate sustainability strategies because they include a pricing mechanism based on the level of sustainability goal achievement (IFC, 2022). These loans encourage businesses to improve their performance not only in environmental but also in social and governance (ESG) areas. By agreeing on the ESG indicators to be monitored within the scope of the loan agreement, lenders and borrowers make performance measurement transparent and traceable (OECD, 2020a). This allows lenders to establish a direct link between environmental impact management and financial returns. Institutional-level research shows that sustainability-linked loans enable companies to adopt sustainability strategies in a more holistic and accountable manner (Flammer, 2021). They also create an attractive investment environment for environmentally sensitive investors and financial institutions. In this respect, SLLs have become the fastest-growing sustainable finance instrument after green bonds (UNEP FI, 2021). Sustainability-linked loans go beyond traditional financing instruments and represent a performance-based credit structure that focuses on environmental and social impact. The importance of such financing instruments in achieving key sustainability goals such as combating climate change, resource efficiency, and social inclusion is increasing.

Carbon Markets & Emissions Trading: Carbon markets are a market-based approach used as an economic tool in the fight against climate change and aim to limit greenhouse gas emissions. This system sets a “carbon quota” by limiting the maximum amount of carbon a company or country can emit in a given period. These quotas are then converted into “emission permits” that can be traded on the market. Thus, companies that emit less can generate financial revenue by selling their surplus rights to those that emit more, while companies that do not comply with the emissions limit are forced to purchase these rights (World Bank, 2022b).

Emissions trading systems generally operate on a "cap-and-trade" model. In this model, a cap is set on total emissions, which are then reduced over time to reduce emissions. The system promotes low-carbon technologies and reduces emissions cost-effectively. Carbon trading provides companies with flexibility to reduce their carbon emission costs and provides economic incentives for the transition to low-carbon technologies. In this respect, carbon markets are considered a key component of green finance mechanisms.

Green Funds and Investment Funds: Green funds and sustainable investment funds are financial instruments that invest in capital markets by focusing on environmental, social, and governance (ESG) criteria. The primary objective of these funds is not only to generate financial returns but also to create environmental and societal benefits. Green funds typically invest in areas such as renewable energy, energy efficiency, clean technology, sustainable agriculture, and environmentally friendly infrastructure projects (UNEP FI, 2021). ESG criteria used in the evaluation of these funds include: environmental carbon footprint reduction, natural resource protection, and climate risk management; social employee rights, occupational health and safety, and social equity; and governance transparency, ethical management, and stakeholder accountability (OECD, 2020a). Green funds have become a rapidly growing investment category globally. The European Union's Sustainable Finance Disclosure Regulation (SFDR) and similar policies, in particular, support the development of these funds by requiring investors to consider sustainability criteria. Thus, green funds both direct capital to a low-carbon economy and become a secure investment vehicle that mitigates long-term risks for investors (Global Sustainable Investment Alliance, 2021).

Grants and Incentives: Grants and incentives are direct financing mechanisms provided by government agencies, international organizations, or development banks for projects targeting environmental sustainability. These types of financial support are particularly used in areas such as renewable energy investments, energy efficiency projects, sustainable agricultural practices, waste management, carbon reduction, and climate adaptation (World Bank, 2022a). Grants are generally non-refundable resources and play a critical role in covering the initial costs of environmental projects. Incentives can take the form of tax breaks, subsidies, low-interest loans, or the redistribution of carbon pricing revenues. These instruments facilitate the private sector's focus on sustainable investments and reduce risks for investors (OECD, 2020b). At the international level, the Green Climate Fund (GCF), the Global Environment Facility (GEF), and funds provided by the European Union under the Green Deal offer significant financing sources for low-carbon and climate-resilient development. Thus, grants and incentives contribute not only to the implementation of environmentally friendly projects but also to increasing social well-being and accelerating economic transformation (UNEP FI, 2021).

Microfinance: Microfinance is a mechanism designed for low-income individuals or small businesses, typically encompassing financial services such as small-scale loans, savings programs, or microinsurance. In the context of environmental sustainability, microfinance plays a critical role in implementing environmentally friendly projects, particularly in rural areas or developing countries (Armendáriz & Morduch, 2010). This type of financing enables smallholder farmers to obtain loans for solar irrigation systems, biogas plants, or organic farming practices, and rural entrepreneurs to receive capital support for recycling, energy efficiency, or sustainable production techniques (UNEP, 2020a). Microfinance is also a tool with high social impact. It promotes inclusive development by increasing access to the financial system for women entrepreneurs and disadvantaged communities. In this respect, microfinance contributes to both poverty reduction and environmental sustainability. Therefore, it is considered an important tool for developing local solutions to combat climate change and supporting the transition to a low-carbon economy at a grassroots level (Schreiner, 2002).

Insurance and Risk Management Products: These products are among the fundamental tools providing economic security against the physical and financial risks posed by climate change. These products help actors operating in climate-sensitive sectors, particularly agriculture, energy, and infrastructure, compensate for losses resulting from sudden climate events or long-term environmental changes. In addition to traditional insurance mechanisms, innovative products such as climate risk-based index insurance, catastrophe bonds, and parameter-based insurance have gained prominence in recent years. These instruments not only enable risk transfer but also promote more transparent risk measurement, pricing, and management within the financial system (Surminski & Thieken, 2017).

These instruments play a critical role in financing the green transition and contribute to the achievement of sustainable development goals. Green finance can not only support economic development but also ensure long-term economic stability. According to the International Energy Agency (IEA, 2021), energy transitions implemented through green finance have the potential to create millions of new jobs worldwide by 2030. Furthermore, according to the OECD's (2021) green recovery database, climate-friendly public spending has a higher economic multiplier effect compared to traditional spending. All these developments demonstrate that green finance is positioned not only as an environmental policy tool but also as an economic development strategy.

BLUE ECONOMY: SEA-BASED SUSTAINABLE DEVELOPMENT

Definition and Principles of the Blue Economy

Their climate-regulating effect within the ecological system, their hosting of wide biodiversity, their production of oxygen, and their being a source of economic and cultural values make oceans and seas an indispensable part of human life. However, climate change, habitat loss, overfishing, oil spills, and pollution deeply threaten the sustainability of oceans and seas. Therefore, the blue economy is gaining importance for preserving the natural structure of oceans and seas, transferring them to future generations, and ensuring their sustainable use (Youssef, 2023, p. 12).

The blue economy represents an approach that prioritizes the sustainability of ecosystem services in the use of marine and ocean resources for economic activities. This concept aims not only for economic growth but also for environmental protection and social inclusion (UNCTAD, 2020). The terms "ocean economy," "blue growth," and "marine economy," which have no clear boundaries in terms of meaning, are often used synonymously for the blue economy (Yıldırım, 2024, p. 1). The roots of the concept date back to the Rio Earth Summit held in 1992 (Voyer et al., 2018, p. 597). The action plan, Agenda 21, an outcome of the Rio Earth Summit, emphasized the protection and sustainability of oceans, seas, coastal areas, and marine living resources, while also highlighting the importance of cooperation and coordination in managing the process (United Nations, 1992). Although the concept of a blue economy was not directly included in the action plan, it laid the groundwork for its development.

The concept of the blue economy was introduced into the literature in 2010 through the work of Gunter Pauli and gained international recognition at the United Nations Conference on Sustainable Development held in Rio de Janeiro in 2012 (Suluk, 2022, p. 134). Gunter Pauli emphasizes that sustainable economic development requires producers to adopt not only profit-oriented business models but also innovative and competitive ones that contribute to environmental sustainability within the scope of the blue economy (Briones-Peñalver et al., 2023, p. 2474). In this context, he emphasizes that profit maximization alone is not sufficient; the efficient use of natural resources is also a necessity, and producers must assume significant responsibilities in this regard.

The blue economy represents an adaptation of the green economy, which aims for environmentally sensitive sustainable development, for the oceans and seas (Cisneros-Montemayor et al., 2022, p. 982). The United Nations Conference on Sustainable Development (UNCSD) report addressed the need for a transition to a blue economy alongside green development efforts. The same report also addressed the sustainability of oceans, seas, and coastal areas, protecting existing biodiversity, reducing pollution and its impacts, and supporting ecosystem-based fisheries (United Nations, 2012). Rather than excluding the costs arising from the misuse of oceans, the blue economy is based on a comprehensive assessment of natural capital, taking this into account in decision-making processes (University of Nairobi Maritime Centre, 2023, p. 5). In this regard, the blue economy aims to optimize the benefits derived from oceans and seas by establishing a link between economic growth and environmental protection (Bari, 2017, p. 5).

The blue economy encompasses not only well-established sectors and areas of activity related to ocean and marine ecosystems, but also new sectors and areas of activity supported by emerging technologies and innovation-focused perspectives. In addition to traditional maritime, fisheries, and tourism sectors, it also encompasses new sectors such as marine-based renewable energy, biotechnology, and carbon sinks (OECD, 2019). While activities and sectors related to the blue economy vary depending on the national vision, for activities and sectors to be considered a blue economy component, they must meet the following criteria (World Bank and United Nations Department of Economic and Social Affairs, 2017, p. vi). These are:

- Ensuring that the benefits derived from ocean and marine ecosystems extend to both present and future generations,
- Restoring the degraded biological richness of ocean and marine ecosystems, increasing their productivity, building resilience, and preserving and maintaining their unique value,
- Building on clean technologies, renewable energy sources, and circular material flows that encourage waste minimization and material recycling.

The blue economy encompasses a wide range of action plans and activities, from products and services derived from sea and ocean resources to maritime transportation, from water management infrastructure to employment creation, from wetland protection to coastal restoration, from water route determination to rainwater use, from graywater recycling to smart water systems (Çoban & Ölmez, 2016, p. 161). The World Bank and the United Nations Department of Economic and Social Affairs have grouped the types of blue economy activities, their subcategories, and the sectors/industries related to these activities under the heading of "components of the blue economy." This grouping is presented in Figure 2.

Figure 2. Components of the blue economy

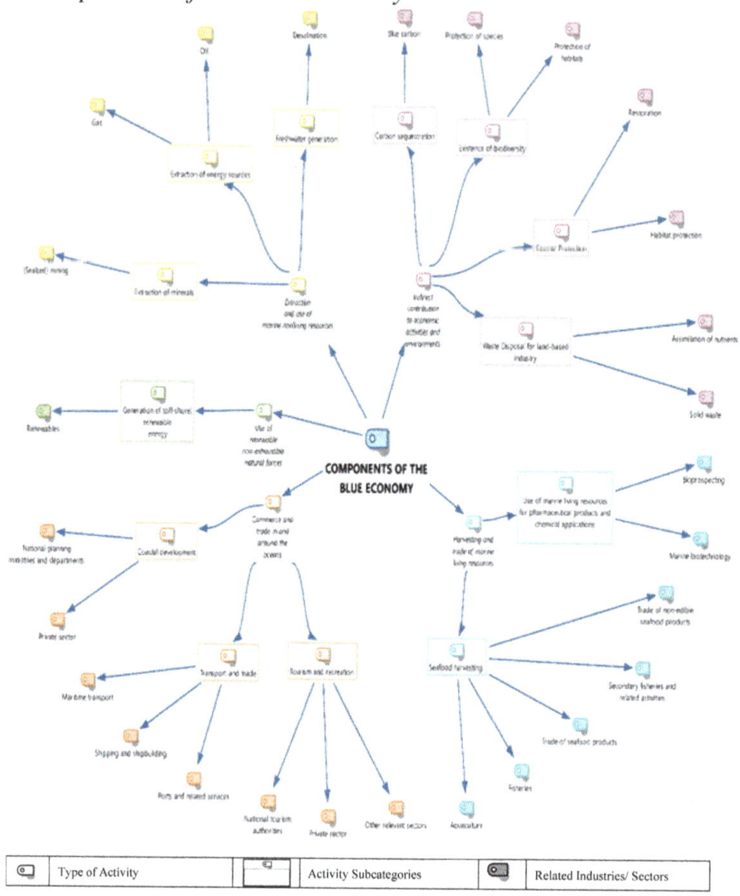

Note: *It was drawn by the authors using the World Bank and United Nations Department of Economic and Social Affairs (2017). The Potential of the Blue Economy: Increasing Long-term Benefits of the Sustainable Use of Marine Resources for Small Island Developing States and Coastal Least Developed Countries. World Bank, Washington DC.*

In addition to developing the blue economy sustainably and transferring ocean and marine resources to future generations, there is a need for generally accepted principles to form the basis of the blue economy and guide stakeholders. Because these principles have been shaped over time by various international organizations, reports, and agreements, there are no universally accepted standards. In this context, the shaping of these principles is influenced by the priorities and perspectives of public and private institutions.

There is a strong connection between Goal 14 of sustainable development, "to protect and sustainably use the oceans, seas, and marine resources," and the blue economy. The blue economy not only provides a framework for achieving Goal 14 of sustainable development but also helps develop action-oriented solutions. The principles of the blue economy align with the criteria set under Goal 14, such as the sustainable use and protection of marine and coastal ecosystems, increasing economic benefits from the seas, preventing marine pollution, improving ocean health, and supporting sustainable fishing practices. In this context, the blue economy should be promoted to achieve sustainable development goals and ensure a prosperous life for future generations.

Within the framework of sustainable development, the scope of applications for the blue economy is quite broad. In the context of the blue economy, sustainability practices contribute to ecosystem protection and biodiversity maintenance in areas such as fisheries, renewable marine energy, marine tourism, maritime transportation, blue biotechnology, and marine mining, while also supporting economic growth. Therefore, the widespread adoption of sustainability practices significantly impacts the long-term success of the blue economy.

Sustainability of the blue economy is possible through good governance, science-based policy design, and technological innovation. Artificial intelligence and big data analytics support decision-making in applications such as monitoring fish stocks, preventing illegal fishing, and monitoring marine pollution (World Bank, 2022b). In this context, the blue economy is strategically important not only for coastal countries but also for global food security, energy transition, and climate goals.

Marine Ecosystems and Sustainable Fisheries

Marine and coastal ecosystems, which provide direct and indirect benefits to human life, include coral reefs, intertidal zones, estuaries, seagrass communities, and coastal aquaculture (Lopez-Rivas & Cardenas, 2024, p. 1). These unique ecosystems possess a vital and unparalleled biodiversity (Netto et al., 2025, p. 1). Marine and coastal ecosystems are under pressure from numerous factors, including human-induced misuse, aggressive coastline development, global climate change, and rising sea levels (Schlacher et al., 2007, p. 556). This multifaceted pressure is also profoundly impacting fisheries, which are both a source of food and livelihood for many people.

The Food and Agriculture Organization of the United Nations (FAO) states that while global fisheries resources were 90% sustainable in 1974, they had largely lost this quality, falling to 62.3% sustainability in 2021 (FAO, 2025). The European Environment Agency (EEA) states that, while an average of 28% of assessed stocks across Europe are sustainable, this rate is 41% in the North-East Atlantic and Baltic Seas, and 9% in the Mediterranean and Black Seas due to regional differences (EEA, 2025).

Factors such as overfishing of fish stocks, illegal or unreported fishing activities, and the catching of immature fish and non-target species; the use of destructive fishing methods such as dynamite, cyanide, and ghost nets; and marine pollution and the destruction of marine habitats such as coral reefs, mangroves, and seagrass meadows lead to serious declines in fish stocks and the endangerment of some species. Furthermore, the lack of clear ownership rights in the seas further exacerbates the problem, leading to resource exploitation (Ozan & Başer, 2024, p. 110). This situation highlights the existence of a serious problem in terms of sustainable fishing and demonstrates the crucial importance of sustainable fishing practices.

Sustainable fishing, a multifaceted concept with economic, ecological, and social dimensions, involves managing fish stocks to ensure the continuity of fishing activities (WWF-Turkey, 2013, p. 11). Sustainable fishing practices encompass a wide range of tools and methods. Stock assessment, quota implementations, seasonal fishing bans, ecosystem-based management, use of targeted fishing gear, designation of marine protected areas, and raising public and local fishermen's awareness are among the most prominent practices and methods.

To achieve positive results from sustainable fishing practices, it is crucial to implement legal regulations, as well as raise public awareness and participation among local fishermen. To this end, detailed and understandable information should be shared on the status of fish stocks, appropriate fishing methods, fishing gear, the harms of overfishing, and the benefits of sustainable fishing practices. Awareness-raising programs should be conducted for local fishermen to ensure the long-term preservation of sustainable fishing practices. In addition to contributing to the conservation of marine biodiversity, sustainable fishing practices offer alternative energy sources that reduce carbon emissions through ocean-based energy production (e.g., tidal and wave energy) (IEA, 2021). Furthermore, blue carbon ecosystems, such as mangrove forests, seagrass meadows, and coral reefs, play a critical role in carbon sequestration in the fight against climate change (UNEP, 2020b).

Renewable Marine Energy

The global economy's heavy dependence on fossil fuels is at risk due to factors such as supply security and climate change (Musa et al., 2018, p. 2281). Reducing risks to energy supply, meeting energy demand, and combating climate change have made diversifying energy sources essential. In this context, the oceans, which cover 71% of our planet, emerge as massive renewable energy resources that can potentially meet global energy demand (Sang et al. 2018, p. 735).

Energy demand is steadily increasing with global developments. Global electricity demand is projected to rise to 65% by 2040, with the majority of this demand coming from developing economies, while global decision-makers aim to increase the share of renewable energy to 28% (Agyekum et al., 2024, p. 2). Ocean energy resources, thought to have a potential of approximately 32,000 GW, exceeding solar and wind energy combined, are estimated to be able to meet 7% of global electricity needs (Segura et al., 2018, p. 528). In this respect, ocean energy offers an advantageous renewable energy source compared to wind and solar energy, thanks to its high energy density, predictability, and consistency (Lehmann et al., 2017, p. 1300).

Ocean energy, in a narrow sense, encompasses energy derived from waves, tides, currents, thermal, and salinity gradient sources, while in a broader sense, it encompasses a richer range of energy sources, including marine biomass, offshore wind, and solar energy (Taveira-Pinto et al., 2020, p. 1160). Wave energy is a type of renewable energy derived from the motion of waves generated by the force exerted by the wind on the sea surface (Küçük, 2025, p. 22). However, utilizing the full energy potential derived from waves depends on technological advancements and consideration of environmental factors.

Tidal energy, a form of ocean energy, occurs in the forms of tidal potential energy and tidal current energy (Soleimani et al., 2015, p. 77). Tidal energy systems generate energy by utilizing water movements created by ocean tides and currents (Khare & Bhuiyan, 2022, p. 1). The high predictability of tidal energy in the long term facilitates its integration into energy systems compared to other renewable sources (Denny, 2009, p. 1914). Although the energy created by tides in shallow seas potentially generates 3000 GW of power, it is largely untapped (Harvey, 2010, p. 316). In this context, large-scale production of tidal energy requires high-capacity turbines operating entirely underwater, replacing the water mills that have been used for small-scale energy production for centuries (Turkenburg et al., 2012, p. 854).

Salinity gradient energy is generated by utilizing the potential difference between waters of different salinity levels (Sharma et al. 2022, p. 1). Salinity gradient energy is released through hybrid processes such as desalination or wastewater treatment plants, or through standalone technologies such as reverse electrodialysis or pressure-delayed osmosis (Azarpour et al., 2022, p. 11). The global energy output of salinity gradient energy in regions where freshwater flows into the sea is estimated at 2.6 TW, and it is thought that this could meet 20% of global energy demand (Post et al., 2007, p. 218).

Ocean thermal energy originates from the temperature gradient between the sea surface and deep water, and ocean thermal energy conversion is used to utilize this energy (Ocean Energy Systems, 2025). The ocean thermal energy conversion process utilizes the temperature difference between the hot water at the ocean's surface and the cold water at depth to drive a low-pressure turbine and generate energy (Faizal & Rafiuddin Ahmed, 2011, p. 1120). It is estimated that approximately 10 TW of power can be generated from ocean thermal energy conversion, which consists of open-loop, closed-loop, and hybrid systems (Üçgül & Elibüyük, 2016, p. 89).

In summary, renewable energy sources such as wave, tidal, and ocean thermal energy hold high promise for meeting future energy needs. To make these technologies commercially viable, intensive research and the development of reliable systems are necessary. Otherwise, even a minor malfunction could lead to the failure of these promising technologies if robust systems that can operate for many years are not created (Harvey, 2010, p. 311).

Blue Finance and Blue Bonds

Blue finance, a new and growing branch of climate finance, plays a significant role in accessing clean water resources, preserving underwater life, and building a sustainable aquatic economy (International Finance Corporation, 2025). Blue finance refers to financial instruments and investments that promote the protection and sustainable use of ocean ecosystems and the growth of the blue economy (Shiiba et al., 2022, p. 1). Blue finance combines the pursuit of economic growth with compliance with legal sustainability standards and the financial support of marine conservation initiatives (Yoshioka et al., 2020, p. 3). In this context, blue finance focuses on raising capital and securing investors to meet the financing needs of ocean-related commercial ventures (Wanta et al., 2023, p. 95).

Blue bonds, a key instrument of blue finance, stand out for their ability to combine financial returns and environmental impact by encouraging private sector investment in sustainable ocean projects. Blue bonds are mostly issued by governments and development banks (Clifford Chance, 2019, p. 2). While the proceeds from blue bonds are channeled into initiatives that support the blue economy, such as biodiversity and sustainable fisheries, these bonds offer financing opportunities similar to traditional debt instruments, with repayments at maturity with interest (Ela & Çoşkun, 2023, p. 1612). Blue bonds are designed in accordance with the Green Bond and Loan Principles to support projects directly related to Sustainable Development Goals 6 (clean water and sanitation) and 14 (underwater life), as well as to provide financing for sustainable initiatives aligned with Sustainable Development Goals 2, 7, 12, 13, and 15 (European Commission, 2025).

The world's first sovereign blue bond was launched by the Republic of Seychelles in 2018 to finance sustainable maritime projects (World Bank Group, 2018). The Republic of Seychelles issued a 10-year bond worth US$15 million to support the expansion of marine protected areas, fisheries management, and the blue economy (United Nations, 2020, p. 4). Since the first blue bond issuance in 2018, blue bonds have been used as a financing tool in various countries. In 2019, the World Bank provided US$10 million from a bond issue aimed at combating marine plastic waste, while the Nordic Investment Bank issued a 2 billion Swedish kronor bond to finance projects aimed at solving the region's water-related problems (United

Nations, 2020, p. 4). In addition, the World Bank raised US$15 million in 2022 by issuing a callable, fixed-interest Blue Development Bond, which draws attention to the problem of marine plastic pollution (European Commission, 2025). Modeled after green bonds, blue bonds have become widespread and have reached a global volume exceeding US$5 billion with more than 26 applications since the beginning of 2023 (March et al., 2024, p. 152). In this context, the use of blue bonds among financial instruments is gradually expanding.

ENABLING MECHANISMS FOR GREEN AND BLUE TRANSFORMATION: GOVERNANCE, MEASUREMENT AND FINANCE

The effective integration of green finance and the blue economy into the sustainable development process is not limited solely to the availability of appropriate financial instruments. The success of this transformation depends on establishing strong governance structures, developing standardized measurement and monitoring frameworks, and ensuring policy alignment. Regulatory fragmentation, inconsistencies in sustainability definitions, and a lack of transparency are among the key obstacles limiting the scalability and credibility of green and blue finance initiatives.

In this context, measurability and accountability are central elements of the transformation process. Developing sustainable finance taxonomies, performance indicators, and impact reporting standards allows policymakers, investors, and other stakeholders to assess the extent to which financial flows contribute to combating climate change, conserving biodiversity, and increasing social inclusion. In the absence of robust measurement mechanisms, green and blue finance practices risk remaining symbolic and increasing the risk of greenwashing.

Evaluating the success of green and blue finance requires monitoring environmental and economic impacts through measurable indicators. In this context, key performance indicators (KPIs) play a critical role in assessing the contribution of investments to climate and sustainability goals. The European Union Taxonomy of Sustainable Finance aims to standardize measurement and reporting processes by defining which economic activities are classified as environmentally sustainable (European Commission, 2021). Among the most commonly used indicators are greenhouse gas emissions prevented, impact on biodiversity, and the investment leverage ratio, which shows the capacity of public resources to mobilize private investment (World Bank, 2022b). In this section, the success of green finance and blue economy practices is evaluated through the key indicators presented in Table 2.

Table 2. Measuring the Performance of Green and Blue Finance: Key Indicators

Dimension	Key Indicator (KPI)	Definition	Policy Relevance
Climate Impact	Avoided GHG Emissions (tCO_2e)	Amount of greenhouse gas emissions reduced or avoided through financed projects	Alignment with Paris Agreement targets
Biodiversity	Ecosystem Restoration Index	Area and quality of restored or protected ecosystems (marine and terrestrial)	Conservation of natural capital
Financial Leverage	Investment Leverage Ratio	Ratio of private capital mobilized per unit of public finance	Efficiency of public funding
Social Impact	Local Employment Creation	Number of decent jobs created in green and blue sectors	Inclusive and just transition
Governance	Taxonomy Alignment Rate	Share of investments aligned with recognized sustainability taxonomies	Transparency and prevention of greenwashing
Resilience	Climate Risk Reduction Index	Reduction in exposure to physical and transition climate risks	Long-term economic resilience

Note: Author's own work.

The measurement framework encompasses six main dimensions: climate impact, biodiversity, financial leverage, social impact, governance, and resilience. While environmental impacts such as greenhouse gas emission reduction and ecosystem protection are monitored at the project level, indicators such as investment leverage ratio and taxonomy compliance ratio allow for the evaluation of the efficient use of public and private sector resources. This approach aims to analyze green and blue financing instruments not only based on their financial size but also on their tangible contributions to sustainable development goals.

Public policies play a guiding and catalytic role in this process. Providing clear regulatory signals, reducing investment uncertainty, and mobilizing private sector capital through mixed financing and risk-sharing mechanisms are critical to accelerating sustainable transformation. However, the multifaceted nature of environmental risks, the diversification of financial instruments, and increasing inter-sectoral interactions are making decision-making processes increasingly complex.

This increasing complexity also necessitates advanced analytical tools for evidence-based policy design and effective investment decisions. Building upon this foundation, which is built upon strengthened governance and measurement infrastructure, artificial intelligence technologies, as discussed in the next section examines the role of artificial intelligence in scaling, monitoring and optimizing green and blue economic systems.

TRANSITION TO A SUSTAINABLE ECONOMY WITH ARTIFICIAL INTELLIGENCE

Artificial intelligence, which acquires the ability to learn, adapt, and autonomously solve problems by processing big data, is a system that enables machines to perform complex tasks by imitating human intelligence (Türkcan, 2025, p. 126). With its inherent capabilities, AI has the potential to synchronize economic, ecological, and social factors. In this regard, artificial intelligence (AI) is emerging as a significant driver of sustainable economic development. AI technologies offer effective solutions in critical areas such as environmental monitoring, energy efficiency, and resource management through big data analysis, machine learning, and automation (Rolnick et al., 2019). Particularly in combating climate change and green transformation strategies, AI-based models provide support for a wide range of applications, from weather forecasting to monitoring greenhouse gas emissions (Vinuesa et al., 2020).

Artificial intelligence, effectively used within industrial processes, not only increases resource efficiency in the industrial sector but also enables green and sustainable economic growth by reducing carbon emissions and the impact of climate change (Qian et al., 2023, p. 16419). In the green economy, AI plays a role in optimizing energy production and consumption, integrating renewable energy sources, and reducing the carbon footprint. In the blue economy, AI is used in areas such as habitat analysis for marine species, preventing illegal hunting, and early detection of marine pollution (Jiang and Li, 2021). These technologies contribute to sustainable policymaking by providing decision-makers with real-time and accurate information.

Artificial intelligence has the potential to bridge the gap between economic development goals and the planet's ecological resilience by acting as a bridge in blue economy applications (Spalding, 2024: 29). By combining massive data from various sources, such as satellites, underwater drones, and IoT sensors, AI generates holistic and meaningful information about ocean conditions (Babu et al., 2023, p. 128). By analyzing large marine datasets and predicting environmental changes, AI offers the opportunity to more effectively understand and protect the oceans through applications such as detecting anomalies such as algal blooms and oil spills, modeling the effects of climate change, and exploring the deep sea with autonomous vehicles (Verma & Sharma, 2025, p. 33). However, the ethical and social dimensions of AI in the sustainability arena also need to be considered. Data privacy, algorithmic bias, and inequalities in access to technology are the primary obstacles to ensuring this transformation is socially inclusive (Floridi et al., 2018).

Consequently, AI is a critical tool that accelerates innovation in the green and blue economy and contributes to sustainable development. Policymakers need to simultaneously plan regulations, education, and infrastructure investments to maximize the potential of AI.

POLICY RECOMMENDATIONS AND FUTURE PERSPECTIVES

The establishment and implementation of effective policy frameworks is crucial for achieving sustainable development. Policies supporting green and blue economies should combine environmental protection with economic growth and encourage innovation and technological advancements (OECD, 2021). In this context, carbon pricing, renewable energy incentives, environmental regulations, and the expansion of green financing mechanisms are among the key strategies (UNEP, 2021b).

Successful examples of implementation are leading the way in both the green and blue economies. The European Union's Green Deal supports sustainable investments and prioritizes green technologies, aiming to achieve carbon neutrality (European Commission, 2019). The European Union Green Deal is a comprehensive policy initiative aimed at restructuring energy, industry, agriculture, and finance policies within a holistic framework, in line with the goal of a climate-neutral economy by 2050. The sustainable finance taxonomy developed under the deal plays a key role in guiding green investments by defining which economic activities are considered environmentally sustainable. Furthermore, it aims to direct private sector investment towards low-carbon projects through green bonds and publicly funded mechanisms. However, differences in financial capacity, institutional infrastructure, and speed of implementation among member states are among the key challenges limiting the effectiveness of the deal (European Commission, 2019). Within the blue economy, countries like Sweden and Canada are protecting ecosystems and increasing economic benefits through innovative policies in sustainable fisheries management and marine conservation (World Bank, 2022b). Furthermore, blue carbon projects developed in Kenya to protect mangrove ecosystems aim to create economic value through carbon markets for coastal ecosystems with high carbon sequestration capacity. These projects contribute to the reduction of greenhouse gas emissions while simultaneously creating alternative income sources for local communities. Blue carbon initiatives offer an integrated policy model by addressing climate change adaptation, biodiversity conservation, and poverty reduction goals together. However, long-term financing sustainability and strengthening local governance capacity are critical to the success of these projects (UNEP, 2020b).

While green and blue economy approaches are often discussed within a normative and optimistic framework, these transformation processes can also bring about significant distributional impacts and social inequality issues. In particular, the financing costs of green investments can place a relatively higher burden on low-income households in the short term, while the unequal emergence of environmental and economic benefits in the long term raises discussions about equitable transition. This situation demonstrates that sustainability policies should be addressed not only from an environmental but also from a social perspective.Furthermore, large-scale blue economy projects can threaten the traditional livelihoods of local communities and lead to social conflicts if adequate participation mechanisms are not established. Therefore, green and blue transition policies need to consider social justice and inclusion dimensions in addition to environmental goals.

The future economic vision requires integrating green and blue economy policies. This integrated approach can offer comprehensive solutions to multidimensional challenges such as climate change, biodiversity loss, and social inequalities. By integrating artificial intelligence and digital technologies into policymaking processes, data-driven decision-making mechanisms will be strengthened and the efficient use of resources will increase (Vinuesa et al., 2020). As a result, to achieve sustainable development goals, policy makers need to increase multi-stakeholder cooperation and international coordination, support innovative financing models and closely follow technological developments.

CONCLUSİON

In today's world, where economics and ecology cannot be considered separately, the concept of sustainability has become central to economic models. Green and blue economy approaches are the cornerstones of a new paradigm that aims to balance economic growth with the protection of natural resources. This paradigm combines multifaceted objectives such as efficient use of resources, reducing environmental impacts, and ensuring social justice.

Combating climate change has moved beyond being merely an environmental imperative to a decisive factor in shaping economic policies. In this process, strategies to reduce the carbon footprint of economic activities, sustainable financing models, and technological innovations are gaining prominence. Technologies such as digitalization and artificial intelligence are enabling the development of more effective, faster, and data-driven approaches to economic decision-making.

As a result, the economic model of the future is taking shape as a flexible and innovative system built on the balanced integration of the environment, economy, and society. This new paradigm is essential for achieving development goals, ensuring the preservation of natural resources, and leaving a livable world for future generations.

REFERENCES

Agyekum, E. B., Khan, T., Ampah, J. D., Giri, N. C., Mbasso, W. F., & Kamel, S. (2024). Review of the marine energy environment-A combination of traditional, bibliometric and PESTEL analysis. *Heliyon, 10*(6), e27771. DOI: 10.1016/j.heliyon.2024.e27771 PMID: 38524577

Armendáriz, B., & Morduch, J. (2010). *The economics of microfinance* (2nd ed.). MIT Press.

Babu, C. V. S., Begum, M. S. A., & Karoldas, S. (2023). Blockchain and AI for advancing sustainability in the blue economy: Innovating marine resource management and environmental stewardship. In Sharma, V., Singh, T., de Bem Machado, A., & Bajcsi-Ábrigo, R. (Eds.), *FinTech for Sustainable Finance and a Well-Blue Economy* (pp. 127–150). Springer.

Bari, A. (2017). Our oceans and the blue economy: Opportunities and challenges. *Procedia Engineering, 194*, 5–11. DOI: 10.1016/j.proeng.2017.08.109

Briones-Peñalver, A. J., Prokopchuk, L., & Samoilyk, I. (2023). Strategic vectors of coastal tourism development as a blue economy component in the international dimension. *Journal of Environmental Management & Tourism, 14*(6), 2473–2496. DOI: 10.14505/jemt.v14.6(70).01

Burke, M., Hsiang, S. M., & Miguel, E. (2015). Global non-linear effect of temperature on economic production. *Nature, 527*(7577), 235–239. DOI: 10.1038/nature15725 PMID: 26503051

Chang, L., Taghizadeh-Hesary, F., Chen, H., & Mohsin, M. (2022). Do green bonds have environmental benefits? *Energy Economics, 115*, 106356. DOI: 10.1016/j.eneco.2022.106356

Cisneros-Montemayor, A. M., Croft, F., Issifu, I., Swartz, W., & Voyer, M. (2022). A primer on the "blue economy:" Promise, pitfalls, and pathways. *One Earth, 5*(9), 982–986. DOI: 10.1016/j.oneear.2022.08.011

Clifford Chance. (2019). *Blue bonds expanding to the oceans*. https://www.cliffordchance.com/content/dam/cliffordchance/briefings/2019/11/blue-bondsexpanding -to-the-oceans.pdf

Columbia Climate School. (2022). *Billion-dollar climate disasters and their economic costs*. Columbia University. https://news.climate.columbia.edu

Denny, E. (2009). The economics of tidal energy. *Energy Policy*, *37*(5), 1914–1924. DOI: 10.1016/j.enpol.2009.01.009

Ela, M., & Coşkun, S. (2023). Blue Sukuk: Bir model önerisi ve potansiyeli [Blue Sukuk: A model proposal and its potential]. *Journal of Abant Social Sciences*, *23*(3), 1609–1620.

European Commission. (2019). *The European Green Deal*. https://ec.europa.eu/info/strategy/priorities-2019-2024/european-green-deal_en

European Commission. (2021). *EU emissions trading system (EU ETS)*. https://climate.ec.europa.eu/eu-action/eu-emissions-trading-system-eu-ets_en

European Commission. (2025). Blue bond. *In Restore our ocean and waters, Horizon Europe*. https://projects.research-and-innovation.ec.europa.eu/en/funding/funding-opportunities/funding-programmes-and-open-calls/horizon-europe/eu-missions-horizon-europe/restore-our-ocean-and-waters/blue-bond

European Environment Agency (EEA). (2025). *Status of marine fish and shellfish stocks in European seas*. https://www.eea.europa.eu/en/analysis/indicators/status-of-marine-fish-and?activeAccordion=ecdb3bcf-bbe9-4978-b5cf-0b136399d9f8

Faizal, M., & Ahmed, M. R. (2011). On the ocean heat budget and ocean thermal energy conversion. *International Journal of Energy Research*, *35*(13), 1119–1144. DOI: 10.1002/er.1885

Flammer, C. (2021). Corporate green bonds. *Journal of Financial Economics*, *142*(2), 499–516. DOI: 10.1016/j.jfineco.2021.01.010

Floridi, L., Cowls, J., Beltrametti, M., Chatila, R., Chazerand, P., Dignum, V., Luetge, C., Madelin, R., Pagallo, U., Rossi, F., Schafer, B., Valcke, P., & Vayena, E. (2018). AI4People—An ethical framework for a good AI society: Opportunities, risks, principles, and recommendations. *Minds and Machines*, *28*(4), 689–707. DOI: 10.1007/s11023-018-9482-5 PMID: 30930541

Food and Agriculture Organization of the United Nations (FAO). (2025). *SDG indicators data portal*. https://www.fao.org/sustainable-development-goals-data-portal/data/indicators/1441-fish-stocks-sustainability/en

Global Sustainable Investment Alliance (GSIA). (2021). *Global sustainable investment review 2020*. GSIA. https://www.gsi-alliance.org/trends-report-2020/

Harvey, L. D., (2010). *Energy and the new reality 2: Carbon-free energy supply*. Earthscan.

Intergovernmental Panel on Climate Change (IPCC). (2023). AR6 synthesis report. *Climatic Change, 2023*, •••. https://www.ipcc.ch/report/ar6/syr/

International Energy Agency (IEA). (2021). *World energy outlook 2021*. IEA. https://www.iea.org/reports/world-energy-outlook-2021

International Finance Corporation. (2025). Blue finance. https://www.ifc.org/en/what-we-do/sector-expertise/financial-institutions/climate-finance/blue-finance

International Finance Corporation (IFC). (2022). *Sustainability-linked finance: Mobilizing capital for a sustainable future*. https://www.ifc.org/en/insights-reports/2022/sustainability-linked-finance

Jiang, X., & Li, B. (2021). Artificial intelligence applications in marine science: A review. *Marine Policy*, *132*, 104664. DOI: 10.1016/j.marpol.2021.104664

Khare, V., & Bhuiyan, M. A. (2022). Tidal energy-path towards sustainable energy: A technical review. *Cleaner Energy Systems*, *3*, 100041. DOI: 10.1016/j.cles.2022.100041

Küçük, A. O. (2025). Artvin Hopa sahil şeridi kısa dönem dalga enerjisi potansiyeli tespiti [Artvin Hopa coastline short-term wave energy potential determination]. *Journal of Cihannuma Technology. Engineering and Natural Sciences Academy*, *1*(1), 21–32.

Lehmann, M., Karimpour, F., Goudey, C. A., Jacobson, P. T., & Alam, M.-R. (2017). Ocean wave energy in the united states: Current status and future perspectives. *Renewable & Sustainable Energy Reviews*, *74*, 1300–1313. DOI: 10.1016/j.rser.2016.11.101

Liu, M. (2022). The driving forces of green bond market volatility and the response of the market to the COVID-19 pandemic. *Economic Analysis and Policy*, *75*, 288–309. DOI: 10.1016/j.eap.2022.05.012

LMA. APLMA, & LSTA. (2023). *Sustainability linked loan principles*. Loan Market Association.

Lopez-Rivas, J. D., & Cardenas, J. C. (2024). What is the economic value of coastal and marine ecosystem services? A systematic literature review. *Marine Policy*, *161*, 106033. DOI: 10.1016/j.marpol.2024.106033

March, A., Evans, T., Laing, S., & Raguain, J. (2024). Evaluating the world's first sovereign blue bond: Lessons for operationalising blue finance. *Commodities*, *3*(2), 151–167. DOI: 10.3390/commodities3020010

Musa, S. D., Zhonghua, T., Ibrahim, A. O., & Habib, M. (2018). China's energy status: A critical look at fossils and renewable options. *Renewable & Sustainable Energy Reviews*, *81*(2), 2281–2290. DOI: 10.1016/j.rser.2017.06.036

Nawaz, F., Khan, M. A., & Khan, S. (2021). Sustainable development goals and climate action: Progress and challenges. *Environmental Science & Policy*, *120*, 23–34. DOI: 10.1016/j.envsci.2021.03.012

Netto, M. P., Freire, P. L. S., Martins, G., Augusto, M., Abude, R. R. S., Cardoso, R. S., Miceli, B. S., & Cabrini, T. M. B. (2025). A decade of ecological awareness on marine coastal ecosystems: University extension engaging basic education students in Brazil. *Marine Pollution Bulletin*, *219*, 118315. DOI: 10.1016/j.marpolbul.2025.118315 PMID: 40544529

Ocean Energy Systems. (2025). *Waves*. https://www.ocean-energy-systems.org/ocean-energy/what-is-ocean-energy/

Organisation for Economic Co-operation and Development (OECD). (2019). *Rethinking innovation for a sustainable ocean economy*. https://www.oecd.org/environment/marine/rethinking-innovation-for-a-sustainable-ocean-economy-bede6513-en.htm

Organisation for Economic Co-operation and Development (OECD). (2020a). *Developing sustainable finance definitions and taxonomies*. OECD Publishing. https://www.oecd.org/en/publications/developing-sustainable-finance-definitions-and-taxonomies_134a2dbe-en.html DOI: 10.1787/134a2dbe-en

Organisation for Economic Co-operation and Development (OECD). (2020b). *OECD business and finance outlook 2020: Sustainable and resilient finance*. OECD Publishing., DOI: 10.1787/eb61fd29-

Organisation for Economic Co-operation and Development (OECD). (2021). *Key findings from the update of the OECD green recovery database*. https://www.oecd.org/environment/green-recovery/

Ozan, S., & Başer, U. (2024). Sürdürülebilir balıkçılık mümkün mü? [Is sustainable fishing possible?]. *Turkish Journal of Agriculture - Food Science and Technology*, *12*(1), 109–115. DOI: 10.24925/turjaf.v12i1.109-115.6186

Post, J. W., Veerman, J., Hamelers, H. V. M., Euverink, G. J. W., Metz, S. J., Nymeijer, K., & Buisman, C. J. N. (2007). Salinity-gradient power: Evaluation of pressure-retarded osmosis and reverse electrodialysis. *Journal of Membrane Science*, *288*(1–2), 218–230. DOI: 10.1016/j.memsci.2006.11.018

Qian, Y., Liu, J., Shi, L., Forrest, J. Y. L., & Yang, Z. (2023). Can artificial intelligence improve green economic growth? Evidence from China. *Environmental Science and Pollution Research International*, *30*(6), 16418–16437. DOI: 10.1007/s11356-022-23320-1 PMID: 36184706

Rolnick, D., Donti, P. L., Kaack, L. H., Kochanski, K., Lacoste, A., Sankaran, K., & Bengio, Y. (2019). Tackling climate change with machine learning. *Nature Communications*, *10*(1), 1–10. DOI: 10.1038/s41467-019-08759-2 PMID: 30602773

Sachs, J. D., Schmidt-Traub, G., Kroll, C., Lafortune, G., & Fuller, G. (2019). *Sustainable development report 2019*. Bertelsmann Stiftung and Sustainable Development Solutions Network (SDSN). https://s3.amazonaws.com/sustainabledevelopment.report/2019/2019_sustainable_development_report.pdf

Sang, Y., Karayaka, H. B., Yan, Y., Yilmaz, N., & Souders, D. (2018). Ocean (marine) energy. In Dincer, I. (Ed.), *Comprehensive energy systems* (pp. 733–769). Elsevier. DOI: 10.1016/B978-0-12-809597-3.00120-6

Schlacher, T. A., Dugan, J., Schoeman, D. S., Lastra, M., Jones, A., Scapini, F., McLachlan, A., & Defeo, O. (2007). Sandy beaches at the brink. *Diversity & Distributions*, *13*(5), 556–560. DOI: 10.1111/j.1472-4642.2007.00363.x

Schreiner, M. (2002). Aspects of outreach: A framework for discussion of the social benefits of microfinance. *Journal of International Development*, *14*(5), 591–603. DOI: 10.1002/jid.908

Sharma, M., Das, P. P., Chakraborty, A., & Purkait, M. K. (2022). Clean energy from salinity gradients using pressure retarded osmosis and reverse electrodialysis: A review. *Sustainable Energy Technologies and Assessments*, *49*, 101687. DOI: 10.1016/j.seta.2021.101687

Shiiba, N., Wu, H. H., Huang, M. C., & Tanaka, H. (2022). How blue financing can sustain ocean conservation and development: A proposed conceptual framework for blue financing mechanism. *Marine Policy*, *139*, 104575. DOI: 10.1016/j.marpol.2021.104575

Smith Azarpour, A., Mohammadzadeh, O., Rezaei, N., & Zendehboudi, S. (2022). Current status and future prospects of renewable and sustainable energy in North America: Progress and challenges. *Energy Conversion and Management*, *269*, 115945. DOI: 10.1016/j.enconman.2022.115945

Soleimani, K., Ketabdari, M. J., & Khorasani, F. (2015). Feasibility study on tidal and wave energy conversion in Iranian seas. *Sustainable Energy Technologies and Assessments*, *11*, 77–86. DOI: 10.1016/j.seta.2015.03.006

Spalding, M. J. (2024). AI and improving the blue economy. *Journal of Ocean Technology*, *19*(2), 28–35.

Suluk, S. (2022). Ekonominin renkleri: Sürdürülebilir mavi ekonomi bağlamında Türkiye'nin değerlendirilmesi [Colors of the economy: An evaluation of Turkey in the context of the sustainable blue economy]. *Dumlupınar University Journal Of Social Sciences*, *74*(74), 132–150. DOI: 10.51290/dpusbe.1123257

Surminski, S., & Thieken, A. H. (2017). Promoting risk reduction: The role of insurance in climate adaptation and disaster risk management. *Nature Climate Change*, *7*(9), 703–706. DOI: 10.1038/nclimate3390

Taveira-Pinto, F., Rosa-Santos, P., & Fazeres-Ferradosa, T. (2020). Marine renewable energy. *Renewable Energy*, *150*, 1160–1164. DOI: 10.1016/j.renene.2019.10.014

Türkcan, B. (2025). Sürdürülebilir kalkınmada yapay zekanın rolü: Amaçlar için etkin politika formasyonu [The role of artificial intelligence in sustainable development: Effective policy formation for achieving the goals]. *Çukurova University Journal of Social Sciences Institute, 34 (Transformation of Civilization: Artificial Intelligence)*, 125-141.

Turkenburg, W., Arent, D. J., Bertani, R., Faaij, A., Hand, M., Krewitt, W., Larson, E. D., Lund, J., Mehos, M., Merrigan, T., Mitchell, C., Moreira, J. R., Sinke, W., Sonntag-O'Brien, V., Thresher, B., van Sark, W., Usher, E., Bilello, D., Chum, H., & Schmid, J. (2012). Renewable energy. In *Global energy assessment: Toward a sustainable future* (pp. 761–900). Cambridge University Press. DOI: 10.1017/CBO9780511793677.017

Üçgül, İ., & Elibüyük, U. (2016). Okyanus termal enerji dönüşüm (OTEC) sistemi [Ocean thermal energy conversion (OTEC) system]. *Erzincan University Journal of Science and Technology*, *9*(1), 87–94.

United Nations. (UN). (2020). *Practical guidance to ıssue a blue bond.* https://ungc-communicationsassets.s3.amazonaws.com/docs/publications/Practical-Guidance-to-Issue-a Blue-Bond.pdf

United Nations Conference on Trade and Development (UNCTAD). (2020). *The blue economy: Policy handbook.* https://unctad.org/webflyer/blue-economy-policy-handbook

United Nations Environment Programme Finance Initiative (UNEP FI). (2021). *Rethinking impact to finance the SDGs*. https://www.unepfi.org/publications/rethinking-impact-to-finance-the-sdgs/

United Nations Environment Programme Finance Initiative (UNEP FI). (2022). *Global guidance on green financial products and services*. https://www.unepfi.org

United Nations Environment Programme (UNEP). (2020a). *Microfinance for ecosystem-based adaptation: Options, costs and benefits*. United Nations Environment Programme. https://www.unep.org/resources/publication/microfinance-ecosystem-based-adaptation-options-costs-and-benefits

United Nations Environment Programme (UNEP). (2020b). *Out of the blue: The value of seagrasses to the environment and to people*. https://www.unep.org/resources/report/out-blue

United Nations Environment Programme (UNEP). (2021a). *Adaptation gap report 2021*. https://www.unep.org/resources/adaptation-gap-report-2021

United Nations Environment Programme (UNEP). (2021b). *Financing a sustainable recovery: Progress and prospects*. https://www.unep.org/resources/report/financing-sustainable-recovery

United Nations (UN). (1992). Agenda 21: *United nations conference on environment & development*. United Nations. https://sustainabledevelopment.un.org/content / documents/ Agenda21.pdf

United Nations (UN). (2012). Report of the United Nations conference on sustainable development (A/CONF.216/16). https://docs.un.org/en/A/CONF.216/16

United Nations (UN). (2015). *Transforming our world: The 2030 agenda for sustainable development*. https://sdgs.un.org/goals

University of Nairobi Maritime Centre. (2023). An assessment of the status of blue economy sectors in Kenya: Synthesis report.UNEP–NC. https://www.nairobiconvention.org/clearinghouse/sites/default/files/Synthesis%20Report%20Popular%20Version%2013062023.pdf

Verma, S., & Sharma, M. (2025). Role of digitalization in a sustainable blue economy. *International Journal of Teaching. Learning and Education*, *4*(3), 618013.

Vinuesa, R., Azizpour, H., Leite, I., Balaam, M., Dignum, V., Domisch, S., Felländer, A., Langhans, S. D., Tegmark, M., & Fuso Nerini, F. (2020). The role of artificial intelligence in achieving the Sustainable Development Goals. *Nature Communications*, *11*(1), 1–10. DOI: 10.1038/s41467-019-14108-y PMID: 31932590

Volz, U. (2018). Financing sustainable development: Implementing the UN Sustainable Development Goals through financial sector reforms and investments. *International Review of Financial Analysis*, *59*, 123–135. DOI: 10.1016/j.irfa.2018.05.002

Voyer, M., Quirk, G., McIlgorm, A., & Azmi, K. (2018). Shades of blue: What do competing interpretations of the blue economy mean for oceans governance? *Journal of Environmental Policy and Planning, 20*(5), 595–616. DOI: 10.1080/1523908X.2018.1473153

Wanta, D., Khomsiyah, K., & Gunawan, J. (2023). Blue finance: Is this supporting SDG 14 financing gap. *International Journal of Social and Management Studies, 4*(1), 93–105.

World Bank. (2022a). *State and trends of carbon pricing 2022*. World Bank Group. https://openknowledge.worldbank.org/handle/10986/37455

World Bank. (2022b). *Blue economy development framework*. https://www.worldbank.org/en/topic/environment/publication/blue-economy-development-framework

World Bank. (2024). *The World Bank impact report 2024: Sustainable development bonds & green bonds (Fiscal year ending June 30, 2024)*. World Bank Treasury. https://treasury.worldbank.org/en/about/unit/treasury/impact/impact-report

World Bank and United Nations Department of Economic and Social Affairs. (2017). *The potential of the blue economy: Increasing long-term benefits of the sustainable use of marine resources for small island developing states and coastal least developed countries*. World Bank.

World Bank Group. (2018). *Seychelles launches world's first sovereign blue bond*. World Bank. https://www.worldbank.org/en/news/press-release/2018/10/29/seychelles-launches-worlds-first-sovereign-blue-bond

WWF-Türkiye (Doğal Hayatı Koruma Vakfı). (2013). *Sürdürülebilir balıkçılık için "ekosistem temelli yönetim" [Ecosystem-based management for sustainable fisheries]*. https://wwftr.awsassets.panda.org/downloads/web_surdurulebilir_balkclk.pdf

Yıldırım, G. (2024). Mavi ekonomi ve turizm. [Blue economy and tourism] In Özgürel, G., & Karadağ, L. (Eds.), *İnsan toplum ve çevre ekseninde turizm* [Tourism in the human, society and environment axis]. (pp. 1–14). Detay Yayıncılık.

Yoshioka, N., Wu, H. H., Huang, M. C., & Tanaka, H. (2020). *Proposing regulatory-driven blue finance mechanism for blue economy development* (ADBI Working Paper Series No. 1157). ADBI.

Youssef, M. (2023). Blue economy literature review. *International Journal of Business and Management, 18*(3), 12–18. DOI: 10.5539/ijbm.v18n3p12

Zhang, D., Zhang, Z., & Managi, S. (2019). A bibliometric analysis on green finance: Current status, development, and future directions. *Finance Research Letters, 29*, 425–430. DOI: 10.1016/j.frl.2019.02.003

ADDITIONAL READING

Bhutta, U. S., Tariq, A., Farrukh, M., Raza, A., & Iqbal, M. K. (2022). Green bonds for sustainable development: Review of literature on development and impact of green bonds. *Technological Forecasting and Social Change, 175*, 121378. DOI: 10.1016/j.techfore.2021.121378

Campiglio, E., Dafermos, Y., Monnin, P., Ryan-Collins, J., Schotten, G., & Tanaka, M. (2018). Climate change challenges for central banks and financial regulators. *Nature Climate Change, 8*(6), 462–468. DOI: 10.1038/s41558-018-0175-0

Khan, M., Serafeim, G., & Yoon, A. (2016). *Corporate Sustainability: First Evidence on Materiality*. Harvard Business School Press.

Maltais, A., & Nykvist, B. (2020). Understanding the role of green bonds in advancing sustainability. *Journal of Sustainable Finance & Investment*, •••, 1–20. DOI: 10.1080/20430795.2020.1724864

Pauli, G. (2015). *The blue economy: 10 years, 100 innovations, 100 million jobs*. Paradigm Publications.

Sachs, J. D., Woo, W. T., Yoshino, N., & Taghizadeh-Hesary, F. (Eds.). (2019). *Handbook of green finance: Energy security and sustainable development*. Springer. DOI: 10.1007/978-981-13-0227-5

Vivas, F. J., De las Heras, A., Segura, F., & Andújar, J. M. (2018). A review of energy management strategies for renewable hybrid energy systems with hydrogen backup. *Renewable & Sustainable Energy Reviews, 82*, 126–155. DOI: 10.1016/j.rser.2017.09.014

Volz, U. (2018). *Fostering green finance for sustainable development in Asia*. Asian Development Bank Institute. DOI: 10.4324/9781315543222-27

Volz, U., Böhnke, J., Knierim, L., & Richert, K. (2020). *Climate Change and Financial Risk*. Routledge.

KEY TERMS AND DEFINITIONS

Artificial intelligence (AI): The ability of computers and software to mimic human-like thinking and decision-making processes.

Climate change: Permanent changes in the Earth's long-term temperature, precipitation, wind, and other climate indicators.

Digital transformation: The process of reshaping the functioning of institutions, businesses, or society through digital technologies.

Ecology: The branch of science that studies the relationships between living things and their physical environment.

Emission: Refers to gases, particles, or other substances released into the environment (atmosphere, water, or soil) from a source. Specifically, in the context of climate change, carbon dioxide (CO_2) and other greenhouse gases released into the atmosphere from human-induced activities such as fossil fuel burning, industrial activities, and transportation are considered emissions.

ESG: It stands for Environmental, Social, and Governance. ESG is an analysis and investment approach that evaluates companies not only on their financial performance but also on their environmental and social impacts.

Sustainable development: The conservation and balanced use of natural resources to meet the needs of future generations in the process of increasing economic growth and social well-being.

Chapter 16

The Contribution of Smart City Applications to Combating Climate Change

Asime Dilara Erdem
https://orcid.org/0000-0002-7422-7473
Çanakkale Onsekiz Mart University, Turkey

Didem Saygın
https://orcid.org/0000-0002-2832-351X
Biga Faculty of Economics and Administrative Sciences, Çanakkale Onsekiz Mart University, Turkey

ABSTRACT

As a result of the urbanization movements that started to increase after the Industrial Revolution, cities cause climate change and are also affected by the climate crisis caused by climate change. Especially overpopulated cities have implemented many smart city applications to combat climate change. The idea of smart cities aims to make cities more livable, resilient and sustainable with the help of technology. In this context, firstly, the concept of smart city is emphasized in this study. The emergence process of smart cities is examined, and it is tried to explain why and in what ways smart city applications are needed. In addition, it is stated which smart city application aims to prevent which aspect of climate change. In this context, it is thought that the study will contribute to the literature in terms of addressing smart city applications from a holistic perspective and matching smart city applications with policies to combat climate change.

DOI: 10.4018/979-8-3373-5238-1.ch016

BACKGROUND

Changing living conditions, coupled with technological advances, have led to population growth, particularly in cities, where employment opportunities are greater than in other settlements. The concept of urbanisation emerged concurrently with the Industrial Revolution, primarily due to employment opportunities. Although the population of cities has now shifted from industry to the service sector, urbanisation has become irreversible. In other words, urban growth raises various issues that could jeopardise the economic, environmental and social sustainability of cities (Neirotti et al., 2014). In particular, the increased use of natural energy resources, along with the growth of global industrial activities, has contributed to global warming (Shuja et al., 2017). The gases emitted from the chimneys of production facilities to ensure the continuity of industrialisation and later the service sector, the increased traffic due to service provision and overpopulation, and the resulting exhaust pollution, and the intensive use of fossil fuels for energy production and heating, among many other factors that have increased and clustered with urbanisation, act as a blanket over the atmosphere, affecting global warming and causing climate change. Climate crises arise due to climate change, and problems such as extreme rainfall or extreme drought can occur.

International organisations, national governments and local authorities are working to implement measures to combat climate change. The United Nations (UN) "2030 Sustainable Development Goals" and the European Union (EU) "European Green Deal" are prime examples of international organisations' efforts to combat climate change. In addition, countries are also participating in the fight against climate change, both by supporting these initiatives and by developing their own policies. However, as cities are where climate change or the climate crisis is most tangibly experienced, it is particularly necessary for local governments to be prepared for the crises created by climate change and to make cities more liveable, resilient and sustainable. In this context, with the advancement of technology, many smart city applications integrated with technology have been created within the framework of the smart city concept.

Examples of smart cities from around the world have been examined, and it has been observed that most smart city applications focus on traffic and public transport, water usage, energy efficiency and citizen awareness, with the aim of reducing greenhouse gases, which are considered to be the root cause of the problem in the context of combating climate change.

1. CONCEPTUAL FRAMEWORK

Cities are considered the main source of greenhouse gas emissions due to the intensity of production and service activities and are also recognised as playing a key role in preventing climate change. For this reason, climate-smart cities, a type of smart city, are gaining importance (Pee & Pan, 2022). Accordingly, it is stated that most actions aimed at mitigating the effects of climate change take place in cities (Contreras & Platania, 2019). There is no universally accepted definition of the concept of smart cities, as it evolves according to the specific needs of each city. However, it generally aims to take action on issues such as sustainability, the environment, energy, and climate change through the use of "information and communication Technologies" (ICT), depending on technological developments. It is emphasised that digital cities should not be confused with smart cities. A digital city refers to a city where all procedures, communication and information are digitised, while a smart city is defined as a digital city that can make high-level decisions based on artificial intelligence through the development of digital applications, i.e. a city with an intelligence layer. In addition, smart cities focus on the practicality of applications and the user experience (Yin et al., 2015). In other words, it is stated that smart cities are built upon the activities of independent and conscious citizens (Giffinger et al., 2007).

Although there are many views on the definition of a smart city, the definitions can be grouped into two main categories. The first defines a smart city as connecting physical, social, information and communication technologies and infrastructures to utilise the collective intelligence of the city (Harrison et al., 2010), while the second view defines a smart city as an intelligent combination of economy, governance, people, mobility, environment and citizen participation in governance (Gülseçen et al., 2013; Peris-Ortiz et al., 2017). Accordingly, a smart city is generally defined as *"an urban development vision that aims to provide technological solutions to problems arising in the field of urbanisation, particularly population growth, environmental issues, energy consumption, transportation, security, education and health"* (Güney & Yalçın, 2014). It has been stated that a smart city has six components: economy, environment, governance, people, mobility, and life (Giffinger & Gudrun, 2010).

It is stated that smart cities will play an important role in the implementation of sustainable economic development models within the framework of the smart economy. This is because the relationship between sustainability and the economy in smart cities is to be able to provide high-quality services at low cost (Vaquero-Garcia et al., 2016). Considering that smart city applications implemented within the scope of combating climate change require significant costs, the connection with the economy becomes even clearer. Smart cities aim to improve the quality of life of city dwellers through sustainable practices (Çetin & Çiftçi, 2019), enabling them to save time and energy and live in a cleaner environment. Furthermore, utilising ICTs to intelligently plan and develop cities forms the basis for data and sustainable development strategies to achieve long-term sustainability goals (Bibri & Krogstie, 2017). It is also stated that for a city to become a smart city, innovation, problem-solving skills and technological development must coexist (Örselli & Akbay, 2019). However, although smart cities are considered "environmentally sustainable", the difficulty of measuring the real impact of many components is seen as one of the problems of smart cities (Yang et al., 2020).

Cities conceptualised as "climate-smart cities" have emerged in the fight against climate change. The aim of these cities is to use information and communication technologies to prepare for the dangers posed by climate change, take measures to reduce carbon emissions, and unite stakeholders by informing them about the factors that cause climate change. More broadly defined, *"climate-smart cities refer to cities where urban development and planning, policies, and services do not harm the environment and where the negative effects of climate change are gradually reduced"* (Tekin, 2023). Another inclusive definition of a smart city is *"an environmentally friendly city where information technologies are used in the provision of urban services, participation is enabled through technology, and individuals and institutions that follow innovations and actively learn are present"* (Gül & Atak Çobanoğlu, 2017). According to this definition, the smart city concept emphasises lifelong learning, as it refers to individuals and institutions that follow innovations and actively learn. Smart cities gain importance in terms of encouraging lifelong learning. According to another definition, *"climate-smart cities are defined as 'climate-priority and climate-resilient, digitally enabled cities that establish a short-term socio-economic development goal linking zero or negative greenhouse gas emissions to citizens quality of life and businesses financial success"* (Pee & Pan, 2022). Accordingly, it can be stated that sustainability is frequently emphasised in the definitions of smart cities and that it is part of a more holistic view based on sustainability (Albino et al., 2015). Furthermore, there is a view that refers to economics in the creation of smart cities. According to this perspective, the concept of smart cities emerged from the idea of using technology to sustainably and efficiently meet unlimited human needs with limited resources (Partigöç, 2023). It is stated that the integration of a city's systems, such as transport, energy, and infrastructure, is the most important component in creating a smart city, emphasising that no system can function independently in complex areas such as cities (Albino et al., 2015). Accordingly, smart city systems should also be integrated with each other, and all components of the system should be equipped with information technologies. In general, it is accepted that smart city applications are also quite important in the context of effective disaster management in the fight against climate change (Bulut & Aslan, 2022). The main reason for addressing climate change issues in cities is the concentration of population in cities, while it is thought that problems such as a decline in air quality, an increase in heat islands and water scarcity may increase in cities due to climate change (Ojo et al., 2024). In terms of problems, it is stated that the potential problems that smart cities need to solve are generally in the areas of transport, energy, water sustainability, health, service security and the environment (Alkan, 2015).

2. COMPONENTS OF A SMART CITY

Certain components are required for a city to be considered a smart city. The most important of these components are the city's systems (including city administrations) and its citizens. Accordingly, in order to evaluate a city within the smart city category, six components are discussed: smart governance, smart economy, smart mobility, smart environment, smart people, and smart living (Giffinger et al., 2007).

Table 1. The Components of a Smart City

Smart Economy (Competitiveness) • Innovative spirit • Entrepreneurship • Economic image & trademarks • Productivity • Flexibility of labour market • International embeddedness • Ability to transform	**Smart People (Social and Human Capital)** • Level of qualification • Affinity to lifelong learning • Social and ethnic plurality • Flexibility • Creativity • Cosmopolitanism/Open-mindedness • Participation in public life
Smart Governance (Participation) • Participation in decision-making • Public and social services • Transparent governance • Political strategies & perspectives	**Smart Mobility (Transport and Ict)** • Local accessibility • (Inter-)national accessibility • Availability of ICT-infrastructure • Sustainable, innovative and safe transport systems
Smart Environment (Natural Resources) • Lack of pollutionof natural conditions • Pollution • Environmental protection • Sustainable resource management	**Smart Living (Quality Of Life)** • Cultural facilities • Health conditions • Individual safety • Housing quality • Education facilities • Touristic • Social cohesion

Source: Giffinger et al. (2007)

The smart economy component is defined as contributing to the city's entrepreneurial and production capacity by utilising emerging new technologies, while the smart people component is interpreted as the city's human capital for creating a smart city. The smart governance component generally emphasises the need for cooperation to achieve sustainable urbanisation through technology (Memiş, 2018). The smart living component refers to the use of smart systems in urban issues that shape the individual's life in a smart city (Memiş, 2018). Examples of this include smart buildings, smart stops, smart energy systems, and smart lighting, which are examples of the city's physical characteristics. Within the scope of this study, the smart environment component is considered important. The smart environment component can be described as the detection, monitoring, and elimination of numerous urban problems such as air and water pollution and waste with the help of technology (Memiş, 2018).

The common characteristics of smart cities are related to smart city components and generally include:

- Energy savings
- More effective and efficient use of resources
- Equipped with technologies that offer cost-reducing solutions
- Technology is effective
- High quality of life
- Low environmental pollution
- Low carbon emissions
- Governance is at the forefront

These characteristics are listed in a way that encompasses the increasing standards developed based on technology (Gürcan & Açıksöz, 2023).

3. CLIMATE CHANGE AND SMART CITY APPLICATIONS

Smart city applications aimed at combating climate change or adapting to climate change; -taking into account city systems and smart city components- have been selected as energy efficiency and the use of renewable energy sources, transport and mobility, urban planning and green spaces, climate monitoring and early warning systems, waste and water management, air quality monitoring, and citizen participation and awareness.

3.1. Energy Efficiency and Renewable Energy Sources

Current energy systems rely heavily on fossil fuels. It is stated that energy use in the industrial sector accounts for 24.2 per cent of greenhouse gas emissions, while energy used in buildings produces 17.5 per cent of greenhouse gases (Ritchie, 2020). Accordingly, electricity and heat energy production are the main causes of greenhouse gas formation (Ritchie et al., 2020). Energy produced using fossil fuels generates various wastes, particularly hazardous gases such as carbon dioxide and compounds consisting of nitrogen and oxygen, leading to the greenhouse effect. The problems it creates are not limited to air pollution or the increase in greenhouse gases; liquid waste released during energy production also disrupts the water ecosystem (Mahmood et al., 2021). Energy efficiency and the use of renewable energy sources are considered important for smart city living (Wang et al., 2021). Similarly, energy efficiency and the use of renewable energy sources can be considered one of the most important applications of smart cities in combating climate change. The reduction in the use of fossil fuels or the use of environmentally friendly renewable energy sources is seen as one of the effective ways to combat global warming and the negative effects of climate change, as it reduces the greenhouse effect (Öztopçu & Salman, 2019). Within the scope of smart cities, the smart energy component stands out in terms of ensuring energy efficiency. The smart energy component refers to technology-based systems that promote the use of renewable energy sources instead of fossil fuels and provide both cost and energy savings (Yimsek & Yakar, 2023). Increased energy efficiency and lower costs are cited as demonstrable benefits of using smart systems in energy management (Wang et al., 2021). In smart cities' energy efficiency and renewable energy source usage, hybrid smart grids that optimise energy consumption and integrate with renewable sources are gaining importance (Khalil et al., 2021). It is thought that the carbon footprint can be reduced in terms of energy thanks to smart grids.

It is believed that smart energy systems, data analytics, and AI-supported energy management can improve energy consumption and prevent unnecessary resource waste (Marinakis, 2020). With the advancement of technology, the use of applications such as artificial intelligence and smart grids in cities ensures energy efficiency. However, the energy consumed by these applications in processing big data must also be taken into account, and there is a need for more studies evaluating the energy consumption of smart city applications in data processing.

3.2. Transport and Mobility

Intelligent transport systems are considered one of the most important components of the smart city under the influence of the fourth industrial revolution (Önder & Ulukavak, 2020). Intelligent transport aims to minimise the time vehicles spend in traffic within the city, manage traffic, plan urban mobility, and encourage public transport.

One of the main causes of climate change linked to greenhouse gases is the gases emitted by vehicles used for transport. According to research, the transport sector accounts for 16.2% of greenhouse gas emissions. Of this share, 11.9% comes from road transport, 1.9% from air transport, 1.7% from shipping, 0.4% from rail transport, and 0.3% from pipeline transport (Ritchie, 2020). Road transport is seen to be the type of transport and transportation that produces the most greenhouse gas emissions. Therefore, it is thought that reducing the fossil fuels used in smart transport vehicles, i.e. the type of fuel and materials, will reduce greenhouse gas emissions (Ünal & Polat, 2023). It is emphasised that, in addition to engineering-based developments such as changes in fuel type, technology-supported applications such as increased usability of additional mobility services to be carried out with the help of smart devices in smart cities, and applications such as car sharing, are also important (Mohieldin & Vandycke, 2017).

Applications such as "green wave applications" are also aimed at reducing carbon emissions from vehicles during transport. It is thought that the green wave application, which can be defined as vehicles always passing through green lights when travelling at a certain speed on a specific road, will result in less carbon emissions as vehicles will stop less often at red lights. It is thought that this will prevent traffic congestion and, as vehicles spend less time in traffic, there will be less carbon emissions, reduced air pollution, and a contribution to combating climate change (Kara et al., 2022).

3.3. Urban Planning and Green Spaces

It is known that urban planning and the creation of green spaces contribute positively to smart cities' efforts regarding climate change and sustainability. Urban green spaces contribute to the urban ecosystem (Türker & Gül, 2022), thus referring to the environmental aspect of smart cities. In this context, green city planning includes environmental sustainability in planning, smart city lighting, air pollution detection systems, etc., while green spaces involve creating carbon sink areas that will reduce carbon emissions in the city, constructing environmentally friendly green buildings, increasing energy efficiency, and prioritising sustainability (Gökalp & Şenol, 2020). It is stated that significant changes must be made in urban planning and management to adapt to climate change issues, increasing extreme weather events, and disaster risks (Tekin, 2023). For example, increasing green spaces as determined by urban planning will create sink areas, which will be effective in ensuring the sustainability and disaster resilience of the city by preventing disasters such as floods that may occur as a result of extreme natural events, such as the heavy rainfall caused by climate change. The integration of these green spaces with technology is also gaining importance as one of the components of the smart city. The fact that smart city applications can ensure the city's open water management can serve as an example of the integration of green spaces with smart city applications.

These efforts, considered green technology, are directly related to climate change and global warming.

3.4. Climate Information and Early Warning Systems

With the advancement of technology, it is stated that it is possible to be resilient to climate change or to adapt more easily to climate change. Technology is crucial for smart cities as it is an element that develops early warning systems, facilitates effective communication in emergencies, and accelerates data-driven decision-making processes. In this context, it is stated that climate-smart and resilient urbanisation can be achieved through technological applications such as "artificial intelligence"" (AI), blockchain, and the "Internet of Things" (IoT) (Pee & Pan, 2022). In particular, meteorological early warning systems provide both local authorities and citizens with the opportunity to take precautions against extreme weather events, thereby averting disasters before they occur (Barış et al., 2023). For this reason, meteorological early warning systems are considered an important smart city component in addressing climate change-related issues. Early warning systems are also important in creating climate-smart cities using the technologies mentioned. Early warning systems are defined as a cost-effective disaster risk reduction and climate change adaptation measure that has been proven to save people, livelihoods and ecosystems from climate-related hazards (UNEP, n.d.). It is stated that climate information and early warning systems can reduce damage by 30% by identifying disaster risks 24 hours in advance. Furthermore, it is estimated that investing just $800 million in "Multi-Hazard Early Warning Systems" could prevent annual losses of between $3 billion and $16 billion (UNEP, n.d.). This demonstrates that establishing early warning systems to take precautions before a disaster occurs can also contribute financially to national economies.

The 2022 "UN Climate Change Conference or UNFCCC Conference" of the Parties (COP27) also outlines the four pillars of the "UN Secretary-General's Early Warning for All Initiative" for the implementation of "Multi-Hazard Early Warning Systems": risk information and management, observation and forecasting, dissemination and communication, and preparedness for response (UNEP, n.d.).

It is also known that artificial intelligence is used in smart cities to determine and predict weather patterns. Thanks to artificial intelligence predictions, the effects of weather-related disasters caused by climate change are reduced, and it becomes easier to take precautions and make decisions (Probowo et al., 2023).

3.5. Solid Waste and Water Management

Smart city applications implemented as part of the fight against climate change also cover waste and water management. Waste and water management is examined in the broadest sense as part of the sustainable resource management component of the smart environment. “Waste management” provides important indicators for assessing whether a city can be considered a smart city (De Novais et al., 2021). Smart waste management refers to the inclusion of technology in the entire process, from waste generation to recycling (Effendy et al., 2023). It is also stated that the use of technology in waste management facilitates the sorting of waste and thus simplifies the recycling process (Effendy, 2023). Smart waste management is also important for protecting public health, reducing environmental pollution, conserving natural resources, enhancing sustainability, and recovering energy (Malik et al., 2023). IoT, is particularly important in waste management. This allows urban technologies to be integrated with each other and processes to be monitored. In addition to these applications, it is thought that energy-efficient waste management can be achieved with technologies such as smart bins (Boubaris et al., 2022). Climate change is affecting the world's water resources by causing extreme weather events such as drought or excessive rainfall. Particularly in the context of drought, water scarcity is anticipated, and accordingly, the importance of managing available clean water is increasing. “Digital water systems” are being evaluated within the scope of smart cities and provide data for water management through IoT technology regarding how much water is used or about wastewater (Figueiredo et al., 2021). It is stated that the “Water Management System” used by smart cities supplies water according to need and facilitates water recycling for specific activities (Gade, 2021). This is thought to enable measures to be taken against droughts that climate change may cause.

In smart waste management, equipping municipal refuse collection vehicles with tracking systems makes it possible to determine whether waste has been collected (Urhan and Güllü, 2023) and to monitor the condition of streets in real time, enabling a faster response to needs.

There are significant outcomes from utilising smart city applications, particularly in solid waste management. Efficient and energy-saving waste management helps reduce carbon dioxide, air pollution and vehicle exhaust emissions. Resources can be used more efficiently, and waste management contributes to the city's budget. The separation of waste through smart city applications and its disposal without harming the environment protects environmental resources such as water, forests, and air. Furthermore, smart waste management reduces the dangers of hazardous waste, such as industrial and medical waste, thereby protecting human health. In terms of evaluating waste as a resource, smart city applications are important both for recycling resources and for creating green energy (Sivasankari et al., 2017).

It is certain that drought is one of the extreme weather events experienced today due to climate change. The need for water is increasing because it is one of the most basic human needs, or indirectly, because it is necessary for food production and industrial activities. The fact that water is now considered a scarce resource has brought water management to the forefront in smart cities. This is because water management is seen as directly related to the development of water resources and environmental protection activities (Polat et al., 2017). When water management is combined with the concept of smart cities, the concept of smart water management comes to the fore (Polat et al., 2017). The smart water management approach aims to ensure the sustainability of water resources through the use of information and communication technologies. Applications of smart water management include smart network integration, monitoring water quality, tracking and reducing unaccounted-for losses, and controlling wastewater (Polat et al., 2017). The importance of urban water management is also recognised by supranational organisations. "The European Union Urban Water Project", which aims to establish smart and web-based urban water management, can be cited as an example of this (Polat et al., 2017). To implement smart water management, flow meters must be installed at water sources (wells, dams, treatment plants, etc.) and controlled with proportional valves so that the water supplied is measured accurately. A smart water meter must be installed, and infrastructure systems must be made smart and digitised. Networks should be controllable with a sufficient number of valves, and all data should be integrated into the system through a system established for water usage or conversion. It has also been determined that personnel involved in this matter should have sufficient equipment (Yıldız & Karakuş, 2019).

The importance of energy efficiency in wastewater treatment plants used in wastewater management is emphasized. In this context, a study analyzing the data obtained shows that in Turkey, the wastewater treatment sector consumes approximately 1.6 TWh of electricity per year. By implementing technology-integrated optimizations, savings of 15-25% are possible, which translates to 0.9-1.5 billion TL in operating costs and 110,000-180,000 t CO_2 equivalent emissions could be achieved (Delikanlı, 2025). Thanks to this study, it has become possible to see the concrete benefits of smart city technologies in wastewater management.

3.6. Citizen Participation and Awareness Raising

Citizen participation is considered one of the most important factors in smart cities' fight against climate change. This is because, fundamentally, the existence of a process triggered by a conscious human resource is discussed for a city to become a smart city (Şahin, 2025). Or, to put it another way, a smart city infrastructure is not merely a city enabled by digitalisation, but is also recognised as a city that incorporates social and relational capital, with people at its core (Nam & Padro, 2011). Because without people, no feature of any city has any significance. Therefore, it is important for citizens to become aware of climate change and to transform their consumption preferences, particularly in ways that reduce greenhouse gas emissions. To achieve this, the importance of mobile applications on smart phones, which everyone uses today, is emphasised. It is stated that the integration of these mobile applications not only ensures the active participation of citizens in protecting the environment but also strengthens their active participation in its conservation (Wolniak et al., 2023). Smart citizens are the main actors in the implementation of smart city applications. Individuals who can collectively monitor the city thanks to the sensing and processing capabilities of these mobile smart devices are defined as smart citizens (Pouryazan and Kantarcı, 2016). In addition, there are studies on the transition from a sustainable smart city to a smart society (Mishra et al., 2022). The role of the smart citizen is crucial in the transition from a sustainable smart city to a smart society. Furthermore, raising citizens' awareness and involving them in governance, i.e., ensuring good governance, is seen as one of the key elements that should be present in smart cities (Almeida et al., 2018). Citizen participation and raising citizens' awareness of technology are seen as important smart city applications in the smart city index in terms of the global competitiveness of smart cities with each other (Göksoy Sevinçli and Turan, 2025). Furthermore, incentive policies should be developed for users to enable smart city systems to become operational and technologies to be widely adopted (Sürmeli, 2025).

4. EXAMPLES OF SMART CITY APPLICATIONS IN THE FIGHT AGAINST CLIMATE CHANGE FROM AROUND THE WORLD

London has set out to reduce carbon emissions and increase urban green spaces as part of its London Environment Strategy. To this end, it has chosen reducing carbon emissions as a key step and has created zero-emission zones in the city centre where only zero-emission vehicles are allowed to enter (Yimsek & Yakar, 2023). "The London Environmental Strategy" strives to create a zero-average return model for greenhouse gas emissions specified in the "2050 Zero Carbon targets" and is recognised as one of the most ambitious smart city projects worldwide in this context (Contreras & Platania, 2019).

A study conducted on South Korean cities indicates that energy efficiency and environmental sustainability are being adopted within the scope of smart cities. For example, urban elements such as smart grids and motion-sensing street lighting are used to ensure energy efficiency (Yang et al., 2020). It can be argued that such applications contribute indirectly to combating climate change and sustainability by reducing factors that cause climate change, such as greenhouse gases emitted during energy production. Or, in New York, the installation of smart thermostats in some buildings to automatically adjust air conditioning to increase energy efficiency can also be considered a contribution of energy-efficient smart city applications to combating climate change (Yang et al., 2020).

The implementation of parking information applications in San Francisco has reduced traffic in the city (Yang et al., 2020). It is thought that such smart city applications, evaluated within the scope of smart mobility, contribute to combating climate change by reducing greenhouse gas emissions associated with vehicle use. In addition, it is stated that the use of 100% LED lighting in street lighting in the city provides energy efficiency (Uçar, 2022).

As part of efforts to combat climate change, waste management in India utilises the IoT to monitor bin fill levels and schedule waste collection times based on these levels, thereby enabling fuel savings for waste collection vehicles (Effendy, 2023). This situation can be considered to contribute to the fight against climate change by both preventing methane gas accumulation and reducing greenhouse gas emissions from waste collection vehicles.

Barcelona also uses IoT systems to monitor air quality, traffic and waste, and it can be argued that the improvement in service quality has also increased the quality of life for the public (Ojo et al., 2024). It is stated that water savings are achieved in Barcelona's parks thanks to sensor-based irrigation systems (Uçar, 2022).

Brazil stands out as one of the countries using smart city applications to reduce greenhouse gas emissions and combat climate change under the "Paris Climate Change Agreement". Brazil, particularly in the transport sector, promotes electricity and renewable energy sources instead of fossil fuels (Lima et al., 2022), which can be evaluated in terms of smart transport or smart mobility, which are components of smart cities, within the scope of combating climate change.

Izmir is one of Türkiye's most densely populated cities and, as such, is among the country's major cities with heavy traffic congestion. "Izmir Metropolitan Municipality" aims to reduce increasing traffic congestion and ensure traffic safety in Izmir through its "Intelligent Traffic Management System" (IZUM). This smart city system has been developed with the ability to monitor and record the city's main arteries throughout the day using sensors, smart cameras and software. IZUM is also used to reduce greenhouse gases, one of the main causes of climate change. The system enables real-time traffic management, and data collected in areas such as traffic light optimisation and public transport management can be used to alleviate traffic congestion and direct drivers to alternative routes. This helps reduce traffic density and congestion (Güney and Yalçın, 2024). In its current form, the system can be considered to contribute to the fight against climate change by reducing greenhouse gas emissions from vehicles in traffic jams. It is also noted that IZUM encourages people to use public transport within the scope of smart mobility (Güney & Yalçın, 2024), which is considered one of the applications that contribute to combating climate change by reducing the number of vehicles on the road.

Los Angeles and San Francisco stand out for their smart waste management. In Los Angeles, there are over 300 locations with solar-powered Big Belly smart bins equipped with Wi-Fi. The bins use sensors to notify waste collectors of their fill levels via text message or email (Herbst, 2015). This relieves workers of excessive workload while preventing waste collection vehicles from spending long periods in traffic. San Francisco implements the "EkoMap" project. The map provides access to waste, energy, and transportation-related emission data, as well as greenhouse gas emission data (Türkiye Ministry of Environment and Urbanisation, 2019).

Toronto has been the subject of studies on solid waste. "The Toronto City Council" has implemented a waste wizard, which informs residents living in flats and apartments where to dispose of rubbish, recycling, organic products and other items appropriately, and has developed an online platform where they can consider donating usable items such as clothes and books (Söylemez, 2018). Furthermore, Google has proposed the idea of monitoring the coastal area with cameras to examine events such as air pollution, water levels, traffic flow, and weather conditions in Toronto (The Guardian, 2018). Although this situation highlights the governance and collaboration features of the smart city, it has been emphasised that constant surveillance is disturbing for people living in Toronto, and it has been stressed that responsibility for such smart infrastructure should not be given to a private company (The Guardian, 2018). Although this situation may be considered detrimental to cooperation in smart cities, it should be borne in mind that people may develop an aversion to smart city systems, particularly due to systems such as cameras, which make them feel constantly under surveillance. On the other hand, in a constantly monitored city, while events such as traffic and weather conditions are being examined, the ability to monitor the city with the same camera systems may reduce crime rates, or camera recordings may be used as evidence against elements that constitute crime, bringing to mind the idea that a smart city can be a safe city.

Seoul, meanwhile, is taking significant steps towards becoming a smart city. Seoul has implemented smart transport, IoT technology, smart street lighting, smart waste and water management, and a smart sewerage network (Erkek, 2023). It is believed that all of the smart city initiatives implemented in Seoul contribute to the fight against climate change.

In Wroclaw, a city in Poland, steps are being taken to implement measures to combat climate change using smart city components by 2030. In this context, the Wroclaw 2030 Strategy has been adopted, which includes smart city applications aimed at improving the environment and urban space quality, such as reducing CO2 emissions by 30 per cent by 2030, investing in environmentally friendly technologies, and investing in environmentally valuable areas (Bednarska-Olejniczak et al., 2019). The Strategy also includes measures related to mobility. The implementation of environmentally friendly practices aimed at mitigating the effects of climate change is planned, such as integrated public transport systems, limiting vehicle traffic in the city centre, promoting sustainable mobility, developing public transport systems, supporting the development of ecological freight transport, and increasing river transport such as environmentally friendly practices aimed at mitigating the effects of climate change are planned (Bednarska-Olejniczak et al., 2019).

In Singapore, the green building initiative aims to be implemented in all buildings across the country by 2030, with the goal of achieving energy savings through this initiative. In Singapore, the "Hackathon application" is also used to monitor public transport systems, with the aim of tracking vehicle occupancy information and terminal times. It is stated that the aforementioned smart city applications will enable savings in both energy and time (Uçar, 2022). As energy efficiency has come to the fore in the fight against climate change, achieving these targets is considered important.

DISCUSSION and CONCLUSION

Cities are the areas most affected by global warming and climate change. Industrialisation, traffic, waste and water usage, air and water pollution, and increased mobility, which have developed alongside population growth resulting from urbanisation, have also led to an increase in greenhouse gases that cause climate change. In this context, the creation of smart cities, particularly through the use of technology, has become important in combating climate change. Cities have been made smart through technologies such as information technology, artificial intelligence and IoT, urban systems have been integrated with each other, and citizens have been enabled to participate in governance.

The study focused particularly on energy in the fight against climate change and determined that energy efficiency can be achieved through the use of renewable energy sources or smart city systems.

Transport and mobility have been major problems in cities due to carbon emissions causing greenhouse gases. Therefore, it has been seen that encouraging the use of zero-carbon vehicles in the context of transport and mobility, integrating these systems with technology, can save both time and energy, while also reducing greenhouse gas emissions.

Climate information and early warning systems encompass preparedness for extreme weather events such as severe droughts or excessive rainfall caused by climate change. It has been stated that early warning systems enable measures to be taken against natural events that could reach disaster proportions. This allows for protection against both material and immaterial damage, which is one of the most important contributions of smart cities in combating climate change.

Waste management and disposal, as well as water management, are also being addressed by smart cities using IoT technologies. Smart city technologies, which are also used in waste recycling, are important in combating climate change because they help prevent the accumulation of greenhouse gases such as methane and protect the environment. With the integration of IoT technologies into smart cities, water can be monitored and water management can be made more efficient. Thanks to smart city technologies, access to data on elements such as clean water access or water accumulation and filtration is becoming easier, and policies appropriate to the current situation are being developed. Particularly in smart cities, switching to smart meters for smart water management, enabling the monitoring of water transfer, detecting water loss, and irrigating urban areas such as parks and gardens with the help of sensor-equipped irrigation tools saves water while demonstrating how water management can be achieved through smart applications.

Citizen participation emerges as the most important component for achieving concrete results from the specified smart city applications. Increasing citizens' technological literacy and their use of city applications on their mobile devices are considered to be the reflection of smart city projects in the field. Therefore, citizens both participate in governance and protect their cities from the effects of climate change, while also being protected from the effects of climate change themselves. Adapting to technological innovations in order to benefit from the advantages of smart cities supports citizens' lifelong learning process.

It is emphasised that governance is important in smart cities, drawing attention to the relationships between local authorities, citizens and other stakeholders. Companies described as technology giants are also contributing to the development of the digital infrastructure of smart cities through cooperation. In cities equipped with technological infrastructure by these companies, many factors such as air quality, water levels and traffic density will be monitored via cameras. However, it is also noted that there are some reservations in this regard. While citizens' greatest concern is being monitored throughout the day, it can be argued that recording cameras throughout the day could reduce crime rates in cities or contribute to the city's security by providing evidence against incidents that could be considered crimes.

It is considered necessary to calculate the cost-benefit ratio by also accounting for the energy used to process big data in smart city applications. Furthermore, while the development of information and communication technologies brings benefits, the growth of the system also makes it vulnerable to cyberattacks (Sağ Örnek and Hasırcı Tuğcu, 2023). Therefore, it has been seen that special attention should be paid to cybersecurity in relation to smart city applications and that this issue needs

REFERENCES

Albino, V., Berardi, U., & Dangelico, R. M. (2015). Smart cities: Definitions, dimensions, performance, and initiatives. *Journal of Urban Technology, 22*(1), 3–21. DOI: 10.1080/10630732.2014.942092

Alkan, T. (2015). Akıllı kentler ya da 21.yy şehirleri. *Bilişim Dergisi*, (182), 71–77.

Almeida, V., Doneda, D., & Moreira da Costa, E. (2018). Humane smart cities: The need for governance. *IEEE Internet Computing, 22*(2), 91–95. DOI: 10.1109/MIC.2018.022021671

Barış, Ş., Zülfikar, A.C., Korkmaz, M. & Tunç, S. (2023). Akıllı şehirlerde afet riski azaltma uygulamaları. *Şura Akademi Dergisi*, (2), 27-41.

Bednarska-Olejniczak, D., Olejniczak, J. & Svobodova, L. (2019). Towards a smart and sustainable city with the involvement of public participation- the case of Wroclaw. Sustainability, 11(2), 332, 1-33. https://doi.org/DOI: 10.3390/su11020332

Bibri, S. E., & Krogistie, J. (2017). Smart sustainable cities of the future: An extensive interdisciplinary literature review. *Sustainable Cities and Society, 31*(25), 1–47. DOI: 10.1016/j.scs.2017.02.016

Boubaris, A., Kantounias, F., Kyriazidis, G., Dasteridis, V., Rigogiannis, N., Papanikolau, N., & Sirakoulis, G. (2022). Smart waste collection system in the context of smart cities. *2022 11th International Conference on Modern Circuits and Systems Technologies (MOCAST)*, Bremen, Germany.

Bulut, Y., & Aslan, M. M. (2022). Doğal afetlere karşı dirençli kentler oluşturmaya yönelik yenilikçi yaklaşımlar: Akıllı kent uygulamaları üzerine bir inceleme. Kartepe Zirvesi 2022- Dirençli Şehirler İçin Vizyoner Yönetim. Hamza Ateş (ed.). 9-24.Kocaeli.

Çetin, M., & Çiftçi, Ç. (2019). Literatüre göre dünya ve ülkemizden örneklerle akıllı kent kavramının irdelenmesi. *Ulusal Çevre Bilimleri Araştırma Dergisi, 2*(3), 134–143.

Contreras, G., & Platania, F. (2019). Economic and policy uncertainty in climate change mitigation: The London smart city case scenario. *Technological Forecasting and Social Change, 142*, 384–393. DOI: 10.1016/j.techfore.2018.07.018

De Novais, M. P., Xavier, V. A., Xavier, L. H., & Hwang, J. (2021). Modelling e-waste management data in smart cities. *2021 2nd Sustainable Cities Latin America Conference (SCLA)*, Medellin, Colombia.

Delikanlı, N. E. (2025). Yapay zeka ve makine öğrenmesi ile arıtma tesislerinde enerji verimliliği ve süreç iyileştirme potansiyelinin değerlendirilmesi. *JOSS Akıllı Sistemler Dergisi*, *4*(1), 42–54.

Effendy, V. A., Wirani, Y., & Sucahyo, Y. G. (2023). Smart waste managment planning to support the implementation of smart city in the national capital city of Indonesia. *2023 10th International Conference on ICT for Smart Society (ICISS)*, Bandung, Indonesia.

Erkek, S. (2023). Citizen participation in a smart city: The Seoul case. *Kent Akademisi*, *16*(4), 2595–2610. DOI: 10.35674/kent.1367710

Figueiredo, I., Esteves, P., & Cabrita, P. (2021). Water wise- a digital water soluiton for smart cities and water management entities. *Procedia Computer Science*, *181*, 897–904. DOI: 10.1016/j.procs.2021.01.245

Gade, D. S. (2021). Reinventing smart water management system through ICT and IoT driven solution for smart cities. [IJAEML]. *International Journal of Applied Engineering and Management Letters*, *5*(2), 132–151. DOI: 10.47992/IJAEML.2581.7000.0109

Giffinger, R., Fertner, C., Kramar, H., Kalasek, R., Pichler-Milanovic, N., & Meijers, E. (2007). *Smartcities – ranking of European medium-sized cities*. Centre of Regional Science.

Giffinger, R., & Gudrun, H. (2010). Smart cities ranking: an effective instrument for the positioning of the cities? *ACE: Architecture. City and Environment*, *4*(12), 7–26.

Gökalp, D., & Şenol, P. (2020). Akıllı kentlerin gelişimi ve kent planlama ilişkisi üzerine bir değerlendirme. *SKETCH*, *2*(2), 1–20.

Göksoy Sevinçli, B., & Turan, M. (2025). Akıllı kent uygulamalarına yönelik vatandaş algısı envanterinin oluşturulması: Ankara/Çankaya örneği. *Dokuz Eylül Üniversitesi Sosyal Bilimler Enstitüsü Dergisi*, *27*(1), 218–243. DOI: 10.16953/deusosbil.1561414

Gül, A. and Atak Çobanoğlu, Ş. (2017). Avrupa'da akıllı kent uygulamalarının değerlendirilmesi ve Çanakkale'nin akıllı kente dönüşümünün analizi. *Süleyman Demirel Üniversitesi İktisadi ve İdari Bilimler Fakültesi Dergisi*, 22(Kayfor 15 Özel Sayısı), 1543-1565.

Gülseçen, S., Özdemir, Ş., Çelik, S., Uğraş, T., & Özcan, M. (2013). *Dijital dünyadan yansımalar: bilgide ve vatandaşlıkta değişim*. XVIII. Türkiye'de İnternet Konferansı 223-227. İstanbul, Turkey

Güney, A., & Yalçın, E. (2024). Belediyelerde akıllı kent uygulamaları: İzmir Büyükşehir Belediyesi. *Uluslararası Yönetim Akademisi Dergisi*, *7*(2), 656–669. DOI: 10.33712/mana.1494407

Harrison, C., Eckman, B., Hamilton, R., Hartswick, P., Kalagnanam, J., Paraszczak, J., & Williams, P. (2010). Daha akıllı şehirler için temeller. *IBM Araştırma ve Geliştirme Dergisi*, *54*(4), 1–16. DOI: 10.1147/JRD.2010.2048257

Herbst, J. (2015). How L.A. works: Smart trash cans. Erişim adresi: https://www.lamag.com/citythinkblog/how-l-a-works-smart-trash-cans/#:~:text=Thc unit isn't all,public Wi-Fi hot spots (18.09.2025).

Kara, H., Sarısoy, G., & Delice, Y. (2022). Yeşil dalga uygulamalarında işıklı kılavuz sistem kullanımının değerlendirilmesi. *Akıllı Ulaşım Sistemleri ve Uygulamaları Dergisi*, *5*(2), 15–32. DOI: 10.51513/jitsa.1096477

Khalil, M. I., Jhanjhi, N. Z., Humayun, M., Sivanesan, S., Masud, M., & Hossain, M. S. (2021). Hybrid smart grid with sustainable energy efficient resources for smart cities. *Sustainable Energy Technologies and Assessments*, *46*, 101211. DOI: 10.1016/j.seta.2021.101211

Lima, M. A., Mendes, L. F. R., Mothe, G., Linhares, F. G., de Castro, M. P. P., da Silva, M. G., & Sthel, M. S. (2022). Renewable energy in reducing greenhouse gas emissions: Reaching the goals of the Paris agreement in Brazil. *Environmental Development*, (33), 1–13.

Mahmood, D., Javaid, N., Ahmed, G., Khan, S. & Monteiro, V. (2021). A review on optimization strategies integrating renewable energy sources focusing uncertainty factor–Paving path to eco-friendly smart cities. *Sustainable Computing: Informatics and Systems, 30*, 100559. 1-19.

Malik, S., Malik, P. K., Kumar, G. R., Singh, R., & Naim, A. (2023). The internet of things and its intervention for waste management in smart cities. *2023 3rd International Conference on Advancement in Electroncis & Communication Engineering (AECE)*, Ghaziabad, India. 171-175.

Marinakis, V. (2020). Big data for energy management and energy efficient buildings. *Energies*, *13*(17), 1555. DOI: 10.3390/en13071555

Memiş, L. (2018). Akıllı teknolojiler, akıllı kentler ve belediye örgütlenmesinde dönüşüm. *Yasama Dergisi*, *36*, 66–92.

Mishra, P., Thakur, P., & Singh, G. (2022). Sustainable smart city to society 5.0: State-of-the-art and research challenges. *SAIEE Africa Research Journal*, *113*(4), 152–164. DOI: 10.23919/SAIEE.2022.9945865

Mohieldin, M., & Vandycke, N. (2017). Sustainable mobility for the 21st century. World Bank Group. Erişim adresi: https://www.worldbank.org/en/news/feature/2017/07/10/sustainable-mobility-for-the-21st-century (18.04.2025).

Nam, T., & Pardo, T. A. (2011). Conceptualizing smart city with dimensions of technology, people and institutions. *Proceedings of The 12th Annual International Digital Government Research Conference: Digital Government Innovation in Challenging Times*, New York: ACM, 282–291. DOI: 10.1145/2037556.2037602

Neirotti, P., De Marco, A., Cagliano, A. C., Mangano, G., & Scorrano, F. (2014). Current trends in smart city initiatives: Some stylised facts. *Cities (London, England)*, *38*, 25–36. DOI: 10.1016/j.cities.2013.12.010

Ojo, B., Nnajiofor, C., Osawe, E., & Aghaunor, C. T. (2024). Akıllı kentsel yapı: İklim değişikliğine karşı daha fazla dayanıklılık için IoT'den yararlanma. *International Journal of Science and Research Archive*, *12*(2), 1355–1364. DOI: 10.30574/ijsra.2024.12.2.1363

Önder, H. G., & Ulukavak, M. (2020). Yerel yönetimler için çok modlu taşımacılıkta akıllı hareketlilik önerileri. *Türkiye Coğrafi Bilgi Sistemleri Dergisi*, *2*(2), 83–89.

Örselli, E., & Akbay, C. (2019). Teknoloji ve kent yaşamında dönüşüm: Akıllı kentler. *Uluslararası Yönetim Akademi Dergisi*, *2*(1), 228–241. DOI: 10.33712/mana.544549

Öztopçu, A., & Salman, A. (2019). Sürdürülebilir kalkınmada akıllı kentler. *Karadeniz Uluslararası Bilimsel Dergi*, *41*(41), 167–188. DOI: 10.17498/kdeniz.476335

Partigöç, N.S. (2023). Sürdürülebilir kentsel planlama süreçlerinde akıllı şehir yaklaşımının rolü. *Çevre, Şehir ve İklim Dergisi*, 2(3), 174-189.

Pee, L. G., & Pan, S. L. (2022). Climate-intelligent cities and resilient urbanisation: Challenges and opportunities for ınformation research. *International Journal of Information Management*, *63*, 1–6. DOI: 10.1016/j.ijinfomgt.2021.102446

Peris-Ortiz, M., Bennett, D. R., & Pérez-Bustamante Yábar, D. (Eds.). (2017). *Sustainable smart cities. ınnovation, technology, and knowledge management.* DOI: 10.1007/978-3-319-40895-8

Polat, N., Bilgiç, E., & Gündüz, O. (2017). Akıllı kentlerde su yönetimi uygulamaları. *4th International Water Congress*, 2-4 November, Izmir-Türkiye. 365-275.

Probowo, O. M., Mulyana, E., Nugraha, I. G. B. B., & Supangkat, S. H. (2023). Cognitive city platform as digital public infrastructure for developing a smart, sustainable and resilient city in Indonesia. *IEEE Access : Practical Innovations, Open Solutions*, *11*(11), 120157–120178. DOI: 10.1109/ACCESS.2023.3327305

Ritchie, H. (2020). Sector by sector: where do global greenhouse gas emissions come from? *Our World in Data.* Erişim adresi: https://ourworldindata.org/ghg-emissions-by-sector (18.04.2025).

Ritchie, H., Rosado, P., & Roser, M. (2020). Breakdown of carbon dioxide, methane and nitrous oxide emissions by sector. *Our World in Data.* Erişim adresi: https://ourworldindata.org/emissions-by-sector (18.04.2025).

Sağ Örnek, R., & Hasırcı Tuğcu, Z. (2023). Akıllı şebeke uygulamlarında derin öğrenme tekniklerinin kullanımına ilişkin kısa bir inceleme. *EMO, 13*(1), 41–61.

Şahin, O. (2025). Akıllı kent olgusu ve sürdürülebilir kentsel gelişim. *Ankara Üniversitesi Sosyal Bilimler Dergisi, 16*(2), 390–405.

Shuja, J., Ahmad, R. W., Gani, A., Abdalla Ahmed, A. I., Siddiqa, A., Nisar, K., Khan, S. U., & Zomaya, A. Y. (2017). Greening emerging IT technologies: Techniques and practices. *Journal of Internet Services and Applications, 8*(1), 1–11. DOI: 10.1186/s13174-017-0060-5

Sivasankari., Shri, B. and Jinila, Y.B. (2017). Smart waste management using WSN and IoT. ICIIECS Conference. Erişim adresi: https://www.researchgate.net/publication/317718894_Smart_Waste_Management

Söylemez, A. (2018). Akıllı kentlerde atık yönetimi ve dünyadan akıllı atık yönetimi üzerine örnekler. *Yasama Dergisi, 37*, 87–100.

Sürmeli, R. Ö. (2025). Karbon nötr geleceğe doğru: Akıllı sistemlerin entegre uygulamalarıyla karbon ayak izi üzerindeki etkisi. *Akıllı Sistemler Dergisi, 4*(1), 55–73.

Tekin, Ö. F. (2023). Global climate change and climate-smart urban development: The case of Hyllie. *Journal of Managment and Economies Research, 21*(3), 473–502. DOI: 10.11611/yead.1247978

The Guardian. (2018). The Guardian view on Google and Toronto: Smart city, dumb deal. Erişim adresi: https://www.theguardian.com/commentisfree/2018/feb/05/the-guardian-view-on-google-and-toronto-smart-city-dumb-deal (18.09.2025).

Türker, H. B., & Gül, A. (2022). Kentsel açık ve yeşil alanlarının niceliksel analizi ve irdelenmesi: Uşak kent merkezi örneği. *Kent Akademisi Dergisi, 15*(4), 2088–2109. DOI: 10.35674/kent.999451

Türkiye Ministry of Environment and Urbanisation. (2019). *Akıllı şehirler beyaz bülteni.* Erişim adresi: https://webdosya.csb.gov.tr/db/cbs/akillisehirler/ (18.09.2025).

Uçar, A. (2022). Akıllı kentler ve su kullanımı. *Journal of Civilization Studies, 7*(1), 1–14.

Ünal, E., & Polat, E. (2023). İklim değişikliği ve akıllı ulaşım teknolojileri. *Th International Congress On Life, Social, and Health Sciences in a Changing World.* 189-195.

UNEP. (t.y). Climate ınformation and early warning systems. Erişim adresi: https://www.unep.org/topics/climate-action/climate-transparency/climate-information-and-early-warning-systems (31.05.2025).

Urhan, O. and Güllü, M. K. (2023). Veriye dayalı akıllı şehir oluşturmada teknoloji trendleri. *Şura Akademi Dergisi*, (2), 19-25.

Vaquero-García, A., Álvarez-García, J. & Peris-Ortiz, M. (2016). *Urban models of sustainable development from the economic perspective: smart cities. sustainable smart cities, 15–29.* DOI: 10.1007/978-3-319-40895-8_2

Wang, C., Gu, J., Sanjuán Martínez, O., & González Crespo, R. (2021). Economic and environmental impacts of energy efficiency over smart cities and regulatory measures using a smart technological solution. *Sustainable Energy Technologies and Assessments, 47*, 101422. DOI: 10.1016/j.seta.2021.101422

Wolniak, R., Jonek-Kowalska, I., & Grebski, W. (2023). Environmental monitoring in smart city- smartphone applications aspects. *Scientific Papers of Silesian University of Technology Organization and Management Series, 2023*(185), 605–615. DOI: 10.29119/1641-3466.2023.185.35

Yang, J., Kwon, Y., & Kim, D. (2020). Regional smart city development focus: The South Korean national strategic smart city program. *IEEE Access : Practical Innovations, Open Solutions*, (9), 7193–7210.

Yıldız, S., & Karakuş, C. B. (2019). Akıllı kentlerde sürdürülebilir su yönetimi. EJONS VI- International Conference on Mathematics-Enginerring- Natural & Medical Sciences. 8-10 March. Adana, Türkiye. 244-252.

Yimsek, F. S., & Yakar, M. (2023). Akıllı kentlere genel bir bakış. *Türkiye Arazi Yönetimi Dergisi, 5*(1), 49–56. DOI: 10.51765/tayod.1270504

Yin, C., Xiong, Z., Chen, H., Wang, J., Cooper, D., & David, B. (2015). A literature survey on smart cities. *Science China. Information Sciences, 58*(10), 1–18. DOI: 10.1007/s11432-015-5397-4

ADDITONAL READING

Albino, V., Berardi, U., & Dangelico, R. M. (2015). Smart cities: Definitions, dimensions, performance and initiatives. *Journal of Urban Technology*, *22*(1), 3–21. Advance online publication. DOI: 10.1080/10630732.2014.942092

Anthony, M. T. (2013). *Smart cities: big data, civic hackers, and the quest for a new utopia*. W. W. Norton & Company, Inc.

Batty, M. (2018). *Inventing future cities*. The MIT Press. DOI: 10.7551/mitpress/11923.001.0001

Batty, M., Axhausen, K. W., Gianotti, F., Pozdnoukhov, A., Bazzani, A., Wachowicz, M., Ouzounis, G., & Portugali, Y. (2012). Smart cities of future. *The European Physical Journal. Special Topics*, *214*(1), 481–518. DOI: 10.1140/epjst/e2012-01703-3

Caragliu, A., Del Bo, C., & Nijkamp, P. (2013). Smart cities in Europe. In *Creating smart-er cities* (pp. 65–82). Routledge.

Harrison, C., & Donnelly, I. A. (2011, September). A theory of smart cities. In *Proceedings of the 55th Annual Meeting of the ISSS-2011,* Hull, UK.

Kim, H., Choi, H., Kang, H., An, J., Yeom, S., & Hong, T. (2021). A systematic review of the smart energy conservation system: From smart homes to sustainable smart cities. *Renewable & Sustainable Energy Reviews*, *140*, 110755. DOI: 10.1016/j.rser.2021.110755

Kitchin, R. (2019). The timescape of smart cities. *Annals of the Association of American Geographers*, *109*(3), 775–790.

Letaifa, S. B. (2015). How to strategize smart cities: Revealing the SMART model. *Journal of Business Research*, *68*(7), 1414–1419. DOI: 10.1016/j.jbusres.2015.01.024

Chapter 17

Need for a Holistic Approach for Green Marketing by Leveraging the Role of Artificial Intelligence

Filiz Eroğlu
https://orcid.org/0000-0002-3223-953X
Canakkale Onsekiz Mart University, Turkey

Elçin Bayraktar Köse
https://orcid.org/0000-0003-4405-9920
Çanakkale Onsekiz Mart University, Turkey

Figen Kılıç
Çanakkale Onsekiz Mart University, Turkey

ABSTRACT

Sustainability awareness has led to green marketing practices, which businesses have either voluntarily adopted or been compelled to embrace by competitors, consumers, and governments. To ensure the effective functioning of this green marketing approach holistically, driven by a firm's own green mission, eco-conscious consumers, green-marketing competitors, and imposed green standards, artificial intelligence (AI) has become the most crucial tool. This study investigates and discusses the driving and assisting role of AI within the complex nexus of green competition, eco-consumers, and green standards as applied in a holistic green marketing approach in businesses.

DOI: 10.4018/979-8-3373-5238-1.ch017

Through its integration with techniques such as algorithms and machine learning, as well as systems like blockchain, AI assumes a supportive and leveraging role in businesses' green marketing strategies.

INTRODUCTION

By 2050, climate change is projected to inflict catastrophic costs, including an extra 14,5 million human deaths and $12.5 trillion in economic losses, according to the 2024 WEF Insight Report. These devastating projections stem from a variety of ongoing environmental and social impacts, such as the droughts and wildfires already affecting some regions and the rising sea levels and floods seen in others, which are all compromising vital resources like water, energy and health. These issues have elevated the concept of sustainability -the principle of satisfying current requirements without jeopardizing future capacity, from a mere idea to a basis element of our lives. The sustainability approach has permeated all aspects of life today, and marketing is one of the primary fields where sustainability-related activities are highly visible. While the concept of sustainability encompasses economic and social dimensions, its environmental component receives the most attention and is the subject of the greatest debate. This has led to green marketing emerging as the most prominent approach within the marketing field.

An increasing number of firms are adopting green marketing initiatives as a core component of their sustainability strategies. They are progressively recognized for their environmental pledges, as these initiatives boost their competitive edge (Chen & Chang, 2013). Firms' pursuit of a competitive advantage is a significant driver within the realm of environmental marketing. In many instances, companies observe and subsequently replicate the environmental initiatives of their competitors in order to preserve their market position (Singal et al., 2013, p. 473). On the other hand, consumer demand also plays a significant role in the transition to green marketing practices, although consumer skepticism towards environmental claims is on the rise. Therefore, as Boztepe (2012, p. 5) noted, it was not expected that businesses would remain unresponsive to environmental awareness. Green marketing integrates ecological considerations into every stage of decision-making, from product design and packaging to corporate culture. By aligning environmental responsibility with consumer expectations, businesses can achieve their profitability goals while simultaneously minimizing their ecological footprint (Gögermiş, 2019, p. 194).

Beyond the companies' desire and consumer demand for green alignment and sustainability, firms are also driven toward green marketing practices by laws and regulations, as well as incentives and penalties imposed by states and international collaborations. Singal et al. (2013) pointed out that while seeking to protect consumers and society, governments are striving to decrease the creation of detrimental products or their by-products, shift the way consumers and industries engage with and use these hazardous items, and manage the amount of harmful refuse businesses generate. Companies need to make major reductions in their environmental impacts across manufacturing, distribution, raw materials, and energy consumption. Otherwise, they risk being unable to compete in the face of new regulations and hardening customer demands (Grant, 2008, p. 25). The purpose of these regulations is also to prevent businesses from making "greenwashing" claims and to ensure that accurate information is provided to consumers. European Regulations like Ecodesign for Sustainable Products Regulation – ESPR, Green Claims Directive, Empowering Consumers for the Green Transition Directive, and international organizations and guidelines like ISO 14000, ICC Framework for Responsible Environmental Marketing Communications, Kyoto Protocol serve as prime examples of such frameworks.

Spurred by a collective focus on the environment and sustainability, businesses are gravitating toward the adoption of green marketing policies as a strategy to attain a competitive advantage and address the growing expectations of consumers and governments. Effectively and efficiently meeting the demands of all parties requires a holistic approach. However, a significantly more advanced tool has become available to them today: Artificial Intelligence (AI). It's powered by technologies like machine learning (ML), deep learning, natural language processing, neural networks, and robotic process automation (Davenport, 2018) impacts marketing strategies and their underlying business models (Davenport et al., 2019). Incorporating the integration of AI into green marketing strategies represents a transformative convergence of technology and sustainability, as AI's ability to provide data-driven insights and enhance operational efficiency empowers businesses to develop targeted marketing strategies for eco-conscious consumers, giving rise to sustainable innovation (Baruno & Indrasari, 2025).

Leveraging AI is instrumental for developing a holistic approach in green marketing, allowing businesses to align their green transformation objectives with competitive demands, consumer preferences, and compliance with strict regulatory standards. It is of paramount importance for businesses to strike a balance among the aforementioned issues. This balance is crucial for ensuring their operations are conducted in an organic, voluntary, and genuine manner. This study examines the necessity of a holistic approach to establishing an effective green marketing system for the benefit of all business partners and explores how AI can be leveraged to optimize the functioning of this approach while discussing the ethical issues about AI.

HISTORICAL EVOLUTION

"Green marketing" was initially established in the academic literature after the American Marketing Association's "Ecological Marketing" workshop in 1975, which brings together numerous bureaucrats, academics, and participants, investigated and analyzed the impact of marketing on the natural environment. This workshop led to the publication of Ecological Marketing by Henion and Kinnear in 1976, which is recognized as the first instance of the term's use. Henion's definition, presented within this foundational text, frames green marketing as creating and executing marketing programs for the eco-conscious consumer base (Kumar et al., 2025). It's a common observation that the term "green marketing" became more notable with the sustainability trend; however, the approach itself is clearly a much older one, having been established long before as part of the green movement, influenced by the ecological and social marketing movements.

This situation is clearly explained by the three periods of green marketing discussed in Ken Peattie's seminal book (1992), which was instrumental in defining and outlining the concept's framework. During the initial ecological period, marketing was used reactively, serving to resist and find solutions for emerging environmental problems. The subsequent period marked a shift towards proactive measures, emphasizing the adoption of cleaner technologies and the designing of eco-friendly products. Currently, the field is in its sustainable period, which emerged in the late 1990s and the early 2000s and remains the prevailing paradigm. As of the 1990s, a widespread and rapid change in direction began to emerge in the promotion of products and services in line with the expectations of environmentally conscious consumers.

Since the 2000s, the field of green marketing has seen a shift where theoretical research is increasingly accompanied by analytical studies of real-world practices. Simultaneously, the rise in industry applications has moved the scholarly debate toward a more practice-oriented focus. In the current landscape, the diverse range of green marketing applications adopted by firms is particularly noteworthy. Patagonia has encouraged the repair and reuse of old clothes, Nike has used recycled products in its products to reduce waste and carbon emissions, Lush cosmetics not only uses natural ingredients but also keeps packaging to a minimum even selling some products without packaging and IKEA uses bamboo and recycled materials in its furniture as well as it sells solar panels (Sharma, 2025). German car manufacturer Mercedes offers its customers driving training that enables them to reduce their carbon emissions by 60%. It actively uses this in its communication campaigns (Yüzbaşıoğlu, 2012). However, some businesses that claim to be sustainable through the concept of greenwashing are actually criticized for the harm they cause to the environment. For example, Walmart stating it will strive to make its stores low-carbon despite the initiative being limited to production and transportation and being insufficient, Banana Boat's claim that it produces reef-friendly sunscreen, despite the product containing chemicals harmful to coral reefs, and Unilever's packaging, which is made from materials that make reuse impossible, contrary to its commitment to collect and recycle more than it sells (Koons, 2024).

LITERATURE REVIEW

The concept of green marketing is expressed through terms such as green business, environmental marketing, ecological marketing, eco-friendly marketing, and green marketing. Therefore, over the years, it has been subject to different interpretations as different topics have come to the fore (Al-Salman, 2016, p. 5). Then it was defined as the adaptation of the 4Ps to green marketing (Lampe & Gazdat, 1995), a marketing philosophy and emphasized the concept of holistically considering the interests of society (Jain and Kaur, 2004), raising awareness of global warming and non-biodegradable solid waste and considering the production and consumption of products within this framework (Mishra & Sharma, 2014), the green transformation of businesses' policies, systems, and processes (Prakash, 2002), how marketing managers attempt to create strategies for consumers who are environmentally conscious (McDaniel & Rylander, 1993, p. 4), and establishing sustainable relationships with consumers, the social environment, and the natural environment (Belz & Peattie, 2009). Fuller (1999), in his more holistic definition, emphasized the need to align both the 4Ps and consumer needs and business objectives with sustainability. It was highlighted the reduction of production's environmental impacts in his definition of green marketing (Peattie, 2001a).

Green Consumers

Within the literature of green marketing, the concept of the green consumer stands as one of the most frequently researched topics. Green consumers are generally defined as individuals who actively or voluntarily consume products that cause the least harm to the environment in order to meet their needs (Peattie, 2001a, p. 129) or utilize the leverage of their consumption choices to foster personal and global preservation (Ali et al., 2011, p. 218). Green consumers are pivotal to the evolution of green marketing because they actively seek to comprehend and address external demands for businesses to enhance their environmental performance (Peattie, 2001b, p. 187).

A comprehensive understanding of green consumers enables marketers to assess if being "environmentally friendly" is a viable selling point for their target audience and to effectively incorporate this into their green marketing mix (Ginsberg & Bloom, 2004, p. 80). Moreover, the same source indicates that the well-known Roper Survey divides consumers into five categories according to their degree of environmental consciousness:

- True Blue Greens: The most committed and active
- Greenback Greens: Those willing to pay the most for green products
- Sprouts: Those undecided about environmental issues
- Grousers: Concerned but less active
- Basic Browns: Not active

This segmentation is based on the environmental concerns of green consumers, who represent a multidimensional structure encompassing knowledge, attitudes, and purchasing behavior. Therefore, it is a mixed model that does not distinguish between green attitudes and green purchasing behavior. Consumers' environmental knowledge, green values, beliefs, and attitudes are the source of the formation of actual green purchasing behavior. However, green attitudes do not always lead to green purchasing behavior. For example, not all consumers with environmental awareness and attitudes are willing to pay the price difference for green products. Therefore, there is a gap or inconsistency between green attitudes and the actual purchasing behavior of green consumers (Fan & Zeng, 2011, pp. 30-31). The limited reach of green marketing necessitates a focus on environmentally conscious segments of the market for effective consumer engagement (Lampe & Gazdat, 1995). On the other hand, growing customer awareness of the seriousness of environmental degradation is driving a rise in ecological consciousness. Consequently, eco consumers are actively seeking out and favoring businesses that demonstrate strong environmental practices by purchasing their eco-friendly products and services (Kalafatis et al., 1999, p. 442). To effectively pinpoint the eco-consumer segment and consistently incorporate their demands into production, enterprises will need to leverage the most advanced technology and digital resources.

Regulations and practices of government on green marketing

To ensure environmental sustainability, governments are actively seeking to guide both businesses and citizens toward responsible production and consumption decisions through specific incentives and restrictions. Green marketing, when considered within the context of sustainability -which is a key agenda item for countries worldwide- emerges as a focal point where governments, along with consumers and companies, play a significant role. In this section, the expectations of governments from businesses in the context of green marketing, the regulations made to realize these expectations, and examples of some of these regulations are included.

Today, governments -as the primary authority in regulating and controlling the amount of harmful waste that businesses release into the environment- operate through laws and regulations to prevent these negative externalities of businesses. According to Polonsky (1994), government regulations on environmental marketing are designed to protect consumers in different ways including: (1) reducing production of harmful goods or by-products, (2) altering individual and industry use and/or consumption of harmful goods, (3) ensuring that all individuals have the ability to evaluate the environmental composition of goods.

While the influence of societal expectations and a stricter legal system focused on environmental protection has led many companies to re-evaluate their approaches and integrate sustainable practices into their production and marketing processes, environmental legislation serves as a regulatory tool that sets legal standards that restrict environmentally harmful activities and encourage more sustainable behavior (Moraes et al., 2025). Governments carry out this regulation and supervision function by imposing environmental taxes, granting green incentives and/or ensuring the implementation of environmental certification and standards.

Environmental (Green) taxes and limitations of governments

Many of the practices that governments impose on businesses, such as environmental taxes, provide businesses with green marketing opportunities while also protecting the environment (Solaiman et al., 2015). In terms of environmental taxes, taxes levied on carbon emissions, energy use and waste generation are practices that affect green marketing (through functions such as production, packaging and distribution) and often impose costs on businesses. These are one of the main tools that governments use to encourage sustainable and green practices, and reduce harmful externalities caused by the operations of businesses and individuals. These taxes include energy taxes, taxes on transportation, pollution taxes and taxes on resource usage (European Commission, n.d., a). One of (and very well-known) these taxes -which has been used extensively in recent years- is the carbon tax, which aims to reduce carbon dioxide emissions of businesses.

While Finland was the first country in the world to implement a carbon tax in 1990, many other countries (Sweden at 1991, Ireland at 2010, Australia at 2012, Chile at 2014.. etc.) began implementing this tax in the following years (Carbon Tax Center, n.d.). In addition, other Asia Pacific countries like China, India, Indonesia, Thailand, Singapore and Vietnam have also included environmental and carbon-related taxes in their policies (Cottrell et al. 2017). In the United Kingdom (England, Scotland, Wales and Northern Ireland), where a carbon tax has been in place since

2013, the tax technically operates as a "carbon floor price" (CPF), which serves as the minimum price that fossil fuel producers (but only those in energy-intensive industries) pay for their CO2 emissions (Carbon Tax Center, n.d.).

To combat climate change and create a cleaner environment within the scope of the European Green Deal, the EU has already declared in 2020 that it aims to reduce greenhouse gas emissions by 55 percent by 2030 and become a climate-neutral continent by (European Commission, n.d., a). The European Union Emissions Trading Scheme (EU ETS) is one of the most important practices implemented by the EU in connection with this goal and which forms the center of its climate policy.

EU ETS, which can be considered as a restriction for reducing carbon emissions other than carbon tax, is a system that imposes costs on energy and industrial facilities based on their carbon emissions, which was firstly implemented in 2005, now includes EU countries, Norway, Iceland and Liechtenstein in the EU Economic Area, and Northern Ireland in the energy sector. Within this system, national authorities in each member state set an emissions quota for each industrial facility; companies whose emissions fall below these quotas can sell their idle quotas to other companies, while companies whose emissions exceed the quota can purchase quotas from other companies (European Commission, n.d., b; Perthuis & Trotignon, 2014; Zhang & Wei, 2010). While all countries participating in the EU ETS price some of their emissions by this system, 15 of the countries additionally have a carbon tax, mostly on emissions not covered by this system (Metcalf, 2021).

According to Fang et al. (2022) considering the damage caused to the ecology by energy consumption, especially fossil energy, the ways in which the energy structure can be improved are seen as the key to achieving the goal of a cleaner future. While studies conducted for developing and developed countries state that energy consumption is the main source of emissions of harmful gases (Shahzad, 2020), it has been observed that it is not clean energy but non-renewable energy that causes environmental degradation (Fang et al., 2022). Aiming to encourage waste recovery and decrease the amount of waste going to landfills, waste (landfill) taxes are a type of tax paid by the landfill owner and reflected in the costs associated with waste acceptance fees (United Nations, n.d.). While the landfill taxes imposed by the 22 EU member states are used in conjunction with bans on the storage of certain types of waste in Europe, there are also incineration taxes imposed by 9 member states (European Environment Agency, 2023).

Green incentives

Another tool governments use to reduce and regulate negative externalities of the businesses is incentives. As governments around the world tighten environmental, fuel economy, and safety regulations, auto companies are accelerating innovation in technology, processes, products, services, and business models (Abo-Khalil et al., 2022); such shifts could present significant opportunities for green marketing efforts, particularly for electric vehicle manufacturers, in areas such as positioning and distribution. One of the leading countries in terms of the use and promotion of electric vehicles is Norway. In Norway, battery electric vehicles and hydrogen vehicles are exempt from vehicle registration tax and value-added tax (VAT) and pay minimal vehicle registration fees; in addition, they receive other incentives, such as exemption from road tolls and ticket fees on ferries in most counties, or free parking in municipal public car parks (Bjerkan et al., 2016).

The Chinese government, which has begun implementing various subsidy policies to encourage businesses to engage in environmentally friendly production, introduced incentives such as VAT exemptions, low-interest loans, and preferential tariffs for new energy vehicle manufacturers and sellers in 2019 as a way to reduce carbon emissions (Nie et al., 2020). Addition to this, a wide range of financial incentives are available in Europe for electric vehicles, including rebates, home charger subsidies, POS tax relief, value added tax (VAT) relief, income tax relief, special depreciation, movement tax relief, company car tax relief, electricity supply subsidies and fee waivers (Münzel et al., 2019). Different countries have different incentives for battery electric vehicles (BEVs) and plug-in hybrid electric vehicles (PHEVs): the Netherlands offers registration tax exemptions or rebates, Korea offers purchase subsidies and tax reductions, the UK offers tax exemptions for company car drivers and point of sale grants, and finally the US tries to encourage ownership and use through national (federal) tax credits and some state rebates (Kong & Hardman, 2019).

One of the important restrictions imposed by states to reduce the use of environmentally harmful products has been related to products containing plastic and plastic derivatives, whose environmental damage has been debated for years. Considering that most plastic use is found in single-use packaging, banning single-use plastic (SUP) products can be considered an easy solution (Hira et al., 2022). With the European Union's Single-Use Plastics Directive -which came into force in 2019, Europe is leading the way in regulating plastic use by banning many single-use plastic products in all member states, including plastic cutlery and plates, plastic straws and stirrers, cotton swabs, oxo-degradable plastics, food containers made of expanded polystyrene, and drinking cups (UKHI, 2025). Canada also banned this category of products in the same year (including plastic bags, straws, and cutlery in 2019). In addition to Canada, many other countries such as Costa Rica, Taiwan, Belize, India, and the United States (California and Florida) also apply this practice (Hira et al., 2022).

Moreover, while SUP bags are still the most heavily regulated by governments across SUP product categories, other SUP products, such as styrofoam products or microbeads, are also among the categories experiencing increased regulation (Knoblauch & Mederake, 2021). Microbead plastics used in personal care products, cleaning products, printer toners, industrial abrasives, and medical applications have also banned in the Netherlands, Tanzania, Australia, Italy, South Korea, New Zealand, the United Kingdom, the United States, and Canada. In addition to bans, governments can adopt punitive measures, such as fines or fees, or incentives, such as container deposit schemes, water refill stations, and free or low-cost paper or cloth bags or coffee cup substitutes (Hira et al., 2022). Choosing more environmentally friendly products over plastic can be helpful for businesses, especially in their green marketing and promotional efforts. For example, some retailers in India sell shopping bags under the Go Green Environment Fund to reduce plastic bag use and support their environmental commitments (Thakkar, 2021).

There is a critical need for the establishment of mechanisms that ensure especially global enterprises can instantaneously assimilate and comply with the full spectrum of evolving environmental regulations, legislation, and policy amendments. This requisite compliance is only feasible through a high-technology instrument capable of rapidly scanning and analyzing vast quantities of data to generate the necessary workflow for enterprises.

AI

Despite the pioneering developments that laid the groundwork for it in prior years, the field of AI was formally established at the 1956 Dartmouth summer workshop, a pivotal event in the history of the discipline. The participants, including John McCarthy and Marvin Minsky, defined the core objective of the project as creating machines capable of exhibiting behavior that would be considered intelligent if demonstrated by a human (McCarthy et al., 1955). De Bruyn et al. (2020: 93) describe AI as machines that exhibit human-like intelligence in tasks such as learning, planning, and problem-solving through the autonomous creation of higher-level knowledge.

According to McCarthy (2004), AI is the science and engineering of creating intelligent machines, particularly intelligent computer programs. Amidst a growing body of scholarly work and application-based examples on the implementation of AI across various domains, it is argued that the technology will lead to paradigm-shifting changes in education (Ahmad et al., 2021), medicine (Ramesh et al., 2004; Reddy, 2022), engineering (Pham & Pham, 1999), economy (Aghion et al., 2017), law (Reiling, 2020) and other fields. Business and industry, including all of their sub-disciplines, have been profoundly influenced by AI. A large body of research highlights the impact of AI on business and industry. This includes both studies that present a general framework on AI's transformative effect (Parunak, 1996; Akerkar,

2019), and research that individually examines its influence on core business functions such as finance (Buchanan, 2019), accounting (Li & Zheng, 2018), human resources (Ahmed, 2018) and marketing (Jarek & Mazurek, 2019).

A body of literature has begun to emerge regarding the impact of AI on marketing in general, and green marketing in particular. Basha's (2023) research, based on interviews with 15 senior marketing executives, revealed that competitive pressure is the most influential factor in the adoption of AI in marketing. This finding suggests that many companies feel compelled by their competitors to implement AI in their marketing approaches. Kumar et al. (2025) contend that advanced technologies such as AI, Big Data, and digital platforms are revolutionizing green marketing via digital transformation. Crucially, AI and ML enable marketers to act on real-time insights, resulting in better-informed decisions.

METHODOLOGY

This study employs a comprehensive conceptual analysis and narrative literature review. This study synthesizes core green marketing research focused on consumer, competitive, corporate, governmental, and international environmental regulations, while also integrating major works on the role of AI in green marketing applications and addressing its potential ethical challenges. The methodology focuses on the intersection of AI-driven digital transformation and green marketing strategies to propose a holistic framework that addresses the demands of companies, consumers, competitors, and governments.

RESEARCH OBJECTIVES

The primary objectives of this chapter are:

- To evaluate the necessity of a holistic green marketing approach in a complex stakeholder environment.
- To analyze the leveraging role of AI in such a holistic green marketing approach.
- To identify the structural requirements and ethical challenges associated with AI-driven sustainability strategies.

THEORETICAL FRAMEWORKS

The theoretical foundations of this study are rooted in the sustainability, green marketing and the integration of technological advancement with ecological responsibility. AI enables marketers to act on real-time insights, revolutionizing green marketing through digital transformation. This study's theoretical framework centers on green marketing theory triggered by sustainability, examining the interplay between firms, competition, consumers, and regulations, while also integrating the role of AI in this entire process. AI's ability to provide data-driven insights allows for sustainable innovation, where technology serves as the infrastructure for organic and genuine green practices.

IMPORTANCE OF THE STUDY

This study, which examines green marketing through a holistic lens involving firms, consumers, competition, and regulations in terms of sustainability, aims to enrich green marketing theory and provide theoretical insights into the nexus of green marketing and artificial intelligence. Moreover, adopting green marketing initiatives is a core component of modern sustainability strategies, boosting a firm's competitive edge (Chen & Chang, 2013). Research demonstrates a positive correlation between green marketing and both brand loyalty (Wu & Li, 2022; Shafiq et al., 2023) and consumer purchasing behavior (Majeed et al., 2022). In an era of intense market competition, businesses replicate the environmental initiatives of rivals to preserve market position. Therefore, leveraging AI becomes instrumental for developing a holistic approach, allowing businesses to align green transformation objectives with competitive demands and consumer preferences.

NEED FOR A HOLISTIC GREEN MARKETING APPROACH

In the wake of the global sustainability movement, businesses have been compelled to increasingly adopt green marketing practices. According to a study by Leonidou et al. (2013, p. 151), green marketing can be a key driver of firm performance. Green product and distribution components are found to enhance product-market performance, while green pricing and promotion positively affect a firm's financial return. The findings also highlight that industry-level environmental reputation plays a moderating role in these links. The main purpose is the aspiration to improve a company's public image and brand value by showcasing its dedication to ecological sustainability (Cowan & Guzman, 2020). This shift is also driven by a confluence of influential factors, including the pressure to conform to internationally standardized regulations, the competitive landscape shaped by rivals who have already embraced green marketing approaches, and the escalating demand from environmentally conscious consumers. Green marketing transcends simple promotional strategies, encompassing a holistic approach from product inception and manufacturing to packaging and distribution, with the fundamental goal of mitigating environmental harm while addressing consumer demands (Fuller, 1999). It is evident that some consumers exhibit intermittent eco-friendly inclinations in their purchasing habits and behavior (Ginsberg & Bloom, 2004, p. 80). As highlighted in the Roper classification mentioned previously (Ginsberg & Bloom, 2004), there is a need to categorize consumers based on their green product purchasing tendencies. With intense market competition, customer loyalty is often challenged as consumers have multiple alternatives and face minimal barriers to switching suppliers so the presence of competitive pressure can serve as a catalyst for organizations to explore and integrate green marketing into their strategies as well (Leonidou et al., 2013).

On the other hand, government regulations, as highlighted by Polonsky (1994) and supported by Moraes et al. (2025), serve as a critical tool for promoting environmental responsibility. These regulations are designed to protect both consumers and the environment by restricting harmful activities and encouraging sustainable behavior. This includes ensuring consumers can evaluate the environmental impact of products and influencing companies to integrate sustainable practices into their production and marketing processes. The implementation of various government initiatives, such as tax incentives for green investments and subsidies for renewable energy, has been a substantial factor in encouraging businesses in many countries to adopt environmentally friendly practices and promote sustainable products (Dangelico & Vocalelli, 2017). By setting clear guidelines and criteria, regulatory frameworks enable businesses to market their products as environmentally friendly in a way that builds consumer trust (Rahman & Nguyen-Viet, 2022).

Based on these points, it is evident that for businesses to develop and implement successful and effective green marketing strategies, they need a holistic approach that considers business interests, consumers, competitors, regulations and standards. Green marketing will only yield productive and sustainable outcomes for all parties if the intersection of all these parties is addressed in a holistic manner. This is because governments and standards, while demanding green marketing practices from businesses, are essentially considering the interests of consumers, the environment, and the economy. Meanwhile, consumer demand for green marketing is shaped by a growing environmental awareness and pressure from government regulations. Consumers also intensify competition, pushing businesses toward adopting these practices. In response, firms implement green marketing to meet consumer demands, avoid falling behind competitors, and gain incentives or avoid penalties from regulations. Ultimately, this results in an improvement in their own performance indicators, demonstrating a convergence between the interests of all parties and the firm's own business goals. Given the broad and holistic nature of this process, the role of a facilitating technology, such as AI, would be invaluable.

THE LEVERAGING ROLE OF AI IN A HOLISTIC GREEN MARKETING APPROACH

Marketing is unquestionably the field where the transformative power of AI is most apparent, as AI is redefining its core applications and strategies. De Bruyn et al. (2020, p. 93) draws on Power (2016) and Murray and Wardley (2014) to argue that the majority of contemporary AI applications in business rely on deep artificial neural networks to address complex predictive challenges that were considered intractable until recently. In marketing, predictive analytics, in particular, empowers practitioners to forecast the outcomes of future marketing initiatives, derive actionable insights to enhance lead generation, facilitate new customer acquisition, and optimize pricing strategies.

There has been extensive research into the different applications of AI in marketing management. AI plays a crucial role in developing sales and marketing strategies that result in substantial gains in corporate performance. It has transformed the marketing environment, helping to update obsolete strategies. This shift presents a significant challenge for organizations, requiring them to proactively plan for the future and provide continuous training for their staff (Basha, 2023). The analysis by Jarek and Mazurek (2019) suggests a widespread adoption of AI in the marketing field, albeit with its applications largely confined to the operational level. Gacanin and Wagner (2019) addressed the practical challenges of implementing autonomous Customer Experience Management (CEM), demonstrating how AI and machine learning (ML) can establish intelligent networks and generate significant business value. AI's capacity to process vast datasets-including numerical, textual, vocal, visual, and facial expression data-enables the extraction of valuable insights from customer and transaction information. Consequently, companies can utilize AI-driven analytics to predict consumer behavior, proactively detect credit fraud, or implement highly targeted, real-time digital advertising campaigns (Davenport et al., 2019). The rise of AI has spurred the development of new analytical methods for managing marketing information and solving related problems (Zhao & Priporas, 2017). Specific applications of AI, such as AI/ML algorithms facilitate more efficient data processing for better decision-making (Maxwell et al., 2011) and Natural Language Processing (NLP)-driven chatbots, have been shown to improve customer experience (Nguyen & Sidorova, 2018). AI is vital for analyzing customer habits and preferences (Chatterjee et al., 2019) and that AI User Interfaces (AIUI) have enhanced Customer Relationship Management (CRM) functions (Seranmadevi & Kumar, 2019). Additionally, in e-commerce, AI's ability to track human senses has been shown to strengthen the consumer-brand relationship (Sha & Rajeswari, 2019) and customer experience management (Nwachukwu & Affen, 2023). AI-powered tools, including chatbots, personalization engines, and predictive analytics, play a crucial role in boosting e-commerce performance (Madanchian, 2024).

AI is also important for green marketing strategies and practices of businesses, which enables businesses to facilitate their green transformation, gain a competitive edge, satisfy the needs of green consumers, and adhere to legislative requirements. Sajili (2024) stated that AI's capacity to process vast volumes of data is pivotal in providing marketers with the actionable intelligence required to tailor their strategies to develop sustainability trends. Leveraging AI allows companies to deploy highly targeted advertising, optimize their supply chain logistics, and enhance the clarity of their sustainability efforts (Baruno & Indrasari, 2025, p. 2). Strategically, AI automates the collection and analysis of data for sustainability reporting, ensuring transparent and timely communication with all partners. AI also enables the implementation of dynamic pricing models that account for a product's environmental impact, helping to align pricing with sustainable values. It also provides the analytical and operational foundation needed to create, execute, and monitor holistic green communication strategies (Ganesh et al., 2024).

Furthermore, AI's capacity for advanced data analytics is central to developing and distributing effective green messaging. By leveraging predictive analytics and semantic analysis, businesses can identify emerging sustainability trends and craft both textual and visual content that is highly relevant to consumers' eco-conscious interests. Furthermore, AI-powered chatbots and gamification platforms can be integrated to foster direct, interactive communication, educating consumers on sustainable practices while incentivizing pro-environmental behaviors. It empowers consumers to authenticate sustainability claims and bolster their confidence in such purchases (John & Varghese, 2024). Simultaneously, AI-driven sentiment analysis allows brands to continuously monitor and respond to public perception regarding their sustainability initiatives, thereby enhancing brand reputation (Ganesh et al., 2024).

According to Chen and Jin (2023), AI's capacity to optimize production processes and tailor consumer experiences through the use of predictive analytics and machine learning ultimately leads to a reduction in carbon emissions. AI technologies are instrumental in optimizing resource efficiency through waste reduction (Muduva & Kiwa, 2025). Furthermore, AI-powered solutions facilitate heightened targeted consumer engagement through the capability to design tailored marketing campaigns that strike a chord with environmentally aware consumers. AI's role in promoting sustainability is evident through its capacity to optimize inventory and supply chain management, which in turn leads to significant waste reduction and more efficient resource allocation (Heins, 2022). AI-driven advancements in product development facilitate the creation of renewable energy solutions, thereby catering to the growing market for environmentally conscious products (Yin et al., 2023).

A report stated that the integration of Earth observation data with AI provides a more effective, efficient way to timely monitor environmental effects and trends. By offering new insights and stronger predictive capabilities, AI can produce key information for environmental management and policy monitoring, ultimately assisting both consumers and businesses in adapting to more sustainable behaviors (Gailhofer et al., 2021, p. 9). AI-driven sustainable finance facilitates a more precise evaluation of ESG metrics, which directly influences the credibility and effectiveness of a firm's green marketing (Marchiori et al., 2022; Azzada et al., 2023). Pulkkinen et al. (2024) contend that AI-empowered participatory budgeting has the potential to foster collaborative innovation within local governments, thereby demonstrating the scalability of AI-enhanced green practices across various sectors. For businesses, it's crucial to adhere to environment-protection regulations and comply with international green standards. Overall, AI will play a positive role by providing tools that better manage the processes of green marketing, leading to its more efficient implementation and maximizing benefits for all relevant parties.

The Synergetic Role of AI in Regulatory Compliance and Green Marketing

As established in the previous sections, the proliferation of instruments such as the EU ETS, carbon taxes, and Single-Use Plastic (SUP) directives creates a complex, multi-layered regulatory environment. Within this framework, AI does not merely function as an operational tool; it serves as a strategic bridge between legal necessity and green marketing success. While governments impose restrictions to mitigate negative externalities (Polonsky, 1994), firms often struggle with the "compliance burden." AI-driven predictive analytics allow businesses to transition from a reactive stance to a proactive one. By analyzing vast datasets related to energy consumption and emission quotas, AI enables firms to optimize their production in real-time to avoid carbon penalties. This efficiency then becomes a core narrative in their green marketing, transforming "compliance" into a "brand promise" of genuine sustainability.

With the introduction of the Green Claims Directive, the legal consequences of greenwashing have become severe. The discussion here centers on AI's role in verification and transparency. AI systems can track raw material origins and supply chain footprints with high precision, providing the empirical evidence required by

Scandinavian and Canadian-style "evidence-sharing" sanctions. Thus, AI acts as a safeguard for brand reputation, ensuring that marketing claims are supported by verifiable, real-time data. Governments offer "Green Incentives," such as VAT exemptions for EVs in Norway or subsidies in China. However, identifying the right consumer segment to utilize these incentives requires sophisticated analysis. AI enhances this process by segmenting "True Blue Greens" and "Greenback Greens" more accurately than traditional models. By aligning government incentives with the right consumer profiles through AI-powered personalization, firms can maximize the ROI of their green marketing campaigns while fulfilling the state's sustainability objectives.

A critical point of discussion is the paradox of AI's own environmental footprint. While AI optimizes resource efficiency, its high energy demand could potentially contradict the goals of green marketing. Therefore, a holistic approach must involve "Green AI" algorithms optimized for low energy consumption. The implementation of regulatory frameworks must eventually address the energy use of the digital tools themselves, ensuring that the technology used to save the planet does not inadvertently accelerate its degradation.

Ethical issues and challenges with AI

While AI offers immense potential, it introduces specific ethical concerns. There is an increasing critical view on the environmental footprint of AI itself, specifically the high energy consumption required to train and run large AI models. As Kelly (2022) points out, our knowledge regarding the environmental impacts of artificial intelligence is still at a limited level. Zhuk's (2023) study highlights that the development and deployment of AI pose significant environmental, ethical, and legal challenges. Key concerns include high energy consumption from non-renewable sources leading to increased carbon emissions, the growing problem of e-waste from improper equipment disposal, and the risk of algorithmic errors causing environmental injustice. Furthermore, AI technologies can disrupt natural ecosystems, threatening wildlife and hindering sustainable ecological progress. Additionally, data privacy issues arise when using AI to track consumer habits for segmentation. For green marketing to remain genuine, businesses must ensure that the use of AI does not contradict the very sustainability goals they promote. The other common ethical issues with AI are transparency, algorithmic bias and accountability challenges (Mensah, 2024). Diminishing transparency in AI systems serves as a major barrier to establishing accountability, thereby undermining the overall reliability of the technology. Also, AI algorithms trained on vast, biased datasets can produce discriminatory outcomes that reinforce existing societal inequalities if left unaddressed.

CONCLUSION

The increasing importance of green marketing is a result of converging pressures from consumers, governments, and businesses themselves. Growing consumer environmental awareness has intensified market competition, while strict regulations and policy measures from governments and international standards have compelled firms to adopt green practices. In response to these external forces, businesses strategically implement green marketing not only to ensure compliance and satisfy consumer demands but also to enhance their brand reputation and gain a competitive advantage. Beyond the apparent pressure on businesses to engage in green marketing, a complex web of reciprocal relationships exists. The literature demonstrates that green marketing initiatives lead to positive outcomes for firms' performance. Moreover, studies highlight the interconnected roles of governments and international standards in influencing consumers, consumers in intensifying competition, and businesses in educating consumers about environmental issues.

Clearly, navigating this complex interplay of stakeholder expectations -from regulators and competitors to consumers- presents a significant challenge. Consequently, the establishment of an effective and integrated green marketing framework and the development of corresponding strategic initiatives is a process that proves exceedingly difficult to execute using conventional methodologies. Undoubtedly, with its applications that simplify a multitude of areas in our lives today, AI will play a significant and effective role in the establishment and implementation of a holistic green marketing strategy.

The application of AI-driven technologies, specifically predictive analytics, machine learning, and automation, can be leveraged to optimize green marketing strategies through the facilitation of precise consumer segmentation and targeting, dynamic green campaign adaptation, and robust sustainability performance measurement. The potential of AI lies in its capacity to enhance consumer engagement, foster trust in sustainable brands, and mitigate implementation challenges across diverse market contexts. AI technologies also enable businesses to monitor emissions, optimize resource usage, and streamline environmental reporting, thereby ensuring compliance with increasingly stringent regulations.

Conversely, addressing the potential environmental drawbacks of AI is essential. The notion that a tool employed to enhance sustainability efforts could ultimately yield a negative impact is untenable. Consequently, keeping a close watch on both academic studies and regulatory developments regarding this aspect of AI is of paramount importance. As long as the potential challenges are mitigated, overall, AI is set to become a key factor in the successful implementation of a holistic green marketing approach, ensuring optimal outcomes for all stakeholders in the present and future.

FUTURE RESEARCH DIRECTIONS

Future scientific studies can demonstrate the role of artificial intelligence within this theoretically established holistic model through quantitative research. Future research also should focus on the quantitative depth of AI's impact on green marketing ROI, in particular. Additionally, the "Net Zero AI" concept, balancing the carbon footprint of AI models with the environmental savings they generate, remains a vital area for academic inquiry. Studies should also explore how AI can specifically help in closing the gap between green consumer attitudes and actual purchasing behavior in diverse global markets.

REFERENCES

Abo-Khalil, A. G., Abdelkareem, M. A., Sayed, E. T., Maghrabie, H. M., Radwan, A., Rezk, H., & Olabi, A. G. (2022). Electric vehicle impact on energy industry, policy, technical barriers, and power systems. *International Journal of Thermofluids, 100134*, 100134. Advance online publication. DOI: 10.1016/j.ijft.2022.100134

Aghion, P., Jones, B. F., & Jones, C. I. (2017). Artificial intelligence and economic growth. NBER Working Paper Series. National Bureau of Economic Research. Retrieved from: chrome-extension://efaidnbmnnnibpcajpcglclefindmkaj/https://www.nber.org/system/files/working_papers/w23928/w23928.pdf

Ahmad, S. F., Rahmat, M. K., Mubarik, M. S., Alam, M. M., & Hyder, S. I. (2021). Artificial intelligence and its role in education. *Sustainability (Basel)*, *13*(22), 12902. DOI: 10.3390/su132212902

Ahmed, O. (2018). Artificial intelligence in HR. [IJRAR]. *International Journal of Research and Analytical Reviews*, *5*(4), 971–978.

Akerkar, R. (2019). *Artificial Intelligence for Business*. Springer Briefs in Business Series. Springer. DOI: 10.1007/978-3-319-97436-1

Al-Salman, F. M. A. M. (2016). Investigation of the factors affecting green purchasing behavior: A study using the concepts of environmental concern, collectivism and self-image. Unpublished Master Thesis, Gazi University Institute of Social Sciences, Ankara.

Ali, A., Khan, A. A., & Ahmed, I. (2011). Determinants of Pakistani consumers' green purchase behavior: Some insights from a developing country. *International Journal of Business and Social Science*, *2*(3), 217–226.

Baruno, A. & Indrasari, M. (2025). Leveraging AI to enhance green marketing strategies. *Jurnal Ekonomi, Manajemen, Akuntansi dan Keuangan, 6*(1), 1-17.

Basha, M. (2023). Impact of artificial intelligence on marketing. [EAJMR]. *East Asian Journal of Multidisciplinary Research*, *2*(3), 993–1004. DOI: 10.55927/eajmr.v2i3.3112

Belz, F. M., & Peattie, K. (2009). *Sustainability Marketing: A Global Perspective*. Wiley.

Bjerkan, K. Y., Norbech, T. E., & Nordtomme, M. E. (2016). Incentives for promoting Battery Electric Vehicle (BEV) adoption in Norway. *Transportation Research Part D, Transport and Environment*, *43*, 169–180. DOI: 10.1016/j.trd.2015.12.002

Boztepe, A. (2012). Green marketing and its impact on consumer buying behavior. *European Journal of Economic and Political Studies*, *5*(1), 5–21.

Buchanan, B. G. (2019). Artificial intelligence in finance. Report, University of Surrey. *Zenodo*. Retrieved from: DOI: 10.5281/zenodo.2612536

Carbon Tax Center. (n.d.). Where carbon is taxed (Some Individual Countries). Retrieved from: https://www.carbontax.org/where-carbon-is-taxed-some-individual-countries/

Chatterjee, S., Ghosh, S. K., Chaudhuri, R., & Nguyen, B. (2019). Are CRM systems ready for AI integration? A conceptual framework of organizational readiness for effective AI-CRM integration. *The Bottom Line (New York, N.Y.)*, *32*(2), 144–157. DOI: 10.1108/BL-02-2019-0069

Chen, Y., & Jin, S. (2023). Artificial intelligence and carbon emissions in manufacturing firms: The moderating role of green innovation. *Processes (Basel, Switzerland)*, *11*(9), 2705. DOI: 10.3390/pr11092705

Chen, Y. S., & Chang, C. H. (2013). Enhance environmental commitments and green intangible assets toward green competitive advantages: An analysis of structural equation modeling (SEM). *Quality & Quantity*, *47*(1), 529–543. DOI: 10.1007/s11135-011-9535-9

Cottrell, J., Ludewig, D., Runkel, M., Schlegelmilch, K., & Zerzawy, F. (2017). Environmental tax reform in Asia and the Pacific. MPDD Working Paper Series WP/17/05, United Nations Economic and Social Commission for Asia and the Pacific (ESCAP). Retrieved from: https://ideas.repec.org/p/unt/wpmpdd/wp-17-05.html

Cowan, K., & Guzman, F. (2020). How CSR reputation, sustainability signals, and country-of-origin sustainability reputation contribute to corporate brand performance: An exploratory study. *Journal of Business Research*, *117*, 683–693. DOI: 10.1016/j.jbusres.2018.11.017

Dangelico, R. M., & Vocalelli, D. (2017). "Green marketing": An analysis of definitions, strategy steps, and tools through a systematic review of the literature. *Journal of Cleaner Production*, *165*(1), 1263–1279. DOI: 10.1016/j.jclepro.2017.07.184

Davenport, T. H. (2018). *The AI Advantage: How to Put the Artificial Intelligence Revolution to Work*. MIT Press. DOI: 10.7551/mitpress/11781.001.0001

Davenport, T. H., Guha, A., Grewal, D., & Bressgott, T. (2019). How artificial intelligence will change the future of marketing. *Journal of the Academy of Marketing Science*, *48*(1), 24–42. DOI: 10.1007/s11747-019-00696-0

De Bruyn, A., Viswanathan, V., Beh, Y. S., Brock, J. K.-U., & Wangenheim, F. (2020). Artificial intelligence and marketing: Pitfalls and opportunities. *Journal of Interactive Marketing*, *51*(1), 91–105. DOI: 10.1016/j.intmar.2020.04.007

European Commission. (n.d.) a. Green taxation- in support of a more sustainable future. Retrieved from: https://taxation-customs.ec.europa.eu/archives/other-taxe/green-taxation_en

European Commission. (n.d.) b. Scope of the EU ETS. Retrieved from: https://climate.ec.europa.eu/eu-action/carbon-markets/eu-emissions-trading-system-eu-ets/scope-eu-ets_en

European Environment Agency. (2023). Economic instruments and separate collection systems - key strategies to increase recycling. Retrieved from: https://www.eea.europa.eu/en/analysis/publications/economic-instruments-and-separate-collection

Fan, H., & Zeng, L. (2011). Implementation of green marketing strategy in China – A study of the green food industry. Unpublished Master Thesis in Business Administration, University of Gävle, Sweden.

Fang, G., Yang, K., Tian, L., & Ma, Y. (2022). Can environmental tax promote renewable energy consumption? An empirical study from the typical countries along the Belt and Road. *Energy*, *260*, 125193. https://www.sciencedirect.com/science/article/pii/S0360544222020837. DOI: 10.1016/j.energy.2022.125193

Fuller, D. A. (1999). *Sustainable Marketing: Managerial-ecological Issues*. Sage. DOI: 10.4135/9781452220611

Gacanin, H., & Wagner, M. (2019). Artificial intelligence paradigm for customer experience management in next-generation networks: Challenges and perspectives. *IEEE Network*, *33*(2), 188–194. DOI: 10.1109/MNET.2019.1800015

Gailhofer, P., Herold, A., Schemmel, J. P., Scherf, C.-S., Urrutia, C., Köhler, A. R., & Braungardt, S. (2021). The role of artificial intelligence in the European Green Deal. European Parliament: Directorate-General for Internal Policies of the Union. Retrieved from: https://data.europa.eu/doi/10.2861/882830

Ganesh, C., Podila, N., Vamsi, G. B. K., Rao, M., & Bhardwaj, N. (2024). AI-enhanced content marketing for sustainability: A theoretical perspective communication strategies. MATEC Web of Conferences, 392, 01045. Retrieved from: DOI: 10.1051/matecconf/202439201045

Ginsberg, J. M., & Bloom, P. N. (2004). Choosing the right green marketing strategy. *MIT Sloan Management Review*, *46*(1), 79–84.

Gögermiş, E. (2019). Yeşil pazarlama. Y. Durmaz & S. Gün (Eds.) Pazarlama Çalışmaları, (191-252), İKSAD Publishing House. Retrieved from: https://www.researchgate.net/profile/Yakup-Durmaz-3/publication/337905699_PAZARLAMA_CALISMALARI_EDITORLER/links/5df1e7ca299bf10bc35464c5/PAZARLAMA-CALISMALARI-EDITOeRLER.pdf

Grant, J. (2008). Green marketing. *Strategic Direction.*, *24*(6), 25–27. DOI: 10.1108/02580540810868041

Heins, C. (2022). Artificial intelligence in retail – A systematic literature review. *Foresight*, *25*(2), 264–286. DOI: 10.1108/FS-10-2021-0210

Henion, K. E., & Kinnear, T. C. (1976). *Ecological Marketing*. American Marketing Association.

Hira, A., Pacini, H., Attafuah-Wadee, K., Vivas-Eugui, D., Saltzberg, M., & Yeoh, T. N. (2022). Plastic waste mitigation strategies: A review of lessons from developing countries. *Journal of Developing Societies*, *38*(3), 336–359. DOI: 10.1177/0169796X221104855

Jain, S. K., & Kaur, G. (2004). Green marketing: An Indian Perspective. *Decision (Washington, D.C.)*, *31*(2), 168–209.

Jarek, K., & Mazurek, G. (2019). Marketing and artificial intelligence. *Central European Business Review*, *8*(2), 46–56. DOI: 10.18267/j.cebr.213

John, A. P., & Varghese, S. (2024). Transformative potential of artificial intelligence in green marketing. *MSNIM Management Review, 2*(1). Retrieved from: https://msnim.edu.in/research/journal/archives/

Kalafatis, S. P., Pollard, M., East, R., & Tsogas, M. H. (1999). Green marketing and Ajzen's theory of planned behaviour: A cross-market examination. *Journal of Consumer Marketing*, *16*(5), 441–460. DOI: 10.1108/07363769910289550

Kelly, B. (2022). Ethical AI and the environment. *The iJournal, 7*(2), 1-10.

Knoblauch, D., & Mederake, L. (2021). Government policies combatting plastic pollution. *Current Opinion in Toxicology*, *28*, 87–96. DOI: 10.1016/j.cotox.2021.10.003

Kong, N., & Hardman, S. (2019). *Electric vehicle incentives in 13 leading electric vehicle markets. Research Report UCD-ITS-RR-19-04. Plug-In Hybrid & Electric Vehicle Research Center, Institute of Transportation Studies*. University of California., Retrieved from https://escholarship.org/content/qt0fm3x5bh/qt0fm3x5bh.pdf

Koons, E. (2024). Greenwashing Examples 2024 [Greenwashing Companies. Retrieved from: https://energytracker.asia/greenwashing-examples-of-top-companies/ .]. *Top (Madrid), 10*, •••.

Kumar, S., Sharma, C., Mishra, M. K., Sharma, S., & Bhardwaj, V. (2025). Smart, sustainable, and green: The digital transformation of green-marketing. *Discover Sustainability*, *6*(1), 388. DOI: 10.1007/s43621-025-01242-5

Lampe, M., & Gazdat, G. M. (1995). Green marketing in Europe and The United States: An evolving business and society interface. *International Business Review*, *4*(3), 303–313. DOI: 10.1016/0969-5931(95)00011-N

Leonidou, C. N., Katsikeas, C. S., & Morgan, N. A. (2013). "Greening" the marketing mix: Do firms do it and does it pay off? *Journal of the Academy of Marketing Science*, *41*(2), 151–170. DOI: 10.1007/s11747-012-0317-2

Li, Z., & Zheng, L. (2018). The impact of artificial intelligence on accounting. [ASSEHR]. *Advances in Social Science, Education and Humanities Research*, *181*, 813–816. DOI: 10.2991/icsshe-18.2018.203

Madanchian, M. (2024). The impact of artificial intelligence marketing on e-commerce sales. *Systems*, *12*(10), 429. DOI: 10.3390/systems12100429

Majeed, M. U., Aslam, S., Murtaza, S. A., Attila, S., & Molnár, E. (2022). Green marketing approaches and their impact on green purchase intentions: Mediating role of green brand image and consumer beliefs towards the environment. *Sustainability (Basel)*, *14*(18), 11703. DOI: 10.3390/su141811703

Marchiori, D. M., Rodrigues, R. G., Popadiuk, S., & Mainardes, E. W. (2022). The relationship between human capital, information technology capability, innovativeness and organizational performance: An integrated approach. *Technological Forecasting and Social Change*, *177*, 1–14. DOI: 10.1016/j.techfore.2022.121526

Maxwell, A. L., Jeffrey, S. A., & Lévesque, M. (2011). Business angel early-stage decision making. *Journal of Business Venturing*, *26*(2), 212–225. DOI: 10.1016/j.jbusvent.2009.09.002

McCarthy, J. (2004). What is artificial intelligence? http://jmc.stanford.edu/articles/whatisai.html

McCarthy, J. L., Minsky, M. L., Rochester, N., & Shannon, C. E. (1955). A proposal for the Dartmouth Summer Research Project on artificial intelligence. August 31, 1955. Retrieved from: http://jmc.stanford.edu/articles/dartmouth/dartmouth.pdf

McDaniel, S. W., & Rylander, D. H. (1993). Strategic green marketing. *Journal of Consumer Marketing*, *10*(3), 4–10. DOI: 10.1108/07363769310041929

Mensah, G. B. (2024). AI ethics. [AJFRA]. *Africa Journal for Regulatory Affairs*, *1*(11), 32–45.

Metcalf, G. E. (2021). Carbon taxes in theory and practice. *Annual Review of Resource Economics*, *13*(1), 245–265. DOI: 10.1146/annurev-resource-102519-113630

Mishra, P., & Sharma, P. (2014). Green marketing: Challenges and opportunities for business. *BVIMR Management Edge*, *7*(1), 78–86.

Moraes, A. S., Mezzomo, A. L., Pacheco, C. S. G. R., Santos, R. P., dos, , Moreira, M. B., Oliveira, M. D. S., de, , Oliveira, D. F., Chagas, S. C. S., Campos, D. F., Bonomo, B., Diniz, M. H. dos S., & Araújo, N. (2025). Environmental Law and Green Marketing: Perspectives for Sustainable Consumption. *Revista De Gestão Social e Ambiental*, *19*(1), 1–9. DOI: 10.24857/rgsa.v19n1-106

Muduva, M., & Kiwa, F. J. (2025). Artificial intelligence-driven green marketing strategies for a circular economy. In Muposhi, A., & Nyagadza, B. (Eds.), *Sustainable Green Marketing Strategies for a Circular Economy*. Palgrave Studies of Marketing in Emerging Economies. Palgrave Macmillan., DOI: 10.1007/978-3-031-81627-7_11

Münzel, C., Plötz, P., Sprei, F., & Gann, T. (2019). How large is the effect of financial incentives on electric vehicle sales? A global review and European analysis. *Energy Economics*, *84*, 104493. Advance online publication. DOI: 10.1016/j.eneco.2019.104493

Nguyen, Q. N., & Sidorova, A. (2018). Understanding user interactions with a Chatbot: A Self-determination Theory Approach. AMCIS 2018 Proceedings. 3. New Orleans. Retrieved from: https://aisel.aisnet.org/amcis2018/HCI/Presentations/3

Nie, D., Li, H., Qu, T., Liu, Y., & Li, C. (2020). Optimizing supply chain configuration with low carbon emission. *Journal of Cleaner Production*, *271*, 122539. DOI: 10.1016/j.jclepro.2020.122539

Nwachukwu, D., & Affen, M. P. (2023). Artificial intelligence marketing practices: The way forward to better customer experience management in Africa (Systematic literature review). *International Academy Journal of Management. Marketing and Entrepreneurial Studies.*, *9*(2), 44–62.

Parunak, H. V. D. (1996). Applications of distributed artificial intelligence in industry. In N. O'Hare and G. O. Jennings (Eds.), *Foundations of Distributed Artificial Intelligence*, (pp. 139-164). Toronto: A Wiley-Inter Science Publication.

Peattie, K. (1992). *Green Marketing*. Pitman Pub.

Peattie, K. (2001a). Towards sustainability: The third age of green marketing. *The Marketing Review*, 2(2), 129–146. DOI: 10.1362/1469347012569869

Peattie, K. (2001b). Golden goose or wild goose? The hunt for the green consumer. *Business Strategy and the Environment*, *10*(4), 187–199. DOI: 10.1002/bse.292

Perthuis, C., & Trotignon, R. (2014). Governance of CO2 markets: Lessons from the EU ETS. *Energy Policy*, *75*, 100–106. DOI: 10.1016/j.enpol.2014.05.033

Pham, D. T., & Pham, P. T. N. (1999). Artificial intelligence in engineering. *International Journal of Machine Tools & Manufacture*, *39*(6), 937–949. DOI: 10.1016/S0890-6955(98)00076-5

Polonsky, M. J. (1994). An introduction to green marketing. *Electronic Green Journal*, *1*(2), 1–10. DOI: 10.5070/G31210177

Prakash, A. (2002). Green marketing, public policy and managerial strategies. *Business Strategy and the Environment*, *11*(5), 285–300. DOI: 10.1002/bse.338

Pulkkinen, M., Sinervo, L.-M., & Kurkela, K. (2024). Premises for sustainability-participatory budgeting as a way to construct collaborative innovation capacity in local government. *Journal of Public Budgeting, Accounting & Financial Management*, *36*(1), 40–59. DOI: 10.1108/JPBAFM-04-2022-0077

Rahman, S., & Nguyen-Viet, B. (2022). Towards sustainable development: Coupling green marketing strategies and consumer perceptions in addressing greenwashing. *Business Strategy and the Environment*, *32*(4), 2420–2433. DOI: 10.1002/bse.3256

Ramesh, A. N., Kambhampati, C., Monson, J. R. T., & Drew, P. J. (2004). Artificial intelligence in medicine. *Annals of the Royal College of Surgeons of England*, *86*(5), 334–338. DOI: 10.1308/147870804290 PMID: 15333167

Reddy, S. (2022). Explainability and artificial intelligence in medicine. *Comment*, *4*(4), 214–215. PMID: 35337639

Reiling, A. D. (2020). Courts and artificial intelligence. *International Journal for Court Administration*, *11*(2), 8. Advance online publication. DOI: 10.36745/ijca.343

Sajili, M., Kinanti, L. A. B., & Rudhan, A. M. (2024). Rise of AI: Transforming data analytics in marketing strategies. *Journal of Economic Business and Accounting (Costing)*, *7*(4), 7216–7722. DOI: 10.31539/costing.v7i4.10179

Seranmadevi, R., & Kumar, A. (2019). Experiencing the AI emergence in Indian re- tail–early adopters approach. *Management Science Letters*, *9*(1), 33–42. DOI: 10.5267/j.msl.2018.11.002

Sha Nazim, S., & Rajeswari, M. (2019). Creating a brand value and consumer satisfaction in e-commerce business using artificial intelligence with the help of vosag technology. *International Journal of Innovative Technology and Exploring Engineering*, *8*(8), 1510–1515.

Shafiq, M. A., Ziaullah, M., Siddique, M., Bilal, A., & Ramzan, M. (2023). Unveiling the sustainable path: Exploring the nexus of green marketing. [IJSSE]. *International Journal of Social Science and Entrepreneurship*, *3*(2), 654–676.

Shahzad, U. (2020). Environmental taxes, energy consumption, and environmental quality: Theoretical survey with policy implications. *Environmental Science and Pollution Research International*, *27*(20), 24848–24862. DOI: 10.1007/s11356-020-08349-4 PMID: 32372356

Sharma, A. (2025). Green marketing examples: 15 inspiring brands promoting sustainability. Retrieved from: https://www.youngurbanproject.com/green-marketing-examples/

Singal, R., Garg, A., & Singla, S. (2013). Green marketing: Challenges and opportunities. *International Journal of Innovations in Engineering and Technology*, *2*(1), 470–474.

Solaiman, M., Osman, A., & Halim, M. S. B. A. (2015). Green marketing: A marketing mix point of view. *International Journal of Business and Technopreneurship*, *5*(1), 87–98.

Thakkar, R. S. (2021). Green marketing and sustainable development challenges and opportunities. *Gap Gyan-. Global Journal of Social Sciences*, *4*(4), 58–65.

UKHI. (2025). A global map of single-use plastic regulations: Navigating the changing landscape. https://ukhi.org/a-global-map-of-single-use-plastic-regulations/

United Nations. (n.d.). Tax on Waste (Landfill Tax). Retrieved from: https://sdgs.un.org/partnerships/tax-waste-landfill-tax

WEF Insight Report. (2024). Quantifying the impact of climate change on human health. Retrieved from: https://www.weforum.org/publications/quantifying-the-impact-of-climate-change-on-human-health/

Wu, L., & Li, Z. (2022). The influence of green marketing on brand trust: The mediation role of brand image and the moderation effect of greenwash. *Discrete Dynamics in Nature and Society, 392172*(1), 6392172. Advance online publication. DOI: 10.1155/2022/6392172

Yin, H.-T., Wen, J., & Chang, C.-P. (2023). Going green with artificial intelligence: The path of technological change towards the renewable energy transition. *Oeconomia Copernicana, 14*(4), 1059–1095. DOI: 10.24136/oc.2023.032

Yüzbaşıoğlu, S. (2012). Yeşil pazarlama altın çağında. Retrieved from: https://www.dunya.com/ozel-dosya/isim-iletisim/yesil-pazarlama-altin-caginda-haberi-193926

Zhang, Y. J., & Wei, Y. M. (2010). An overview of current research on EU ETS: Evidence from its operating mechanism and economic effect. *Applied Energy, 87*(6), 1804–1814. DOI: 10.1016/j.apenergy.2009.12.019

Zhao, S., & Priporas, C. V. (2017). Information technology and marketing performance within international market-entry alliances. *International Marketing Review, 34*(1), 5–28. DOI: 10.1108/IMR-01-2016-0024

Zhuk, A. (2023). Artificial intelligence impact on the environment: Hidden ecological costs and ethical-legal issues. *Journal of Digital Technologies and Law, 1*(4), 932–954. DOI: 10.21202/jdtl.2023.40

ADDITIONAL READING

Biggi, G., Iori, M., Mazzei, J., & Mina, A. (2025). *Green intelligence: The AI content of green technologies*. Eurasian Business Review., DOI: 10.1007/s40821-024-00288-1

Bukhari, S. S. (2011). Green Marketing and its impact on consumer behavior. *European Journal of Business and Management, 3*(4), 375–384.

Chen, S. (2024). The influence of artificial intelligence and digital technology on ESG reporting quality. *International Journal of Global Economics and Management, 3*(1), 301–310. DOI: 10.62051/IJGEM.v3n1.36

Kilbourne, W. E. (1998). Green marketing: A theoretical perspective. *Journal of Marketing Management, 14*(6), 641–655. DOI: 10.1362/026725798784867743

Sharma, R., & Singh, A. (2025). Integrating AI, green marketing, and influencer strategies: Shaping sustainable consumer behavior in the digital age. Z. Hussain, M-N. S. Sharipudin, A. Albattat, A. Khan (Eds.), Strategic Brand Management in the Age of AI and Disruption, (325-342), IGI Global.

Sohaib, O., Alshemeili, A., & Bhatti, T. (2025). Exploring AI-enabled green marketing and green intention: An integrated PLS-SEM and NCA approach. *Cleaner and Responsible Consumption*, *17*, 100269. DOI: 10.1016/j.clrc.2025.100269

KEY TERMS AND DEFINITIONS

Artificial Intelligence (AI): The ability to perform tasks associated with human intelligence.

Eco/Green Consumer: A person who is sensitive to the environment and adopts a consumption policy that will cause minimal harm to natural assets and the physical environment.

Environmental Awareness: Being aware of the impact of our actions on the environment.

Environmental Legislation: Laws and regulations established to protect the environment and ensure the use of natural resources with a sustainable consumption approach.

Environmental/Eco Taxes: These are taxes levied to encourage the refraining from activities deemed to be harmful to the environment.

Environmentally Friendly Marketing Practices: Marketing practices that consider environmental protection and sustainable use of natural resources in all product development, pricing, promotion and distribution functions.

Green Incentives: A benefit given to a person or organization in the form of a reward for the sustainable use of natural resources that protect the environment.

Green Marketing: An environmentally sensitive marketing approach where product development, pricing, promotion and distribution activities are carried out based on the sustainable use of natural resources with minimal environmental damage.

Greenwashing: Marketing communications that mislead consumers by a business to appear more environmentally friendly than it actually is.

Holistic Green Marketing Approach: A holistic approach where green marketing is implemented by encompassing all relevant parties and relationships and is operated in the most efficient manner in the eyes of the business, consumer, competitive environment and legal regulations.

Chapter 18
Innovative Strategies for Achieving Goals:
Policy, Technology, and Multisectoral Collaboration

Fadhil Afiq Badruzzaman
https://orcid.org/0009-0003-8190-166X
Universitas Esa Unggul, Indonesia

Binastya Anggara Sekti
https://orcid.org/0000-0001-5489-4888
Universitas Esa Unggul, Indonesia

Qori Halimatul Hidayah
https://orcid.org/0009-0008-7567-4102
Universitas Esa Unggul, Indonesia

ABSTRACT

The transition to a sustainable ocean economy requires innovative approaches that integrate environmental protection with technological advancement. This chapter investigates the transformative role of Artificial Intelligence (AI) and climate-smart policies in advancing the Blue Economy and supporting the achievement of the United Nations Sustainable Development Goals (SDGs), especially SDGs 13, 14, and 9. It highlights the use of AI in marine biodiversity monitoring, climate risk modeling, and sustainable resource management. The discussion includes real-world examples from Indonesia and small island developing states (SIDS), emphasizing multisectoral collaboration and ethical challenges such as data inequality, algo-

DOI: 10.4018/979-8-3373-5238-1.ch018

rithmic bias, and the digital divide. By outlining strategic pathways for AI adoption in ocean governance, the chapter offers policy and innovation recommendations to promote inclusive, eco-digital futures in the maritime sector.

INTRODUCTION

The ocean is fundamental to the stability of the planet, regulating climate systems, sustaining biodiversity, and providing essential resources for more than three billion people. Despite this importance, marine environments are increasingly threatened by climate change, pollution, habitat loss, and unsustainable exploitation of resources. The Blue Economy concept has gained global attention as a framework that supports economic growth while ensuring that marine ecosystems remain healthy and productive. However, the operationalization of this concept frequently encounters barriers such as fragmented governance, inconsistent environmental policies, and limited monitoring capabilities, especially in regions with constrained technological infrastructure. These challenges underscore the need for innovative, data-driven approaches that can strengthen marine management and policy implementation. The blue economy approach connects existing opportunities and challenges to advance integrated planning and management by considering local knowledge systems that empower effective implementation (Hoareau, 2025).

In the short term, policymakers should prioritize investments in ocean data infrastructure, AI literacy programs, and regulatory sandboxes for marine innovation. Over the long term, national blue economy strategies should institutionalize AI governance frameworks, promote regional data-sharing agreements, and align climate-smart AI deployment with SDG monitoring and reporting mechanisms.

Artificial Intelligence is emerging as a central tool for addressing these limitations by transforming how we observe, analyze, and manage ocean systems. Through applications such as automated satellite analytics, predictive modeling, and real-time environmental sensing, AI enhances our ability to detect illegal fishing, monitor coral reef health, forecast climate impacts, and assess marine pollution. These technologies have the potential to close long-standing knowledge gaps and provide rapid, high-resolution data that conventional marine monitoring fails to deliver. AI also supports climate-smart decision-making by enabling early warning systems, adaptive marine spatial planning, and low-carbon strategies that are particularly crucial for vulnerable coastal communities. The main drivers of digital transformation in this rapidly developing region include a strong emphasis on technological advancements, artificial intelligence (AI), cybersecurity, biotechnology, and the green and blue economies (Voronkova et al., 2024).

The AI technologies discussed in this chapter encompass a range of concrete tools, including satellite-based computer vision for marine biodiversity monitoring, machine learning models for coral reef health prediction, AI-driven climate risk modeling using oceanographic time-series data, and decision-support systems for sustainable resource management. These technologies translate abstract policy goals into operational, data-driven interventions. The integration of machine learning (ML) and artificial intelligence (AI) has transformed traditional processes, creating smarter, more efficient, and adaptable production systems (Sekti, 2025c).

While AI presents transformative opportunities, its integration into marine governance introduces a new set of ethical and operational challenges. Data inequality, limited access to digital tools, algorithmic bias, and insufficient institutional capacity may reinforce disparities between highly digitalized nations and developing coastal states. Without inclusive policy frameworks, there is a real risk that AI solutions will benefit only those with technological and financial advantages, further widening gaps in the implementation of the Blue Economy. Ensuring that AI contributes to environmental justice requires robust rules for data sharing, transparency, accountability, and the involvement of local communities in the governance process. Observing marine life has emerged as a pivotal catalyst for the growth of the Blue Economy (Han et al., 2024).

Ethical tensions in AI-driven Blue Economy initiatives are closely linked to data inequality, algorithmic bias, and the digital divide. These challenges can be mitigated through inclusive policy design that prioritizes open data frameworks, capacity-building programs for coastal communities, and participatory AI governance models that integrate local and indigenous knowledge into algorithmic decision-making.

Given these dynamics, multisectoral collaboration is essential to align technological innovation with climate-smart policies and the Sustainable Development Goals, particularly SDG 13 on Climate Action and SDG 14 on Life Below Water. This chapter examines how an integrated approach that connects AI, climate adaptation, and marine policy can create sustainable and equitable eco-digital futures for the Blue Economy. Through case studies, strategic frameworks, and practical recommendations, it outlines how governments, industries, civil society, and scientific communities can jointly build digital infrastructures, develop inclusive ocean governance, and support climate-resilient economic growth. The goal is to provide a comprehensive foundation for leveraging AI to advance marine sustainability in a way that is both technologically advanced and socially just. The blue economy is an emerging concept that emphasizes the protection of marine life and ecosystems while promoting economic growth through the sustainable utilization of ocean resources (Saraswati, Hananto, Prananda, Mahramhafiz, & Pennesi, 2025).

Effective multisectoral collaboration has emerged as a critical enabler of innovation in ocean governance. Successful models demonstrate coordinated partnerships among governments, private technology firms, academic institutions, and civil society organizations, where public agencies provide regulatory direction, academia ensures scientific rigor, industry scales technological solutions, and civil society facilitates community engagement and accountability.

THE BLUE ECONOMY IN THE CONTEXT OF THE SUSTAINABLE DEVELOPMENT GOALS (SDGS)

The Blue Economy refers to the sustainable use of ocean resources for economic growth, improved livelihoods, and ecosystem health. It emphasizes the integration of economic development with marine conservation, recognizing the ocean not only as a source of wealth but also as a shared natural heritage that underpins planetary stability. As the global community accelerates its efforts to achieve the 2030 Agenda for Sustainable Development, the Blue Economy has become central to advancing multiple Sustainable Development Goals (SDGs) particularly those related to climate, biodiversity, poverty reduction, and innovation. Figure 1 shows the blue economy and the Sustainable Development Goals.

Figure 1. The Blue Economy and The SDGs

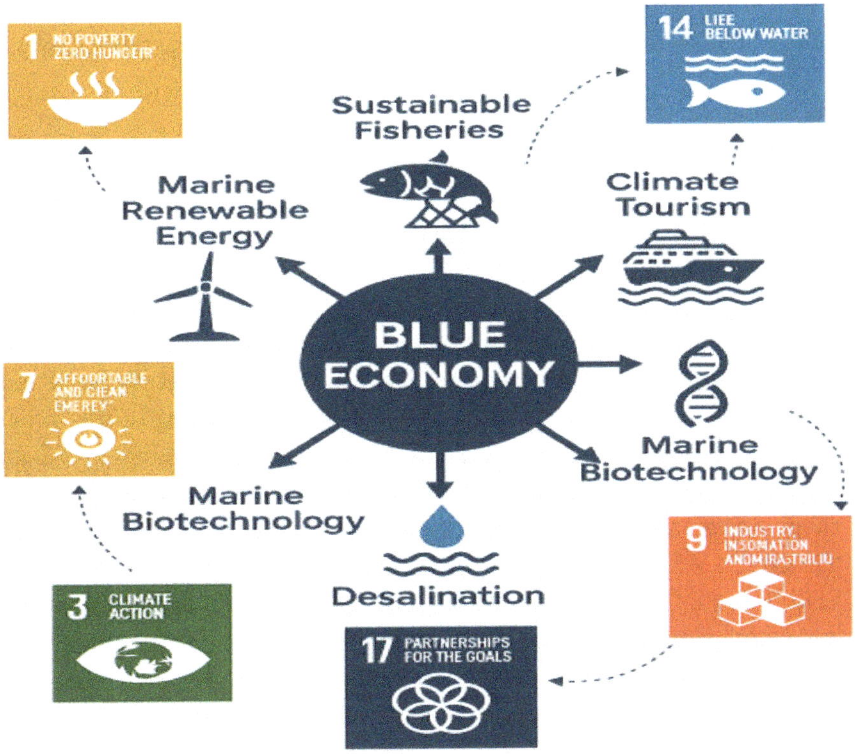

Blue Economy: A Multidimensional Framework

Unlike traditional ocean economies, which focus on exploitation of resources such as fisheries, shipping, and oil extraction the Blue Economy emphasizes long-term sustainability, low carbon pathways, and inclusive development. It spans across various sectors:

- o Sustainable fisheries and aquaculture
- o Marine renewable energy (e.g., offshore wind, tidal, wave energy)
- o Ecotourism and coastal development
- o Marine biotechnology
- o Desalination and water services
- o Shipping and port activities

This framework aligns economic activities with the preservation of marine ecosystems, while ensuring that coastal communities, especially in Small Island Developing States (SIDS) and developing countries, benefit equitably(Onyena & Sam, 2025).

Alignment with Key SDGs

The Blue Economy directly contributes to several SDGs:

- SDG 14 – Life Below Water: The core of the Blue Economy aligns with SDG 14, which calls for the conservation and sustainable use of the oceans, seas, and marine resources. Targets under this goal include reducing marine pollution, restoring marine ecosystems, and regulating harvesting of fish stocks.
- SDG 13 – Climate Action: Oceans play a vital role in climate regulation, absorbing over 90% of excess heat and nearly a third of carbon dioxide emissions. Sustainable ocean strategies are essential for climate mitigation and adaptation.
- SDG 1 & SDG 2 – No Poverty & Zero Hunger: Fisheries and aquaculture are major sources of income and nutrition for millions. Sustainable practices in these sectors can ensure long-term food security and economic empowerment for marginalized populations (Sekti, 2025a, 2025b).
- SDG 7 – Affordable and Clean Energy: The growth of marine renewable energy contributes to reducing reliance on fossil fuels and expanding energy access, particularly in coastal and island regions.
- SDG 9 – Industry, Innovation, and Infrastructure: The digitization of ocean activities, including smart ports and AI-enabled logistics, supports modern infrastructure and industrial innovation (Sekti, 2025c).
- SDG 17 – Partnerships for the Goals: Implementing Blue Economy strategies requires global cooperation across governments, businesses, and civil society, reinforcing SDG 17's emphasis on partnerships and shared knowledge.

Gaps in Current Implementation

Despite the growing policy attention toward the Blue Economy, its real-world application faces several challenges:

- Fragmented governance: Many countries lack cohesive national ocean strategies, resulting in overlapping mandates, policy silos, and inefficient resource use.

- o Data and technology limitations: Accurate ocean data is scarce, particularly in the Global South. This limits the ability to monitor progress, enforce regulations, and design adaptive policies.
- o Inequality and exclusion: Women, indigenous peoples, and small-scale fishers are often excluded from Blue Economy planning and decision-making processes.
- o Short-term economic interests: In many cases, short-term profits from extractive industries such as deep-sea mining or industrial fishing overshadow long-term sustainability goals.
- o Climate vulnerability: Sea-level rise, ocean warming, and increased frequency of marine hazards continue to threaten coastal economies, particularly those without access to risk-reducing technologies.

The Role of AI in Accelerating SDG Progress via Blue Economy

As nations seek to close these gaps, AI offers a strategic advantage. By improving marine monitoring, enabling predictive modeling, and facilitating integrated ocean planning, AI can:

- o Track SDG indicators more efficiently and accurately
- o Optimize marine spatial planning to reduce resource conflicts
- o Support real-time enforcement of marine protected areas
- o Enable climate adaptation strategies for vulnerable coastal zones

AI thus becomes not just a tool but a catalyst for systemic change, enabling more dynamic, data-driven, and inclusive pathways to achieving the SDGs through the Blue Economy.

A Digital Turning Point for Ocean Sustainability

The convergence of the Blue Economy and AI represents a digital turning point for ocean sustainability. However, this transformation will only succeed if guided by ethical standards, inclusive governance, and international cooperation. Nations must ensure that digital innovation serves the needs of both people and planet, and that AI is deployed not as a replacement for local knowledge but as a complement to community-led stewardship (Andika et al., 2025).

CLIMATE-SMART POLICIES: CONCEPTS AND GLOBAL APPLICATIONS

In recent decades, the escalation of climate-related marine threats—rising sea levels, ocean acidification, coastal erosion, and ecosystem collapse—has underscored the urgent need for integrated, adaptive policy frameworks. Climate-smart policies aim to simultaneously address climate mitigation, adaptation, and development goals. Within the marine context, these policies are crucial to reducing vulnerability and increasing the resilience of ocean-dependent communities and ecosystems. However, in an increasingly complex and data-driven world, traditional policy approaches alone are insufficient. This is where Artificial Intelligence (AI) becomes a key enabler of climate-smart transformation in ocean governance(Sekar et al., 2025).

Understanding Climate-Smart Policies in the Marine Context

Climate-smart policies in the marine domain are defined by their ability to anticipate climate impacts, reduce greenhouse gas emissions, promote adaptation strategies, and support sustainable development. These policies cover multiple dimensions, including:

- The protection and restoration of critical habitats (e.g., mangroves, coral reefs)
- Regulations to decarbonize maritime industries
- Investment in early warning systems and disaster risk reduction
- Integrative frameworks that link marine biodiversity with economic planning

The innovation gap, however, lies in the operationalization of these policies. Many governments lack the resources or tools to implement them effectively. AI, when embedded into policy cycles, can transform how risks are detected, how policies are shaped, and how outcomes are monitored over time.

AI as a Catalyst for Policy Intelligence

AI's strength lies in its predictive capacity, pattern recognition, and ability to manage complexity. In climate-smart marine policy, this enables a shift from reactive to proactive governance. AI-powered systems can:

- Predict coastal flooding or sea-level rise scenarios using machine learning models trained on satellite imagery and historical ocean data

- Model the effectiveness of marine protected areas (MPAs) by simulating biodiversity responses under different climate futures
- Optimize zoning decisions for marine spatial planning by integrating ecological, social, and economic layers
- Continuously monitor policy compliance using autonomous drones, computer vision, and acoustic AI systems

These AI tools provide decision-makers with real-time intelligence and dynamic feedback loops, making policy design more agile and evidence-based. Table 1 shows the Examples of AI-Enhanced Climate-Smart Policies in Marine Systems.

Table 1. Examples of AI-Enhanced Climate-Smart Policies in Marine Systems

Country/Region	AI Application in Climate-Smart Policy	Policy Outcome
Indonesia	AI-based mangrove mapping using drone and image recognition	Informed restoration targets in national coastal resilience plan
European Union	AI for shipping emission tracking and smart fuel routing	Supported carbon-neutral maritime transport policy
Pacific Island States	Deep learning for cyclone and storm surge prediction	Enabled adaptive early warning and evacuation systems

AI-Driven Early Warning and Risk Modeling Systems

One of the most impactful contributions of AI to climate-smart marine governance is its use in disaster forecasting and risk modeling. In regions like Southeast Asia and the Pacific Islands, where coastal communities are highly vulnerable, AI is increasingly being used to model the behavior of tropical cyclones, wave surges, and coastal erosion patterns. For instance, convolutional neural networks (CNNs) have been used to analyze satellite-derived ocean temperature anomalies and correlate them with coral bleaching events(Deaza-Fernández et al., 2025). Recurrent neural networks (RNNs) have also been applied to long-term tide and storm data to generate real-time alerts. These predictive models not only save lives but also inform adaptation funding, insurance schemes, and climate action prioritization. AI's capacity to handle non-linear, uncertain, and multidimensional datasets is especially valuable in climate policy, where decision-making is often clouded by incomplete or delayed information(Alka et al., 2025).

The Role of AI in Policy Co-Creation and Multilevel Governance

Beyond technical modeling, AI is beginning to reshape how policies are formulated and negotiated. Natural Language Processing (NLP) algorithms, for example, have been used to analyze public sentiment, stakeholder submissions, and legislative texts. This enables governments to identify patterns in public concerns, align marine policies with social priorities, and detect regulatory overlaps or contradictions(Luo & Wang, 2025).

Moreover, AI facilitates cross-jurisdictional governance by connecting local data with national and international policy frameworks. For instance, AI tools can map how local fishing patterns, biodiversity zones, and carbon sink areas interact across borders—creating more coherent regional strategies, especially under frameworks like the Paris Agreement or the Kunming-Montreal Global Biodiversity Framework(Hong et al., 2025).

Addressing the Limitations of AI-Enhanced Climate Policies

Despite its promise, AI-enhanced climate-smart policymaking is not without limitations. The quality of AI outputs is only as good as the data fed into them. In many coastal nations, data on seabed composition, marine life distribution, or water quality are sparse, outdated, or inaccessible. Additionally, AI systems risk reinforcing existing power asymmetries, particularly when designed without transparency or accountability.

To mitigate these risks, policies must be developed that:

- Ensure open access to marine data while respecting local knowledge systems
- Promote capacity-building in AI literacy across ministries and coastal stakeholders
- Institutionalize ethical frameworks that govern how AI is applied, especially where human rights and indigenous knowledge are concerned

Toward a Climate-Smart, AI-Enabled Marine Policy Future

Integrating AI into climate-smart marine policymaking offers a way to bridge the gap between knowledge and action, between data and impact. It helps governments and communities move beyond fragmented, static policy approaches toward dynamic, intelligent, and adaptive governance. As we face an increasingly volatile climate future, it is not enough to simply respond to change—we must anticipate it. AI provides the cognitive infrastructure to do so. But the success of this transition will depend on political will, ethical stewardship, and international collaboration.

The next chapter will explore how these tools are being used in one of the most critical sectors of the Blue Economy: sustainable fisheries and aquaculture, where AI is transforming everything from harvesting to supply chains in the face of climate uncertainty.

ARTIFICIAL INTELLIGENCE FOR OCEAN GOVERNANCE

The governance of the ocean has traditionally relied on legal treaties, field monitoring, and manual bureaucratic processes. Frameworks such as the United Nations Convention on the Law of the Sea (UNCLOS) and regional fisheries management organizations have shaped global maritime rules, but their operational capacity often falls short of the growing complexity and scale of marine threats in the 21st century. With the climate crisis accelerating sea-level rise, biodiversity collapse, and ocean warming, new tools are urgently needed to enhance transparency, responsiveness, and efficiency in ocean governance. One of the most transformative of these tools is Artificial Intelligence (AI)(Torres-Ceron et al., 2025).

AI introduces a paradigm shift in how oceans can be governed not merely as a data-processing technology, but as a governance infrastructure capable of anticipating risks, identifying violations, and optimizing the use of marine space and resources. By integrating machine learning, computer vision, satellite imagery analysis, and predictive modeling, AI enables governments, institutions, and civil society to monitor and manage marine ecosystems in real time, at scales that were previously unmanageable. The result is not only a smarter ocean but a more accountable and inclusive policy environment(Han et al., 2025).

One of the most prominent applications of AI in ocean governance is the use of machine learning to detect illegal, unreported, and unregulated (IUU) fishing. Through the analysis of vessel Automatic Identification System (AIS) signals and behavioral patterns, platforms like Global Fishing Watch have enabled authorities to identify suspicious activity far beyond what conventional patrol boats can detect. In Indonesia, AI is being deployed through a system known as Ocean Eye, which uses drone surveillance and image recognition to detect unauthorized activity in marine protected areas. Meanwhile, the European Union's Copernicus Marine Service leverages satellite-based AI to monitor ocean pollution, oil spills, and plastic leakage, providing actionable data for regulatory response.

In Indonesia, the deployment of AI-enabled mangrove mapping and coastal risk analytics has resulted not only in improved restoration targeting but also in measurable reductions in shoreline erosion and enhanced community-based adaptation planning. However, implementation challenges remain, particularly related to data continuity, inter-agency coordination, and limited AI capacity at the local governance level. Similar patterns are observed in Small Island Developing States, where AI-supported early warning systems have improved disaster preparedness but face constraints due to infrastructure gaps and reliance on external technical support. Aquaculture impacts the economic standard of living in Indonesia and the effects of urbanization and industrialization on human health (Jamilah et al., 2025).

Beyond enforcement, AI is also transforming marine spatial planning (MSP), a key component of ocean governance that determines how ocean areas are allocated for different uses such as fisheries, conservation, tourism, and shipping. Traditional MSP relies on stakeholder negotiation and environmental assessments, which are often slow and fragmented. AI enables planners to simulate multiple zoning scenarios, integrating ecological, economic, and climate data to produce optimized outcomes. Reinforcement learning and multi-objective optimization algorithms allow for predictive planning under various climate futures, helping to minimize user conflict and environmental degradation. For example, in East Africa's coastal regions, AI-enhanced MSP has been used to resolve conflicts between artisanal fishing communities and expanding tourism zones while protecting marine biodiversity hotspots(Jiang et al., 2025).

AI's ability to enhance ocean governance is not limited to national governments. Civil society organizations and citizen scientists are increasingly using open-access AI tools to hold authorities accountable. In the Philippines, for instance, local NGOs use AI-processed satellite feeds to track encroachment in protected areas and push for policy enforcement. Natural Language Processing (NLP) models are being used to analyze public feedback on marine legislation in countries like Canada and members of the European Union, making it easier for policymakers to align regulations with public sentiment and emerging social concerns. These developments show that AI is not just a tool for top-down control, but also for bottom-up participation and transparency. Table 2 summarizes several high-impact initiatives where AI has strengthened ocean governance. Figure 2 shows the AI Application in the Blue Economy.

Table 2. High-impact initiatives where AI has strengthened ocean governance

Initiative / Region	AI Application	Governance Outcome
Global Fishing Watch (GFW)	Machine learning on vessel AIS data	Detection of illegal, unreported, and unregulated (IUU) fishing
NOAA (USA)	AI for coral reef monitoring and threat detection	Dynamic allocation of conservation funding and enforcement efforts
Indonesia's Ocean Eye	AI-driven drone surveillance of MPAs	Strengthened monitoring of protected zones and illegal logging
EU Copernicus Program	Satellite-based ML for pollution detection	Policy response to oil spills and plastic leakage

Figure 2. AI Application in the Blue Economy

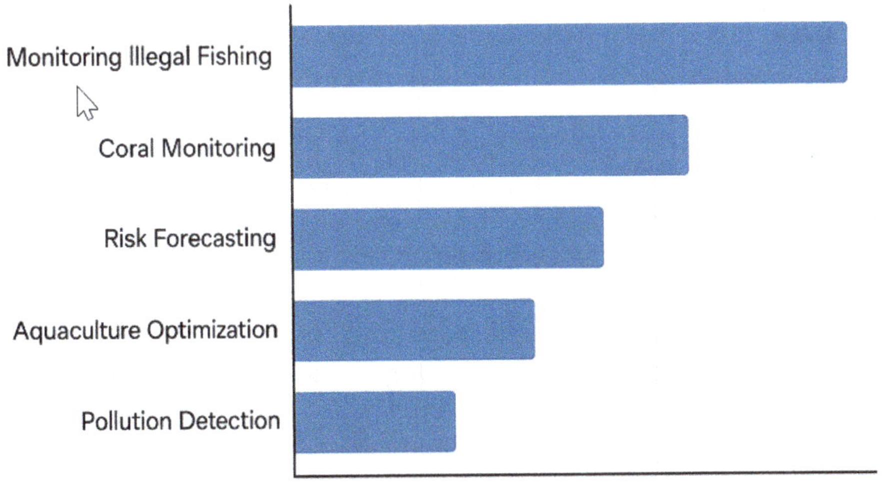

However, the use of AI in ocean governance also raises new questions and limitations. Much of the world's ocean data—such as detailed bathymetry, biodiversity mapping, and pollution monitoring—remains incomplete, inconsistent, or inaccessible, especially in developing countries. This uneven data landscape limits the effectiveness and equity of AI systems. Moreover, algorithmic bias, lack of model transparency, and cybersecurity vulnerabilities pose real risks, particularly when AI is used to guide enforcement or allocate resources. The growing reliance on automated systems must also be balanced with human judgment, local knowledge, and democratic oversight, especially when policies affect the rights and livelihoods of coastal communities(Obisi et al., 2025).

To navigate these challenges, a hybrid model of ocean governance is needed—one where AI acts as an enabler of policy innovation but not a substitute for human accountability. Capacity-building programs must equip marine policymakers with AI literacy, ethical training, and tools to ensure transparency. Open data platforms and international cooperation are also essential to ensure that the benefits of AI are shared globally, not just concentrated in technologically advanced states. Most importantly, AI must be developed and deployed with the active participation of diverse stakeholders, including indigenous peoples, fisherfolk, women, and youth who are often marginalized in high-tech decision-making processes.

In conclusion, AI holds the potential to transform ocean governance into a more adaptive, anticipatory, and inclusive system. It enables authorities to see farther, act faster, and govern more fairly in the face of marine uncertainty. But realizing this potential requires a conscious effort to embed ethical principles, human rights, and climate resilience into every layer of the AI-governance interface. As the ocean becomes a space of digital governance, the challenge is no longer just how to use AI—but how to ensure that AI serves the ocean and all who depend on it.

Eco-Digital Tools for Sustainable Fisheries and Aquaculture

The fisheries and aquaculture sectors are central to food security, livelihoods, and economic development, particularly in coastal and island nations. Yet, overfishing, habitat destruction, and climate-induced changes such as ocean acidification have put immense pressure on these systems. Traditional approaches to fisheries management—like quota systems, seasonal bans, and field monitoring—while important, are often insufficient due to delayed data collection, limited surveillance capacity, and weak enforcement. This is where Artificial Intelligence (AI) and eco-digital tools are proving to be transformative, offering real-time, scalable, and predictive capabilities that align fisheries and aquaculture with sustainability goals under the SDGs, particularly SDG 14 (Life Below Water) and SDG 2 (Zero Hunger)(Kurmis et al., 2025).

AI-powered tools are increasingly used to enhance the monitoring, control, and surveillance (MCS) of capture fisheries. For example, deep learning algorithms are applied to vessel tracking data (AIS and VMS) to identify anomalous behavior that suggests illegal fishing activity. This technology has enabled countries like Indonesia, Peru, and Ghana to detect and deter IUU (Illegal, Unreported, and Unregulated) fishing with greater accuracy and at a fraction of the cost of traditional patrol systems. These AI models learn from vast datasets of vessel movement patterns and environmental conditions, enabling not only real-time alerts but also forecasts of future violations.

In aquaculture, AI is being utilized to optimize feeding strategies, water quality management, and disease detection. Using underwater cameras and computer vision, AI systems can monitor fish behavior and growth, adjusting feed delivery to reduce waste and improve yield. For instance, Norwegian salmon farms have adopted AI-enabled systems that detect when fish are no longer interested in feed, automatically stopping the process to avoid overfeeding. Similarly, predictive algorithms based on environmental sensors can anticipate algal blooms or disease outbreaks, allowing proactive intervention. This integration of machine learning and IoT (Internet of Things) devices leads to both environmental and economic benefits—minimizing chemical use, improving animal welfare, and increasing profitability.

Another promising area is the use of AI for stock assessment and ecosystem-based fisheries management. Traditionally, stock assessments rely on costly field surveys and historical catch data. AI models, trained on satellite imagery, oceanographic data, and species population trends, can produce dynamic assessments of fish stock health, even in data-poor regions. In West Africa, researchers have developed neural network models that predict the biomass of small pelagic species based on temperature and chlorophyll levels. These models are critical for supporting sustainable harvest limits and adapting fisheries policy to climate variability(Eneh et al., 2025).

In terms of sustainability certification and traceability, AI plays a growing role in ensuring that seafood products are responsibly sourced. Natural Language Processing (NLP) and blockchain technologies, combined with image recognition tools, can verify the authenticity of seafood products across the supply chain. For example, the WWF's Blockchain Seafood Traceability Project integrates AI to verify fishing logs, vessel identities, and supply routes. This helps consumers and regulators distinguish between legally caught fish and illicit catch, reducing market incentives for unsustainable practices. Table 3 shows the summary of AI applications in fisheries and aquaculture.

Table 3. A summary of AI applications in fisheries and aquaculture

Area	**AI Application**	**Sustainable Outcome**
Capture Fisheries	AIS-based anomaly detection	Reduction in IUU fishing
Aquaculture	Computer vision for feed control	Improved feed efficiency and reduced environmental impact
Stock Assessment	Predictive modeling using oceanographic variables	Data-driven management and dynamic quotas
Traceability & Certification	NLP & image recognition for product verification	Transparency and accountability in supply chains
Disease Prevention	Predictive analytics on water quality and behavior	Early intervention, reduced antibiotic use

Despite these advances, significant challenges remain in scaling AI-driven tools in low-income and small-scale fishing communities. Many such communities lack the digital infrastructure, reliable electricity, and data literacy required to effectively implement AI solutions. Moreover, the risk of digital exclusion—where high-tech tools are adopted mainly by large industrial operations—may further widen the gap between industrial and artisanal sectors. To address this, inclusive innovation models are necessary. Open-source AI tools, mobile-based platforms, and partnerships with local cooperatives can help democratize access. For example, apps like Abalobi in South Africa allow small-scale fishers to record catches, track income, and access market prices through smartphones, using lightweight machine learning algorithms to provide insights(Kwon et al., 2025).

Integrating AI into fisheries and aquaculture requires a governance framework that balances technological innovation with equity, sustainability, and ethical responsibility. National authorities must establish data standards, ethical guidelines, and AI procurement policies that support inclusive and climate-smart resource management, while international cooperation is essential for managing transboundary and migratory marine resources to avoid data fragmentation and unequal advantages. When responsibly deployed, AI and eco-digital tools can transform fisheries and aquaculture from extractive and reactive systems into regenerative, anticipatory, and resilient governance models, provided that technological robustness is matched by social inclusion and alignment with long-term ocean health and community livelihoods.

AI-Enabled Marine Biodiversity Monitoring and Protection

The degradation of marine biodiversity is one of the most urgent challenges facing the global ocean system. Climate change, overfishing, habitat loss, marine pollution, and invasive species have led to significant declines in ocean biodiversity, threatening the resilience of marine ecosystems and the livelihoods of millions who depend on them. Traditional monitoring methods such as manual surveys, diving expeditions, and static sensors while valuable, are often expensive, labor-intensive, and limited in spatial and temporal coverage. Artificial Intelligence (AI), when integrated with sensor networks, satellite data, and autonomous systems, offers a transformative approach to biodiversity monitoring and protection, enabling faster, more precise, and adaptive conservation strategies(Shen et al., 2025).

AI-based image and acoustic recognition technologies have significantly enhanced marine biodiversity monitoring by enabling large-scale, near real-time analysis of visual and audio data. Convolutional neural networks can process imagery from underwater cameras, ROVs, and AUVs to identify marine species, assess population dynamics, and detect ecological changes such as coral bleaching and reef degradation. Complementing this, AI-driven acoustic monitoring systems analyze vast underwater sound recordings to distinguish species-specific vocalizations, track migratory patterns, and detect noise pollution from human activities. Together, these approaches provide non-invasive, data-rich tools that strengthen conservation planning, support evidence-based policymaking, and improve the monitoring of marine ecosystems at unprecedented spatial and temporal scales.

Beyond detection, AI is now contributing to predictive biodiversity modeling. Using satellite data, climate variables, and ecological indicators, machine learning models can forecast the distribution of vulnerable species under various climate scenarios. This is particularly important in planning Marine Protected Areas (MPAs), where conservation efforts must anticipate future ecosystem shifts. For example, dynamic ocean management platforms use AI to recommend temporary closures or shifts in fishing zones based on predicted movements of endangered species like sea turtles or blue whales. This allows for adaptive management that protects biodiversity while reducing unnecessary restrictions on marine users(Sofuoglu et al., 2025).

However, the deployment of AI in biodiversity protection is not without ethical and technical concerns. First, the availability of high-quality, labeled datasets remains a bottleneck, especially for lesser-known marine species. Collaborative platforms and citizen science initiatives such as Naturalist Marine and OceanEyes are helping to bridge this gap by crowdsourcing data collection and validation. Second, data privacy and sovereignty, especially in international waters, raise questions about how biodiversity information is used and by whom. Establishing clear data governance standards and ensuring equitable access to AI tools are critical for avoiding exploitation and ensuring inclusive conservation(Gabr et al., 2025).

Finally, while AI provides unprecedented capabilities, it is not a replacement for human expertise and local ecological knowledge. Integrating AI with community-based monitoring and traditional conservation practices creates a more holistic and resilient approach to biodiversity stewardship. By engaging indigenous communities, fishers, and coastal residents as co-creators and users of AI systems, conservation becomes not only technologically advanced but also socially rooted.

In conclusion, AI has become a vital ally in the protection of marine biodiversity. It enables continuous, scalable, and intelligent monitoring, supports predictive and adaptive conservation policies, and enhances restoration and pollution control efforts. To maximize its potential, AI must be ethically designed, inclusively deployed, and guided by a shared commitment to the long-term health of ocean ecosystems. The future of biodiversity conservation will not be shaped solely by algorithms, but by how we choose to harness these tools in service of nature and society(Y. Chen et al., 2025).

DATA-DRIVEN DECISION-MAKING AND GOVERNANCE IN THE BLUE ECONOMY

As the Blue Economy continues to expand encompassing sustainable fisheries, offshore renewable energy, marine biotechnology, and ocean-based tourism the complexity of managing ocean resources grows in tandem. Traditional governance frameworks, often fragmented across national agencies and regional jurisdictions, struggle to address the fast-evolving dynamics of marine environments. In this context, Artificial Intelligence (AI) and data-driven decision-making present powerful opportunities to enhance governance transparency, effectiveness, and inclusivity. Through the integration of real-time data streams, predictive analytics, and algorithmic decision support systems, AI is transforming the way ocean policies are designed, implemented, and adapted.

At the core of AI-enhanced governance is the ability to process vast volumes of ocean-related data from satellite observations and sensor networks to economic reports and citizen science inputs. These diverse datasets, often referred to as "big ocean data," are increasingly being harmonized through cloud-based platforms and digital twins of the ocean. AI algorithms can uncover hidden patterns, anomalies, and trends from this data, providing marine authorities with actionable insights. For instance, machine learning models can identify illegal, unreported, and unregulated (IUU) fishing activities by analyzing vessel behavior patterns using Automatic Identification System (AIS) data. This has empowered coast guards and regulatory agencies to respond more proactively to violations and enhance maritime security(Li et al., 2025).

AI also plays a growing role in ocean finance and investment governance. Predictive algorithms assess the economic and environmental performance of Blue Economy projects, informing risk assessments and green financing decisions. For example, AI can estimate the carbon sequestration potential of mangrove restoration projects or model the profitability of sustainable aquaculture under future climate scenarios. This supports financial institutions and development agencies in channeling capital toward projects aligned with both economic returns and SDG targets(Qiao et al., 2025).

Importantly, AI-driven governance must be transparent and accountable. The use of opaque or biased algorithms can reinforce existing inequities or exclude local knowledge systems. Therefore, there is a growing push toward "explainable AI" (XAI) in marine governance. XAI frameworks aim to make AI decisions understandable to non-experts, including policymakers, stakeholders, and affected communities. This is especially important in managing sensitive ocean areas where socio-economic livelihoods are directly impacted by governance choices(Dagkinis et al., 2025).

Ethical data governance is also crucial. Ocean data often crosses national boundaries, raising issues of data sovereignty, privacy, and equitable access. Small Island Developing States (SIDS), which are highly dependent on ocean resources, may lack the digital infrastructure to collect, process, or benefit from AI-derived insights. International cooperation is essential to build digital capacities in under-resourced regions and ensure that AI technologies serve the global ocean commons rather than reinforcing digital divides.

Furthermore, AI contributes to adaptive governance by enabling feedback loops between monitoring, evaluation, and policy adjustment. For example, real-time dashboards powered by AI can track the effectiveness of marine protected areas, blue carbon initiatives, or pollution control measures. When indicators deviate from desired outcomes, AI systems can trigger alerts or recommend policy revisions. This continuous learning cycle supports agile governance in a world where ocean conditions and human impacts are constantly shifting. Citizen engagement is another area where AI strengthens governance. Chatbots, digital assistants, and participatory apps powered by natural language processing (NLP) and geospatial

AI allow citizens to report environmental violations, participate in monitoring efforts, and access ocean literacy resources. These platforms democratize access to information and encourage more inclusive decision-making. As such, AI doesn't just make governance smarter it makes it more participatory(Fegert et al., 2025).

In summary, data-driven governance underpinned by AI is reshaping how decisions are made in the Blue Economy. By integrating real-time monitoring, predictive analytics, and participatory tools, AI empowers ocean policymakers to make faster, fairer, and more informed decisions. The success of these systems, however, depends on ethical frameworks, strong institutional capacities, and a commitment to transparency and inclusivity. With the right safeguards, AI can be a force multiplier for good governance, ensuring that ocean development aligns with ecological sustainability and social equity.

CROSS-SECTORAL COLLABORATIONS AND AI ECOSYSTEMS FOR SDG 14 IMPLEMENTATION

The implementation of Sustainable Development Goal 14 "Life Below Water" requires far more than isolated policy interventions or singular technological innovations. It demands a deeply interconnected ecosystem involving governments, academia, the private sector, civil society, and international organizations. As Artificial Intelligence (AI) emerges as a powerful enabler in achieving SDG 14 targets, its effectiveness is increasingly tied to cross-sectoral collaboration and the development of inclusive AI ecosystems tailored to ocean sustainability(Badriyah & Wardi, 2026).

Unlike conventional digital systems, AI thrives in multi-stakeholder environments where access to quality data, domain expertise, and computational resources are pooled. In the context of ocean governance, this means forging strong alliances between marine scientists and AI engineers, local communities and data providers, regulators and private innovators. These partnerships are essential not only for training accurate AI models but also for ensuring that these models are ethically aligned, socially accepted, and contextually appropriate.

One notable example of this collaborative dynamic can be found in projects like Global Fishing Watch, where partnerships between NGOs (such as Oceana), tech companies (Google), and intergovernmental organizations (like FAO) have enabled the use of AI and satellite data to track global fishing activity in near-real time. This has improved transparency in the seafood supply chain and empowered regulatory bodies to combat illegal fishing. The project demonstrates how open data infrastructures and collaborative governance models are key to scaling AI applications for marine sustainability(W. Chen et al., 2025).

Similarly, AI-powered coral reef monitoring systems developed by joint efforts between universities and marine conservation organizations use underwater image recognition and machine learning to detect coral bleaching events and track reef health. These tools are often co-designed with local divers and marine rangers, ensuring usability and grounding the technology in local knowledge. This participatory approach strengthens trust and increases adoption, showing how bottom-up innovation can complement top-down policy goals. In the private sector, startups focusing on marine robotics, satellite oceanography, and sustainable aquaculture increasingly rely on partnerships with research institutions and ocean data initiatives. These AI-powered enterprises often lack the in-house scientific depth to model complex ocean processes but gain insight and validation through joint ventures with oceanographers and ecologists. Governments and development agencies can facilitate such linkages through innovation hubs, blue tech incubators, and public-private R&D programs(Alam et al., 2025).

AI ecosystems also benefit from regional and global platforms that facilitate cross-sector collaboration. Initiatives like the Ocean Data and Information System (ODIS) led by UNESCO-IOC, or the Decade of Ocean Science for Sustainable Development (2021–2030), create enabling environments for data exchange, knowledge co-creation, and AI innovation. These platforms promote interoperability of datasets, common AI ethics guidelines, and shared capacity-building programs that reach across sectors and regions. Moreover, cross-disciplinary education and training are essential to sustain AI ecosystems. Marine biologists must gain fluency in data science, while computer scientists must understand oceanographic processes and sustainability imperatives. Universities and research centers play a vital role by designing transdisciplinary curricula that foster this hybrid skillset and produce the next generation of "ocean AI stewards." A major challenge in cross-sectoral collaboration, however, lies in aligning incentives and timelines. While research institutions may prioritize long-term ecological outcomes, private tech firms often seek faster returns and market scalability. Meanwhile, policymakers operate within budgetary and electoral cycles. Bridging these differences requires strong coordination mechanisms, clear shared goals (like SDG 14 indicators), and collaborative governance frameworks that distribute risks and benefits equitably(Ramezani et al., 2025).

Another barrier is the digital divide, especially in low-income coastal nations and Small Island Developing States (SIDS), where AI literacy, data infrastructure, and computational capacity may be limited. International support mechanisms such as digital capacity-building grants, open-source toolkits, and inclusive AI design processes are essential to ensure that no region is left behind in the ocean AI revolution. In sum, the successful application of AI to SDG 14 depends less on standalone innovations and more on the strength and inclusivity of AI ecosystems.

By building bridges between sectors, disciplines, and geographies, cross-sectoral collaboration enables the co-development of robust, fair, and scalable AI solutions for ocean sustainability. In this collective effort, AI serves not merely as a tool, but as a platform for co-creation, bringing together diverse actors to safeguard marine life and ensure the long-term health of our blue planet(Galeboe et al., 2025).

RISKS, BIASES, AND ETHICAL CONSIDERATIONS IN APPLYING AI TO OCEAN GOVERNANCE

While Artificial Intelligence (AI) offers transformative potential in enhancing ocean governance and advancing Sustainable Development Goal 14 (SDG 14), its application is not without significant risks, biases, and ethical concerns. In fact, the very features that make AI powerful its ability to process massive datasets, automate decision-making, and reveal patterns invisible to human observers can also lead to harmful consequences if not carefully managed within ethical and equitable frameworks. One of the primary risks involves the quality and representativeness of ocean data used to train AI models. Many marine ecosystems, particularly in the Global South, suffer from a lack of consistent data collection due to limited funding, poor infrastructure, or geopolitical restrictions. When AI systems are trained predominantly on data from well-mapped and monitored regions often in wealthier nations they may generate biased outputs that do not accurately reflect local conditions elsewhere. This could result in inappropriate conservation recommendations or flawed predictive models when applied across diverse contexts(Wang et al., 2025).

Furthermore, the black-box nature of many AI algorithms, especially deep learning systems, poses serious concerns in regulatory and legal settings. For example, an AI system used to allocate marine protected areas (MPAs) or monitor illegal fishing must be transparent and auditable, particularly if its outputs affect livelihoods or sovereign maritime claims. Without explainability, stakeholders may not trust AI recommendations, or worse, may be unable to challenge decisions that are technically complex but ethically contentious. Another ethical concern arises from the uneven access to AI technologies and the asymmetric distribution of benefits. High-income nations and multinational tech firms often dominate the development and deployment of AI tools for marine resource management, while coastal communities who are most directly affected by ocean policies are rarely included in the design and decision-making process. This creates a power imbalance and may exacerbate digital colonialism, where technological solutions are imposed without local participation, consent, or capacity building(Mostafa et al., 2025).

There is also the potential for algorithmic harm. For instance, AI systems that prioritize efficiency or economic optimization in fisheries management might recommend quotas or practices that disregard ecological complexity or marginalize small-scale fishers. In the absence of human-centered AI design, such systems could promote short-term profits at the expense of long-term sustainability or social equity. Privacy and surveillance are growing ethical frontiers in ocean AI. The increasing use of autonomous underwater drones, satellite tracking, and facial recognition for fish stock monitoring raises questions about data ownership, consent, and oversight. Who owns the data collected from ocean spaces? Can AI monitoring be weaponized or used for geopolitical advantage? These are not hypothetical questions they are urgent issues, especially in contested maritime regions or areas with fragile governance.

In addressing these risks, a robust ethical framework is necessary, grounded in international human rights principles, environmental justice, and open science. Key recommendations include:

- Transparency and explainability: AI models used in ocean governance must be interpretable and auditable, especially when linked to regulatory enforcement or public policy.
- Data inclusivity: Efforts should be made to collect diverse, high-quality data across underrepresented regions, ensuring that AI models reflect the ecological and cultural diversity of ocean systems.
- Participatory design: Involving coastal communities, indigenous peoples, and local fishers in the co-creation of AI tools fosters trust, relevance, and ethical alignment.
- Capacity building: Global cooperation must prioritize training, infrastructure, and digital literacy in low- and middle-income countries to close the AI access gap.
- Environmental impact assessment of AI systems: The computational resources used by AI models, including energy consumption and e-waste, should be monitored and minimized in line with broader sustainability goals.
- Policy alignment and regulatory oversight: AI deployment in marine governance should be guided by international frameworks, including the UN Convention on the Law of the Sea (UNCLOS), the UNESCO Ocean Decade roadmap, and emerging AI ethics guidelines (e.g., OECD, UNESCO, EU).

The ethical implementation of AI in ocean governance is not merely a technical challenge—it is a political and moral imperative. Misuse, neglect, or uncritical adoption of AI can deepen inequalities, undermine trust, and damage fragile ecosystems. Conversely, by embedding ethical considerations from the ground up, AI can become a catalyst for more just, transparent, and sustainable ocean governance. Ultimately, safeguarding AI integrity in the context of SDG 14 is a collective responsibility. It requires not only engineers and data scientists but also ethicists, ecologists, indigenous leaders, policymakers, and civil society to engage in continuous dialogue and co-governance of AI futures for the ocean.

FUTURE DIRECTIONS AND RECOMMENDATIONS FOR AI-DRIVEN BLUE ECONOMY TRANSFORMATION

The integration of Artificial Intelligence into the blue economy presents a transformative opportunity to enhance ocean sustainability, restore marine ecosystems, and accelerate progress toward Sustainable Development Goal 14. To realize this potential, future AI-driven initiatives must be designed with foresight, equity, and resilience in mind. This requires not only advancing technological capabilities but also addressing data access, governance, and social inclusion to ensure that AI benefits are broadly shared across regions and communities.

A central priority is data democratization and the development of AI models suited to the complexity of marine systems. Open, standardized, and interoperable marine data platforms are essential to prevent data silos and enable inclusive innovation. At the same time, investment in domain-specific AI models that integrate oceanographic, ecological, and socioeconomic data will improve the relevance and reliability of AI applications in fisheries, aquaculture, conservation, and climate risk management. Public-private partnerships, blue technology hubs, and targeted funding mechanisms can help scale these solutions while supporting sustainable ocean-based industries.

Equally important are governance, capacity-building, and ethical safeguards. AI must be embedded into national ocean strategies, marine spatial planning, and disaster risk reduction frameworks in a transparent and participatory manner. Expanding AI literacy and training for coastal communities and local institutions is critical to avoiding new digital divides. In parallel, context-specific AI ethics guidelines and accountability mechanisms are needed to address algorithmic bias, data sovereignty, and emerging cyber risks. By aligning AI innovation with environmental justice, inclusive governance, and long-term stewardship, the blue economy can evolve toward a resilient eco-digital future that supports healthy oceans and sustainable livelihoods.

POLICY INNOVATION AND DIGITAL INFRASTRUCTURE FOR EQUITABLE AI ADOPTION

To ensure that the benefits of AI in the blue economy are widely shared and sustainably realized, policy innovation must go hand-in-hand with digital infrastructure development. As AI systems become more deeply embedded in ocean governance, maritime industries, and conservation efforts, the enabling environment legal, institutional, and infrastructural must evolve accordingly. This final section explores how forward-thinking policies and inclusive digital infrastructure can lay the groundwork for equitable, scalable, and ethical AI deployment in service of SDG 14 and the broader 2030 Agenda. At the heart of this effort is the recognition that AI is not neutral. Its impact is shaped by who builds it, who accesses it, and how it is governed. Therefore, policies must aim not only to promote innovation, but to distribute the benefits of AI fairly across nations, sectors, and social groups. In the context of the blue economy, this means prioritizing small island developing states (SIDS), coastal indigenous communities, artisanal fishers, and marine-dependent populations who are often left behind in digital transformations(Rajaperumal & Columbus, 2025).

One key area of policy innovation involves establishing AI regulatory sandboxes for ocean technology. These controlled environments allow innovators to test AI applications such as autonomous underwater vehicles, marine pollution detection systems, or predictive fishery models under real-world conditions with adaptive regulatory oversight. By encouraging responsible experimentation, governments can reduce regulatory uncertainty, speed up learning, and identify potential risks early. Governments must also address the digital infrastructure gap that prevents equitable AI adoption. In many coastal and developing regions, basic connectivity, cloud storage, data centers, and computational power are lacking or prohibitively expensive. Public investment and international cooperation are needed to expand submarine internet cables, satellite coverage, and high-performance computing (HPC) infrastructure tailored for ocean monitoring and AI analysis. Projects like Africa's Blue Cloud or ASEAN's digital marine corridors can serve as blueprints for regional coordination and digital capacity building.

Policy frameworks should further support the interoperability and transparency of AI systems. This means adopting open standards for marine data exchange, promoting explainable AI (XAI), and requiring algorithmic documentation for public sector tools. For example, if AI is used to allocate fishing quotas or delineate marine protected areas, affected stakeholders must be able to understand the rationale behind those decisions and contest them if necessary. Such transparency builds trust and reduces the risk of technocratic overreach. Another crucial dimension is data governance, especially for data collected from oceans that cross multiple jurisdictions or involve indigenous territories. Policies must respect data sovereignty while encouraging responsible data sharing for the global good. Multilateral agreements can enshrine principles such as fair data benefit sharing, informed consent, and ethical data commercialization in maritime AI applications.

To ensure that these innovations are sustained and not fragmented, countries should embed AI strategies within broader National Ocean Policies, Blue Economy Roadmaps, and Digital Transformation Agendas. A unified policy architecture will align technological, environmental, and social objectives maximizing synergies and avoiding duplication. Inter-ministerial coordination (e.g., between Ministries of Fisheries, Environment, and Digital Affairs) and stakeholder co-creation processes are key to crafting coherent, future-ready policies. The private sector also has a role to play. Tech companies developing marine AI solutions should be encouraged or required to adhere to Environmental, Social, and Governance (ESG) principles, conduct AI impact assessments, and partner with local institutions. Governments can use public procurement, subsidies, and outcome-based funding models to incentivize responsible innovation that advances both economic growth and ocean sustainability.

International organizations and development banks must mobilize funding and technical assistance to help low- and middle-income countries develop the policy and infrastructure foundations for AI adoption. Grants, concessional finance, and South-South cooperation mechanisms can support national AI roadmaps, digital literacy programs, and maritime AI centers of excellence in under-resourced regions. In closing, the success of AI in transforming the blue economy will not be determined by algorithms alone but by the policies and infrastructures that shape how those algorithms are developed, governed, and used. By centering equity, inclusivity, and ocean stewardship, policy innovation and digital infrastructure can unlock a truly eco-digital future one in which AI serves as a catalyst for resilience, justice, and a sustainable marine world.

SOLUTIONS AND RECOMMENDATIONS

A sustainable and inclusive Blue Economy requires integrating Artificial Intelligence (AI) with climate-smart policies through strengthened data ecosystems, enhanced capacity-building, ethical governance, and global cooperation. Countries must prioritize high-quality, real-time, and interoperable ocean data to support AI-driven monitoring and predictive modeling. This includes expanding remote sensing systems, deploying IoT-enabled marine buoys, and developing open-access databases that promote transparency and innovation. At the same time, capacity-building efforts are essential, particularly in low- and middle-income countries where climate vulnerability intersects with digital inequality. Training programs in AI and marine sciences, along with community-based digital empowerment, should be supported through coordinated South-South and North-South collaborations. The deployment of AI must follow climate-smart principles, enabling early warning systems, sustainable marine spatial planning, and the advancement of low-carbon industries such as offshore wind, seaweed farming, and eco-tourism while avoiding technologies that reinforce extractive or exploitative practices.

Achieving eco-digital transformation further depends on strong multisectoral collaboration and innovative financing mechanisms. Governments, researchers, technology developers, and civil society should work through cross-sectoral innovation hubs to co-create AI solutions tailored to real-world marine challenges and to pilot scalable models for broader adoption. Financial strategies must evolve to support long-term digital and climate resilience, using climate adaptation funds, technology funds, green and blue bonds, and blended finance to mobilize public, private, and philanthropic investments. By implementing these recommendations, stakeholders can harness AI not only for improving efficiency and governance, but also for advancing climate resilience, protecting marine ecosystems, and promoting social inclusion — core principles of a sustainable and equitable Blue Economy aligned with global development agendas.

CONCLUSSIONS

This chapter emphasizes that achieving sustainable development goals in the blue economy requires close integration between visionary policies, the use of AI-based technologies, and inclusive cross-sectoral collaboration. AI and eco-digital tools offer new capacities to improve marine ecosystem monitoring, coastal resource management, and climate change resilience, especially in support of SDGs 13, 14, and 9. However, the positive impact of this technology can only be realized if it is supported by strong governance frameworks, data-driven policies, and ethical approaches that ensure social justice and ecosystem protection.

Furthermore, this chapter highlights that multisectoral collaboration between government, academia, the private sector, and civil society is key in translating technological innovations into real action on the ground. Case studies and strategic discussions show that the success of the transformation towards a sustainable blue economy depends not only on technological sophistication, but also on institutional capacity, community participation, and cross-regional learning. By adopting an adaptive policy approach, investing in capacity building, and responsible AI governance, the blue economy can thrive as an inclusive, resilient, and aligned growth pillar with long-term sustainability.

REFERENCES

Alam, T. M., Stoica, G. A., & Özgöbek, Ö. (2025). Asking the classroom with technology: A systematic literature review. *Smart Learning Environments*, *12*(1), 7. Advance online publication. DOI: 10.1186/s40561-024-00348-z

Alka, T. A., Raman, R., & Suresh, M. (2025). Critical success factors for successful technology innovation development in sustainable energy enterprises. *Scientific Reports*, *15*(1), 14138. Advance online publication. DOI: 10.1038/s41598-025-98725-2 PMID: 40269233

Andika, A., Perdana, T., Chaerani, D., & Utomo, D. S. (2025). Transitioning towards zero waste in the agri-food supply chain: A review of sustainable circular agri-food supply chain. *Sustainable Futures : An Applied Journal of Technology, Environment and Society*, *10*, 100917. Advance online publication. DOI: 10.1016/j.sftr.2025.100917

Badriyah, L., & Wardi, M. (2026). Policy analysis of Islamic educational institutions in facing the challenges of society 5.0: Innovation, learning, and technology-based infrastructure. *Multidisciplinary Reviews*, *9*(1), 2026012. Advance online publication. DOI: 10.31893/multirev.2026012

Chen, W., Cao, Q., Cao, B., & Jin, B. (2025). An innovative coverage optimization method for smart information monitoring in agricultural IoT using the multi-strategy Pelican optimization algorithm. *Scientific Reports*, *15*(1), 12634. Advance online publication. DOI: 10.1038/s41598-025-95885-z PMID: 40221446

Chen, Y., Gao, Y., Zhou, J., Zhang, S., Li, T., Zhang, Z., Gong, X., & Liu, Y. (2025). Vagus nerve stimulation: Innovative applications in the field of hand function rehabilitation. *Cognitive Neurodynamics*, *19*(1), 98. Advance online publication. DOI: 10.1007/s11571-025-10284-4 PMID: 40568488

Dagkinis, I. K., Spyrou, S., Georgantis, G. K., Psomas, P. M., Platis, A. N., & Tsoulfas, G. (2025). Impact of innovative technologies on quality and safety in surgery. *Journal of Robotic Surgery*, *19*(1), 31. Advance online publication. DOI: 10.1007/s11701-024-02192-2 PMID: 39707082

Deaza-Fernández, M. P., Sarmiento-Soto, D., & Quintanilla-Carvajal, M. X. (2025). Innovative Approaches in the Design of Flavonoid-Based Functional Foods for Pediatric Nutrition. *Applied Food Research*, *5*(2), 101114. Advance online publication. DOI: 10.1016/j.afres.2025.101114

Eneh, S. C., Obi, C. G., Ekwebene, O. C., Edeh, G. C., Awoso, O., Udoewah, S. A., Onukansi, F. O., Ikhuoria, O. V., Okoli, I. A., & Ojo, T. O. (2025). Eliminating malaria in Nigeria: Insights from Egypt's success and pathways to sustainable eradication. *Malaria Journal*, *24*(1), 183. Advance online publication. DOI: 10.1186/s12936-025-05391-w PMID: 40490749

Fegert, J. M., Gottschalk, G., Schneider, R., Sitarski, E., Sounderajah, V., & Graham, G. (2025). Navigating life transitions and mental wellbeing in the digital age: A call for stakeholders to embrace innovation and collaboration. *Child and Adolescent Psychiatry and Mental Health*, *19*(1), 67. Advance online publication. DOI: 10.1186/s13034-025-00932-2 PMID: 40517237

Gabr, M. A. M., Sleem, W. F., & El-wkeel, N. S. (2025). Effect of virtual reality educational program on critical thinking disposition among nursing students in Egypt: A quasi-experimental pretest–posttest design. *BMC Nursing*, *24*(1), 874. Advance online publication. DOI: 10.1186/s12912-025-03488-w PMID: 40624646

Galeboe, K. A., Moalosi, R., Rapitsenyane, Y., & Ruele, V. (2025). What is the impact of using design and technology pedagogy to support the attainment of 21st-century skills? *Discover Education*, *4*(1), 175. Advance online publication. DOI: 10.1007/s44217-025-00604-5

Han, C., Wu, C., Hu, M., Li, J., & Chen, H. (2024). C2F-SemiCD: A coarse-to-fine semi-supervised change detection method based on consistency regularization in high-resolution remote sensing images. *IEEE Transactions on Geoscience and Remote Sensing*, *62*, 1–21. DOI: 10.1109/TGRS.2024.3370568

Han, S., Li, J., Jiang, F., & Shi, Q. (2025). Mapping the landscape of metabolic reprogramming research in lung cancer: A bibliometric and visualized analysis. *Discover. Oncology*, *16*(1), 583. Advance online publication. DOI: 10.1007/s12672-025-02327-3 PMID: 40257698

Hoareau, K. (2025). Knowledge, Power and the Blue Economy: Breaking down Barriers. *Marine Policy*, *182*, 182. DOI: 10.1016/j.marpol.2025.106883

Hong, J., Bae, E.-Y., Cha, S., & Lee, J. (2025). The value-for-money assessment and funding arrangements for high-priced drugs in an era of uncertainty: A comparative analysis of national health technology assessment agencies in South Korea, England, Australia, and Canada. *BMC Health Services Research*, *25*(1), 74. Advance online publication. DOI: 10.1186/s12913-025-12207-9 PMID: 39810177

Jiang, B., Cheng, T., Tsou, M.-H., Zhu, D., & Ye, X. (2025). Advancing translational human dynamics research: Bridging space, mind, and computational urban science in the era of GeoAI. *Computational Urban Science*, *5*(1), 12. Advance online publication. DOI: 10.1007/s43762-025-00171-3

Kurmis, A. P., Lustig, S., Zambianchi, F., & Chen, Y. (2025). Editorial: Advances in artificial intelligence and robotics in joint arthroplasty. *Arthroplasty (London, England)*, *7*(1), 17. Advance online publication. DOI: 10.1186/s42836-025-00302-5 PMID: 40176158

Kwon, J.-Y., Kaur, M. N., Elsman, E. B. M., Mehdipour, A., Lo, L. S. H., Osman, A. M. Y., Herbelet, S., Ng, C.-A., & van der Weijst, L. (2025). Research trends among new investigators at ISOQOL: A bibliometric analysis from 2019 to 2023. *Journal of Patient-Reported Outcomes*, *9*(1), 54. Advance online publication. DOI: 10.1186/s41687-025-00878-1 PMID: 40366537

Li, H., Wu, H., Li, Q., & Zhao, C. (2025). A review on enhancing agricultural intelligence with large language models. *Artificial Intelligence in Agriculture*, *15*(4), 671–685. DOI: 10.1016/j.aiia.2025.05.006

Luo, Q., & Wang, J. (2025). Cross-regional collaborative innovation in the era of data openness: A factor mobility perspective. *Technology in Society*, *83*, 102989. Advance online publication. DOI: 10.1016/j.techsoc.2025.102989

Mostafa, S. K., Abdel-Rahman, R. H., Mansour, A. K., & El-Sherbeny, M. A. (2025). Fostering collaborative practice: A pilot study on interprofessional education through simulation-based team-oriented learning sessions among pharmacy and medical students. *Discover Education*, *4*(1), 19. Advance online publication. DOI: 10.1007/s44217-025-00407-8

Obisi, J. N., Abimbola, A. N. J., Babaleye, O. A., Atidoglo, P. K., Usin, S. G., Nwanaforo, E. O., Patrick-Inezi, F. S., Fasogbon, I. V., Chimezie, J., Dare, C. A., Kuti, O. O., Uti, D. E., & Omeoga, H. C. (2025). Unveiling the future of cancer stem cell therapy: A narrative exploration of emerging innovations. *Discover. Oncology*, *16*(1), 373. Advance online publication. DOI: 10.1007/s12672-025-02102-4 PMID: 40120008

Onyena, A. P., & Sam, K. (2025). The blue revolution: Sustainable water management for a thirsty world. *Discover Sustainability*, *6*(1), 63. Advance online publication. DOI: 10.1007/s43621-024-00631-6

Qiao, J., Zhan, S., Ren, M., Wang, H., Huang, Y., Zhang, Z., Luo, Y., Yin, H., Jia, Z., Huang, W., Zhou, H., Liu, J., Liu, X., Sun, Q., Li, X., Bai, J., Liu, F., Liu, Y., Jin, Y., & Xu, M. (2025). A new phase of China-ASEAN health cooperation: The China-ASEAN Beijing Declaration on Cooperation in Innovation of Health Products and Technologies. *Global Health Research and Policy*, *10*(1), 3. Advance online publication. DOI: 10.1186/s41256-024-00401-x PMID: 39810277

Rajaperumal, T. A., & Columbus, C. C. (2025). Transforming the electrical grid: The role of AI in advancing smart, sustainable, and secure energy systems. *Energy Informatics*, *8*(1), 51. Advance online publication. DOI: 10.1186/s42162-024-00461-w

Ramezani, M., Bakhtiari, A., Daroudi, R., Mobinizadeh, M., Fazaeli, A. A., Olyaeemanesh, A., Rabiee, H. R., Mostafavi, H., Sazgarnejad, S., Bordbar, S., & Takian, A. (2025). Applications of artificial intelligence and the challenges in health technology assessment: A scoping review and framework with a focus on economic dimensions. *Health Economics Review*, *15*(1), 46. Advance online publication. DOI: 10.1186/s13561-025-00645-4 PMID: 40461901

Saraswati, R., Hananto, P. W. H., Prananda, R. R., Mahramhafiz, M., & Pennesi, L. (2025). The Role of Indonesia as a Presidency for ASEAN and Blue Economy's Driver: SDG's Issues and Legal Perspective. *Law Reform: Jurnal Pembaharuan Hukum*, *21*(1), 180–202.

Sekar, S., Jeyachandran, S., Giri, J., & Aman, M. (2025). Advancing sustainable agriculture: The potential of seaweed-derived bio pesticides from marine biomass. *Bioresources and Bioprocessing*, *12*(1), 22. Advance online publication. DOI: 10.1186/s40643-025-00849-w PMID: 40120004

Sekti, B. A. (2025a). "Information Security Awareness." Pp. 111–45 in *User-Centric Cybersecurity Implications for Sustainable Digital Transformation*.

Sekti, B. A. (2025b). "IT Service Information Security Management." Pp. 61–93 in *Cases on Information Systems Service Management*.

Sekti, B. A. (2025c). "Machine Learning and Artificial Intelligence (AI) in Manufacturing." Pp. 255–92 in *Integrating Digital Innovation and Integrated Frameworks in Manufacturing*.

Shen, S., Qi, W., Liu, X., Zeng, J., Li, S., Zhu, X., Dong, C., Wang, B., Shi, Y., Yao, J., Wang, B., Jing, L., Cao, S., & Liang, G. (2025). From virtual to reality: Innovative practices of digital twins in tumor therapy. *Journal of Translational Medicine*, *23*(1), 348. Advance online publication. DOI: 10.1186/s12967-025-06371-z PMID: 40108714

Sofuoglu, Z., Başer, A., Sofuoglu, T., & Sönmez, Ö. F. (2025). First responders' experiences with major earthquakes in Türkiye: A qualitative study of innovation needs and challenges. *BMC Emergency Medicine*, *25*(1), 60. Advance online publication. DOI: 10.1186/s12873-025-01217-9 PMID: 40229688

Torres-Ceron, D. A., Amaya-Roncancio, S., Fuentes-Gandara, F., Restrepo-Parra, E., Bohorquez-Santiago, L., & Velasquez-Tamayo, J. P. (2025). Scientometric analysis of computational calculations on hydrogen adsorption. *Global Journal of Environmental Science and Management*, *11*(1), 295–320. DOI: 10.22034/gjesm.2025.01.18

Voronkova, V., Nikitenko, V., Sobol, Y., Oleksenko, R., Zelenin, Y., Kravchenko, O., Levchenko, L., & Yukhymenko, N. (2024). The Digital Transformation Of The Most Dynamic Region In The World (China, Malaysia, India) As A Key Factor In Economic Development Within The Context Of INDUSTRY 5G: Trends, Challenges, And Strategies. *Pakistan Journal of Life and Social Sciences*, *22*(1), 1522–1531. DOI: 10.57239/PJLSS-2024-22.1.00105

Wang, H., Guo, J., Zhang, Y., Fu, Z., & Yao, Y. (2025). Closed-loop rehabilitation of upper-limb dyskinesia after stroke: From natural motion to neuronal microfluidics. *Journal of Neuroengineering and Rehabilitation*, *22*(1), 87. Advance online publication. DOI: 10.1186/s12984-025-01617-9 PMID: 40253334

KEY TERMS AND DEFINITIONS

Artificial Intelligence (AI): A branch of computer science focused on building systems capable of performing tasks that normally require human intelligence, such as learning, decision-making, pattern recognition, and adaptation.

Blue Economy: The sustainable use of ocean and marine resources for economic growth, improved livelihoods, and ecosystem health. It encompasses sectors like fisheries, maritime transport, coastal tourism, and renewable ocean energy.

Climate-Smart Policies: Strategies and frameworks that integrate climate adaptation and mitigation into development planning. These policies are designed to enhance resilience, reduce emissions, and support long-term environmental sustainability.

Data Inequality: Disparities in the availability, access, and quality of data between different regions or social groups, which can affect the fairness and effectiveness of AI systems.

Digital Divide: The gap between those who have access to modern information and communication technologies and those who do not, often influenced by geography, socioeconomic status, and infrastructure limitations.

Eco-Digital Futures: A vision of development where ecological sustainability and digital technologies converge to support inclusive and adaptive systems, especially in the context of climate change and natural resource management.

Marine Spatial Planning (MSP): A strategic process that guides where and how ocean space is used to balance ecological, economic, and social objectives.

Multisectoral Collaboration: Cooperative efforts involving various sectors such as government, private industry, academia, and civil society to solve complex problems and create integrated solutions.

Predictive Analytics: The use of statistical algorithms and machine learning techniques to analyze historical data and predict future outcomes. In marine contexts, it supports climate forecasting and early warning systems.

Sustainable Development Goals (SDGs): A set of 17 global goals adopted by the United Nations to address major challenges such as poverty, inequality, climate change, and environmental degradation by 2030.

About the Contributors

Altuğ Günar is an Associate Professor of European Union Politics and International Relations at Bandırma Onyedi Eylül University. Dr. Günar received a BSc in political science and international relations from Yeditepe University (YTÜ-Türkiye), an MSc in European Union politics and international relations from Kocaeli University, and PhD in European Union studies from Istanbul University. Since 2019, Dr Günar has been the head of the Department of European Union Politics and International Relations and Director of the Mediterranean Policy Applications and Research Center. He has numerous academic works, including articles, book chapters, books and book editorials in the field of the European Union politics, and economy. His main research areas are the European Union studies, International Relations and Economics, and Schumpeterian Economics. Dr Gunar is currently the Director of the Mediterranean Policy Applications and Research Center, Lecturer in the Department of International Relations, and head of the Department of the European Union Politics and International Relations. His most recent book is entitled "Economic and Political Consequences of AI: Managing Creative Destruction", Green Transition in Europe: Strategies, Regulations and Instruments", "Creative Destruction, Crisis, and European Union: The 2008 Crisis within the Framework of the Schumpeterian Approach European Union and Globalization" and "Future of The European Union Integration: A Failure or A Success? Future Expectations". Dr Günar takes an active part in one Jean Monnet Module project as a researcher, and he served as a trainer in European Union literacy-expertise training organized by Konrad Adenauer at the national level and has participated as a speaker in numerous programs.

Didem Saygın is an Associate Professor and member of the Department of Public Administration, Biga Faculty of Economics and Administrative Sciences at Çanakkale Onsekiz Mart University. She completed her PhD in European Union at Istanbul University. From 2012 to 2013, she was a PhD researcher at the Institute of Advanced Legal Studies, University of London. She has several studies in the field of the European Parliament, The European Union-Türkiye Relations, Regional

Development Agencies in the European Union, European Union law and Decision Making.

Çağatay Başarır works as a Professor at Bandırma Onyedi Eylül University, International Trade and Logistics Department. He has studies on financial markets, stock exchange markets, commodity markets and precious metals in the field of time series analyses, multi-criteria decision analysis and performance measurement. He instructs financial management, financial analysis, financial markets, international finance and portfolio management lessons both in graduate and undergraduate degree.

***

Melike Aktaş Bozkurt brings 26 years of professional expertise, including 10 years of experience in the banking sector, with a specialization in corporate finance, risk management, and compliance with anti-money laundering and counter-terrorism financing regulations. Currently, she serves as an Assistant Professor at Ostim Technical University, where she actively contributes to research, teaching, and professional development in finance and accounting. Dr. Aktaş Bozkurt holds multiple professional certifications, including those as an Independent Auditor, Chartered Accountant (CPA), and Licensed Real Estate Appraiser. Her extensive career spans managerial roles in private sector companies, where she oversaw accounting and financial operations, provided strategic consultancy for manufacturing firms, and developed comprehensive financial systems. In academia, she focuses on bridging the gap between theoretical frameworks and real-world applications in areas such as financial reporting, corporate governance, and investment strategies. Her contributions include publications in academic journals, books, and participation in international conferences.

Arzu Alvan is a Professor of Business Administration and Dean of the Faculty of Business, Economics and Social Sciences at Cyprus Science University. An economist by training with a Ph.D. from Eastern Mediterranean University, she works at the intersection of productivity and growth, sustainable development, innovation economics, and political economy. Her scholarship spans SSCI/Scopus-indexed journals, international book chapters with leading publishers, and ongoing projects on the economics of energy, AI-driven transformation, and urban air mobility. Over two decades in academia, Arzu has taught a wide spectrum of courses—from micro and macroeconomics to econometrics, managerial economics, and quantitative research at MBA and PhD levels—while serving in key academic leadership roles including department chair, vice dean, senate member, and graduate institute vice director. She has supervised master's and doctoral research, reviewed for prominent journals such as Applied Economics and Computational Economics, and contributed to international conferences as author, chair, and scientific committee member. Motivated by bridging rigorous analysis with real-world impact, Arzu's recent work explores CBDCs and global monetary dynamics, green and digital economies, and the implications of AI for energy systems. She communicates in Turkish and English, has experience lecturing in Italy, and is committed to fostering data literacy and evidence-based policy thinking in business and economics education.

Rüya Ataklı Yavuz is an Associate Professor of Economics at Canakkale Onsekiz Mart University. Dr. Ataklı Yavuz received a BSc in economics from Balıkesir University, a MSc and a PhD in economics from Canakkale Onsekiz Mart University. Her areas of specialization are sustainable development, economic growth, unemployment, economic policy etc... Her most recent article is entitled "Unemployment hysteresis by gender and youth in the EU: evidence from new factor PANIC unit root tests with breaks", "Economic and Financial Developments in the European Union After the 2008 Financial Crisis", "European Union's Tourism Policies and Sustainability in the Shadow of Covid-19", "2030 Global Goals and The European Union" and "From GATT to WTO, Where to Now? The World Economic Order in the Midst of the Trade Wars".

Fadhil Afiq Badruzzaman is an IT Engineer with expertise in information systems, digital infrastructure, and emerging technologies supporting data-driven and intelligent solutions. His professional experience spans system integration, network and cybersecurity support, and the development of scalable IT architectures for organizational and research environments. He has been actively involved in projects that bridge information technology with applied research, contributing to the implementation of secure, efficient, and reliable digital platforms. His interests include cybersecurity, artificial intelligence applications, cloud computing, and the role of IT engineering in supporting sustainable and innovative technological ecosystems.

K. Balaji is an Assistant Professor in Business and Management, Christ University, Bangalore. He did Ph.D. from KLU Business School, KL University,Vijayawada. He is having 12 years of teaching experience. He has to his credit 23 Research Articles published in reputed National and International Journals. He has participated and presented more than 12 papers in both National and International Seminars/Conferences and also attended more than 50 workshops/ FDPS organized by various reputed organizations. He has also participated in more than 50 seminars / webinars organized by various reputed organizations. He has also attended in 6 Short Term Training programs.He has successfully conducted 3 webinars, one 5 day FDP and one International Conference on Reimaging marketing in New Normal. He is a ratified as Assistant Professor by both JNTUH, JNTU Anantapur and Sri Venkateswara University,Tirupati. His specialized areas include, Accounting and Financial management, Marketing Management, Retailing Management, Consumer Behavior studies, Entrepreneurship and Business Laws.

Elçin Bayraktar Köse author graduated from the Department of Business Administration in 2008. Between 2018 and 2021, she worked as a research assistant at Çanakkale Onsekiz Mart University and completed her PhD at the Department of Business Administration of the Çanakkale Onsekiz Mart University School of Graduate Studies in 2021. The author, who started teaching on fields of study of marketing and business management in 2022, continues to serve as a faculty member in the Business Administration Program of the Çanakkale Vocational School of Social Sciences as of 2023 and conducts scientific studies on consumer behavior.

Lokeshwari D. V. M.Com, K-SET (MBA, Ph.D.) is an Assistant Professor in the Department of Commerce and Management at Dayananda Sagar College of Arts, Science, and Commerce, Bengaluru, with over seven years of teaching experience in undergraduate and postgraduate programs. She has previously served at Surana College -Autonomous and VET First Grade College, Bengaluru, handling a wide range of commerce and management subjects. Alongside her teaching, she has taken up key academic and administrative roles, including AICTE Coordinator, Placement Officer, BBA Program Coordinator, IQAC Committee Member, and PG Department In-Charge (M.Com). Prolific researcher and author, with 29 publications, 16 conference presentations, 4 authored books, 9 book chapters, and 2 patents to her credit. Her research areas include financial literacy, sustainable finance, artificial intelligence in financial inclusion, behavioral finance, and education technology. She is also a Reviewer for IGI Global (Scopus Indexed) and the Asian Journal of Economics, Finance, and Management (Publons – Web of Science), and has been invited as a Resource Person, Guest Speaker, and Conference Chair at various academic forums. With a strong commitment to research-driven teaching and academic excellence, continues to contribute to the advancement of commerce education through innovation, mentorship, and interdisciplinary learning.

Asime Dilara Erdem completed her bachelor's degree in Public Administration at Çanakkale Onsekiz Mart University in 2015, her master's degree in Public Administration at Çanakkale Onsekiz Mart University in 2019, and her doctorate in Public Administration at Çanakkale Onsekiz Mart University in 2026. The author's academic interests include disasters, disaster migration, resilient cities, disaster policies, urbanization, local governments, and the European Union. She has national and international publications in the aforementioned fields.

Filiz Eroğlu After completing her undergraduate education in Business Administration, the author worked in the private sector in sales, marketing and foreign trade. She completed her master's degree in International Trade. In 2008, she started her Marketing PhD program at Istanbul University, and in 2009, she started working as a Research Assistant at Çanakkale Onsekiz Mart University. The author, who was in Canada between 2012-2015 and completed her doctorate during this period, has been working as a faculty member at Çanakkale Onsekiz Mart University Biga Faculty of Economics and Administrative Sciences, Department of Business Administration since 2016. The author, who received the title of associate professor of marketing in 2023, continues her scientific studies in the field of marketing, especially on consumer behavior, sustainable consumption and sustainable marketing.

Nalan Gelirli is an Assistant Professor in the Aviation Management Department at İstanbul Bilgi University in İstanbul, Turkiye. She began her career in 1995 as a Data Control and Analysis Specialist at Turkish Airlines A.O. She later served as a Revenue Specialist in the Production Planning and Revenue Management Department, and subsequently as a Lead Auditor in the Flight Safety and Quality Assurance Department. Her professional career in airline operations and regulatory processes continued at Turkish Airlines until the end of 2011. In the following years, she worked in leading aviation organizations such as MY Technic MRO, AtlasGlobal Airlines, and Pegasus Airlines. Between September 2018 and September 2025, she also held several senior academic administrative positions in Northern Cyprus, including Department Chair, Head of Department, and Director of the School of Civil Aviation (HMYO). Gelirli completed her Ph.D. at Istanbul University and her undergraduate studies at Marmara University. Her primary research interests lie in airline management, from theory to design to implementation. As a researcher in the social and economic sciences, she has collaborated actively with colleagues from multiple disciplines. She places particular importance on producing publications in aviation studies, a field still scarce within the Social Sciences. Gelirli was born on June 20, 1967, in Berlin, Germany, and grew up in Istanbul, Turkey.

Leonardo Goliatt Bachelor degree in Civil Engineering from the Federal University of Juiz de Fora and a Ph.D. in Computational Modeling from the Brazilian National Laboratory for Scientific Computing. Currently, Leonardo is an associate professor in the Department of Applied and Computational Mechanics and a permanent member of the Graduate Program in Computational Modeling at the Federal University of Juiz de Fora (UFJF). Leonardo has been head of the Department of Applied and Computational Mechanics and is a member of UFJF's Self-Assessment Commission (CPA), serving as its president. He has coordinated the Graduate Program in Computational Modeling at UFJF and currently coordinates research programs under UFJF's Office of Graduate Studies and Research. His research focuses on Artificial Intelligence and Data Science, with involvement in innovative and effective research and development projects that address current challenges through sustainable, data-driven solutions.

Tiago Silveira Gontijo PhD in Production Engineering from UFMG in the research line "Stochastic Modeling and Simulation". Participated in the Academic Doctorate for Innovation (DAI) Program from CNPq. Master's and Bachelor's degrees in Production Engineering. Bachelor's degrees in Administration and Economics. Interrupted PhD in Public Health from the Faculty of Medicine at UFMG. Has 10 years of teaching experience and in educational management. Currently an adjunct professor at the Federal University of São João del-Rei (UFSJ). Developed teaching activities at institutions such as UFOP, PUC-MG, UNIBH, UNA, and Izabela Hendrix. Has articles published in various journals. Has experience in data analysis and applied statistics, with emphasis on biostatistics, renewable energies, systematic literature review, and time series. Proficient in the R language.

Gustavo de Souza Groppo Graduated in Economic Sciences from the Federal University of Viçosa (1997), Specialization in Strategic Management in Finance from the Federal University of Minas Gerais (2000), Master in Sciences (Applied Economics) from the University of São Paulo (2004) and a doctorate in sanitation from the Federal University of Minas Gerais (2019). It has experience in consulting in business management, strategic planning, elaboration of macroeconomic scenarios, teaching, as well as developing research on financial and forecasting employing Machine Learning acting on the following subjects: macroeconomic variables, stock markets, ibovespa, monetary policy variables, future contracts, co-interpretation, causality of granger, demand forecast.

Jipson Joseph has a PhD in Ethics from DVK Bangalore. Currently, he is a Research Scholar at School of Law, Christ University, Bangalore, India. His research interests are in the fields of ethics, gender studies, constitutional law, human rights, human security, environmental justice, SDGs, developmental policies, and global poverty. He has published 6 books, co-edited one book, and contributed scholarly articles to many international and national journals and edited volumes including IGI Global Scientific Publishing's edited books.

Harmanpreet Kaur is Assistant Professor in university school of business having 10 years of experience in Academics and reserch. She published her reserch work in SCOPUS, ABDC journal and UGC care along with 6 patents by her name.

İhsan Erdem Kayral graduated from the Department of Economics at Hacettepe University as the top student of the faculty and department. He received his MSc. and Ph.D. degrees in the field of Finance in the Department of Business Administration at the same university. Completing his studies, Assoc. Prof. İhsan Erdem Kayral started to work in TUBITAK (The Scientific and Technological Research Council of Turkey) as a Chief Expert in 2010 and he served there for over 8 years. He worked in the Department of Economics and Department of International Trade and Business Administration at Konya Food and Agriculture University for approximately 4 years. Assoc. Prof. Kayral worked as a faculty member in the International Trade and Finance Department (served as Director of Graduate School of Social Sciences almost one year) at OSTIM Technical University for 2 years. Currently, he works as Assoc. Prof. in the Department of Business Administration at Başkent University. He has written several papers and book chapters about finance, clean energy and forecasting the macroeconomic data. Assoc. Prof. Ihsan Erdem Kayral has lots of awards in various fields. He also receives a National Tubitak Master's Degree Scholarship and Turkish Economic Association Achievement Grant at Bachelor's Degree Level.

Anant Prakash Awasthi is a distinguished data scientist and educator with a Ph.D. (Pursuing) in Statistics. Born in India, he has built a stellar career as the Lead Data Scientist at Optum, specializing in industries such as Banking, Financial Services, Insurance (BFSI), operations, and healthcare. Anant is deeply passionate about demystifying complex concepts and bridging the gap between data science theory and practice. His first book, Explainable Artificial Intelligence (XAI): Concepts and Applications, caters to practitioners in the field of data science, aiming to make cutting-edge AI technologies accessible and transparent. Beyond his professional achievements, He has made significant contributions to education and mentoring. He dedicates his time to teaching and training aspiring data scientists, particularly those from underprivileged backgrounds. As a visiting faculty member, he has taught "Data Science for Managerial Decisions" in the executive management program at IIT Patna and contributed to several learning and development programs at reputed institutions. With a mission to empower and educate, Anant continues to inspire the next generation of data scientists and make meaningful contributions to the evolving world of artificial intelligence.

Figen Kılıç, a lecturer at Çanakkale Onsekiz Mart University, Biga Vocational School of Higher Education. holds a Bachelor's degree in Business Administration and a Master's degree in Economics. She gained private sector experience in the banking sector. Currently pursuing her doctoral degree in the Business Administration department at Çanakkale Onsekiz Mart University, she has scholarly studied on economics and consumer behavior, with a current focus on sustainable and green marketing, which is the subject of her doctoral dissertation.

Fulya Kıroğlu holds bachelor's degrees in two different fields: Economics from Anadolu University (2009) and Labor Economics and Industrial Relations from Çanakkale Onsekiz Mart University (2014). She completed her master's degree in Labor Economics and Industrial Relations in 2019 and her PhD in Economics in 2023, both at Çanakkale Onsekiz Mart University. The author's main research areas include precarity in working life and social inequalities, as well as the economic and social impacts of innovative models such as green finance and the blue economy within the context of sustainability. She conducts her research as an independent scholar.

Sina Kısacık is an Assistant Professor Doctor at the Department of International Relations under the Faculty of Economics, Administrative and Social Sciences in Cyprus Aydın University, located at Ozanköy-Girne (Kyrenia)-Turkish Republic of Northern Cyprus (TRNC) since October 2021. Before that, he was a lecturer in the energy law module within the Public Law MA Programme (with thesis) at the Social Sciences Institute of Özyeğin University, İstanbul, Türkiye between January 2020 and June 2021. Before teaching at Özyeğin University, he taught "Current Issues in International Relations" course at Political Science and International Relations Department of the Social Sciences Institute at İstanbul Sabahattin Zaim University between September 2019 and January 2020. He also worked as an assistant professor at the Political Science and International Relations Department of İstanbul Gelişim University between February 2019 and July 2019. Here, he taught courses on scientific research methods (in Turkish and English), Public Administration and Türkiye and Balkans. He holds a Ph.D. in Political Science and International Relations Department from Yeditepe University, İstanbul, Türkiye in July 2017 by the dissertation titled "The Ukraine Policy of the Russian Federation and The Paradigms in European Energy Security." He got his MA degree in European Union studies from Galatasaray University and BA on international relations from Bilgi University, İstanbul, Türkiye He has co-authored six books and has authored/ co-authored several book chapters and papers in various national and international academic journals in English and in Turkish. His research interests include: energy geopolitics and security, Turkish foreign policy, Russian foreign policy, Caucasia, Central Asia, Eurasia and European Union. He is a columnist at Uluslararası Politika Akademisi and also with the editorial board and academic advisory board of peer-reviewed UPA Strategic Affairs and some other academic journals.

Pawan Kumar, PhD, is a prominent academic and researcher with over 20 years of expertise in business studies. A professor at Graphic Era Deemed to be University, Dehradun, India, he has supervised six PhD scholars and published over 60 research papers, including 51 in SCOPUS/ 06 in Web of Science, and 14 books, including eight indexed in Scopus. His works, featured by prominent publishers like Emerald, IGI Global, Nova Science, and Cambridge Scholars, have gained recognition. Dr. Kumar earned his PhD in Commerce from Kurukshetra University in 2007 and holds an M.Phil. and M.Com. Degrees. Specializing in finance, banking, and securities markets, he has authored books on green finance, AI, and fintech. His innovative research includes five patents and has served as a guest editor for Scopus-indexed journals. In 2022, he received the prestigious Researcher Award from Chandigarh University. Dr. Kumar's impactful work on sustainable finance, blockchain, and fintech is globally recognized. His active participation in high-impact conferences and faculty development programs reflects his commitment to advancing research, teaching, and innovation in commerce and management.

Biswajit Mondal Biswajit Mondal is a Senior Data Science professional with 9 years of experience, including 6 years in applied machine learning. He builds scalable, cloud-native AI solutions across healthcare and finance, with expertise in MLOps, deep learning, and statistical modeling. His work has driven cost optimization, efficiency gains, and robust production systems. He is now focused on responsible AI innovation using LLMs, RAG, and Agentic AI.

Mesut Savrul is a faculty member in the Department of Econometrics at Çanakkale Onsekiz Mart University, Biga Faculty of Economics and Administrative Sciences. He earned his doctorate in Economics from Istanbul University. His research focuses on international economics, economic growth, economic integration, and regional development.

Burak Şakir Şeker is an Associate Professor at Ankara Hacı Bayram Veli University, a retired Navy Lieutenant, and an industrial engineer. He previously served in the Turkish Naval Forces and the Turkish Armed Forces Headquarters in various roles, including operation officer, project officer, and commander. During his career, he represented the Turkish Delegation at the International Maritime Organization (IMO) Maritime Safety Committee in London, contributing to the Maritime Security Working Group. He was also awarded the United Nations Staff Officer Certificate (Peacebuilding Process/Peacekeeping Operations) by the Swedish Armed Forces. Following his military service, Dr. Şeker worked as a postdoctoral researcher and guest lecturer at the World Maritime University (WMU) in Sweden, a United Nations institution. He has also been invited by the NATO Multinational Maritime Security Centre of Excellence (MARSEC COE) and Sciences Po Grenoble – Université Grenoble Alpes in France to deliver lectures on international relations and security. His research and commentary have appeared in The Conversation, Al Jazeera English, Cairo Review of Global Affairs, ORIENT (Deutsche Orient-Stiftung), Newzroom Afrika, and Sveriges Radio (SR), as well as in peer-reviewed journals and international publishers. His academic focus spans global maritime geopolitics, the Middle East and Africa, the cyber environment and international politics, space security, global actors and great power competition, comparative politics and political regimes, and refugee and international issues.

Binastya Anggara Sekti has been a lecturer at the Faculty of Computer Science, Esa Unggul University, Jakarta, since 2016. He is actively involved in the Artificial Intelligence and Information Security Study Group. He holds a Bachelor's degree in Informatics Engineering from the Bandung Institute of Technology and a Master's degree in Information Systems Management from Trisakti University, Jakarta. Prior to entering academia, he accumulated over 25 years of professional experience as an IT practitioner in various companies across Jakarta. His work focused on developing data centers for internet service providers and delivering IT solutions for the manufacturing and oil & gas sectors. His research interests include artificial intelligence, cybersecurity, and the development of IT infrastructure.

Amrik Singh is working as Professor and HOD in the School of Hotel Management and Tourism at Lovely Professional University, Punjab, India. He obtained his Ph.D. degree in Hotel Management from Kurukshetra University, Haryana, India. He has published 115 Scopus-indexed publications. He has published 12 patents, and 01 patent has been granted in the inter-disciplinary domain. Dr. Amrik Singh participated and acted as a resource person in various national and international conferences, seminars, research workshops, and industry talks. He has published 20 books with IGI Global Publication, Emerald Publication, Bentham Publication, Nova Publication etc., on trending areas such as Food waste, Artificial Intelligence, Mountain Tourism, Virtual tourism, Healthcare Tourism, Guest behavior in tourism and hospitality. His area of research interest are Accommodation management, sustainability, Green practices, Human Resource Management in Hospitality, Artificial Intelligence, AR VR in hospitality, Hotel pricing, Employee Disability, Work life balance, women empowerment etc. He is currently guiding 8 Ph.D. scholars, and 5 Ph.D. scholars have been awarded Ph.D.

Niraj Kumar Singh Dr. Niraj Kumar Singh has received his Ph. D. in Statistics from Banaras Hindu University and presently he is working in Amity University, Noida as an Assistant Professor. Dr. Niraj has published more than 30 research papers in the area of Mathematical Demography, Applied Statistics and Biostatistics. He has also served as a reviewer for many peer reviewed and reputed journals.

Murat Yorulmaz is an assistant professor of international relations with Trakya University in Edirne, Turkey. He has published widely on regional studies centered on security and energy.

Index

A

B

C

D

E

F

G

H

I

J

L

M

N

O

P

R

S

T

W

www.ingramcontent.com/pod-product-compliance
Lightning Source LLC
LaVergne TN
LVHW081318300926
849161LV00008B/1803
9798337352381